RELIGION

IN SOCIOLOGICAL PERSPECTIVE

Fifth Edition

RELIGION
IN SOCIOLOGICAL PERSPECTIVE
Fifth Edition

KEITH A. ROBERTS
Hanover College

DAVID YAMANE
Wake Forest University

SAGE | PINE FORGE

Los Angeles | London | New Delhi
Singapore | Washington DC

Los Angeles | London | New Delhi
Singapore | Washington DC

FOR INFORMATION:

Pine Forge Press

An Imprint of SAGE Publications, Inc.

2455 Teller Road

Thousand Oaks, California 91320

E-mail: order@sagepub.com

SAGE Publications Ltd.

1 Oliver's Yard

55 City Road

London EC1Y 1SP

United Kingdom

SAGE Publications India Pvt. Ltd.

B 1/I 1 Mohan Cooperative Industrial Area

Mathura Road, New Delhi 110 044

India

SAGE Publications Asia-Pacific Pte. Ltd.

33 Pekin Street #02-01

Far East Square

Singapore 048763

Acquisitions Editor: David Repetto

Editorial Assistant: Maggie Stanley

Production Editor: Eric Garner

Copy Editor: Megan Markanich

Typesetter: C&M Digitals (P) Ltd.

Proofreader: Susan Schon

Indexer: David Yamane

Cover Designer: Gail Buschman

Marketing Manager: Erica DeLuca

Permissions Editor: Karen Ehrmann

Cover Art Credits: ©iStockphoto.com/anandkrish16; ©iStockphoto.com/Leontura; ©iStockphoto.com/urbancow; ©iStockphoto.com/syagci; Medioimages/Photodisc/Thinkstock.

Printed in the United States of America

Library of Congress Cataloging-in-Publication Data

Roberts, Keith A.

Religion in sociological perspective / Keith A. Roberts, David Yamane.—5th ed.

p. cm.

Includes bibliographical references (p.   ) and index.

ISBN 978-1-4129-8298-6 (pbk.)

1. Religion and sociology—Textbooks. I. Yamane, David. II. Title.

BL60.R58 2012

306.6—dc22

2011011899

CONTENTS

PREFACE

Religion is a complex phenomenon. It involves a meaning system with an inter-related set of beliefs, rituals, symbols, values, moods, and motivations. Each of these interacts in diverse and complex ways with one another, sometimes being mutually supportive and sometimes conflicting. Religion is also a structural system with established statuses, organizational patterns, and even bureaucratic dilemmas. This structural system is itself diverse and multifaceted, characterized by both conflicts over self-interests and strains toward coherence and integration. Finally, religion is composed of a belonging system, with friendship networks, group boundaries, and informal norms that may be quite independent of the formal structure or official meaning systems. These three subsystems of religion are themselves *interdependent*, forming a larger system that is in some ways coherent and in some ways in tension and discord. Further, religion is part of a larger social system, and as such it both affects and is affected by this larger system. It is precisely this complexity of religion, including the complexity of its relationship to the larger society and to the world system, that we explore in the following pages.

This book is designed as an introduction to the sociology of religion. Our intent has been to present and illustrate the basic theories sociologists use to understand the social dimensions of religion. First and foremost, we seek to help students understand the *perspective* from which sociologists view religion. By the time students have finished this book, they should understand the central theories and methods of research in the sociology of religion, and they should have an idea of how to apply these analytical tools to new groups they encounter. The goal of this text is to be *illustrative* rather than all-encompassing. Insofar as it is adopted as a text in a university or seminary course, we have assumed that it will be complemented with monographs or anthologies that explore the specific groups or specific processes the instructor chooses to emphasize.

Theoretically, the 5th edition continues to draw on a wide range of perspectives. We seek to help students recognize the contributions of various theoretical perspectives and the blind spots of each theory. Conflict, functional, social constructionist, and rational choice paradigms are used throughout the text, and various middle-range theories are utilized to explore specific processes. Despite the effort to introduce many perspectives, however, we have made an effort to enhance integration of the text by using one framework throughout the book. That perspective, which stresses both structure and dynamic process, is the open systems model.

We have also tried in this book to provide students with an understanding of the relationship between research methods and findings. Without belaboring the issue of methods, we have attempted to make students aware of the relationships between accurate data, operationalization of ideas, correct use of the language of causality, and support of one's generalizations. Our hope is that students sense that doing research really can be exciting work. In this 5th edition, we have continued to provide features on "Doing Research on Religion" to intrigue readers with the exciting process of inquiry. "Critical Thinking" questions are also provided throughout to stimulate engagement with the material and more active reading and learning.

Each chapter of this text has been student-tested for readability and clarity among nonsociologists, and all revisions have been made with an eye to making the material ever more accessible to the nonspecialist. Photos have been selected with great care to illustrate concepts and to heighten interest. At the outset of each chapter, there are inquiry questions to heighten reader curiosity and to help students become more "active readers." While we have made every effort to enhance clarity and to amply illustrate abstract ideas, we have also tried to resist appeals from some quarters to lessen the theoretical sophistication of the text.

The first four editions of this text were a solo "labor of love" by the first author. When we began to look at a new edition of *Religion in Sociological Perspective*, a major obstacle was in the way. The original author was so involved in continual revisions of an introductory book (*Our Social World*, coauthored with Jeanne Ballantine) that he was unable to find the time necessary to complete the necessary revisions. The solution was to bring on board a coauthor. The five selection criteria for finding a coauthor were as follows: (1) effective writing skill in communicating with lay audiences, (2) solid familiarity of the issues in the scholarship of teaching and learning, (3) cutting-edge knowledge of the scholarship in the sociology of religion, (4) a balanced, nonideological approach to the discipline (with no ax to grind on a particular theoretical paradigm), and (5) being less fossilized than the now-aging first author. The two authors had worked together on the editorial board for the American Sociological Association (ASA) journal *Teaching Sociology*, and David Yamane was also the editor of one of the leading journals in this subfield: *Sociology of Religion*. It was an obvious choice, and it has been a superb

choice. We have worked together in this revision, and David Yamane will increasingly take the lead in any future revisions of this book.

Many of the changes in the 5th edition are matters of updating data and including recent findings, but we also take advantage of the recent availability of some very important new data sets on religion. These include the National Congregations Study, Hartford Institute's Megachurches Today project, National Survey of Youth and Religion, and the Pew Research Center Forum on Religion & Public Life's Religious Landscape Survey. These are not just updated data but cover new ground in religious life. Therefore, these data play a major role in several chapters. Moreover, the "hot issues" in sociology of religion have changed somewhat. This is reflected in new topical coverage and substantial reorganization of several chapters. Part III of the book has been expanded from one to two chapters, with new material on religious socialization, religion and youth, the life course, and identity all incorporated into Chapter 5. Part IV of the book, by contrast, has been reduced from three to two chapters, but much of Chapter 8 on denominationalism and congregationalism is new. The major dimensions of inequality in society—class, race, gender, and sexuality—sadly remain unchanged in society and therefore remain at the heart of Part V. Each of these four chapters have been augmented by perspectives developed and studies published since the 4th edition. For example, Chapter 9 talks about "Prosperity Theology" in relation to the "Protestant ethic." Chapter 10 adds new data on racial differences in religious affiliation and a new section on racial segregation in congregations. Chapter 11 discusses theories of risk tolerance and aversion as explanations for the gender gap in religious practice and also how women negotiate sexism in religious institutions in everyday life. The newly constituted Part VI has just three chapters instead of four, but we think we do more with less in this edition. Chapter 13 still addresses secularization but brings in more of Yamane's perspectives (he coined the term *neosecularization* as a participant in these scholarly debates) and considers the issue of "civil religion" in relation to secularization processes. Chapter 14, "Religion Outside the (God) Box," combines the best of two chapters from the 4th edition with new material on religion and the media (especially the Internet) and sport. The final chapter still addresses globalization but adds new considerations of the process of "glocalization" (how the global is local and the local is global) and the expanding number of transnational connections in the world. Some of the material that has been dropped in this edition—witch hunts as religious gynocide and religious insight as creativity, for example—is now available on this book's website at **www.pineforge.com/rsp5e.** Those who may still wish to use it can do so via the web.

There is substantial value to texts that have interchangeable chapters, for they allow the instructor flexibility in designing the course. What a text gains in flexibility, however, it often loses in cumulative development of analytical complexity. While it is possible to vary the order of the chapters in this text, the text is designed to be cumulative. Later chapters are more intricate in their analysis than earlier ones and attempt to stretch students to greater levels of analytical sophistication. Considerable deliberation has been given to building greater analytical complexity as one moves through the book. If an instructor does choose to use these chapters in a different sequence, we recommend that the first three chapters be covered first as a foundation. Chapters 10 and 11 will also make more sense after Chapter 4 has been read. Other chapters may be selected as one wishes.

RESOURCES FOR INSTRUCTORS

This textbook is unique among sociology of religion texts in providing an instructor's manual and a test bank, as well as ancillary materials for students. All of these are available online. The instructor's manual is designed to provide ideas for creative teaching and active learning, written by someone who has won both the Hans Mauksch

Award for Teaching and Learning (given by the ASA Section on Teaching and Learning) and the ASA Distinguished Contributions to Teaching Award (in 2010). Roughly a dozen nonlecture teaching strategies are provided for each chapter, including several simulation games, in-class writing exercises, simple class surveys, small group work ideas, film ideas, and discussion questions. In total, there are roughly 240 instructional strategies for teaching sociology of religion. Our hope is that some of the ideas provided on the webpage will serve as a stimulus to each instructor's own creativity. To obtain access to the instructor's manual, please contact your SAGE/Pine Forge sales representative or visit the website for this book at **www.pineforge.com/rsp5e.** At the website, you will be able to download a Word version of the manual and other support materials for this text.

The test bank includes both essay and multiple-choice test questions for each chapter and essay ideas for a comprehensive final examination. There are 18 to 40 multiple-choice questions and 8 to 12 essay questions per chapter (approximately 500 questions total). Instructors will need a password from your sales representative to gain access to the test bank. If instructors need help with this, call 1-800-818-7243.

People interested in creative teaching strategies for the sociology of religion may also want to contact the ASA's Teaching Resources Center to acquire a copy of *Syllabi and Instructional Materials for the Sociology of Religion.* This monograph has an excellent annotated film guide, possible term projects, innovative teaching strategies, and sample syllabi from many people who teach this course. We will not repeat here all the interesting term paper ideas in that monograph, but they are well worth your consideration. There are suggestions for term projects involving students in number crunching and data analysis, content analysis of hymns, field experiences in participant observation of worship services, an observation exercise on religion and the courts, a project for creation of a new civil religion for one's state (which just seceded from the union), an exercise in hypothesis testing in the sociology of religion, ideas for extensive use of the Internet, and a host of other ideas that involve students as active learners. There are also essays on teaching sociology of religion in various social contexts (large state universities, denominational seminaries, and so forth). We are convinced that providing a set of choices on term projects increases the quality of the projects. None of us writes well on a task that we do not undertake with some volition, and our students are no different. So we recommend provision of a diversity of active course projects from which students can choose. We know of no better source of ideas than this ASA monograph. *Syllabi and Instructional Materials for the Sociology of Religion* is available from

ASA Teaching Resources Center

1430 K Street, NW—Suite 600

Washington, DC 20005

202-383-9005

www.asanet.org/teaching/resources.cfm

A UNIQUE PROGRAM SUPPORTING TEACHING OF SOCIOLOGY

The first author has been instrumental in the founding of a unique program to support and enhance the quality of college teaching. A new award program—the SAGE/Pine Forge Teaching Innovations & Professional Development Award—is designed to prepare a new generation of scholars within the teaching movement in sociology. People in their early career stages (graduate students, assistant professors, newer PhDs) can be reimbursed $500 each for expenses entailed while attending the daylong ASA Section on Teaching and Learning workshop. The workshop is the day before ASA meetings. By 2011, 95 people had already benefited because of donations from royalties of SAGE/Pine Forge authors and the generosity of SAGE/Pine Forge. This book will also be a cosponsor of this program with contributions from book royalties.

ACKNOWLEDGMENTS

While only two names go on the book cover, a project of this nature is enriched by the labor and support of many people. Keith Roberts is deeply appreciative of the assistance offered by the librarians of Bowling Green State University and Hanover College. Reviewers for the first four editions deserve special comment for their incisive analysis, helpful criticism, constructive suggestions, and encouragement in this project. The reviewers for earlier editions were Charles Bonjean, David Bromley, Dennis J. Cole, James D. Davidson, Barbara J. Denison, Roger Finke, Sharon Georgianna, John W. Hawthorne, Tom Kearin, Fred Kniss, James R. Koch, Les Kurtz, Harry LeFever, Richard Machalek, Perry McWilliams, Edgar W. Mills Jr., Wade Clark Roof, Frank Sampson, John S. Staley, and David Yamane. Reviewers for this edition were as follows:

Samuel Inuwa Zalanga, Bethel University

Martin Laskin, Southern Connecticut State University

Marty Laubach, Marshall University

Christine O'Leary-Rockey, Central Pennsylvania College

Liane Pedersen-Gallegos, University Of Colorado–Boulder

Matthew Vos, Covenant College

Phil Zuckerman, Pitzer College

Keith also had several student editorial assistants: Jennifer Crye, Rachael Ernst, Emily McDarment, Sarah Opichka, Elise Roberts, and Sarah Sheagley. An exceptional reference librarian, Kelly Joyce, has helped with finding and documenting obscure sources. David Yamane benefited from the assistance of Rebecca Dore, Anna Durie, Kim McIntee, and Glen Thomas. His spring 2010 Sociology of Religion class at Wake Forest University also helped critique and update the 4th edition of this book. His administrative assistant, Joan Habib, also kept him pointed in the right direction on a daily basis. SAGE/Pine Forge has been wonderful to work with, including Dave Repetto (acquisitions editor), Maggie Stanley (editorial assistant), Eric Garner (production editor), and Megan Markanich (copy editor). While all of the persons just listed have contributed tremendously to the finished product, they are certainly not responsible for its flaws.

Several colleagues who teach at universities with very large classes necessitating objective examinations have shared with us the multiple-choice test banks they have generated for this book. Many of the multiple-choice questions that appear in the test bank originally had been used in classes taught by Roger Finke, David Bromley, and Jeffrey Hadden. Several undergraduate students—Jeremy Castle, Rachel Ernst, Emily McDarment, Sarah Opichka, and Elise Roberts—have also served as assistants in the production of the instructor's manual and test bank. We are indebted to these junior colleagues for their help. Samuel Inuwa Zalanga (Bethel University) has also added a great deal to the instructor's manual by providing additional video suggestions.

Finally, Keith must express his deep appreciation to his wife, Judy, and children, Justin, Kent, and Elise. Judy Roberts has been enormously supportive and has offered helpful comments and criticism at untold times in this project. Keith's progeny contributed to earlier editions through provision of delightful interludes away from the intensity of the project. Now that they are all young adults, their critique of Keith's work and stretching of his thinking has been stimulating, encouraging, and insightful. Elise has also provided some of the photographs for this book from her extensive international travels.

David is blessed with three wonderful children, Paul, Hannah, and Mark, and a solid group of friends in the tennis community of Winston-Salem. All of these significant others keep David from taking himself too seriously. He dedicates his work on this book to his tennis teammates who mercilessly harass him for "only working 3 days a week" and "having summers off"—even as he writes this acknowledgment at 11:41 pm on a Saturday night in August.

If you have suggestions for this text or for the instructor's manual, we would be happy to hear them. We can be reached at the addresses below:

Keith A. Roberts
Hanover College
robertsk@hanover.edu

David Yamane
Wake Forest University
yamaned@wfu.edu

ABOUT THE AUTHORS

Keith A. Roberts is professor of sociology and department chair at Hanover College in Indiana. Previously he has taught at a 2-year campus and at a PhD-granting research university. In 2000, he was awarded the Hans O. Mauksch Distinguished Contributions to Teaching Award, by the ASA Section on Teaching and Learning, and in 2010, his work in the scholarship of teaching and learning was recognized through the ASA Distinguished Contributions to Teaching Award.

David Yamane is associate professor and chair of the Department of Sociology at Wake Forest University. He has published in the sociology of religion and in the scholarship of teaching and learning. He recently completed a 4-year term as editor of a leading journal in the field, *Sociology of Religion: A Quarterly Review*. In 2007, he was chosen by Wake Forest students to receive the Kulynych Family Omicron Delta Kappa Award for Contribution to Student Life.

PART I

INTRODUCTION TO THE SOCIOLOGY OF RELIGION

Before we delve too deeply into analysis of our topic, we must establish some common assumptions and understandings. The reader need not agree with the authors, but at least the reader should know how the authors are using key concepts and approaching various topics. That is the purpose of these first two chapters; we are laying the groundwork for a shared investigation.

In Chapter 1, we explore what we mean by religion and the importance of different definitions of "religion" as different "ways of seeing" a complex, multifaceted social phenomenon. Then in Chapter 2, we examine what it means to take a *social scientific* approach to the study of religion, including the unique perspectives and methodological assumptions of sociology as a discipline.

1

WHAT DO WE MEAN BY THE TERM *RELIGION*?

Here are some questions to ponder as you read this chapter:

- What is religion? What makes something "religious"?
- Why might one's definition of religion create blinders that cause one not to include important phenomena?
- How are religion and magic similar or different?
- How are religion and spirituality similar or different?
- Do religion and science serve the same functions or different ones?
- What is critical in my own idea of what is or is not a form of "religion"?

We begin our venture together with a difficult and complex problem: the definition of our topic of study. To students who have never studied the sociology of religion, the definition of religion may seem clear. Certainly, everyone knows what religion is! Let's get on with more important matters! Yet we dare not be so hasty. Some definitions are so narrow and specific as to exclude Buddhism as a religion. Other definitions are so broad and inclusive that many social behaviors may be considered forms of religion—including patriotism, systematic racism, or any other core set of values and *beliefs* that provides an individual or community with a sense of worth and of meaning in life.

We must begin our analysis, then, by exploring the question of what it is we intend to study. What, after all, is religion? We begin to answer this question by recognizing that the way we define our subject matter sets boundaries on what are and are not legitimate topics or groups for analysis—on what will be included in our studies of "religion" and what will be excluded. In this sense, definitions are "ways of seeing" a complex, multifaceted social reality, as the literary theorist Kenneth Burke (1935) noted, "Every way of seeing is also a way of not seeing" (p. 70).

An important implication of this approach is that definitions are not mirrors of reality to be judged as "true" or "false" but are *tools* that can been seen by those who use them as more or less *useful* (Berger, 1967). As you read and think about the following ways in which social scientists have defined religion, think about which definitions you find more or less useful and why.

SUBSTANTIVE DEFINITIONS

Many sociologists employ what has been called a *substantive* definition. This approach hinges on identification of the "substance" or "essence" of religion. Edward B. Tylor (1873/1958) used this approach in 1873 when he defined religion as "belief in Spiritual Beings" (p. 8). For most of us, a reference to God or gods does seem to be an essential element in religion. The reason Tylor used the term *spiritual beings* is that many non-industrialized people worship and/or fear their deceased ancestors. They have little or no concern about gods, as such, but their world is peopled with many unseen beings. Hence, spiritual beings seemed to Tylor a more inclusive term than belief in gods. Some contemporary scholars have reaffirmed Tylor's insistence that religion involves a belief in a Being or beings that are not encountered in normal empirical processes (Spiro, 1966).

Trying to define the essence of religion is a difficult task, but it becomes more difficult if our definition is to be applied cross-culturally. In the Western world, we tend to feel that religion is essentially a matter of belief. In fact, some social scientists have attempted to measure the religiosity of people—the extent of their "religiousness"—by determining how orthodox they are. (An orthodox person is one who believes the traditional doctrines of a religious tradition.) However, as R. R. Marett (1914) suggested, in many cultures, religion is "not so much thought out as danced out" (p. xxxi). That is to say ritual and emotion are primary to religion, and belief is only secondary.

Sam Gill's (2004) study of traditional Native American religions has convinced him that these faiths are expressed through tribal practices, prayer, and religious objects, not creeds, dogmas, or theologies. Scholars studying Orthodox Judaism have also consistently pointed out that a focus on behavior, rather than on beliefs and attitudes, is characteristic of that faith (Cohen, 1983; Moberg, 1984; Sklare & Greenblum, 1979). Anthropologists studying other cultures have also insisted that emphasis on belief is a bias of the Western world that causes investigators to miss the underlying thrust of many non-Western religions (Kluckhohn, 1972). For example, several observers have insisted that any concepts of a deity or superhuman beings are peripheral to official Buddhism (Benz, 1964; Herbrechtsmeier, 1993; Zaechner, 1967). On the other hand, most common folks around the world who identify themselves as Buddhists do believe in superhuman beings (Herbrechtsmeier, 1993; Orru & Wang, 1992; Spiro, 1978). So a definition that emphasizes a belief in superhuman beings leaves doubt about whether Buddhism is a religion. Strictly speaking, many Buddhist gurus (who are not concerned with superhuman beings) would not be considered to be practicing religion. Does religion refer only to those who hold a specific kind of belief?

Another definitional approach that tries to capture the essence of religion but that avoids the requirement of a specific belief was first suggested in 1912 by Emile Durkheim. Durkheim was fascinated with the cleansing exercises and the change of attitude that in many cultures were necessary before one could enter into religious ritual. Durkheim (1912/1995) maintained that a recognition of the division of life into sacred and profane realms allows us to identify religion in any culture. People around the world seem to undergo a psychological shift when encountering holy objects or engaging in religious ritual. This shift involves a sense of awe, a feeling of fear and majesty. The attitude differs from anything one encounters in the everyday life of these people.

Durkheim recognized that not all experiences of awe or sacredness by an individual are religious in character. Religion, he maintained, is a communal activity. It involves a social group: "In all history we do not find a single religion without a Church" (Durkheim, 1912/1995, p. 59). The sacred attitude must be fundamentally a group experience if it is to be identified as religion. Durkheim's (1912/1995) formal definition, then, is "A religion is a unified system of beliefs and practices relative to sacred things, that is to say, things set apart and forbidden— beliefs and practices which unite into a single moral community called a Church, all those who adhere to them" (p. 62). Since the late 1950s, Mircea Eliade and his students have reasserted the importance of the sacred–profane distinction in defining religion (Eliade, 1959).

This approach is helpful in a great many cases, and it avoids the problem of deciding which specific belief is intrinsically or inherently religious. Yet social scientists who have used this approach have often implied (if not asserted) a dualistic *worldview*. That is to say, life has a religious dimension and a nonreligious dimension. For example, Durkheim (1912/1995) insisted that

> the religious life and the profane life cannot coexist in the same unit of time. It is necessary to assign determined days or periods to the first, from which all profane occupations are excluded.
>
> . . . There is no religion, and, consequently, no society which has not known and practiced this division of time into two distinct parts. (p. 347)

Mircea Eliade (1959) concurred and wrote this regarding space: "For religious [persons], space is not homogeneous; he [or she] experiences interruptions in it; some parts of space are qualitatively different from others" (p. 20).

While it is true that many people organize their life experience into separate categories, not all do. The United Society of Believers in Christ's Second Appearing (the Shakers) attempted to sustain an attitude that all of life is sacred. More recently, a deeply religious utopian society called the Bruderhof (with communes in New York, Pennsylvania, and Connecticut) has also attempted to make *all* life hallowed and has de-emphasized sacraments and rituals. All of one's life is to be lived in the spirit of worship, and the community works to sustain this attitude. Benjamin Zablocki (1971) wrote, "The Bruderhof is concerned with bearing witness in the simple, everyday acts of living. . . . There are no activities, however trivial, that cannot be permeated by the divine spirit" (p. 31). This community even refuses to erect a church building lest religion be identified with a distinct time and place. Groups such as these do not seem to bifurcate life into sacred and profane realms. In fact, Kingsley Davis (1949) has suggested that while this distinction is useful in studying nonindustrial societies, it erects a false dichotomy in contemporary society and religion.

In Islam the inside of the mosque is holy ground, and Muslims recognize this extraordinariness and sacredness by removing their shoes when they enter.

Andrew Greeley has employed this criterion of sacredness as a defining characteristic of religion but has avoided the dualism of many writers. Greeley (1972) suggested that any being, social process, or value that gives meaning and purpose to life tends to become a source of reverence or profound respect. When Greeley referred to a sacred attitude, he suggested that it is not totally unlike a secular outlook but is a matter of intensified respect. Hence, Greeley's reference to a sacred attitude does not entirely preclude the study of nationalism, worship of the free-enterprise system, or any other example of profound loyalty as a form of religion. However, he did insist that not all experiences of transcendence or profound reverence are religious. He did not say how one distinguishes religious transcendence from nonreligious transcendence, but the distinction appears to pivot on the degree to which the object of awe is supernatural.

An underlying question in this whole debate, then, is whether religion by definition includes only that which has an otherworldly or supernatural dimension. What about the person whose ultimate value and deepest commitment is to the United States? He or she has a deep sense of loyalty to the flag of the United States and will even give his or her life to defend it. The American Way of Life and the American Dream provide a sense of meaning, purpose, and value in life. National holidays are celebrated with devotion, and a tear is shed when the national anthem is played. This individual may belong to several organizations for the preservation and promotion of patriotism and the glorification of America. The person loves this country above all else, according it highest value and deepest loyalty. Is this religious behavior? Can nationalism be a form of religion? It is not otherworldly, and it is not essentially supernatural (although the belief structure may include a God who overlooks, blesses, and judges the nation). Certainly the individual feels a sort of sacredness toward the nation and its primary symbol, the flag. Yet this sacredness does not involve the fear and trembling that Rudolf Otto (1923) and Durkheim (1912/1995) describe as part of the sacred attitude. How does the feeling of awe and reverence toward a nation differ from the awe and reverence toward a supernatural being or realm? Is this difference significant enough to call one experience religious and the other not? These are not easy questions to answer. Some scholars feel that nationalistic behavior as described above *is* religious in character and that a broader definition of religion is appropriate. This has even caused one prominent scholar to suggest that we simply focus on the sociology of the sacred, even if the behavior is not "a religion" in the strictest sense, since anything that is considered "sacred" is likely to interest the sociologist of religion (Demerath, 2000).

The major criticism of the substantive definitions is that they tend to focus the researcher's attention solely on traditional forms of religion. Some writers feel that people in complex and changing societies such as ours are religious in new ways. The substantive definitions are felt to be too narrow and too tradition-bound, hence blinding researchers to these new modes of religiosity.

FUNCTIONAL DEFINITIONS

Milton Yinger (1970) has offered a more inclusive definition of religion. He suggested that we focus not on what religion essentially *is* but on what it *does*. He proposed that we define a social phenomenon as religious if it fulfills the manifest function of religion.[1] He followed Max Weber in asserting that meaning in life is a basic human need (although the nature and intensity of that need will vary among individuals). The theologian Paul Tillich (1957) has described religion as that which is one's "ultimate concern," and

---

[1] Manifest functions are the *conscious* and *intended* functions of a social pattern or institution. Latent functions are unconscious and unintended (Merton, 1968).

Yinger also drew on Tillich's treatment in developing his own definition. The underlying conviction is that a fundamental concern of human beings is to understand the purpose of life and the meaning of death, suffering, evil, and injustice. In line with this conviction, Yinger (1970) wrote, "Religion, then, can be defined as a system of beliefs and practices by means of which a group of people struggles with these ultimate problems of human life" (p. 7). Religion helps individuals cope with these perplexities by offering an explanation and by providing a strategy to overcome despair, hopelessness, and futility.

Using this type of definition, the range of phenomena that we consider under the heading *religion* is considerably expanded. Yinger insisted that nontheistic and even nonsupernatural systems of belief and practice can be appropriate social patterns for the sociologist of religion. "It is not the nature of *belief*, but the nature of *believing* that requires our study" (Yinger, 1970, p. 11). Wherever one sees a closing of the gap between fact and hope, wherever one sees a leap of faith that allows a person to assert that suffering and evil will somehow, someday be defeated, there one sees the manifestations of religion. Even a secular faith that science and technology will ultimately solve all our problems is, by this definition, a religious or quasi-religious phenomenon. Yinger (1970) wrote, "A term that already includes, by common consent, the contemplations of a Buddhist monk and the ecstatic visions of a revivalist cult member, human sacrifice, and ethical monotheism may have room in it for science *as a way of life*" (p. 12). Intense faith in nationalism, in capitalism, and in many other objects of deep loyalty may become grist for the student of religion if the object of loyalty and belief is expected eventually to solve the ultimate human perplexities over the purpose of life and the meaning of death, injustice, and suffering. Yinger argued that if a narrower definition is utilized, one may misunderstand and misidentify religion in a society, particularly in societies undergoing cultural change. This definition assumes that, to some extent at least, all people are religious. Yinger (1970) wrote, "To

me, the evidence is decisive: human nature abhors a vacuum in systems of faith. This is not, then, a period of religious decline but is one of religious change" (p. vii). The assumption underlying the *functional* definition of religion does not really invite the question of whether a society is becoming less religious but rather asks what new forms religion is taking. The sociologist adopting this approach is less likely to overlook nontraditional or alternative forms of religion (see Chapter 15) or new developments in the ways that people practice religion such as the way young people today approach religion (see Chapter 5).

Yinger, without making judgments about the truth or falsity of belief, also suggested that all people have a god or gods that give meaning to life (although the intensity of the need for meaning and the relative consistency of belief patterns vary widely among individuals). The task of the social scientist is to discover what it is that gives meaning to people's lives, for that is their religion. Many scholars who study religion agree with Yinger that most humans are more or less religious. Others, like Clifford Geertz (1968), think that the assumption of an intrinsic and universal need for a transcendental meaning in life is not justified. Social scientists differ, then, in their judgments about whether there is an innate religious tendency in humans.

Another well-known functional definition of religion is Robert Bellah's (1970c): "a set of symbolic forms and acts that relate [people] to the ultimate conditions of [their] existence" (p. 21). Like Yinger, Bellah's view of religion was influenced by the theologian Tillich's view of "ultimate concern." One problem is that "ultimate concern" or "ultimate conditions of existence" are difficult phenomena to identify and are even more difficult to measure. Nevertheless, Yinger asserted that any system of belief and action that fails to address the fundamental questions of meaning in life is not a religion.

Some scholars have argued rather persuasively that a supernatural dimension to the belief system is a fundamental characteristic of religion (Stark & Bainbridge, 1996; Stark & Finke,

2000). They argue that if a definition of religion does not include a supernatural dimension, the term *religion* may become so inclusive that it is virtually meaningless. Yinger rejected the idea that a supernatural dimension is necessary.

> *Critical Thinking:* Consider your own presuppositions: Is a belief in the supernatural necessary when you use the term *religion*? Alternatively, is it the quality of one's convictions and the depth of commitment that is essential?

Having discussed Yinger's (1970) approach to defining religion, it is appropriate to conclude with his full definition:

> Where one finds awareness of and interest in the continuing, recurrent, *permanent* problems of human existence—the human condition itself, as contrasted with specific problems; where one finds rites and shared beliefs relevant to that awareness, which define the strategy of an ultimate victory; and where one has groups organized to heighten that awareness and to teach and maintain those rites and beliefs—there one has religion. (p. 33)

A SYMBOLIC DEFINITION

Anthropologist Clifford Geertz developed a symbolic definition of religion that is somewhat more detailed in describing what religion does. *Symbols*—objects, behaviors, or stories that represent or remind one of something else—are powerful forces in human behavior, and they are central to religion. Given the abstract nature of the focal point of religion, symbols become its indispensable medium. Symbols include objects

(e.g., the cross, the Star of David), behaviors (e.g., genuflecting before the altar; baptizing a convert; touching the mezuzah on the doorpost of a Jewish home before entering; kneeling, facing Mecca, and praying five times a day), and myths or stories (e.g., the creation story, the story of Jesus washing the disciples' feet, legends of ancestors). (Myths and rituals will be discussed in more detail in Chapter 4.) Geertz was impressed with the way in which various levels of meaning can be communicated through symbols. Moreover, symbols are more accessible to observation than a subjective experience of "ultimate concern." Hence, he used symbols as the starting point for his definition of religion. Not all symbols are religious, of course. A handshake, a kiss, and a wave of the hand are all symbolic behaviors, but they do not normally have religious connotations.

Religious symbols are distinct from nonreligious ones in that the former are macro symbolic. *Macro symbols* are those that help one interpret the meaning of life itself and that involve a cosmology or worldview: a cross, circumcision, a loaf of bread and a cup. Because they serve this important function, they tend to acquire a sense of sacredness or profound respect. Many nonreligious symbols are *micro symbols*—that is, symbols that affect everyday interaction with others and that enhance daily communication and cooperation: a handshake, a wink, a smile. Micro symbols do not claim to explain the purpose of life and do not suggest values and beliefs that claim highest priority in one's life.[2]

Geertz's definition is so fully and carefully developed that it deserves a close examination. Geertz (1966) wrote,

> Religion is (1) a system of symbols which acts to (2) establish powerful, pervasive, and long lasting moods and motivations in [people] by (3) formulating

---

[2]Geertz did not stress the macro and micro difference, but this distinction is implicit in his discussion and seems to be more helpful than distinguishing "sacred symbols" from "nonsacred" ones.

conceptions of a general order of existence and (4) clothing these conceptions with such an aura of factuality that the moods and motivations seem uniquely realistic. (p. 4)

Religion is a "system of symbols which acts" in that the symbols provide a blueprint for understanding the world. These symbols provide a model of the world by helping people understand what the world and life really are. Many people believe, for example, that life is actually a testing ground in which God determines one's fitness to live in the heavenly kingdom. These individuals live their lives with reference to this understanding. Other religious perspectives offer alternative views of what life really is. In any case, these symbols not only suggest a model *of* the world but they also propose a model *for* the world. The symbol system describes what life is and also prescribes what it *ought* to be. Not only do many Christians assert that life is a testing ground but they claim access to the answers that will help them pass the test.

This *system of symbols*, Geertz (1966) continued, acts to "establish powerful, pervasive, and long lasting moods and motivations" (p. 4) in people. In other words, acknowledgment of the symbols affects one's disposition. Religious activity influences two somewhat different types of dispositions: (1) *moods* and (2) *motivations*. Geertz suggested that moods involve depth of feeling, whereas motivations provide a direction for behavior. Moods vary in intensity, and they affect our total outlook on life, but they are not aimed at any particular goal. One simply experiences a mood; one does not gain a feeling of obligation about a specific goal to be attained from a mood. Some born-again Christian groups emphasize that to be a Christian is to be joyful, even in the face of adversity. The emphasis is on a pervasive mood that is to characterize the believer, regardless of the specific circumstances.

Some religions may emphasize moods as primary (in Buddhism the focus is on mystical experience), while other religions stress motivations and a system of ethics (the Unitarian–Universalist society illustrates this latter focus). Nonetheless, Geertz suggested that in all religions the symbol system produces moods that intensify commitment and motivations to act in specified ways. In another context, Geertz (1958) referred to the moods and motivations together as the *ethos* of the religion.

Not only do the symbol systems enhance a particular disposition but they also act to "formulate conceptions of a general order of existence." A distinguishing characteristic of religion is that it provides a worldview, a cognitive ordering of concepts of nature, of self, of society, of the supernatural. Religion creates not only intense feelings but also establishes a cosmology that satisfies one's intellectual need for reasonable explanations. Geertz emphasized that not all intense feelings of awe are religious. One may be overwhelmed by powerful emotions (moods) in viewing the Grand Canyon, but such feelings may be either purely aesthetic or deeply religious. If no explanatory perspective or overview of the meaning of life is involved, the experience is not religious.

Geertz properly pointed out that religion involves an intellectual ordering. Some people claim that their religious experience comes from a walk in the woods rather than from going to a house of worship. Geertz suggested that a religious sense of awe must include a reaffirmation and commitment to a particular view of the world, a particular mode of interpreting the meaning of suffering, pain, death, and injustice. A walk in the woods may be refreshing and may involve an intense aesthetic feeling, but in most instances it does not change the way one thinks about the world or conceives of the overall purpose of life: "A man may indeed be said to be 'religious' about golf, but not merely if he pursues it with passion and plays it on Sundays: he must also see it as symbolic of some transcendent truth" (Geertz, 1966, p. 13).

There are three major challenges that seem to belie the meaningfulness of life, and it is these that a religious worldview must resolve: (1) a sense of coherence and reasonableness of events of life; (2) a sense of meaning in suffering so that it becomes sufferable; and (3) a sense of moral order in which evil will be overcome and that virtue, goodness, and justice will somehow, someday prevail.

Symbol systems, then, attempt to "account for, and even celebrate, the perceived ambiguities, puzzles, and paradoxes in human experience" (Geertz, 1966, p. 23). The worldview represents an intellectual process by which people can affirm that life makes sense, that suffering is bearable, and that justice is not a mirage—that in the end, good will be rewarded.

Geertz continued his definition by attempting to answer the question of how a particular worldview or set of concepts comes to be believed. The symbols act to "clothe those conceptions in such an aura of factuality that the moods and motivations seem uniquely realistic" (Geertz, 1966, p. 23). How is it, he asked, that despite common sense, everyday experience, and empirical evidence, people will come to believe irrational and unsupportable things? What compels a Christian Scientist to deny the reality of illness, even though the person experiences the symptoms of influenza? Why does a Mormon believe that a new revelation was written to Joseph Smith on golden plates, even though no one could read them but Smith? Why do members of the Unification Church assert that their leader, the Reverend Sun Myung Moon, is the Christ, even though other observers see him as a rather unexceptional man? Why do Christians continue to affirm that Jesus is the son of God who ushered in God's kingdom, even though he died in the manner of a criminal nearly 2,000 years ago? Geertz pointed out that religious ritual often creates an aura in which a deeper reality is said to be reached. Truths are experienced or understood that are more profound than everyday experience provides.

Religion is communal in character and often involves intense emotional experiences. The photo depicts a worship service among evangelical Christians. The intense emotional experience acts to clothe the religious concepts in a mystique that makes these concepts seem uniquely realistic.

As we will see, Geertz described a very important aspect of religion. He pointed out that religion is able to provide a foundation for social values that has its authority outside of empirical verification and therefore cannot be invalidated (Geertz, 1968). Values and perspectives come to be enshrouded in sacredness and unquestioned certainty. We will explore this issue further in our discussion of plausibility structures in Chapter 8.

Many students have difficulty understanding Geertz's heavy emphasis on symbol systems (which are discussed in more detail in Chapter 4). Let it suffice here to say that meaning is commonly "stored" or encapsulated in symbols. They are powerful factors in the lives of people. They are also more concrete and observable than an "ultimate concern."

Geertz's definition is long, abstract, and quite elaborate. (His explanation of the definition runs 46 pages.) This makes it difficult to translate it into concrete research procedures. However, his definition does offer a contribution to the debate over what it is that distinguishes religion from other cultural phenomena. His central contributions are that religion must have a macro symbol system that acts to reinforce both a worldview and an ethos and that has a built-in system of believability or plausibility.

Researchers who like to employ the participant observation method of research find Geertz's definition useful, for it identifies the general properties of religion and helps the observer know what to observe. Yet, it does so without specifying the content of religious beliefs. Those researchers who engage in quantitative studies, such as sample surveys, find Geertz's generalizations so broad and encompassing that his identification of religion is unhelpful. How does one quantify (turn into countable units) a worldview or an ethos? The broadness and inclusiveness of Geertz's treatment is frustrating to those who seek unequivocal precision in categorizing types of behavior, but it is this same generality that attracts those who undertake cross-cultural investigations.

Geertz's analysis is really more than a definition. It is an essay on how religion "works" to reinforce itself and on what religion "does" in the society. Because of this focus on what religion does, the symbolic definition may be considered as one type of functional definition (Berger, 1974).

Critical Thinking: How do you think symbols, moods, and motivations play a role in religious commitment? What symbols elicit strong moods and motivations for you?

THE CONCEPT OF RELIGION AS EMPLOYED IN THIS TEXT

Our underlying interest is in the way people generate and sustain new systems of meaning in the midst of social change. Moreover, we are interested in how individuals create their own systems of meaning. Usually, meaning systems involve a synthesis of official religious doctrine with other cultural beliefs. Rather than dichotomizing religion from nonreligion, we seek to explore anything that provides meaning and purpose in the lives of people. We tend to ask *how* people are religious rather than *whether* they are religious. Hence, the perspective of this book will be most compatible with the functionalist definitions of Yinger and Geertz, each of whom was interested in religion as a cultural system. While we incorporate the research and the insights of those who use a more narrow substantive definition of religion, we also address the broader issues of meaning in a changing culture.

Let us synthesize the debate over definitions by simply highlighting our view of the distinguishing characteristics of religion. First, religion is a social phenomenon that involves the grouping of people around a faith perspective. Faith is an individual phenomenon that involves trusting in some object, event, principle, or being as the center of worth and the source of meaning in life. We sympathize with Yinger's insistence that the "nature of believing" is probably more indicative of religion than the "nature of belief" itself. Hence, a profound commitment to Marxism, intense nationalism, or faith in science and technology as the ultimate solution to our human predicament could be considered at least quasi-religious phenomena. Yet religion is also viewed here as a social phenomenon—involving a group of people with a shared faith or a shared meaning system. Beyond being just a social phenomenon, religion has to do with that assortment of phenomena that communicates, celebrates, internalizes, interprets, and extrapolates a faith. These phenomena include *beliefs* (myths[3]), *rites* (worship),

[3]The term *myth* does not refer to a belief that is untrue. Myth refers to any belief that helps people understand and interpret the events in their own lives (see Chapter 4). It is a macro symbolic system of meaning.

an *ethos* (the moods and moral values of the group), a *worldview* (the cognitive perspective by which the experiences of life are viewed as part of a larger and ultimately meaningful cosmology), and a *system of symbols* that serve to encapsulate the deepest feelings and emotion-packed beliefs.

The criteria identified here should be specific enough to distinguish religion from many other cultural phenomena. For example, we have suggested that a worldview and an ethical system are intrinsic aspects of religion. A feeling of awe or sacredness is not considered religious in this text unless it includes a cognitive pattern that helps persons make sense out of life and helps explain the meaning of suffering, death, and injustice.

On the other hand, we hope that the criteria for identifying religion are sufficiently broad so that we do not miss the religious significance of nontraditional groups. We will be studying Methodists, Muslims, Mormons, and Moonies, but this approach also allows us to explore scientific humanism, Transcendental Meditation (TM), and American nationalism as religious or quasi-religious movements.

To summarize, we believe that religion is an interdependent system by which a community of people are bonded

- by a shared meaning system (a faith or a worldview);
- by a set of myths (beliefs), rituals, and symbol systems that sacralize the meaning system for the members;
- by a sense of belonging to a reference group;
- by a system of ethics or values that is directive in the lives of the members; and
- by a set of routinized social expectations and patterns.

Those phenomena in society that have most, but perhaps not all, of these characteristics will be explored as possible "invisible religions" or "quasi-religions" that can impact traditional religion and that may well be emerging as new religions in the society.

> **Critical Thinking:** Think about how the authors have summarized the core elements of religion. Do you agree that the authors have identified the core of what makes something a religion? Why or why not?

DISTINCTION BETWEEN RELIGION AND MAGIC

When studying religion, social scientists often find themselves investigating a closely related phenomenon—magic. The two are often found together, but because they serve rather different functions, students of religion should understand the distinction between them.

Many criteria have been used to discriminate between religion and magic. In the listing that follows, some of the differences are juxtaposed. Various scholars have emphasized one or more of these factors (Titiev, 1972).

These seven distinguishing factors make a nice, neat picture of magic and religion, but in reality the distinction is not so precise. There are many worship experiences in which the members may not find the ritual meaningful. Many members may attend worship only at critical times in their lives (e.g., funerals, baptisms, marriages, or times of emotional stress for the individual), and the minister may emphasize utilitarian reasons for belonging (e.g., God's blessing on one's business or promises of good health). Yet the phenomenon hardly seems to be magic rather than religion. Furthermore, the factors listed do not show a high correlation with each other. The distinction between magic and religion remains somewhat foggy. Nonetheless, one factor seems worthy of our attention.

The primary difference between religion and magic has most commonly been defined in terms of the attitude toward the transcendent: Is it considered of intrinsic or utilitarian value? The religious perspective views the object of worship as being of inherent, categorical worth.

Religion	*Magic*
1. There is a sense of a "group" of common believers: a church, temple, or mosque.	1. No faith community or "group consciousness" is involved.
2. Moral ethos, or a system of ethics, guides behavior.	2. No moral ethos or systematic pattern of ethics is present.
3. Rites are meaningful: They reinforce patterns of belief.	3. Rites are not necessarily meaningful; they are used to cast a spell or make something happen.
4. Rites occur calendrically (on a regular basis each week, month, and/or year).	4. Rites occur at critical (crisis) times.
5. It functions for both the individual and the structure.	5. It functions only for individuals, not for social structure.
6. Participation is open; a leader leads the entire group in performance of ritual.	6. The leader is the only one to know ritual and how to perform it; others present are passive.
7. The worship of a transcendent Being or Power as intrinsically worthy of one's attention occurs.	7. Manipulation of impersonal, transcendent power occurs for utilitarian reasons.

It is not worshiped primarily because of favors to be returned for performance of the ritual. God is worshiped in the Jewish, Christian, and Islamic traditions—or at least is supposed to be worshipped in those traditions—simply "because God is God." In religion, the object of worship is the center of worth or the epitome of value.

In the case of magic, the world is believed to be influenced by supernatural forces that control one's destiny. The objective is to get those forces working for you and not against you. Magic involves manipulation of supernatural forces. In a discussion of magic in American culture, George Gmelch (1971) pointed out that many baseball players will often follow such patterns as rising in the morning at a specific time, eating only certain foods prior to a game, consistently taking the same route to the stadium, getting dressed according to a specific preset pattern (e.g., always putting on the right sock first), and always stepping on third base on the way to the outfield. The players who followed these rituals seemed to feel that they would be jinxed if they broke them. Somehow, the unspoken forces of fate would betray them if the proper (nonrational) steps were not taken.

While this may be an unusual example, the worldview of expecting supernatural and impersonal forces to control one's destiny is characteristic of what anthropologists have called magic. The supernatural forces are not personal beings to whom one prays or with whom one develops a relationship. Rather, they have no will of their own. They simply represent laws of the universe that must not be violated and that might be used to one's own advantage. Hence, magic in tribal societies is a supra-empirical means of controlling one's chances of success in any endeavor. Whenever the desire for success is high but the chances of failure are great, people may look for any means possible to ensure success. Bronislaw Malinowski (1931) wrote, "Magic is to be expected and generally to be found whenever man comes to an unbridgeable gap, a hiatus in his knowledge or in his powers of practical control, and yet has to continue in his pursuit" (p. 638).

Some of the functions of magic are the same as those of religion. In a study of water witching (using a forked twig to locate a source of underground water), Evon Vogt (1952) found some important functions being served. In Homestead, New Mexico, the cost of drilling was high, the chance of failure great, and success in finding water critical. Hence, people turned to a dowser (water witch). The persons investing money in drilling were satisfied that they were doing *something* to increase the likelihood of success. Because the water witch had confidence, the anxiety of the people was reduced, and they were willing to invest more in any given drilling site. They agreed to drill deeper because they were assured that water was there. Vogt insisted that there is no scientific evidence to support the claim that dowsers can actually find water and that mere chance is actually a more reliable means of hitting water than employing a dowser. Nonetheless, farmers continue to use them because doing so lessens their anxiety. The worldview of those farmers is magical in that they assume some supernatural force is at work in helping the witch to locate water. The behavior of people who forward chain letters out of fear of "bad luck" in the next week if they fail to do so—or in expectation of some "blessing" in the next 4 days if they do pass it along—is another example of a "magic" worldview.

Magic provides a concrete action to *control* adverse events. In actuality, magic serves as a form of primitive technology or as a complement to empirical techniques. If we want to control events, but we do not have at our disposal rational means to do so, then we may turn to supernatural methods. In the case of magic, one looks for ways to manipulate supernatural forces for one's own benefit. One does not have to confess sins or otherwise earn a right relationship with supernatural beings. The twig will work regardless of

the moral righteousness of the dowser. Magic has no ethos and no "moral community."

It is true that religion is often concerned with mundane adversities, but the method of resolving the problem calls for moral purification, confession, or some other transformation of the person involved. Faith healing, for example, is usually expected to be effective only after the person is in "right relationship with God." If a person is ill and resorts to magic for a cure, it is believed that healing will come if the words are said correctly and the ritual is performed properly. If healing is not forthcoming, the ritual was not performed correctly. However, if a person turns to a religious faith healer for help and the ailment is not cured, the explanation is frequently that the afflicted person or the healer is not in harmony with God.

When studying any one group, the question of whether one is encountering magic or religion may not be as clear and categorical as suggested here. Nevertheless, the primary appeal of magic is to manipulate the world. Science and technology are more rational means of controlling one's environment. As science and technology allow persons to have increasing control over their lives, magic tends to decline. The amount of magic in a society tends to be inversely correlated to the amount of science and technology.[4] As Malinowski said, we do not find magic wherever the pursuit is certain, reliable, and under the control of rational methods.

It seems reasonable to suggest that magic will be replaced in large part by science, technology, and the modern secular worldview. The primary functions of religion, however, are not to manipulate one's environment. There are other functions that are central to religion; explaining the meaning of life and preserving central values and ethical codes in the culture are among those. Technology may help us control our environment

---

[4]This does not mean that magic is the primitive substitute for science. Anthropologists have shown that even the simplest society has some empirically validated knowledge (Fabian, 1992; Malinowski, 1948), and technical skill is never replaced by magic. Malinowski (1931) wrote, "Magic, therefore, far from being primitive science, is the outgrowth of clear recognition that science has its limits and that a human mind and skill are at times impotent" (p. 637). Only when technology can ensure success in a task does magic cease.

and can make life more comfortable, but it can never explain why life is meaningful in the first place. Max Weber (1946) quoted Tolstoy to this effect in his famous lecture on "Science as a Vocation": "Science is meaningless because it gives no answer to our question, the only question important for us: 'What shall we do and how shall we live?'" (p. 143).

This is not a defense of religion. Often the answers to these questions offered by a given religion may be bizarre. However, it is crucial to distinguish between *empirical facts* and moral values. Science can only answer the question of how something happens in terms of material causality; religion attempts to answer the question of why something happens in terms of values and ultimate meanings.

> **Critical Thinking:** Do science and religion address different functions? Why do you think as you do?

SPIRITUAL BUT NOT RELIGIOUS?

Another issue that emerges is whether private systems of belief are to be called religion. After all, many individuals have patterns of belief that solve the ultimate meaning issues for them but that are not necessarily shared with others. Yinger (1970) insisted, as do most sociologists of religion, that religion is a "social phenomenon: it is shared and takes on many of its most significant aspects only in the interaction of the group" (p. 10). Privately held patterns of meaning may have religious aspects, but they are not religion per se.[5]

Indeed, today some people consciously reject organized religion in favor of more individualized forms of "spiritual" belief and practice. It is increasingly common to hear people utter the phrase, "I am spiritual, not religious." Spirituality in this sense is seen as a quality of an individual whose inner life is oriented toward God, the supernatural, or the sacred. Spirituality is considered primary, more pure, more directly related to the soul in its relation to the divine, while religion is secondary, dogmatic, and stifling, often distorted by oppressive sociopolitical and socioeconomic forces. Some scholars have argued that in the new millennium there is a "divorce" between spirituality and religion with more personal forms of spirituality destined to replace traditional, organized forms of religion (Cimino & Lattin, 2002). However, the relationship between spirituality and religion is not quite as simple as that.

Robert Wuthnow (1998) argued that, "At its core, spirituality consists of all the beliefs and activities by which individuals attempt to relate their lives to God or to a divine being or some other conception of a transcendent reality" (p. viii). There is nothing in this definition of spirituality that makes it inherently antithetical to religion. To the contrary, spirituality has long been historically connected to religion. Even though it is a social phenomenon, individual forms of piety such as prayer, meditation, or other devotions (often with a mystical component) have long been part and parcel of many major religious traditions. Sufism in Islam; Kabbalah in Judaism; and Benedictine, Franciscan, and Dominican spirituality in Roman Catholic Christianity are well-known examples. Given the historical connection between traditional religion and spirituality, it may be better to use the term Eva Hamberg (2009) suggested—"unchurched spirituality"—to refer to religious beliefs and practices that exist outside of traditional religious institutions.

A second important point to consider is that "unchurched" does not mean "not social."

---

[5]The faith of individuals, even if not celebrated in a group context, is also of interest to sociologists of religion. Such individualized faith systems are sometimes referred to as *invisible religions* (Luckmann, 1967). These will be discussed in Chapter 14. However, individual meaning systems will be referred to henceforth as personal religiosity or as faith; religion will be defined as a social phenomenon.

Wuthnow (1998) pointed out that "spirituality is not just the creation of individuals; it is shaped by larger social circumstances and by the beliefs and values present in the wider culture" (p. viii). That is, we construct our spirituality out of the "toolbox" of cultural resources that is available to us at the time we are living. Many of the most significant cultural resources for spirituality, as we have already noted, come from the major historical religious traditions. In the end, although it is conceptually distinct, individual spirituality is never far removed from religion.

This is borne out, at least in the U.S. context where organized religion remains strong, in some empirical studies of the relationship between spirituality and religion. In a survey of baby boomers—individuals born between 1945 and 1963 who are supposed to be on the leading edge of the revolution in spirituality in American society (see Chapter 5)—Wade Clark Roof (1999, p. 178) asked individuals to identify themselves as "religious" and/or "spiritual." The pattern of responses to these two identity questions was as indicated in Figure 1.1 below.

Nearly 60% of the respondents are *both* spiritual *and* religious. Only 15% of the respondents answered that they are spiritual but not religious, and an equal number rejected the designation of spiritual even though they self-identify as religious.

More recently, *Newsweek* and the website Beliefnet polled a random sample of Americans and found a similar overall pattern—with some important variations (see Figure 1.2a). A majority of older Americans, including baby boomers (the 40- to 59-year-old category in 2005), continue to see themselves as religious *and* spiritual, but among those under 40 years of age, a minority (48%) say that best describes them (see Figure 1.2b). Like the baby boomer generation, nearly 30% of these younger adults characterize themselves as spiritual but not religious—nearly twice the proportion of 60-year-olds and up who describe themselves that way (See Figure 1.2c). These younger adults differ from the baby boomers, however, in their willingness to deny being either religious or spiritual—11% of 18- to 39-year-olds

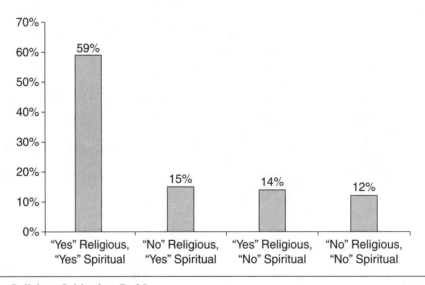

Figure 1.1 Religious, Spiritual, or Both?

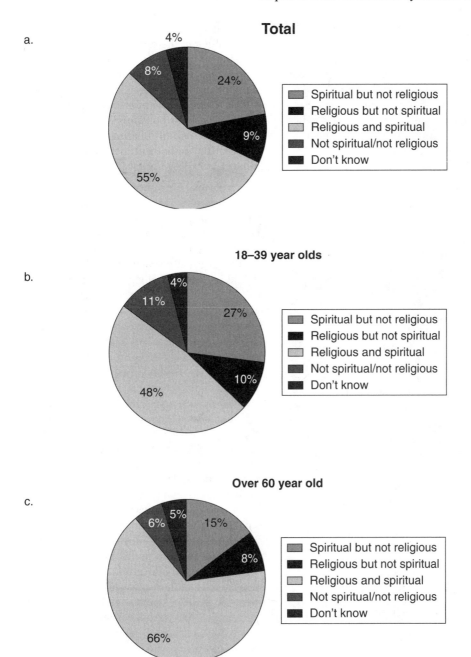

a.

Total

- Spiritual but not religious
- Religious but not spiritual
- Religious and spiritual
- Not spiritual/not religious
- Don't know

b.

18–39 year olds

- Spiritual but not religious
- Religious but not spiritual
- Religious and spiritual
- Not spiritual/not religious
- Don't know

c.

Over 60 year old

- Spiritual but not religious
- Religious but not spiritual
- Religious and spiritual
- Not spiritual/not religious
- Don't know

Figures 1.2 Relationship Between Being Spiritual and Being Religious Among American Adults, Comparing Age Groups

SOURCE: Beliefnet.com. (2005, August). Newsweek/Beliefnet poll results. Retrieved from http://www.beliefnet.com/News/2005/08/Newsweekbeliefnet-Poll-Results.aspx

describe themselves this way, compared to just 5% of 40- to 59-year-olds.

These important generational changes notwithstanding, these data also clearly demonstrate that the majority of Americans do not see religion and spirituality as a zero-sum game where one must be one *or* the other, a fact that other studies have born out (Marler & Hadaway, 2002).

A FINAL WORD ABOUT DEFINITIONS

One's definition of religion is important, for it specifies what is and what is not appropriate data for study. The discussion in this chapter was designed to help the reader understand the differences in the ways religion has been defined by scholars. We hope that the discussion has stimulated you to think through your own criteria for identifying religion. A consensus among us would be convenient, but a lack of agreement need not cause problems (Lechner, 2003). The purpose of this text is not to convert readers to the authors' theoretical persuasion but to help them think more clearly about the relationship between religion, culture, and society. Before going further, it would be helpful to consider (1) your own assumptions regarding the definition of religion, (2) the defining criteria used by the social scientists discussed in this chapter, and (3) the perspective of the authors. As we noted at the outset of this chapter, and as Yinger (1970) had written,

> Definitions are tools; they are to some degree arbitrary. . . . They are abstract, which is to say they are oversimplifications. . . . We must relinquish the idea that there is any one definition that is correct and satisfactory for all. (p. 4)

The definition we each use tends to "slice up life" a little differently and causes us to focus on slightly different phenomena as most important. Hence, we have begun by making our assumptions about religion conscious and explicit.

Although there is no consensus on the definition of religion, there is agreement among sociologists that any investigation of religion must be based on empirical methods of investigation. In the next chapter, we explore what it means to take a social scientific approach to studying religion.

SUMMARY

Definitions of religion are usually one of two types: (1) substantive (which focus on the substance or essence of religion) and (2) functional (which focus on what religion does). Substantive definitions usually emphasize a specific belief, such as in spiritual beings or in a supernatural realm, or they stress the distinction between sacred and profane realms of experience. Substantive definitions tend to focus attention on the traditional forms of religiosity. Functional definitions identify religion as that which provides a sense of ultimate meaning, a system of macro symbols, and a set of core values for life. They invite an investigation of any profound loyalty or ultimate meaning system as a form of religiosity. Social scientists who are interested in cultural change and the new forms of meaning that emerge in times of cultural transition tend to favor functional definitions. Because they are not overly focused on traditional forms of religiosity, they often view religion as changing rather than as declining.

This text is based on the definition of religion as an interdependent system by which a community of people are bonded (a) by a shared meaning system (a faith or a worldview); (b) by a set of myths (beliefs), rituals, and symbol systems that sacralize the meaning system for the members; (c) by a sense of belonging to a reference group; (d) by a system of ethics or values that is directive in the lives of the members; and (e) by a set of routinized social expectations and patterns. It also views religion as different from, though complexly related to, magic, science, and spirituality.

2

A SOCIAL SCIENTIFIC
PERSPECTIVE ON RELIGION

Here are some questions to ponder as you read this chapter:

- What perspective do sociologists take in analyzing religion? What is their major focus?
- If sociologists want to find out if something is true, how do they gather data?
- What is considered "evidence" in sociology—that is, what is considered sound enough to reach a conclusion?
- What kinds of observations are not considered dependable evidence?

A highly respected sociologist once said of religion, "There are few major subjects about which people know so little, yet feel so certain" (Yinger, 1970, p. 2). This seems to be true of both those who are sympathetic to religion and those who are hostile toward it. In light of this, we suggest two characteristics or attitudes that will be particularly helpful in approaching our topic of investigation: a healthy dose of humility and a corresponding openness to new ideas. No one has all the answers on religious behavior, but by listening to one another and by paying close attention to the empirical evidence we have, we can each broaden our understandings.

In this text, we will be presenting a sociological perspective on religion and explaining some of that discipline's findings and theories. Sociology does not offer the whole truth any more than any other discipline or perspective, but it does offer insights that can be helpful to those interested in religion—whether they be believers or skeptics. Our point is that we seek understanding here; a posture of defensiveness—with each person seeking only to preserve his or her own preconceived notions—is counterproductive. The 17th-century philosopher Baruch Spinoza's (2009) summary of the intellectual life applies equally to our work here: "Not to laugh, not to lament, not to curse, but to understand" (I §4). Among other things this means that we seek neither to dissuade believers from their faith nor to convince skeptics of the efficacy of religion. Our goal is to gain a new perspective on religious behavior and thereby expand our comprehension of religion.

Sociology offers only one of many possible vantage points from which to view religion. Consider the following analogy: Many people may interact with the same child and yet have quite different perspectives and understandings of that child. An artist may try to capture the child's charm by focusing on his or her unique physical properties such as facial features; body proportions; and shades of coloration in the eyes, skin, and hair. A physician is interested in the physiological needs of the child and the requirements for good health—concentrating on such characteristics as height and weight, immunizations, and the family's history of congenital diseases. A developmental psychologist may study the child not because of an interest in the characteristics of this child but as part of a broader investigation of childhood development. Parents, of course, are interested in the uniqueness and specialness of this *particular* child. Their concern is not one of detached analysis, for their emotional attachment to the child influences their perception. As the persons responsible for the child's social and emotional development,

they try to be concerned about the child's overall welfare. Because they are so close to the child, some important *patterns* of behavior may go unnoticed. The findings of each of the previous observers may be of interest to the parents as they come to a fuller appreciation of their child and a better understanding of their parental responsibilities. Finally, the child is a unique self-experiencing individual with his or her own self-understanding. The subjective experience is unique and important but also has its own biases and limitations.

It would be foolish to ask which of these persons has the "true" understanding of the child. No one perspective is total and complete. In fact, the insights of the psychologist may influence the socialization practices of the parent, or the prescription of the physician (e.g., the child needs to wear a back brace) may affect the child's behavior and experiences. In each case, the "objective" view of an outsider may differ from that of the parent or of the child, and that perspective may lead to changes. In the long run, those changes may be beneficial to both the parent and the child.

Just as people from many fields have unique perspectives on a child—none of which contains the whole truth about that child—so also can religion be understood from many angles. The psychologist analyzes religious experience as a mental and emotional experience of the individual. The concern is not with religion as ideas or beliefs that may hold eternal truth but with the effect of religion on the human psyche. The philosopher of religion approaches the subject by comparing, contrasting, and analyzing beliefs of various faiths, focusing on the ideas of life, death, suffering, and injustice among the many religions of the world. The systematic theologian formulates doctrines about God and about God's relationship to the universe and to humanity, placing what is believed about God (in a particular tradition) into a logically comprehensive and coherent framework. The religious ethicist attempts to define moral responsibility of religious persons, or at least to clarify moral discourse and identify moral dilemmas for members of a faith. The faithful follower understands his or her religion through yet another lens—that of personal commitment. The person's faith is viewed as a source of ultimate truth and personal fulfillment. Members of the clergy also view religion from a vantage point as committed followers, but they are also leaders who seek understanding of religious processes so they can be more effective in guiding others. Hence, they may use the insights of the social scientist, the philosopher, the theologian, and the ethicist in order to understand more fully both their faith and their leadership responsibilities. The sociologist, as we will see, offers a unique perspective that differs from these others and that can contribute to a holistic understanding of this multidimensional phenomenon we call religion.

THE SOCIOLOGICAL PERSPECTIVE

The sociological approach focuses on religious groups and institutions (their formation, maintenance, and demise), on the behavior of individuals within those groups (e.g., social processes that affect conversion, ritual behavior, or decision to defect to another group), and on conflicts between religious groups (such as Catholic vs. Protestant, Christian vs. Muslim, mainline denomination vs. cult). For the sociologist, beliefs are only one small part of religion.

In modern industrial society, religion is both a set of ideas (values, beliefs) and an institution (structured social relationships). We will be looking at both in order to understand how they affect human behavior. We will investigate differences in beliefs not because we expect to prove their truth or falsehood but because beliefs—regardless of their ultimate veracity—can influence how people behave and how they understand the world.

Religious institutions, however, can also affect behavior quite independently of beliefs. In fact, religious institutions sometimes entice people to behave contrary to the official belief system of that religion.

Later in this book, we will discuss the fact that religious organizations may contribute to racism and combat it at the same time. While the belief systems of most mainline Christian denominations proclaim prejudice to be wrong, the institutional structures of the church unwittingly permit it—and sometimes even foster it. Furthermore, religious beliefs themselves can have contradictory effects. While most religions teach that antipathy toward others is wrong, certain other beliefs have contributed—often unconsciously and unintentionally—to racial prejudice, sex bias, and anti-Semitism. For example, some 1st-century Christians believed that women were incapable of being saved, unless they were first transformed into men. These people reinforced the accepted cultural view of that time that women are defective human beings. In this case, beliefs shaped how people viewed the worth of themselves and others. For example, between AD 1400 and 1700, the Christian churches (both Protestant and Catholic) were involved in burning between 500,000 and 1 million women as witches (Ruether 1975; Nelson 1975; Trevor-Roper 1967). However, scholars have discovered that this massive gynocide[1] was due much less to religious beliefs than to changing sex roles in the society. Secular conflicts (over the proper role of men and women) were expressed in "religious" activities (burning witches as infidels). Religious behavior can be either a cause or an effect of other social processes. There is no assumption in our analysis that religion is always the consequence of other forces, but we do not ignore that possibility either (Stark, 2000a).

In short, there are many ways in which religious groups, religious values, and secular social

---

[1]Genocide is the annihilation or attempted annihilation of an entire people (such as the Nazi holocaust in Germany). Gynocide (*gyno* meaning "women") is the attempted annihilation of the female sex.

processes can be interrelated. Beliefs are not always at the heart of religious behavior. Social scientists have found that persons sometimes become committed to new religious groups with little knowledge of the group's beliefs. They become committed through group pressures and social processes (discussed in Chapter 5). Sociologists are convinced that knowing what a group believes provides insights only into one aspect of this complex phenomenon we call religion. For a fuller understanding, one must comprehend the social processes as well.

Most Americans believe that the central differences between religious groups have to do with their beliefs, but there are many interesting and important variations in style of worship, authority structures, and psychological appeal of religious groups. These differences can have a real impact on how one views life, others, nature, and their own experiences.

Sociology, then, focuses on the social dimensions of religion—including the manner in which religion affects society and the ways the society influences religion. Like the developmental psychologist who studies a child to discover the stages of personality development in all children, we will be looking for the common patterns—the general rules—rather than for unique characteristics of each religion. When we do look at unique characteristics, it is to find how those characteristics affect behavior in special ways.

Beyond its particular theoretical perspectives, sociology as a discipline aspires to take a *scientific* approach to the study of religion. A scientific approach is characterized by two fundamental principles: (1) objectivity and (2) methodological empiricism. In the first case, the sociologist tries to be objective. Objectivity does not mean that the sociologist claims to be above error or to have the whole truth; we have already pointed out that no discipline can claim omniscience. Objectivity means that the sociologist tries to prevent personal beliefs, values, or other biases from affecting the study. The social scientist is committed to the search for truth wherever that search leads. Although sociologists as human beings have preferences and commitments, they seek to be open to the data and to avoid prejudgment of any particular group or any particular religious process. A sociologist may not agree with the views of a group being studied but makes every effort to understand the group on its own terms and to avoid bias in interpreting the processes of the group. Even if they do not agree with the activities of the group in question, sociologists try to base their judgments only on the data in front of them.

Sociologists seek objectivity as a rule of behavior in exercising their discipline. However, this objectivity is not always easy to achieve. Philosopher Leszek Kolakowski once observed that no well is so deep that leaning over it one does not discover at bottom one's own face (Burawoy, 1987, p. 59). There is no standpoint ("Punctum Archimedis") of perfect objectivity from which an observer can see the totality of reality without bias. A sociologist who is active in a church or temple or is otherwise a "believer" is perhaps more likely to be sympathetic to "believers" than is a sociologist who is a committed atheist. As in other areas of study in sociology, those who choose to study religion often do so out of their personal interest in the subject.

The authors of this textbook include an active church member with a theological education (Boston University School of Theology) who was ordained in the United Methodist Church (Roberts) and an adult convert to Roman Catholicism (Yamane). We recognize that our personal commitments may unconsciously affect our perspectives on religion, but we do not abandon the quest for objectivity because of this. To the contrary, we redouble our efforts to be aware of and counter possible biases we bring to our studies. The important point here is that the first step toward objectivity is to identify as clearly and thoroughly as possible any feelings and biases one might have about the object of inquiry. This is not a simple matter of declaring one's group memberships as we just have. It involves an ongoing soul-searching into previously unrecognized feelings and biases and a constant effort to understand each group on its own terms.

In this, we are no different than feminists who study gender in society or Marxists who study class inequality, second generation Americans who study immigration, or African Americans who study race relations. These personal connections to one's subject matter are not problematic in themselves. Our approach mirrors that of Max Weber's (1949) position in early methodological debates in the social sciences over the place of values and the possibilities of objectivity. From a social scientific perspective, values can influence the topics sociologists choose to study and the questions they ask, but they should not affect the data collection, analysis, or conclusions that are drawn.

Again, this is to some extent idealistic given that sociologists are human beings, but it is a valuable ideal nonetheless. Sociologist Kristin Luker (2008) put it well when she wrote, "I see the search for objectivity in the social sciences as something along the lines of Zen enlightenment: I don't personally expect to achieve either of them, but I do find the pursuit worthwhile" (p. 6).

This leads us to the second fundamental principle of a scientific perspective: methodological empiricism. This principle helps meet the challenge of objectivity just discussed by emphasizing the need to make claims about the social world based on systematically gathered data.

Reliance on empirical data means the sociologist as a social scientist considers only data that are observable through the five senses. A sociologist does not assume that a report of supernatural influence means that supernatural influence is "really real," though it can be real in its influence on the beliefs and behavior of the informant. Like other scientists, the sociologist deals in facts that can be measured, observed, and tested. If a sociologist claims that belief X causes behavior Y, empirical studies can be set up and data gathered to support or refute the hypothesis. If politicians and members of the public believe that devotees of religious cults are more likely than nondevotees to be mentally unstable, that hypothesis can be substantiated or disproved through empirical investigation. If members of certain religious groups or persons with particular religious beliefs are thought to be more racially prejudiced than others, only empirical investigation can establish the validity of that claim. The sociologist is not satisfied with general impressions but seeks concrete, verifiable data to verify or disprove any generalization.

This is not to say that sociology will ever achieve the precision of measurement or explanatory power of a natural science like physics or chemistry. The object of study in sociology has an inherently subjective element—human beings capable of exercising "agency" (the independent capacity to act)—that the object of study in the natural sciences does not. As many sociologists will tell you, this makes natural science a comparatively "easy" science and sociology the truly "hard" science.

> *Critical Thinking:* What are your own religious convictions or your assumptions about religion? How might these ideas—or your own gender, socioeconomic, ethnic, or national status—create a possible blind spot for you in understanding religious groups that differ from your own point of view?

SOCIOLOGICAL METHODS OF STUDYING RELIGION

An important aspect of sociologists' claim to being a (social) science is our use of systematic and rigorous methods of data collection and analysis. Unlike the journalist who builds a story based on those willing to talk to her, or the partisan who only listens to those whom he agrees with, or the novelist who is free to create her own alternative realities, sociologists (ought to) restrict their claims about social life to those for which they have good empirical evidence. It is appropriate therefore to discuss the methods by which social scientists gather their data.

Survey Research and Statistical Analysis

Since the middle of the 20th century, the dominant mode of research in sociology generally has been statistical analysis of data from closed-ended surveys of individuals. This is often called "quantitative" research, because the focus is on numbers. Some estimate that 85% of all articles published by sociologists are statistical analyses of survey data. The same is true for sociologists of religion, especially as access to large-scale data sets has expanded in the study of religion.

The key components of this approach are the survey instrument (what questions are asked) and the survey sample (who is asked the questions). In contrast to interviews that tend to be less structured and more flowing, surveys tend to have closed-ended questions—that is, questions that have prespecified answer categories. Here is an example:

How often, if ever, do you pray privately? (Please check one.)

- More than once a week
- Once a week
- Two or three times a month
- Once a month
- Less than once a month
- Almost never
- Never

Responses to these closed-ended questions are easily quantifiable, making survey data amenable to statistical analysis. For example, we can create descriptive statistics like the percentage of people who pray weekly or more. Alternatively, if we have a second question, such as frequency of religious service attendance, we can cross-tabulate and determine the percentage of people who both pray weekly or more and attend religious services at least weekly.

The most common way of analyzing survey data in sociology is *multivariate inferential statistics*. This simply means that the researcher attempts to infer the causal effect of one or more *independent variables* on some *dependent variable*. Independent variables are sometimes called causal variables or explanatory variables; they are believed to cause or explain something else that occurs as a result, which we call a dependent variable. These statistical studies that try to study causality will also include "control variables"—variables that the investigator is not directly interested in but which could influence the dependent variable and cause a mistaken assumption about the links between independent and dependent variables. Control variables, therefore, need to be held constant in order to assess the real influence of the main variables being studied. Common variables used in sociological studies of religion—either as control, independent, *or* dependent variables—include age, gender, race, region, education, and income. Causality is central to scientific theories, which is why it is so important to identify independent and dependent variables and to control for other influences. The use of causal language is also important in understanding causality, as indicated in more detail in the "Illustrating Sociological Concepts" feature.

Creating the questions that will go on a closed-ended survey must be done very carefully, because once the survey instrument is administered to the respondents, the researcher cannot easily change it. The questions should get at the underlying concept in a way that makes sense both to the researcher and (more important) the individuals being surveyed. Social scientists call this process *operationalization*. When we "operationalize" a concept in survey research, we translate conceptual abstractions—like "religiosity"—into specific questions to be answered, such as How often do you pray?

ILLUSTRATING SOCIOLOGICAL CONCEPTS

Sociology and the Language of Causality

As in any other science, sociology attempts to identify social processes that cause, or effect, other social processes. To fully understand subtle differences that social scientists make in the importance of one variable in shaping another, it is critical to understand the language of causality.

First, a variable can be a *contributing factor*, which means that it is one of many factors. If we were to explore the variables that cause persons to become juvenile delinquents, for example, we might find that coming from a single-parent family greatly increases the likelihood—the *probability*—that one will become a delinquent.

This does not mean that all children who come from single-parent families will be deviants, nor does it mean that this is the only factor leading to deviance. It simply means that five or six circumstances together may incline one toward this form of behavior and that being from a single-parent family may be one of those factors. However, finding that single-parent family membership contributes to deviance is not the same as saying that it is necessary.

A *necessary factor* is one that must be present. It may not be the only factor that may cause a behavior pattern, but if the factor is absent, the behavior will not occur. Some social scientists believe that social instability and unrest are necessary conditions for the rise of a witch-hunt hysteria, but social turmoil is not enough by itself to cause witch-hunts. A statement that a variable is "necessary" is a much stronger statement of causality than one asserting a variable is a "contributing" factor.

If a factor is *sufficient* to cause a certain social behavior, then that factor alone can cause a particular behavior. It does not need to combine with other factors. *A* alone is enough to cause *X*. However, various other factors may cause *X* as well. The combinations of *B* and *C* or of *C*, *D*, *E*, and *F* may also cause *X*. Although variable *A* may be sufficient to cause *X*, it may not be a necessary condition. The diagram below is a very simple model of causality depicting several factors that contribute to religious commitment. The solid line leading from C indicates that this is a necessary factor. It is must be present for a high level of religious commitment to occur, but it is not sufficient by itself.

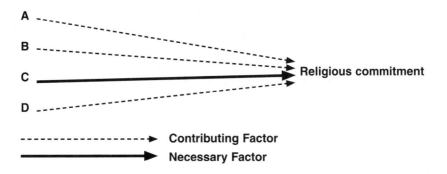

(Continued)

(Continued)

The strongest statement of causality is that a factor or a particular set of factors is necessary *and* sufficient. This means that the behavior will not occur unless these variables are present, and these variables will cause certain behaviors regardless of whether any other conditions are present. This is the strongest and most direct statement of cause, but due to the tremendous complexity of the social world, it is extremely rare that a sociologist can assert with assurance that a factor is both necessary and sufficient. Indeed, sociology is often referred to as a probability science because it usually points to various contributing factors. Statements about the necessity and/or sufficiency of one factor in causing another are usually made with many cautions and are normally stated tentatively rather than as firm conclusions.

Indeed, scientists and especially social scientists almost never use the word *proved* in their conclusions. You want to jettison this word from your vocabulary when you are in a social science class studying humans. First, the variables that affect humans are vast and they are related in complex ways. Second, scientists always set up their studies in a way that tries to disprove their assumptions at the outset. If they fail to disprove a causal relationship it does not mean proof has been established. It means that there is still more evidence to support the hypothesis or the theory. However, we cannot assume that every possible variable has been properly tested. We always remain open to the possibility that something new may be learned in the future. Indeed, that is why we keep doing research. Now some propositions have been tested many times using a wide range of research methodologies. If all of these point to the same conclusion, we may be 98% certain that a proposition or a causal relationship is true. Since there is a miniscule possibility that another study might reveal something no one had thought to test before, we simply do not say we have proved something. The same is true in biology, psychology, political science, history, and communications fields.

In reading this book or attending to any other commentary on social behavior—newspapers, magazines, textbooks, talk shows, or lectures—students would be well served to be sensitive to these distinctions. Unfortunately, some writers and some speakers assert "necessity" and "sufficiency" when the evidence requires a much more modest conclusion. Scholars must learn early to be precise in their use of causal language.

The conclusions of sociologists of religion are not etched in stone, and we often recognize multiple paths of causality in the formation of any behavior pattern. The fact that we are very cautious about the use of causal language does not mean that the findings are ambiguous. Indeed, caution allows for greater precision in describing the social world accurately. The unwillingness to make simple and absolute statements of cause is due to the complexity of our subject matter, not to the confusion of sociologists.

Critical Thinking: How we operationalize religiosity in a study is sometimes linked to our research method, but it is also influenced by the way we define religion (discussed in the previous chapter). If you visit www.pineforge.com/rsp5e, you can see some of the ways scholars have operationalized religion. Which of these seems to you to get at the heart of what it means to be religious? Which ones seem to have missed something important about religion as you understand it? Why do you think so?

Because most survey research seeks to generalize from the individuals surveyed to some broader population, the process of determining who will be given the survey is very important. This is known as the *sampling* technique. The ideal for making generalizations is a simple random sample of the population. In this technique, every member of the population has an equal chance of being selected into the sample.[2] Because of studies done on this approach to gathering data, the researcher knows within a certain margin of error that the conclusions drawn from the sample represent the entire population. For example, suppose we want to know the religious beliefs and practices of all adult Americans. We do not have to ask every American; we can ask 1,500 or 3,000 of them, if we use the right sampling technique. Many of the studies we will discuss in this textbook are based on representative samples of the American adult population.

Because such studies are very expensive, however, most surveys involve a smaller population base: a sampling of college students at a particular type of institution (such as state universities or at private liberal arts colleges), a sampling of residents in a particular city or set of cities, or a sampling of members of a particular denomination. Although the units of analysis in survey research are typically individuals, it is also possible to sample organizations. Sociologist Mark Chaves did this in the "National Congregations Study." By asking a representative sample of Americans to name the mosque, church, or temple they attend (if any), Chaves and his team were able for the first time to generate a random sample of congregations in America (Chaves, Konieczny, Beyerlein, & Barman, 1999).

Although it is the most common method employed in sociological research, statistical analysis of survey data is not without problems. One difficulty is that although statistical analysis tells us on a large-scale basis which religious characteristics can be found in a population and how they are correlated with which other social attributes, it does not necessarily tell us which factor *causes* which. Does a lower level of education lead one to hold fundamentalist religious beliefs or vice versa? If we observe a correlation between frequency of religious service attendance and frequency of reporting mystical experiences, how can we know which caused which (Yamane & Polzer, 1994)? In a cross-sectional survey—that is, a survey of individuals only at one point in time—a great deal of interpretation enters at the point of suggesting causality. One way of addressing the issue of causality empirically is to do a longitudinal study, surveying individuals at multiple points in time. This helps determine which variable occurs first, though most survey data is not longitudinal because of the high cost of following the same group of individuals over time.

A second difficulty is the fact that the data produced by large-scale surveys are characterized by correlations of hundreds of answers to specific questions, which also means that the interpretations of the meaning of the responses for the respondents themselves is sometimes lost. Even a straightforward question—like How often do you pray?—can be hard for a person to answer using the closed-ended response categories given. Furthermore, survey data do not demonstrate the process that an individual goes through; they are static or nonhistorical. They do not reveal, for example, the vicissitudes of an individual's prayer life over time, even from day-to-day or week-to-week.

Another difficulty is that sociologists sometimes assume that negative responses to certain questions mean that the respondent is "less religious" than other respondents. The questions frequently do not allow people to express alternative modes of religiosity and do not account for the fact

---

[2]Because there is no list of all adult Americans from which research can draw a simple random sample, most large scale surveys employ what are called "stratified" random samples. The "sampling frame" in these studies are segments or strata of the population, which taken together are thought to represent the population as a whole. That is why these are often called "representative samples" even when they are not the ideal simple random samples. For reasons too complex to explain here, by sampling from these strata, researchers believe that they can generalize to the broader population with a high degree of confidence (Kalton, 1983).

that what is a core tenant in one sect of a religious community may be insignificant or even ridiculous in another sect of the same religion. The presuppositions of the researcher (the filtering processes) are sometimes at work in formulating and interpreting the questions. The critical point in this type of research, then, is the objectivity of the questions and accuracy in interpreting the answers.

A final problem with survey studies is that sometimes what people say is quite different from what they do (Deutscher, 1966, 1973). For example, for years major surveys of American adults including the Gallup Poll found that about 40% reported attending religious services in the previous week. When sociologists went and counted individuals actually attending services on any given Sunday, however, they found a much smaller percentage of people actually in church (Hadaway, Marler, & Chaves, 1993). The reasons for the discrepancy are not entirely clear, but some believe that it is due to individuals giving socially preferred responses. That is, people believe that going to Sabbath worship is expected of "good people," so when they are asked in a survey context if they attended services last week, they say "yes" whether they did or not. This is particularly problematic if different social groups are more likely to give socially preferred responses than others because it would incorrectly make it appear that there is a correlation between two variables. For example, if evangelical and mainline Protestants are equally likely to have premarital sex, but the evangelicals are more likely to *say* they have not had premarital sex (because they believe that premarital sex is a sin), it will appear in a survey that evangelicals have premarital sex less often than mainline Protestants. Thankfully for survey researchers, some scholars who study these issues have found that even on sensitive issues like sex and sexuality there is no systematic relationship between religion and likelihood of inaccurately self-reporting behavior (Regnerus & Uecker, 2007).

More generally, survey information does not involve a direct study of religious experience itself but focuses on reports of religious experience or, more frequently, on the consequences of religious experience. It is a common finding in the social sciences that people often operate on assumptions and respond to symbols of which they are only partially conscious. Hence, the value of survey information is affected by how "hard" (how verifiable, objective, and unchangeable) the data are. Statements about one's religious preference or one's political party affiliation provide rather hard data. In such cases, the answers given to researchers are more reliable (e.g., if you asked the same person 6 months later, you would likely get the same answer). On the other hand, to determine the relevance or the importance of religion to respondents, sociologists sometimes ask the following questions: How significant is your religion to you? How important is religion in your everyday life? The information gathered by such questions is much "softer" in nature. A person may say that his religion is very significant to him, and may want it to be so, but it may actually have little influence on the individual's everyday life. Another respondent may report that religion has very little influence, yet her childhood moral and religious training may significantly affect her daily decisions in unconscious ways. Moreover, a question on the "significance of religion"—as one question among two dozen—does not measure the importance to that person of any specific belief with accuracy. Individuals may say that they do believe in life after death or in the existence of the Devil and that they feel religion is very significant or definitely affects their everyday life. Yet this does not reveal the importance of those particular beliefs. The meaning of the correlations needs to be interpreted with great care. Fortunately, any bias or questionable assumption by the researcher is made manifest in the questions and is accessible to other social scientists to recognize and correct. (In other methods of research, the assumptions of the researcher are usually less explicit.)

Interviewing

Interviewing as a research method has some similarities with survey research. Indeed, some people call interviews "open-ended" or "semi-structured" surveys. In both cases, a sample of individuals is asked a set of questions, the questions operationalizing some concepts. However, there are some important differences as well, and as with research methods generally, the weakness of one can be the strength of another.

As noted, a weakness of survey research is that the data produced by it do not speak clearly to meaning and process. For example, if a person answers a closed-ended question—for example, How important is religion in your life?—by checking the box "somewhat," what exactly does that mean? Here, interviews are a much better method of data collection.

As with surveys, interviews begin with questions, but they do not prescribe a set of responses from which the respondent must choose. Good interview questions get respondents to begin talking about the topic of interest and then specify certain follow-up questions (called "probes") that the interviewer should use to encourage the respondent to elaborate. The list of questions the interviewer is going to cover with the respondent is called the "interview schedule." An interview schedule with very general questions and few probes is called an open-ended, and one with more specific questions and more numerous probes is called semistructured.

Consider as an example sociologist Robert Wuthnow's "Spiritual Journals Interview," which was an important part of the empirical basis for a couple of his books: *After Heaven* (Wuthnow, 1998) and *Growing Up Religious* (Wuthnow, 1999). The following sample question is a good illustration of a semistructured interview question:

Give me a sense, then, of your involvement in religious organizations. Again, start with childhood and work up to the present.

For each one:
  Did you like being part of that organization?
  What did you like best?
  What didn't you like?
  So how old were you when you were involved in this?
  Then what were you involved in next?
  How did you get involved in that?
  Did you shop around? What did you look into?
  Why did you get involved in this?

So, were there times when you weren't involved at all?

One way to learn in-depth about how people understand their faith or their spirituality is to conduct semistructured interviews.

Have religious organizations, then, been an important part of your spiritual journey, or not?
  In what way have they been important?
  In what way haven't they been important?

This is but one of 32 questions in Wuthnow's interview schedule, which runs over 13 printed pages.

Most often interviews are tape-recorded and then transcribed. The transcripts from research interviews can be quite lengthy, from several pages to several dozen pages in length. This makes data analysis challenging in an interview study. In analyzing these data, the researcher will seek to identify patterns in the data, often in the types of language individuals use to describe their experiences and understandings. That language is not unique to each individual but is drawn from larger social patterns (often called cultural "frames" or "tool kits"). As Kristin Luker (2008) put it, "The point of interviews . . . is not what is going on inside one person's head, but what is going on inside *lots* of people's heads" (p. 167). Returning to Wuthnow's example, one of the patterns he identified in his spiritual journeys work was the rise of a new "spirituality of seeking" in contemporary society: an emerging way of thinking about being spiritual that is flexible and less dogmatic.

This sort of research is often called "qualitative"—in contrast to statistical analyses of survey data, which are quantitative—because the focus of the former is nonnumeric. The richness of the data produced in qualitative research is a clear strength, but there is a downside of this method as well. Wuthnow (1998) reported that the spiritual journeys interviews took an average of 2 hours to complete (p. 201), and although he does not say, the transcripts made from the interviews must have been quite long. As a result, an interview-based study usually has under 100 respondents, as compared to the thousands that are often found in survey-based studies. Furthermore, those respondents are usually drawn using what is called "convenience sampling" rather than representative sampling techniques, making generalizations from the sample of respondents to a broader population problematic.

Another limitation of interview-based studies is that there is a double interpretation that stands between the reality being investigated and the findings of the study. In the first place, the individual is interpreting her understanding to the researcher. For example, sociologist Christian Smith (2005) interviewed adolescents about their religious beliefs and practices, asking questions such as "What is God like?" The youths' responses to this question—for example, "Um, good. Powerful"—reflect an interpretation (p. 135). In the second place, the researcher has to interpret those understandings in the process of establishing the broader pattern. In Smith's (2005) case, he argued that the pattern of responses in his study reflected a new form of religion operative in the lives of American youth: "moralistic therapeutic deism" (pp. 118–171). So interviews can speak more to process than surveys, but that is still not the same as directly observing social processes. To address this particular weakness, we have to look to still another sociological research method: participant observation.

Participant Observation

Also called ethnography—the primary method of data collection employed by anthropologists—participant observation research involves immersing oneself in the situation being studied. The period of immersion can vary from days to months, and the commitment can range from visiting for a few hours at a time to actually living among the people being studied. By participating in the group, a researcher can observe the religious beliefs and behavior of people in a concrete social context. This can be done overtly (with subjects aware that the observer is studying them) or covertly (with subjects believing the observer is a new member). The latter type of study is considered unethical and will not be approved by the human subjects committees unless there are compelling reasons for the secrecy and assurance is given that no one will be hurt by the study.

There are three related advantages of this research method. In the first place, the data that are produced are very "thick" (Geertz, 1973). If survey-based research tells you a little bit about a lot of people, the strength of ethnography is the opposite: It tells you a lot about a few people.

Second, the data that are collected in participant observation research are social rather than individual. Although sociologists are explicitly concerned with groups and the social dimension, much research done by sociologists is methodologically individualistic: deriving the data from surveys or interviews with individuals, then drawing conclusions about groups, organizations, or social systems from that. Participant observation research entails observation of the groups and organizations in the first place.

Third, unlike surveys and interviews, which rely on people's self-reporting, in this method sociologists directly observe the phenomena being studied. This allows sociologists to see social processes in action, rather than just hearing people talk about them. Sometimes there is a gap between what people say and what they do.

Consider the case of multiracial congregations as an example. We know that most religious congregations in the United States are racially segregated, so how is it that some congregations manage to integrate racially? Sociologist Gerardo Marti is an ethnographer of religious life who has sought to understand this question. Rather than simply asking people what they think about religious racial integration, he spent considerable time in multiracial congregations observing how people *do* religious racial integration. Through his fieldwork, he eventually identified a three-stage process by which members came to identify with the multiracial religious community (Marti, 2009). Although Marti, like most ethnographers, also conducted interviews with those he studied, it is safe to say that a survey or interview alone would have been unlikely to identify these sequential processes. They had to be observed in the situation.

Participant observation allows the researcher to observe subtleties that would not be revealed through responses to a questionnaire or through a 1-hour interview. The quality of the data is rich, but it is also limited to the researcher's observations. This presents two problems with this research method. First, the data are normally limited to one case (one congregation, a single new religious movement [NRM], or the clergy in one denomination). This limits the ability of the researcher to make generalizations. What is true in the one case that is being studied may not be true in others. Case studies permit the researcher to go into great depth in gathering data, but breadth is thereby sacrificed. A large number of case studies is needed before generalized patterns can be identified, and then they are often cataloged by persons who did not have firsthand experience in each group.

A second problem is that the reported data are bound to be somewhat biased by the observer's own "filtering" system. First, not all observers are likely to notice or be interested in the same patterns. Hence, there is the issue of unconscious filtering of data. Second, what the observer chooses to write up cannot possibly include all those observations made. The researcher must consciously choose which data are relevant. This provides a second filtering of data. The same sort of filtering process occurs in all research, but in other sorts of research, it is easier to identify the sources of possible bias by analyzing the nature of questions asked of informants or by checking other available data. Because it frequently happens that only one person is undertaking participant observation on a group at any one time, that observer's data are sometimes the only written record of that particular phase of a group's development. This subjective element in observational studies is a serious hazard.

One way to focus the research using this approach and to lessen the random distractions that any given observer might be prone toward is to have a research guide. The "Doing Research on Religion" feature provides a model of how to do participant observation—what to notice. This one was created by a veteran and award winning ethnographer.

DOING RESEARCH ON RELIGION

Participant Observation of Local Congregations

So you want or have to observe a congregation? Simply wandering into a synagogue, temple mosque, or church and looking around is not advisable. Although many congregations welcome visitors, not all do, and some welcome visitors only under certain circumstances. Although not written by or for sociologists or their students, the book *How to Be a Perfect Stranger: The Essential Religious Etiquette Handbook* (Martins & Magida, 2003) is an excellent guide to understanding the "dos" and "don'ts" of attending the religious services of various religious traditions. Also, observing

(Continued)

(Continued)

religious services without planning ahead and knowing what one is looking for is likely to yield subpar data. Nancy Ammerman, a leading sociologist of religion studying congregations, has written "Observing Congregations: A Guide for First Visits (and Beyond)" to help students sociologically observe congregations in a productive way.

Before You Go

- Think about how much you do and don't know about what you may see. How much of a stranger are you? If this group is new to you, consult a source such as *How to Be a Perfect Stranger* (Martins & Magida, 2003), and/or ask a more knowledgeable friend to brief you or go along.
- Think about how visible you will be because of ethnic or other differences. The smaller and more intimate the group, even if you look like them, the more you will stand out. Think about how you will explain yourself, and pay attention to the effects of your presence on the group.
- Be sure you are clear about any expectations the group may have about proper dress and limitations on the participation of outsiders.
- Double-check the service times and location.
- Take along a small notebook and pen for jotting notes, as appropriate.

Getting There

- Plan to arrive in the area early.
- Observe modes of transportation. How easy is it to get there via available modes of transport?
- Observe the neighborhood. Walk or drive around. What sorts of businesses, homes, parks, etc., are present? Who, if anyone, do you see? What does the area feel like? How does the congregation's meeting space fit into its surroundings? Does it look "at home" in the neighborhood, or does it stand out? How?
- Observe the outside of the building. What is your impression as a newcomer—is it attractive and/or welcoming? Is it accessible? Are there signs, and what do they say? What, if any, visible religious symbols are there? Do you know what the symbols mean? Would most people?
- Don't forget to take notes—discreetly. Make sketches for yourself, and make notes about anything you see but don't understand.

Going Inside

- Unobtrusively find a place where you can see as much of the action as possible, while still being part of it. This will usually be near the back or the side of the congregation, but beware of ending up where you can't see or hear what is happening at the front (or being so near the front you can't see the people).
- If people greet you or ask who you are, be open about why you are there. You can say, for instance, that you are doing a class assignment, that you plan to do research on religious groups, or that you are just curious and wanted to visit. Whatever your explanation, be sure to express your appreciation to them for allowing you to visit.
- Read the bulletin. If you are handed a program of any sort, read it.

- Observe what is done where. As you are able, sketch a "function map" in your notes. Where do people mill around and visit? Where do they position themselves to participate in the service (and do different kinds of people—by gender, age, etc.—occupy different spaces)? What spaces do leaders use? Are there multi-use territories? Visible barriers?
- Observe what is displayed. What do the banners and bulletin boards say? Pick up any flyers or other literature that may be available.

Observing the Service

- Observe time. How long does the service last, and does it start on time? Are there clear "segments," and how long are they? Do people seem aware of time?
- Observe leaders. What are they wearing? What are the leadership activities (e.g., singing, announcing, directing, preaching, conducting rituals)? Do different kinds of people (by gender, age, ethnicity, etc.) do different leadership things?
- Observe participants. How are they dressed? What is their demeanor? What do they do at each stage of the service? What physical activities are called for?
- Observe groups and divisions. Who is most involved and who is least involved? Who gathers with whom?
- Observe "artifacts." What special decorations, furniture, clothes, and implements are used? Is there a printed program? Pay attention to any object that does not seem to be "ordinary." If you don't know the meaning or use of something you see, find someone to ask.
- Observe feelings. What emotions do people express? What are your own emotional responses to the service?
- Listen to what is said. Pay attention to announcements and prayers, as well as sermons. Make special note of the picture they paint of their deity, what they say about how people should behave, what they think is good and bad about the world. Listen for what is assumed about people's everyday lives, concerns, and social circumstances. Note also what is said about the congregation's activities and projects, what they are trying to accomplish.
- Take notes, if you can. Be as inconspicuous as the situation will allow. Writing is often possible during the sermon, but avoid writing during prayers or other important ritual moments.

After the Service

- Strike up conversations with anyone who will talk, and pay attention to who seeks *you* out. Remember to be open about who you are and why you are there. Stick around as long as it seems polite to do so.
  - You might start by asking a person if he or she is a member or regular attendee and how long they have been coming.
  - Ask how they found out about the congregation and why they first came.
  - Ask what they especially like about this congregation.
  - Ask if there is a part of the service they especially like and why.
  - Ask if they participate in other activities sponsored by the congregation.
  - Ask them to explain things you didn't understand.

(Continued)

(Continued)

- Look around. Either before or after the service, ask someone to show you around other parts of the building. This might be an usher or greeter or perhaps someone you've struck up a conversation with. Note how much space they have, what they do with it, and what sort of condition it is in. Ask about other organizations that use the space.

Back Home

- Go back through the notes you jotted, and make additional notes immediately to remind yourself of the things you couldn't jot down at the time. Start by completing your map of the space and a chronology of the event.
- As soon as humanly possible, preferably before you sleep, type up a full set of notes. Your jottings will be good for jogging your memory for no more than a couple of days; you will need to put flesh on those notes if you need to be able to retrieve the memories weeks and months from now.
- At the end of your notes, jot questions you have for future exploration, and reflect on the particular analytical questions that your project or assignment required.

Despite shortcomings, the method has proved to be extremely valuable in the sociological study of religion. In fact, readers may find it enriching to do some participant observation of religious groups different from their own as they read this text. While you will not be trained in participant observation methods, you may find that your observations will provide illustrations and/or questions regarding concepts discussed in this book.

Critical Thinking: How is observational research different from just being a visitor and looking around? What makes it more scientific and empirical? What are its strengths and weaknesses relative to other social scientific methods?

Content Analysis

Still another method used by sociologists is content analysis. The name is descriptive of the process: analyzing the content of cultural "texts," where texts are broadly understood to include written documents, spoken words, or other media (including mass media—newspapers, songs, movies, TV programs, cartoons, the World Wide Web). In this instance, the researcher tries to ferret out underlying religious themes or unarticulated assumptions by analyzing written materials. The secular message of American Protestantism can be identified through an analysis of the ways that the parable of the Prodigal Son is preached in different congregations (Witten, 1995), or the relative importance of religion and science can be examined by looking at the content of television commercials (Maguire & Weatherby, 1998).

Content analysis has been useful, but one difficulty is its assumption that the texts being analyzed accurately represent the views of the people. Sermons may not reflect the attitudes and values of the people in the pews. Television commercials may or may not appeal to the actual values and interests of the viewers. Those doing content analysis have to be cautious in generalizing from the texts under consideration to wider views in the population.

Historical–Comparative Analysis

Sociologist Kristin Luker (2008) made a humorous analogy to help understand historical–comparative analysis when she observed the following:

> As methods go, "historical-comparative methods" are something like that drawer in your kitchen where you put all the useful stuff that doesn't logically go in other drawers. Mind you, I'm not saying that it's the *junk* drawer—quite the contrary. . . . The things in this special drawer are unique, and therefore of high value because they do what they do better than anything else in the kitchen. So it is with historical-comparative methods. (p. 190)

What historical–comparative methods do better than others is to help us understand what events in the past shaped the present situation in which we are living and why things turned out differently here than in other places.

Max Weber's famous studies of the rise of capitalism in Western Europe exemplify these strengths. Weber sought to understand the connection between religious beliefs and economic practices in parts of the world dominated by different religious traditions. Examining Confucian, Hinduist, Buddhist, Christian, Islamic, and Judaic religious ethics, he found that only Protestant Christianity provided the motivation necessary to facilitate the rise of modern industrial capitalism. In this, he explained both the seeds of the world we live in today and why those seeds flowered first in the West rather than some other part of the world (see more in Chapter 9).

Working very much in a Weberian tradition, Philip Gorski (2003) tried to understand the role

that religion played not in the rise of modern capitalism but the modern state (the word social scientists use to describe the entire apparatus of government in a nation). Comparing the Dutch Republic and Brandenburg-Prussia, Gorski identified the key role played by religious beliefs unleashed by the Protestant Reformation in creating the disciplined order needed for the modern state to flourish. As with Weber's study of the spirit of capitalism, Gorski found these beliefs most clearly in Calvinism (as compared to Lutheranism and Catholicism), again helping to explain how and why our contemporary world came into being.

It is important to note that when sociologists use historical material, they have a different emphasis than historians. Historians normally seek to offer a detailed description of historical situations and perhaps to elaborate the specific circumstances that seemed to have caused, or resulted from, a particular set of events. Sociologists are likely to be interested in whether a particular social situation is usually accompanied by or followed by some other "typical" situation or circumstances (Nottingham, 1971). The sociologist is normally looking for a pattern—a general rule—in the relationship between social events and religious characteristics. The goal is to develop a generalization or theory that explains the relationship—not just in that particular circumstance but in most cases. This approach has led to the development of typologies of religious groups (Chapter 8) and to the development of theories about the evolution and decline of religion in modern society (Chapter 13).

Historical analysis is important, but the danger is a tendency to impose one's own pattern on the data and thereby distort history. The goal of this method is to recognize and uncover historical patterns that are relevant today, but a nonbiased execution of this method is far from simple to accomplish. The "Doing Research on Religion" feature provides some insight into the excitement a researcher can experience when doing historical research.

DOING RESEARCH ON RELIGION

Statistical Analysis Employing New Procedures on Historical Materials

Research often utilizes several methods in a single study. Professor Rodney Stark and his graduate students, for example, used sophisticated statistical methods to analyze some very old data in new ways. Professor Stark told how exciting the discovery of research can be for a scholar—and also revealed the unanticipated way in which a scholar may discover a new line of research.

* * * * * * * *

One day I was sitting in my office at the University of Washington when Kevin Welch, one of my graduate students, brought me a two-volume set of books, published by the Bureau of the Census, titled *Religious Bodies, 1926.*

"What do you know about these?" he asked.

"I've never heard of them," I confessed. "Where did you get them?"

"In the census section of the library. There are sets for other years too: 1890, 1906, 1916, and 1936."

As I leafed through the volumes I was stunned at their range and depth. They provided the most extraordinary historical, doctrinal, and statistical information on 256 separate religious bodies, including very tiny ones as well as those outside the Judeo-Christian tradition. As I grew excited about the possibilities the volumes presented, I also grew increasingly embarrassed. I had been publishing studies in the sociology of religion for more than fifteen years and I had never known such books existed. How could I have been so uninformed? Then I recalled Petersen's paper, which I had read during my second year in graduate school. How could he not have known? How could generations of scholars have remained ignorant of what was clearly a massive census undertaking? Indeed, how could demographers, and especially their graduate students, fail to notice a whole shelf of books of religious statistics located alongside the regular census volumes? In fact they didn't overlook them—the books weren't missed, they were dismissed. I have since talked with many demographers who noticed these volumes while in graduate school, but in each case they were informed by faculty members that these statistics were nothing but junk that the Bureau should never have bothered with. Why? Because they were not based on tabulations of individual responses to a question on the regular census form. Instead, they were based on reports prepared by individual pastors or boards of elders. It was claimed, such people were not to be trusted. The figures would necessarily be hopelessly inflated.

Perhaps these demographers knew little about American religion and never bothered to examine the data with care. Hopelessly inflated statistics are precisely what are obtained when individuals are asked their religious affiliation. Ever since the start of public opinion polling in the late 1930s, surveys have always found that virtually everyone has a religion. I know of no national survey in which as many as 10 percent answered "none" when asked their religious affiliation. But the nation's churches cannot possibly seat 90 percent of the population. More careful investigation reveals that for many, their claim to a religious affiliation amounts to nothing more than a vague recollection of what their parents or grandparents have passed along as the family preference. Consider that substantial numbers of students at major universities commit the most unlikely spelling errors when filling out a questionnaire item on their denomination. Can students who think they go to the Pisscaple Church have ever seen the word Episcopal? What of Presditurians? If one wishes to know which churches have how many members (that is, people sufficiently active to have their names on the membership rolls), one needs to visit churches and count the rolls. Alternatively, one can ask how many names are on the rolls. But it takes very careful and elaborate research techniques to calculate membership by summing the responses given by individuals.

Rather than being hopelessly inaccurate, then, there are strong prima facie grounds for thinking that these old census statistics are relatively accurate. But there are strong indications of their accuracy as well. The first of these is that the numbers claimed by the churches are modest. The national rate of religious adherence based on the 1890 data is only 45 percent. Second, the data are extremely stable over space and time. Had there been substantial local misrepresentation, then the data ought to jump around between nearby communities, and they do not. By the same token, there should have been a great deal of inconsistency from one decade to the next, and there is not. Finally, the Bureau of the Census was very concerned with accuracy and provided extensive, sophisticated, and persuasive evaluations of its procedures.

* * * * * * * *

After this rather accidental discovery of these existing sources, Stark teamed up with Roger Finke in a new research project. Applying several statistical procedures to control for errors and to aid in analysis, the two developed a new body of evidence and a fresh interpretation of long-term trends in American religious affiliation (reported in part in Chapter 13). Sometimes innovative lines of research are spawned by such chance encounters. Despite their cause, scholars find such moments of insight and opportunity exhilarating. Doing sociological research can be akin to solving a mystery: one starts with a question and seeks ways to solve the puzzle.

SOURCE: Finke, Rodney, & Stark, Roger. (1992). *The churching of America: Winners and losers in our religious economy* (pp. 7–8). New Brunswick, NJ: Rutgers University Press. Reprinted by permission of Rodney Stark.

Experimentation

The most powerful tool of the social sciences in terms of ability to control variables and therefore isolate causal influences is experimentation. In a pure experimental model, research subjects are randomly assigned to "treatment" and "control" groups. Random assignment to these groups is key because the systematic influence of any confounding variables should be nullified. The only systematic difference between individuals in the two groups is the exposure of the treatment group to the causal force in question. Experimental designs are seen most often in medical research, where members of a treatment group are given a drug and those in the control group a placebo, and the effect of taking the drug can thereby be isolated.

Although it has great explanatory power, experimentation is also the least used research method by sociologists. Experimental research in sociology has been undertaken primarily by social psychologists. Experimental research on religion—even by social psychologists—has been almost nonexistent. Why?

Obviously, one cannot use control groups and treatment groups to experiment on the factors operative in conversion to a new religious organization. We could not, for example, experiment on students with several techniques for converting people to a new religion. In fact, most forms of experimentation regarding people's religiosity would be considered a gross violation of social norms. The public would be outraged—and rightfully so. Religion is so intensely personal and deeply felt by so many people that manipulating it for purposes of study would not be tolerated. Institutional Review Boards or Human Subjects Committees (which approve research projects involving the study of human beings) are unlikely to endorse such a study unless participants agreed to the experiment and were fully aware of its nature. On the other hand, informing participants of the full nature of the research would bias the outcomes. Thus, true experimentation on religious behavior has been limited

Instead, sociologists of religion have tended to employ various types of "quasi-experimental" designs. These are research projects that employ the logic of treatment and control groups without the actual random assignment of individuals to groups. For example, J. M. Darley and Daniel Batson (1973) conducted a quasi-experiment with seminary students in which they tried to measure effects of biblical stories on subsequent behavior. They staged a scene in which they placed a groaning, coughing, sinister-looking man who was apparently in need of help in an alley through which the seminary students passed en route to a location where they were to deliver speeches. Some of the students had recently read the Good Samaritan story and were to speak on that parable. Others were assigned to speak on the role of theological education for nonpastoral occupations. The purpose was to see whether those who had recently read the Good Samaritan story and were to speak on it were more likely to stop and help. (They were not more likely to help—although other factors of religious orientation did affect the likelihood that the students would help.) In such experiments, the controls on variables are not as strong as in most

experimental research, and that is why Batson insists they are only quasi-experimental.

Other quasi-experimental designs may also be used if one allows a natural event to be the manipulator of variables. For example, Daniel Batson (1977) suggested that one could do pretests and posttests on psychological characteristics and values of people before and after a revival meeting. While such a procedure would not conform to all of the normal standards of an experiment, it could offer an important means for establishing the validity or inadequacy of certain theories regarding the effects of religious rituals and celebrations. Most people would probably not object to such tests, for the experimenter would not be attempting to manipulate an individual's faith directly but only to measure the effectiveness of other "manipulations." Further, research ethics committees do not usually object as long as the research project does not put subjects at risk of harm.

The fact remains that most research on religion must proceed without this powerful tool of the social scientist. Although the lack of much experimentation on religious behavior is fully understandable, it does limit the research tools available to the sociologist of religion.

Triangulation

To some extent, the weaknesses of each research design can be compensated for and overcome. However, no one research design is entirely adequate in itself. The variety of approaches allows scientists to check the accuracy of their theories from a variety of data sources. Sociologists often quarrel over which research design is most adequate, and most researchers tend to prefer one approach over others. However, our ability to gather data in several ways provides checks on the weaknesses inherent in any one approach—a process called *triangulation* of the data. Together, these approaches allow social scientists to substantiate or dismiss various generalizations about religious behavior. Examples of each of these methods of research—except for the experimental method—will be incorporated in this text.[3]

> **Critical Thinking:** One issue in empirical research involving humans is that of ethics. Researchers must ensure that no one will be hurt—including damage to a person's self-concept or self-esteem. Also, the privacy of subjects must be protected so that information that is given voluntarily does not later jeopardize their social status or their reputations in the community. What kind of ethical problems might be encountered in the study of religion for each of the following research methodologies: Experiment? Sample survey? Observational studies?

SUMMARY

Sociology offers a unique vantage point for viewing religion and religious behavior. While sociology certainly does not claim to offer the whole truth about human behavior, it does provide insights that other approaches may fail to recognize. The sociological perspective focuses on religious groups and institutions or the behavior of individuals within those groups and on conflict between groups. While religious beliefs are seen as important, they are not the exclusive focus of sociologists. In fact, beliefs are viewed as one variable of religion among many—and often beliefs are found to be the effect of other social behaviors rather than the cause. Sociology of religion does not attempt to prescribe how religion ought to work. Rather it attempts to describe accurately the social underpinnings of religious groups and to generalize about common patterns and apparent causal correlations.

Two fundamental principles characterize this sociological approach: (1) objectivity and (2) reliance on empirical data (methodological empiricism). The sociologist is not satisfied with general impressions but seeks concrete, verifiable data to confirm or disprove any empirical generalization about a group. Although sociologists have preferences and commitments of their own, they seek to be open to the data and to avoid prejudgment of any particular group or any particular religious process. Such objectivity is not easy to maintain and serves as an ongoing goal for each sociologist.

Sociology is commonly referred to as a "probability science" because the statements about causality are normally statements about contributing factors. It is important for readers to be cognizant of the language of causality and to recognize that the complexity of the social world makes it very rare to be able to identify necessary and sufficient factors. Our conclusions about the social world are not etched in stone, and we often recognize multiple paths of causality in the formation of any behavior pattern. The fact that we are very cautious about the use of causal language does not mean that the findings are ambiguous. Indeed, caution allows for greater precision in describing the social world accurately. The unwillingness to make simple and absolute statements of cause is due to the complexity of our subject matter, not to the confusion of sociologists.

In order to gather empirical data, a number of different types of research methods have been employed. Each has its own advantages and disadvantages. Collectively, they can provide checks and corrections on errors that any single method might make. Use of several methods to be sure the data from several sources all point toward the same conclusions is called triangulation. However, due to the nature of religion as a topic of study, some special problems do arise. This is

---

[3]Experimental studies of religion have been undertaken on only a very few occasions and those have dealt with very specialized areas of theoretical interest—areas that we will not be investigating in this text.

especially true in the case of experimentation. Ethical considerations prevent many aspects of religious behavior from being tested and analyzed through experimentation. Any other method of research may also be forbidden by research ethics committees if there is any likelihood of someone being hurt or there is no consent given by the subjects.

PART II

THE COMPLEXITY OF RELIGIOUS SYSTEMS

Integration and Conflict

Religion is a complex phenomenon. It plays a diversity of roles in society and is itself composed of an intricate interplay of symbols, myths, rituals, mystical experiences, and social interactions. In Chapter 3, we take an overview of the multifaceted relationship between religion and the larger society. To do so, we first need to understand two major theoretical perspectives employed by sociology and how each contributes to our overall understanding of societies and social processes. At the end of the chapter, we will explore a theoretical model that attempts to integrate insights from both perspectives. Then in Chapter 4, we turn our attention to a micro perspective and to the internal complexities of a religious system. In this chapter, we will examine and apply another critically important sociological theory: social constructionism. In both chapters, we will discover ways in which religion can serve as a source of social cohesion or as a major contributor to conflict and divisiveness.

3

Religion in the Larger Society

Macro Perspectives

Here are some questions to ponder as you read this chapter:

- How do different macro level theories analyze the place of religion in a *society*?
- What are the functions and dysfunctions of religion? How might a faith have positive consequences for society and negative consequences for individuals (or vice versa)?
- How might conflict impact a religious group?
- Is religious belief and behavior shaped by the self-interests of the adherents?
- How does open systems theory contribute to our understanding of religion?

The relationship between religion and the larger society is multidimensional and complex. To explore this relationship, we will be engaging in a macro level analysis. By *macro*, we simply mean an attempt to explain the relationships and dynamics of the society at large. Macro theories look at the *structure* of societies, the overall patterns of organization, rather than at the minutiae of everyday interaction between individuals. The two primary macro perspectives used by sociologists are (1) functional and (2) conflict theories. Each of these has a rather different approach to understanding social behavior, focusing on slightly different aspects of society and illuminating certain elements in the relationship between religion and the larger society. Although the theories are sometimes viewed as incompatible because of these differences, they can complement each other and together can broaden our understanding.

FUNCTIONAL THEORY AND THE FUNCTIONS OF RELIGION

Our first macro perspective is *functionalism*. This perspective emerged out of concern by early social scientists to explain why a few social patterns exist in all societies. Religion, for example, was believed to be a cultural universal, to exist in some form in every culture in the world. Why is this the case? Functionalism sought to answer this question about religion, and to explain social organization and behavior in general, in terms of the consequences it had for society; that is, how it satisfied basic needs.

Although functionalism can be first attributed to Emile Durkheim, it was developed as a systematic method of analysis by anthropologists Bronislaw Malinowski and A. R. Radcliffe-Brown. One principle of their approach was that any social pattern or institution that does not serve a function will cease to exist. Furthermore, any pattern found among all people is believed to have its basis in innate human needs. Before focusing on the functions of religion, let us review the development and controversies within functional analysis itself, especially the divergences between Malinowski and Radcliffe-Brown.

Functional and Structural Functional Analysis

Malinowski maintained that all basic human needs or drives must be satisfied in a way that does not cause social chaos. The hunger drive, the sex drive, and the need to relieve oneself of bodily wastes are all satisfied in ways controlled by society. One cannot satisfy one's sex drive with just anyone and everyone. Certain norms specify which persons are or are not potential sexual partners. Sex without consent by the partner (rape) or sex with a close family relationship (incest) are only two of the more common prohibitions societies set down. Similarly, the society as a whole has a stake in controlling how one procures food and where one deposits body waste. Hence, all societies are made up of institutions—regularized patterns for the satisfaction of human need. Human needs, or drives, are thereby satisfied without creating social conflict or chaos (see Figure 3.1).

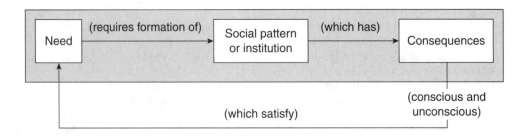

Figure 3.1 Satisfaction of Human Needs Through Social Institutions

Malinowski believed that all institutions and regularized patterns of behavior satisfy a basic human need; otherwise those institutions and behaviors would cease. Therefore, the task of the social scientist is to identify which needs, conscious or unconscious, each social pattern serves (Malinowski, 1944).

Malinowski maintained that evil and misfortune cause people to feel helpless. Religion allows one to feel that there is a source of power and hope that is greater than one's own resources. Especially important for humans is the need to cope with the anxiety and personal disorganization caused by the death of a loved one. Although the various religions deal differently with death, all religions establish some belief and ritual that functions to reduce anxiety about death.

Malinowski traced the function of religion and that of many institutions to basic human needs, drives, and emotions. In this sense, functionalism à la Malinowski tends to be rather individualistic and psychological in character. The bottom line in understanding many social patterns is the needs of individuals (Malinowski, 1944).

Radcliffe-Brown developed a different approach to functionalism. He saw the function of most social patterns as traceable not to individual needs but to needs or requirements of the society as a whole. For example, according to Radcliffe-Brown (1939), fear of hell and damnation—and perhaps fear of death itself—would not exist if the teaching of religious groups did not instill such fear. Religion generally functions not to resolve anxiety but to create, foster, or heighten it.

Why would religion act to enhance anxiety? Some sociologists answer that social stability is important for the survival of the society. Fear of hell, anxiety over offending one's gods, or fear of being bewitched by evil spirits tends to ensure social conformity. Radcliffe-Brown insisted that by making one anxious about breaking cultural rules, religion functions to discourage deviant behavior, and this enhances stability. The function served is not an individual but a societal need. For Radcliffe-Brown the need for social integration and stability is a driving force behind most institutions. Because his emphasis has been on the way social patterns meet societal or structural needs, Radcliffe-Brown's brand of functionalism has been called structural functionalism.

This structural approach was also used by Emile Durkheim. In 1912, Durkheim (1912/1995) maintained that God stands in the same relationship to worshipers as does a society to its members. God transcends the individual in power and scope, is immanent within the individual, and occupies a world that is fundamentally different from the world of the individual. Furthermore, the divine has priority over the individual: Human need must always give way to God's demands. Similarly, society transcends the individual in power, scope, and longevity. Society is contained within the individual in that each member of society has internalized the values and norms of his or her group. The demands and prerequisites of the society have moral priority over the desires of individuals (Durkheim, 1912/1995).

A thoroughly socialized person is one who wants to do that which is necessary for the society to survive. Persons must be motivated to *want* to behave in ways that are good for the society, even when the action may contradict their own desires. If the only argument presented to a person for not engaging in extramarital liaisons was that social stability would be adversely affected, most persons would probably not suppress their desires. However, when the prohibition is presented as a moral principle that is based on divine command, people will more likely internalize that norm. Similarly, people may not curb their desires if they are simply told that incest is wrong because it creates intrafamilial competition, undermines the socialization role of parents, and confuses the lines of inheritance and authority. These may be the functional reasons for the incest taboo, but many people honor the taboo because incest is an unthinkable sin. It is a moral absolute, the violation of which would seem to undermine the laws of the universe.

Durkheim's approach has been referred to as "metaphoric parallelism" (Winter, 1977). He believed that the sacred world is a world that parallels the mundane world. Behavior patterns that would cause social chaos are prohibited by fear of sanction from the supernatural realm. That which is structurally dysfunctional is simply made taboo. According to Durkheim, the term *God* is a metaphor for society; worship of God is really worship of one's own society. People are not aware of this projection process, and taboos and moral models become unquestioningly absolute and binding.

Functionalists do not always assume, as Durkheim did, that the belief in God is a mirage. For most functionalists, the existence or nonexistence of God is beyond the capacity of empiricism to prove or disprove. The concern of the functionalists, regardless of the truth or falsity of a belief, is how a belief or ritual operates in the society. What needs does it meet?

> **Critical Thinking:** Do you find that you tend to identify with the functionalists or with the structural functionalists? Does religion *alleviate* anxiety about death, or does it *create* anxiety about death in order to assure social conformity? Readers might find it instructive to attend funeral services in several different religious traditions. Notice comments made by the clergy who officiate and listen carefully to the prayers. What emotions do they seem to create? During regular worship services, do prayers and sermons create anxiety about death? If so, how is this accomplished?

Durkheim and Radcliffe-Brown believed the most important functions of religion are structural, while Malinowski focused on individual needs like personal anxiety. George Homans (1941) offered a synthesis that attempts to avoid the either/or choice, pointing out instead that both processes may be at work. He used the term *primary anxiety* to refer to the anxiety an individual may feel as a result of the loss of a loved one. Some rituals are designed to alleviate this anxiety or grief; they serve an individual function. On the other hand, rituals may also create anxiety—called *secondary anxiety*; these rituals ensure that society's members will conform to social norms. Hence, they serve a structural function.

It is possible for a single ritual to relieve primary anxiety while creating secondary anxiety. For example, a Protestant funeral service may emphasize that the departed loved one has gone on to the next world and is in the loving care of God. The virtues of the individual are emphasized. The minister may suggest that the bereaved family members will be united with "Grams" (as she was known) when they, too, pass on. "The best way to honor her is to live our lives with faithfulness to the values she taught and lived," they are assured. Comments that produce fond memories, assurance that the deceased is with us in spirit, affirmation that the bereaved will again see the deceased, and confidence that the deceased is dead only in physical form are all reassuring to the survivors. They help to lessen primary anxiety: grief.

At the same time, persons attending the funeral may be reminded that their own lives have not been as noble or righteous as the deceased. The thought of meeting God and receiving one's just reward may heighten their anxiety. The challenge to remember Grams by living a righteous life like hers may increase the listener's sense of humility and inadequacy: How could I ever be as generous and wise as Grams? (One tends to recall primarily the most *positive* qualities of the deceased at that moment.) The funeral service may also cause one to reflect on one's own life and its brevity. This secondary anxiety may cause the person to reaffirm conformity to the mores of that religious group. Hence, the same ritual may reduce primary anxiety and create secondary anxiety. As Homans's synthesis illustrates, the analysis of the function of any institution may be a complex process. Figure 3.2 illustrates the complexity of relationships between individual and structural functions, manifest and latent functions, and dysfunctions. Let us turn now to some of the specific functions of religion and see how modern functionalism is applied.

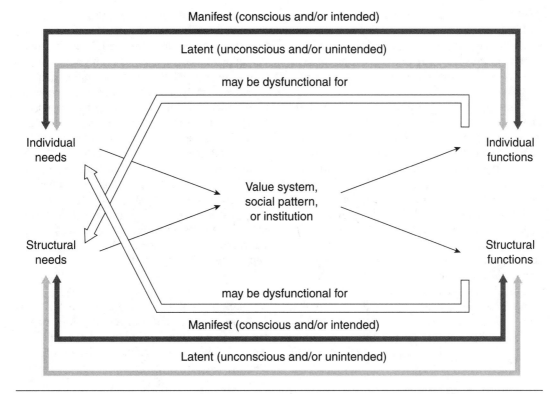

Figure 3.2 Model of Functional Analysis

The Functions of Religion

From our discussion thus far, it should be apparent that religion performs a variety of functions. Religious faith and religious organizations serve a number of needs of individuals and the social structure. However, it can be misleading to offer a general list of the functions of religion, for the way in which a religion functions will vary somewhat depending on the social structure, the culture of the society, and the specific characteristics of the religion itself.

For a group that has immigrated to a new country, religion may take on increased significance as a source of identity. For a group experiencing great suffering, religion may offer a supra-empirical explanation that makes the suffering bearable. In a society experiencing rapid social change, religion may provide a feeling of security and assurance. However, without overgeneralizing it is possible to point to four types of functions that religion typically serves: (1) meaning functions, (2) belonging and identity functions, (3) cultural functions, and (4) structural functions.

Individual Functions

Religion functions for individuals in two primary ways: (1) provision of meaning and (2) provision of a sense of identity and belonging.

Meaning Functions The function most explicitly associated with religion is the provision of a sense of meaning in life. Religion provides a worldview by which injustice, suffering, and death can be seen as ultimately meaningful. Clifford Geertz said when suffering or death has meaning, it becomes sufferable. Friedrich Nietzsche said the same thing when he insisted that "He who has a *why* to live can bear almost any how."

To meet the meaning function, however, religion must include more than a set of ideas or notions about the world. Abstract philosophical ideas seldom satisfy this function for the masses. Meaning involves both concept (idea) and demand (imperative) (Kelley, 1972). The worldview must be presented to the prospective believer in such a way that the person seems to be held by the belief rather than voluntarily holding the beliefs. Philosophical systems of thought seldom address people's emotions in a way that impels them to believe: There is no demand. The communication of concepts through rituals and symbol systems, on the other hand, incorporates both affective *and* cognitive dimensions; this makes the religion seem compelling.

The desire to make sense out of the world seems to be nearly universal. People may be willing

A core aspect of religion is providing a sense of belonging and identity. Worship is usually done in groups, and one's worship community or prayer group often becomes a central reference group. This photograph is of nuns praying at the start of the day at Mother Teresa's orphanage in India.

to admit that their own worldview is less adequate than another, but they are not willing to give up their interpretation for none at all. They are simply not willing to say that human events are meaningless. It seems that bafflement (lack of explanation) is an extremely anxiety-producing experience; religion acts to combat it. The religious response to "why?" is answered primarily in terms of values: What does a particular event *mean* in understanding the ultimate purpose or goal of one's life? Another way to say this is that religion locates a specific experience, event, or observation within a larger context of experiences, events, or observations. The larger context is attributed with *ultimate* meaning, and the specific event is viewed as having significance because of its relationship to the "big picture" (Wuthnow, 1976b).

Belonging and Identity Functions A less often recognized aspect of religion is its importance for the sense of identity of the believers. Andrew Greeley (1972) argued that the reason denominationalism is so strong in the United States is because of the function religion has played in the lives of immigrants (pp. 108–126). Before coming to the United States, many Italians (for example) did not identify themselves primarily as Catholics. Placed in a new environment, however, with different norms and values, many Italian Americans came to identify strongly with the Catholic Church. Many Italian Catholic congregations became community centers that helped members preserve their sense of roots. Similarly, other immigrant groups have shown increased denominational loyalty and intensified religiosity after coming to the United States. The denomination, in effect, became a source of ethnic identity and a bastion of cultural stability for those facing culture shock (Marty, 1972). As decades pass, religious services eventually come to be conducted in English, assimilation occurs, and ethnic loyalty starts to fade; religion then becomes the source of identity in and of itself (Warner, 1993).

Along similar lines, in Poland, Ukraine, and other Eastern European countries that were dominated by the Soviet Union, the Roman Catholic tradition became a source of identity and solidarity for those opposing foreign control. In Poland, the church contributed a sense of commonality for people of several ethnic groups and gave sacred sanction to their resistance to a government that they viewed as an illegitimate puppet government controlled by Moscow. Unity of the Polish people—based on a common *religious* identity—was an important component of their survival and ultimate success (Tamney, 1992).

Even beyond the cases of immigrants or of victims of colonialism, a sense of religious belonging often affects individuals' understanding of whom and what they are. A teenager at a party may forgo alcohol or may refuse to participate in certain activities. The individual may identify herself strongly as a Christian, and according to her definition of what that means, certain activities are unacceptable. Likewise, a young Mennonite, faced with his draft papers for the U.S. Army, may conclude that his identity as a Christian conflicts with his identity as an American. As a Mennonite, he may accept the prohibition against war as categorical. His petition for status as a pacifist may be a profound statement of identity as a Christian. Interestingly, Peek (2005) found an increasing religious identification among some young Muslims in the United States after the 9/11 attacks, a time when one might think that religious identity would be downplayed. This suggests the importance of the belonging function of religion for individuals. Greeley emphasized that while meaning functions may be primary (in the sense of being closer to the manifest purpose of religion), the belonging functions may actually be prior to meaning in terms of chronology. Greeley (1972) wrote, "With some exaggeration we may say that instead of Americans belonging to churches because they believe in religion, there may be a strong tendency for them to believe in religion because they belong to churches" (p. 115).

When we first read that statement by Greeley, we thought he was mistaken. After

doing a good deal of research on the social psychology of religion, we have come to believe that he is essentially correct. In a rapidly changing society, especially in a society with a high degree of geographic mobility, a sense of belonging is a critical need. If one's kin all live 500 miles away, the need for emotional support during a time of crisis can be intense. Religious groups often serve to satisfy this need. This is true for both mainline congregations and new religious movements (NRMs) (McGaw, 1979; Roof, 1978).

Societal Functions

Religion serves the society as a whole by bolstering both the culture and the social structure.

Cultural Functions In most societies, religion functions to sacralize cultural values. Of course, generalizations about the relationship between religion and morality, or religion and everyday values, must be cautiously made and must take into account the researcher's definition of religion. They are most accurately made with reference to a specific religion in a specific society. Nonetheless, we would concur with Geertz in asserting that for most individuals in most religions in most societies, religion and everyday morality are closely linked. Geertz (1958) wrote,

> The tendency to desire some sort of factual basis for one's commitments seems practically universal; mere conventionalism satisfies few people in any culture. However its role may differ at various times, for various individuals, and in various cultures, religion, by fusing ethos and worldview, gives to a set of social values what they perhaps most need to be coercive: an appearance of objectivity. In sacred rituals and myths values are portrayed not as subjective human preferences but as the imposed conditions for life . . . . (pp. 426–427)

A moral code may not seem entirely compelling to people if it is enforced by nothing more than social tradition. One of the important functions of most religions, then, is to provide a metaphysical basis for the moral order of the social group and to reinforce obedience to norms. Thomas O'Dea's (1966) comment seems to be accurate for most cases: "By showing the norms and rules of society to be part of a large supraempirical ethical order, ordained and sanctified by religious belief and practice, religion contributes to their enforcement" (p. 6).

By providing a metaphysical foundation for the culture's values, moral codes, and outlooks on life, a religion helps to combat the confusion, disorientation, and deviance that anomie can generate. This aspect of religion enhances stability for the culture.

Structural Functions Religion also serves the organizational structure of society. Durkheim emphasized that religion served as a sort of glue to bond together people who otherwise had diverse self-interests; it helped them to define themselves as a moral community with common values and with a common mission in life. This unifying and self-defining function is especially true of religion in nonindustrialized, homogeneous societies. In the pluralistic society of the United States, no single traditional religion can claim that role.

As indicated earlier, religion also enhances social stability by sacralizing (making sacred) the norms and values of the society. This persuasive power of religion can have a number of important consequences. If certain behavior is taboo (so heinous and so dangerous that it is unthinkable), people may conform for fear of the consequences of breaking the taboo. Many Innuit (Eskimos) believe that Sedna, the goddess of the sea, determines one's success or failure in hunting seals. Such believers are not likely to risk offending her by violating one of her rules. Hence, no Innuit kills more than is needed; waste is unthinkable. This conservation ethic is functional for Innuit society, yet it is enforced by religious rather than legislative sanction. In most societies, religion creates an environment of moral obligation to norms that benefit the social structure.

Dysfunctions of Religion and the Diversity of Consequences

Any institution or behavior pattern is likely to have a multiplicity of consequences in a society; it may be both functional and dysfunctional at the same time. In Western culture, religion has often rejected scientific knowledge. Copernicus's claim that the earth revolves around the sun was condemned as heresy, and Galileo was tried by the Inquisition for asserting that he had proved Copernicus correct. Much later, biblical literalists condemned Charles Darwin's theories and insisted that such heresy not be taught in the public schools. The criterion of truth was the Bible, and alternative theories or ideas were to be suppressed. Such an attitude is certainly dysfunctional for scientific inquiry. This may have had positive effects for some individuals by creating an unchanging and stable worldview, but it also stifled creativity and advancement of human knowledge through scientific inquiry.

The stabilizing function of religion itself has more than one consequence. It is interesting that Karl Marx and Emile Durkheim basically agreed on the way in which religion contributes to society. Religion tends to unite people around common values and beliefs. This occurs even when the self-interests of members are contrary. Durkheim admired the way religion functioned for social unity; Marx was appalled by it and referred to religion as the "opiate of the people." Marx claimed that religion unites people under a "false consciousness," a false sense of common interests. By claiming that injustice would be rectified in an afterlife or in a reincarnation, religion has served to keep oppressed people in bondage. Religious support of the existing social system may benefit the system while being dysfunctional for the disfranchised members of that society.

We have seen that the sense of belonging that a religious community provides can serve important psychological needs. However, this sense of belonging and strong group-identification may lead to extreme parochialism, bigotry, and ethnocentrism. Religion frequently breeds narrow-mindedness and strong group boundaries.

The key point is that we must always be cautious about broad generalizations regarding religion. In order to be precise, we must discuss the functions and dysfunctions of a specific religion for specific individuals or a specific structure in a specific society. We must acknowledge that religion may be functional in one way and dysfunctional in another.

> *Critical Thinking:* How might a religious behavior or religious norm be functional for society but dysfunctional for an individual? How might a religious belief or behavior be functional for an individual but have harmful consequences for the larger society?

Problems With Functional Analysis

Functional analysis is but one lens through which any social process can be viewed and understood. There are, however, distortions in this lens.

1. Functionalism tends to err in the direction of over-emphasizing social stability and underemphasizing conflict and change. In so doing, functionalists have often assumed that societies are quite well-integrated systems; the positive functions have been stressed more heavily than the dysfunctions.

2. In stressing functions, one also loses sight of the historical process by which any particular religion established its present character. The contemporary emphasis on historical sociology (or diachronic analysis) in the study of religion is an important corrective to this oversight (Geertz, 1968).

3. By emphasizing the critical needs that religion fulfills, one might assume that the traditional forms of religion are indispensable. However, much research in the past three decades has been devoted to discovering new forms of religion or "religion surrogates" (Bellah 1970b, 1975; Bellah, Madsen, Sullivan, Swidler, & Tipton, 1985; Luckmann, 1967; Roof, 1999;

Wuthnow, 1976b; Yinger, 1969, 1977). Hence, it is important to avoid the trap of thinking that traditional forms are indispensable.

4. Functionalism has sometimes evolved into a social philosophy rather than a tool of empirical research. As a social philosophy, functionalism has sometimes used circular reasoning to posit and then prove a point. Certain basic needs, whether biological, psychological, or social, have occasionally been set forth as basic and inherent to human society. Any existing social arrangements are said to be *created by* those needs and, in turn, to have the *effect* of satisfying them. This makes a neat theory because it cannot be disproved. The answer is presupposed in the question; satisfaction of some posited need is both cause and effect.

Most sociologists do not accept many of the assumptions of functionalism as social philosophy, but functionalism continues to provide an important methodology.[1] By this, we mean that the basic questions of functional analysis continue to guide much sociological inquiry: What individual and social needs does a particular social pattern serve in a particular society at a particular time? How does a social pattern or institution affect the lives of individuals and influence other social patterns and institutions? What are the social forces that contribute to social integration? As a method of analysis, these questions are asked as open-ended queries and with the realization that the amount of social integration in a society is never total and is always subject to change.

5. Structural functionalists have often evaluated social functions as being primary, while seeing individual dysfunctions as necessary evils. Therefore, some functional theories have operated with a conservative bias. This need not be an inherent problem with functionalism as a methodology, but it has been a tendency of functionalism as a social philosophy. For functionalism as a method of analysis, social order is not intrinsically good or evil but a variable to be assessed in each society.

---

[1]For a more detailed treatment of the history, criticisms, and logical problems of functionalism, see Jonathan Turner and Alexandra Maryanski (1979). The authors also provide an excellent discussion of the difference between functionalism as an empirical *methodology* and functionalism as a social philosophy.

As a method of analysis, functionalism has certain common concerns with conflict theory, for both seek to discover the social consequences of a particular belief, behavior, or structure. However, many sociologists are principally interested in issues of change, conflict between interest groups, and inequalities in power. Conflict theory addresses these issues more adequately than does functional theory, and it is to this second macro-level theory that we now turn.

CONFLICT THEORY AND RELIGIOUS CONFLICT

Marx maintained that the fundamental reality of history and modern society is a conflict between the classes. The "haves" use every tool available, including coercion and ideology, to sustain their advantageous position over the "have-nots." According to this view, understanding modern industrial society does not necessitate an analysis of cultural values and beliefs. The basic issue is economic conflict. Hence, Marx is often identified as the father of modern conflict theory. He maintained that values and beliefs basically operate (after the fact) to justify the self-interests of various groups. Accordingly, Marx viewed religion as an ideology that justifies the current social arrangements. It serves as a tool of the upper classes and helps maintain stability. Marx, like Durkheim, viewed religion as a force for social integration. However, for Marx, this had a tragic consequence; religion served to maintain an unjust status quo. Religion acts to unite people of various classes with one another when, according to Marx, the lower classes should be uniting against all those in the upper class. In fact, the ideology that promised rewards in an afterlife for conformity in this world has had as much of a pacifying effect as opium (hence, his comment about religion being the "opiate of the masses").

Certainly there are many examples that would lend credence to the Marxian interpretation. The Hindu belief in reincarnation has led many lower caste Indians to conform to the laws of *dharma*. Only by conforming to dharma (which reinforces caste lines) can one expect to be reincarnated in a higher position. Those who violate these laws can expect to be reincarnated in some lower animal form. This sort of belief system tends to undermine any impetus to rebel against the social system. Christian beliefs in otherworldly salvation sometimes act in a similar way to pacify the poor and the disfranchised (see Chapter 10).

Conflict as a Source of Social Disruption

Not all conflict theorists emphasize that religion integrates and stabilizes society. In fact, a number of them emphasize that society is not well integrated at all. Society comprises interest groups, each of which seeks the fulfillment of its own self-interests. They believe there is no consensus over values and beliefs that serves to unite the society; rather, modern society is characterized by conflict, coercion, and power plays by various groups. When stability does occur in a society, it is because (a) a temporary balance of power exists between groups or (b) one group has gained enough power to control the others. Stability lasts only so long as the distribution of power remains the same. In this scenario, religious groups are simply viewed as one more set of interest groups in society. Hence, Marx's principle—that self-interest is the key factor in shaping social relationships—is a central emphasis of all conflict theory.

Interreligious Conflicts

Some conflicts involve hatred of those who adhere to a different faith, and this disrupts the working of the society as a whole. Conflicts between Christians and Jews provide a vivid example. Christians and Jews have coexisted in the Western world for nearly 2,000 years. Yet because Christianity has been the dominant religious force in Europe since late Roman times, Christian leaders have often determined the nature of the relationship. At some points in history, Jews have been enticed to come to predominantly Christian cities because they brought needed skills and services. For example, in AD 1084, the bishop of Speyer attracted Jews to that city because of their professional skills and because they would provide loans, which Christians would not do because they believed that usury (lending money for interest) was immoral. Jews were quite willing to lend money and thereby provided a needed service for the community. As part of the enticement, Jews were given their own section of the city, a section the Jews called a "ghetto."[2] In fact, they willingly purchased a charter and paid a lease for the privilege of having their own ethnic enclave. However, as Christian mores changed, usury became an acceptable Christian enterprise, Christians moved into the professions Jews had occupied, and conflict between the groups began to intensify. By 1555, Pope Paul IV had made the Jewish ghetto compulsory rather than voluntary, and Jews became objects of severe persecution (Berry & Tischler, 1978).

The record of such conflict is long and consistent. The day before Columbus first set out for America, all Jews were ordered to leave Spain. The same thing had occurred in England in 1290 and in France in 1306 (Berry & Tischler, 1978). Although the United States and Canada have not expelled their Jewish populations, the pattern of discriminatory treatment has continued in North America. Many Christians blame Jewish victims for their plight, insisting that some characteristic of the Jews causes them to be persecuted. Despite this assumption, the evidence is overwhelming that the primary cause of

discrimination is the desire to gain an edge in a conflict over scarce resources (e.g., jobs, the best housing, the best educational opportunities). For many decades in the United States, universities and professional schools had quotas for Jews. Only a limited percentage of Jews would be admitted each year, regardless of superior qualifications of Jewish applicants (Karabel, 2005). Christians have often used their power as the dominant group to place Jews at an economic disadvantage. Ironically, those same Christians have then labeled the Jews as being the ones who are devious, manipulative, and driven by economic interests. There were 1,352 reported incidents of anti-Semitism in the United States in 2008. These incidents included vandalism to Jewish institutions and property (such as arson and bombings of synagogues or Jewish homes and the painting of Nazi swastikas on synagogue walls) and harassments, threats, and assaults to Jewish individuals (beatings, bomb threats, harassing hate calls on the phone, etc.). Although this number represents a decline of nearly 16% compared to a decade earlier, Jews are still the most frequently targeted religious group in the United States for hate crimes (Anti-Defamation League, 2009).

Intrafaith Conflicts

Religio-economic conflict has certainly not been limited to that between Christians and non-Christians as the "Illustrating Sociological Concepts" feature demonstrates. In the United States, Protestants have used their dominant numbers and established positions of power to oppress Catholic immigrants. Because they were here first, Protestants were well established before Catholic immigrants of Irish, Italian, or Hispanic background came to this

---

[2]A ghetto is not necessarily a poor area of the city. A poor area is a slum; a ghetto is an ethnic enclave. Some ghettos in America are also slums, but the terms are not synonymous.

Illustrating Sociological Concepts

Conflict and Coveting: The Norm in Jerusalem's Church of the Holy Sepulchre

Roman Catholics visiting the church of the Holy Sepulchre in Jerusalem are often led by brown-robed Franciscan monks as tour guides. The group may pause at what is believed to be Christ's tomb and offer a short prayer. But they have little time to tarry, for black-hooded Armenian priests are on their way with another group—accompanied by chanting choir boys swinging incense burners. Each group has its turn, and the schedule is rigidly enforced.

Five religious groups "share" the most sacred church in Christendom, but here Christian religions mix "about as well as holy oil and holy water." Every inch of the church is carefully divided between the groups and is jealously guarded. Protestants and other groups get nothing, and those who do "own" a share spend much of their time coveting their neighbor's share. A chapel in the church is built over a mound that some people believe is the historic Golgotha—where Jesus was crucified. The floor in that chapel is made up of tiny marble stones, and each stone can be identified as belonging to a specific group: Greek Orthodox, Coptic, Armenian, Syrian, or Roman Catholic. A priest who polishes a stone that does not belong to his own group may be in serious trouble!

Members of each of the groups have been known to throw stones at one another and to switch the locks during the night so others cannot get into the church. Sometimes the combat is limited to verbal exchanges. According to a *Wall Street Journal* article, highly educated clergy—who claim to worship the same Messiah—have been fussing with each other for years about their relative portions of a pillar supporting the roof.

On Easter Day, times for services are calculated so precisely that a twenty-one-page booklet is published to inform worshipers *who* may pray, *when* they may do so, and *where* this will occur. Yet not even this level of cooperation is always achieved. Repairs on the church, which was built during the time of the Crusades, are a serious problem. The groups are unable to agree on how to maintain and restore the church. If one group goes ahead and whitewashes the walls or repairs damaged floors, zealots from the other group may tear up the repairs and even retaliate by destroying some other part of the church identified with the sect that initiated repairs.

Similar conflicts exist in other holy sites. In the Church of the Nativity in Bethlehem, Greek and Armenian Monks fought with fists and brooms at Christmas time a few years ago. The fight broke out over who would dust a cornice. Hatred between the various groups of Christian devotees in the Holy Land is sometimes intense, despite the teachings of their master. Forget the admonition to love thine *enemies;* these folks can't stand their religious kinsfolk. The tenth commandment, which advises that "Thou shalt not covet," is obviously passe—or at least irrelevant when such important issues are at stake!

SOURCE: Rosewicz, Barbara. (1985, April 5). At Jerusalem church, people often ignore tenth commandment. *Wall Street Journal*, pp. 1, 5.

country. The Know Nothing Party and the Ku Klux Klan are two examples of American movements that were intensely anti-Catholic and that limited membership to white, American-born Protestants. Some analysts insist that the temperance movement was also essentially an anti-Catholic phenomenon in which Protestants sought to force their own definitions of Christian morality on Catholics and Jews (Gusfield, 1963). The Anti-Saloon League, formed in 1896, specifically attacked the symbol of the urban, ethnic, Catholic, lower-class leisure lifestyle. For many Protestants, the saloon was a symbol of Irish Catholics, who had immigrated in large numbers, were taking jobs, and were beginning to upset the economic and political advantage Protestants had enjoyed. In 1844, the conflicts between Protestants and Catholics actually deteriorated to armed combat in Philadelphia (Shannon, 1963).

According to this analysis, Prohibition was largely an attempt to define ethnic Catholic lifestyles as illegal and thereby to label such persons as deviants. Protestants continued to dominate the top positions in business and finance into the 1950s, with Catholics—regardless of competence or credentials—effectively shut out. For example, Charles Anderson (1970) reported that in the mid-1950s, 93% of the top executives in manufacturing, mining, and finance were Protestant and that Protestants held 85% of the highest positions in the 200 largest corporations. This was despite the fact that Catholics comprised approximately 25% of the American population (Greeley, 1974; Stark & Glock, 1968). Fortunately, since then, levels of anti-Catholicity among Protestants and anti-Protestant sentiments among Catholics have declined. Still, Protestants have clearly been an interest group that used a position of power and influence to ensure an economic advantage (see Chapter 9).

Hostility against "those others" varies in large measure with the extent to which lines of religious affiliation are coextensive with ethnic and class lines. When ethnic ties and economic interests act to create social solidarity, religious differences may serve as one more symbol of differentiation.

American and Canadian societies are increasingly characterized by many cross pressures and countervailing forces. People may have group loyalties, friendships, and business partnerships that involve alliances with members of other religious groups, social classes, and ethnic groups. Insofar as this is true, the multiplicity of crossed alliances provides the basis for integration and social stability. Interreligious conflict is less

The most extreme expression of antipathy against a religious group occurred in the 1940s, when an estimated 6 million Jews were gassed by the Nazi government in Germany. The stated goal of Hitler was to exterminate all Jews. This scene is from the concentration camp at Buchenwald. (Photo used by permission of the United States Holocaust Memorial Museum.)

severe in those settings where religious, ethnic, and class lines are not coextensive and where countervailing forces can provide structural integration to the society. In this circumstance, there is less tendency to view members of other faiths as enemies.

Conflicts Regarding Religious and Secular Authority

Religion may prove divisive in another instance: recognition of what constitutes proper authority. The central value system of a religion may come into conflict with the secular legal system. In this case, people must choose which set of values and norms they will respect. Often this boils down to the issue of which value system is authoritative or which leader (religious or political) is attributed with proper authority. In the late 1960s and early 1970s, a number of religiously motivated individuals violated federal laws in protest of the Vietnam War (Hall, 1990). For example, two Roman Catholic priests, Daniel and Phillip Berrigan, broke into draft board offices and poured blood on files. They maintained that their Christian conscience would not permit them to remain idle while young men were being drafted for war. Likewise in the 1980s, evangelical Christians bombed and sabotaged abortion clinics, despite the illegality of the act in the civil law (Blanchard & Prewitt, 1993), while liberal Christians provided sanctuary to Central American immigrants who—fleeing the war in that region—were in the United States illegally (Smith, 1996b). In these and other such cases, a person's theological understanding may generate norms that differ from those the civil laws uphold.

The Amish regulation that their children must not go to school beyond the eighth grade provides another example of conflict between religious and state authorities. When this norm conflicted with state law compelling all children to stay in school until graduation or age 16, some Amish parents in the 1950s and 1960s chose to go to jail rather than obey the law. In 1972, the Supreme Court settled the issue in *Wisconsin v. Yoder*, ruling that a state is in violation of the First and Fourteenth Amendments to the Constitution if it forces the Amish to send their children to high school (Hostetler, 1993).

The Mormons also encountered intense conflict with the federal government in the 19th century. In this case, the issue was over which authority, the church or the state, would decide how many wives a man may have. In a 5 to 4 decision, the Supreme Court ruled against the Mormons, and later Latter-day Saints leaders revised their doctrine in conformity with federal law. However, conflict between the Mormon Church and the government had nearly escalated to a small-scale civil war.

In each of these cases, religious norms have conflicted with secular law. The result has been considerable disruption of social harmony and unity. Whether the disruption was good or bad is not our concern here; the point is that religious loyalties sometimes result in discord and conflict in the larger society. Religiously motivated people often march to the beat of a different drummer. Religion, then, may contribute to either consensus or dissension.

Conflict as a Source of Integration

While religious conflict may bring disruption to the larger society, it may also be a source of internal unity for the religious group. Some groups seem to cultivate conflict with the out-group because conflict is functional to the group's internal solidarity. As we will see in Chapter 5, groups with high boundaries usually sustain stronger member commitment and retention than do those with low boundaries. Some Amish groups shun their deviant members—especially those who marry outside the group or who adopt the lifestyle of the outside. The shunning, or *meidung*, involves a refusal to interact with persons so labeled. In some cases, Amish parents whose child marries outside the group have considered the child to be dead, have refused to speak the child's name, and have refused to acknowledge the child's presence when in the same room. The interesting point sociologically is that those groups that are very strict about practicing meidung have greater retention of members than do more liberal Amish groups (Hostetler, 1993; Kraybill, 2001).

Conflict produces internal cohesion for several reasons. First, repression and hostility by outsiders tend to create a feeling of common plight and common destiny. The more animosity that neighbors direct toward the Amish, the more inner unity is created within Amish communities. In the late 1970s, there were a series of incidents of youths attacking Amish buggies, and at least one person was killed—an Amish infant whose skull was crushed by a rock. Such an attack bonded together Amish communities that previously had been in conflict.

Second, a common rejection of something helps articulate one's own beliefs. Actually, it is usually easier for members of a group to agree that they reject something than it is to formulate a constructive statement about what they do believe. A shared disgust at the actions of a deviant member may provide unity for the conformists (Erikson, 1966). Likewise, rejection of worldliness or of some specific religious movement (e.g., satanic cults) may be a significant source of group harmony. For this reason, social conflict and social integration must not be seen as opposites but as different sides of the same coin. One can see the same phenomenon at work at the national level. When the United States was at war with Iraq over Saddam Hussein's invasion of Kuwait, the sense of patriotism in both countries was greatly strengthened. The same was true following the September 11, 2001, terrorist attacks on New York and Washington D.C. Conflict in one situation often creates integration and unity in another. Moreover, the most threatening conflict occurs within the context of a meaningful or important relationship with others. When conflict occurs, it is within a larger system of interrelatedness. Social harmony and social conflict must be understood together (Coser, 1954, 1967).

Conflict as a Source of Change

Conflict is often a source of change. Clearly, religion is capable of contributing to conflict and to social change. One aspect of religion that most interested Weber (1922/1963, 1946, 1947) was the role of the charismatic leader. He found charisma to be fundamentally contrary to social stability and a major source of change. As a person who is attributed with divine authority, the charismatic religious leader is able to challenge social mores where there is otherwise little room for social and political dissent. (This will be treated further in Chapter 7.)

In the eyes of those who defend the status quo, such conflict is disruptive and therefore dysfunctional. Such conflict is indeed disruptive, but disruption may be needed. Religiously motivated abolitionism was thought to be dysfunctional by slave owners. In terms of the interests of blacks and in terms of the structure of contemporary post-industrial society, abolitionism was clearly a beneficial disruption of the status quo. Religion is capable of disruption because religious values sometimes define social relationships differently from the way the encompassing culture does. If a religious ideology maintains that all people are equal before God and if the society is rigidly stratified, a dissonance may be created for those who take the religious ideology seriously. By affirming the innate personal worth of blacks, the African American church in the United States provided a foundation for these oppressed people to assert their rights and to insist on social change. It was no mere coincidence that Martin Luther King Jr. was both a civil rights leader and a Baptist minister. Whether the contribution is direct or indirect, religion can contribute to the modification of the larger society, for it can provide an alternative worldview and an alternative set of values.

On the other hand, conflict and change in the society itself can also be a major source of religious change. Rapid social change is one factor in the rise of NRMs. Societies that are stable and highly integrated are not conducive to the emergence and expansion of religious cults, but loosely integrated, changing societies do experience such phenomena with great frequency (Stark & Roberts, 1982).

The broad sweep of social change in the direction of secularization has also brought modifications in religion. The advent of rational, scientific, empirically oriented culture has caused changes in the worldviews of many people. Scientific interpretations of such issues as the origins of humanity have induced liberal theologians to reformulate religious ideas so that they are compatible with scientific ones. This, of course, has involved conflicts between modernists and conservatives within religious bodies. Hence, the growth of science as a major institution—an institution autonomous from religious authority—has led to changes and conflicts within religious groups. As will be amply illustrated throughout this text, changes and conflicts in the larger society often result in changes and conflicts within religious bodies.

Conflict as a Pervasive Element in All Social Life

Marxian theory suggests that conflict is pervasive in human life; it can arise not only between groups but also within them. Social behavior is always guided by individuals and groups protecting their self-interests. For Marxian theorists, social animosity and inequality are always the result of a struggle for power, privilege, and prestige. Religious groups experience these processes no less than other groups.

For example, men have dominated the leadership roles in most denominations. In Orthodox Jewish groups, only male members are counted as part of the necessary quorum for prayer meetings. Until recent decades, most Protestant congregations excluded women from the ordained ministry or even from lay positions of leadership (such as offices of deacon or elder in

Congregational or Presbyterian churches). Some conflict theorists have viewed these practices as evidence that men are an interest group—whose members cling tenaciously to their positions of authority and power (see the discussion in Chapter 11).

Intrachurch conflicts between clergy and laity or between theological liberals and conservatives may also be rooted in socioeconomic competition and class conflict. Peter Berger (1981) suggested this conflict is actually between two elites in American society that are struggling for power, privilege, and prestige. One is the business elite—a class of people managing industrial production and manipulating business enterprise. This group is more attuned to conservative theology, as is evidenced in the conservatism of most laity. Berger claims that conservative theology tends to justify the self-interests of the business elite.

The "new elite" are those who manipulate symbols and words and who manage the production of ideas. This group includes intellectuals, educators, members of the helping professions, media people, and various social planners and bureaucrats. Highly educated clergy of the mainline denominations are also part of this group, which advocates federal support of education, social welfare, minority rights, and environmental protection. This new class is highly represented in government-supported work, and its members argue for increased business regulation. Hence, this new elite really seeks its own self-interests. Berger insisted that proclamations by the National Council of Churches normally reflect the current interests of this new class. Clergy and modernists are much more likely to support the National Council than are the laity in general and conservatives in particular. Producers of ideas and producers of material objects are viewed as two diverse social classes in a postindustrial society. The conflict between such groups as the New Christian Right and the National Council of Churches is interpreted as part of a deeper struggle for access to power and privilege.

Even the conflict over creationism versus science can be viewed as being, in part, a conflict of self-interests. Christians who are well educated, whose professions are related to scientific investigation, or whose business depends on scientific advances tend to be the ones who dismiss creationism and claim that theology should be reformulated to be consistent with science. In other words, theological modernism can be self-serving. On the other hand, biblical literalism is adhered to largely by the lower- and working-class people—folks whose jobs are sometimes threatened by technological innovations. Furthermore, working-class Americans may be rankled by a perception that many of the persons who occupy high-paying and high-prestige positions—physicians and scientists—are foreign born. Antiscientific views may be interpreted—through a conflict theory analysis—as veiled attacks on those who threaten the jobs and prestige of lower-class Protestants. Although such an analysis certainly does not tell the whole story, it does raise some interesting questions about the extent to which our behavior—including religious behavior and beliefs—is influenced by our self-interests.

A second contribution of conflict theory is that it focuses attention on issues of change. Religious conflicts can contribute to social change, and social conflicts can cause religious change. Those changes may create social dissension and disruption, but depending on the nature and location of the conflict (external versus internal), they are also capable of contributing to the integration and cohesion of a group.

Conflict theory's tendency to use historical analysis is another strength. In using historical perspective and data, it offers an important corrective to more static theories that ignore the specific causes of a contemporary social pattern.

Most important, perhaps, conflict theory illuminates the many ways in which self-interests affect perceptions and behavior—including religious ideas and (supposedly) religiously motivated behavior.

Critical Thinking: How might your own self-interests have shaped your religious beliefs? Do you think it is true that our economic self-interests may bias or even completely blind us as we seek eternal truths? Why or why not?

Problems With Conflict Analysis

Although conflict theory offers important insights and correctives to functional analysis, it is not without its problems. If functional theory often errs in overemphasizing consensus and harmony, conflict theorists often see only social stress, power plays, and disharmony. Although conflict theory is helpful in illuminating the causes of change, it is less complete in explaining social cohesion and cooperation.

Perhaps the most important criticism of conflict theory is its tendency to view all behavior as motivated by self-interest. For those who believe that religious commitment may call a person to genuine unselfish action, the appeal to self-interests as the ultimate motivator of all behavior is unsatisfactory (Smith, 2003). The response among conflict theorists is that appeals to altruistic motives are simply ways of mystifying or hiding the true motives. However, the willingness of deeply religious persons to sacrifice even their lives for others or for their faith raises questions about this unicausal interpretation. Persons may be influenced by a wide range of motives. Conflict theory correctly points out that self-interests can be pervasive and can influence a wide range of behavior and attitudes. However, when it insists that *all* behavior is determined by self-interests, conflict theory commits the error of reductionism.

Another problem with this emphasis on self-interests is that interests are often interpreted only in economic terms. Although economic self-interests are more encompassing than we once realized, other sorts of self-interests can also profoundly affect behavior. For example, a calculation of one's spiritual self-interests—such as a desire to attain a heavenly afterlife and avoid hell—could cause persons to behave in ways that are contrary to their economic self-interests. Religious moods and motivations are, to quote Geertz, "powerful, pervasive, and long-lasting." They are capable of influencing behavior in ways that are insensitive to economic consequences. Conflict theory does not normally take into account such forces.

Conflict theory stresses conflict and dissension and the role of self-interests in shaping behavior. It de-emphasizes the harmony, consensus, and interrelatedness that so intrigue functionalists. There is no shortage of data on conflict within religious groups, but given all of the conflict that does exist, most groups cohere surprisingly well.

> *Critical Thinking:* Why is the more historical approach of conflict theory important as a complement to the rather nonhistorical or "snapshot" analysis of functionalism?

TOWARD SYNTHESIS: AN OPEN SYSTEMS MODEL

Some sociologists believe that no synthesis of the structural functional and conflict perspectives is possible. They do offer very different perspectives on society and on human behavior. However, one effort to combine the insights of both perspectives is promising. That approach is called open systems theory. Variations of systems theory have been used to understand units as small as nuclear families and processes as large as the "world system" of international economic interdependence (Shannon, 1989; Wallerstein, 1974, 1984). Interestingly, systems theory—especially world systems theory—is often claimed to be an outgrowth of conflict theory, but structural functionalists also claim that the systems approach is an extension of their perspective. Open systems theory is not a panacea; it does not

solve all of the problems or discrepancies between the two dominant macro perspectives, but it does offer some overview of how an entire system may be characterized simultaneously by interdependence and by conflicts of interest.

Contemporary systems models stress the idea that societies are both stable and dynamic, depending on whether one focuses on structure or on processes. Structure refers to the organized aspects of a society; this dimension of a social system is resistant to change. The structure can be diagrammed on an organizational chart: statuses, roles, departments, and organizational units each in relationship to one another. Religious denominations have stable sources of income, officers who fill certain roles and responsibilities, and clergy who are responsible to bishops or district superintendents. Local faith communities have commissions and committees that make decisions regarding certain parts of the life of that church or temple: worship, religious education, fundraising, care of the building, and so forth. These offices or interrelated positions may evolve over a period of time, but generally they remain with similar job descriptions and a consistent set of responsibilities. This is structure.

On the other hand, the process aspect of a social system is characterized by action; nothing is ever fixed or final. The members of committees change; the clergy retire or move to new churches or temples and are replaced in the old ones; and the values and ideas of the congregation are transformed in light of national events (such as a recession), international crises (such as a war in which some members are called to serve and others feel compelled to oppose), or local incidents (such as a hate crime against a homosexual in the community). Such happenings may change the value orientations of members, bring a shift in perceived interests of various groups, or induce a transfer in the allocation of resources within the organization. While it is true that organizations are somewhat stable, it is equally true that they are always becoming something new as they interact with the larger environment.

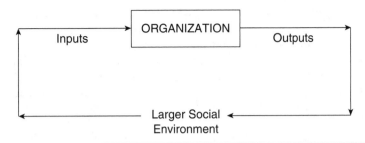

Figure 3.3 Open Systems Model

Structure and process are both necessary parts of any social unit; they cannot be separated. Structure provides the context for action; but action brings the structure to life. Without process, an organization would wither and die; without structure, action would be disorganized and aimless (Ballantine & Hammack, 2008; Olsen, 1978).

Figure 3.3 illustrates one dimension of the systems perspective on how organizations work.[3] Any organization or institution exists within a field of forces, which it tries to influence and which try to influence it. The model provided here is simplified for our understanding, but it is important to recognize that the "organization" in this diagram might be a nation (e.g., Canada), a national institution (e.g., a particular denomination), a local organization (e.g., a congregation), a unit within a local religious community (e.g., the religious education committee of the congregation), or a subsystem of the education program (e.g., an adult scriptural study class).

One of the first points that we must comprehend is that all organizations have *outputs*. For schools, the outputs are graduates, new ideas, new skills and ways of doing things, and developing cultural values and outlooks. For religious organizations, the outputs or "products" that influence the larger culture include the following: values, attitudes, benevolence or "mission" programs, published instructional materials, broadcast programming, movements to influence the larger society (e.g., peace organizations, anti-abortion groups, and movements to have "creation science" taught in the science curriculum of the public schools). Nearly all organizations want to influence, at least in some marginal way, the larger social environment in which they exist. Their outputs are the means through which they do this. Even if influencing the larger culture is not an explicit goal, religious groups are bound to have some impact on the secular society around them as these groups struggle for resources necessary for survival. Also, religious groups typically feel an obligation to engage in benevolent acts, or what they interpret as benevolent acts, for those "less fortunate." This requires outputs beyond the boundaries of their own group.

As the diagram suggests, outputs of any organization will enter and influence the environment—that is, the dominant culture or larger society. The extent of influence varies with the goals and purposes of the group and with the amount of power the group wields. Regardless of the intent of the religious groups themselves, segments of the larger society are being influenced by the religious organization and may even be competing with it for resources (including the time, energy, and monetary donations of members) and for the ability to shape central values and beliefs of the culture. Therefore, those in the social environment want to

---

[3]This model is an adaptation from Ludwig Von Bertalanfly's (1962) general systems theory and is effectively applied to educational systems by Jeanne H. Ballantine and Floyd Hammack (2008).

influence the religion itself. So the final dimension of the process is *inputs* into the organization: new members, secular values and beliefs (including nationalistic pride, scientific theories, and definitions of morality), attempts by governments to legislate religious groups (such as financial disclosure requirements to prevent fraud), new ideas (e.g., use of the latest business principles to manage a large congregation), and even new technologies (including computerization of mailing lists and sophisticated technology to record religious services).

> ***Critical Thinking:*** How might a global event influence a religious community? How could other institutions (e.g., economics, the family, government, sports) affect a religious group? How might a religion influence or impact other institutions in society?

The more the organization is an *open system*, the freer is the flow of both information and resources to and from the surrounding society. In theory, a totally open system would have a completely open interaction with the larger society. In reality, in all environments there are some people who will resist influence from the organization because their own interests are threatened. Further, all systems like to control the input from the community. In fact, in order to develop in a controlled manner, any organization must maintain an appropriate ratio of input and output. Internal needs of the organization demand some filtering of the flow of information (Olsen, 1978).

At the other end of the spectrum, some groups are *closed systems*, organizations that try to exclude outside influence as much as possible and whose desire to shape the larger culture is minimal. It is as if the group built an enormous barricade around the organization. That barricade is what sociologists call a social boundary. Especially important for some groups is the building of a wall preventing inputs, often to protect the self-interests of the organization. We

will see examples of this resistance to input from the surrounding society later on in this book. No organization, however, is totally closed, for all must live in an interdependent relationship with the environment. For effective adaptation to new circumstances, any organization must have some flow of information into the organization from the environment, but the important point here is that each group is selective about—or at least *tries* to control—what penetrates the organization. Religious liberals and religious conservatives each try to filter the inputs (with varying degrees of success), but both allow inputs of one kind or another.

No one diagram can fully explain all of the complexities of the systems approach. Figure 3.4 illuminates another dimension of the social systems view of society. As Marvin Olsen (1978) put it, "Organizations overlap and interlock with each other, forming a gigantic social web, the totality of which is human social life" (p. 99). Any organization or system is made up of a number of interdependent parts. A family is made up of the various members who are linked together with a variety of bonds and obligations. Imagine that in Figure 3.4, each dot represents a member of your family and the lines between them represent the bonds of love and concern. If one person is undergoing some stress (e.g., chemical dependency, unemployment or job stress, or poor health), the rest of the family might well support that individual by taking up some of his or her roles, by offering emotional support, and by helping in other ways. This stressor on one person might then cause stressors on other family members as they try to help out and "make things work" during this time of special tension.

Religious groups are also characterized by interdependent roles and statuses, each dependent upon the other for effective functioning. For the religious education committee to do its job, the stewardship committee must solicit enough money from the congregation to meet the budget. Likewise, the stewardship committee may not be able to be successful if the

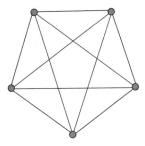

Figure 3.4 Systems Composed of Subsystems

worship committee and the minister do not function effectively and members find worship unsatisfying and empty.

Yet systems are not just dependent upon each other; each system is related to other systems. The family is linked to and dependent upon many other institutions: health care, the workplace, schools, the religious community, the criminal justice system, the banking system, and so forth. If any one of those systems is not functioning properly, the family experiences pulls and tugs on their own life. If the economy is in recession and the workplace lays off family members, or if the agents of law enforcement are harassing members of your ethnic group, or if your local congregation is wracked with angry conflict and division, the family will likely feel the stress. Likewise, if families are not socializing their children with values consistent with those of the temple or church (or are not functioning at all), or if large numbers of the membership are out of work and unable to pay their pledges because of local industry closings, the congregation and even the denomination will feel the stress.

All of these institutions (and others) comprise an interdependent system that make up the nation. However, our nation is only one system that is interrelated to many others: Great Britain, Canada, Iraq, Germany, Russia, Somalia, Pakistan, Israel, El Salvador, Nigeria, and so forth (see Figure 3.5). If we have trade relations or diplomatic concerns with that system (which

we most assuredly do), then a malfunction in that country or that region of the world may cause stresses on our nation. If war is declared in the Persian Gulf or Afghanistan, stresses are generated on the resources of our country, and they impact families (whose members may be called into combat duty), the economy, the churches (which may have some members advocating peace and others wanting the congregation to support the war effort), and so forth. Likewise, an economic recession in Europe or Japan can influence trade and therefore negatively impact certain businesses in this country; a destruction of rain forests can have devastating effects on the health care community as certain medicines become scarcer.

Some institutions would be influenced directly and profoundly by stresses in the global system; others would experience only indirect impact. The important point is that social systems are interrelated in a variety of ways; this brief discussion greatly oversimplifies the complexity of the ties, but our purpose here is to illustrate the point that social systems are layered. Subsystems exist within subsystems, which exist within subsystems. Sometimes subsystems are parts of two or more overlapping macro systems (Olsen, 1978). One example would be a military chaplaincy in which the chaplain is both an officer in the military (and answerable to that hierarchy) and an ordained member of a specific denomination (and responsible to a bishop and/or to denominational policies and values).

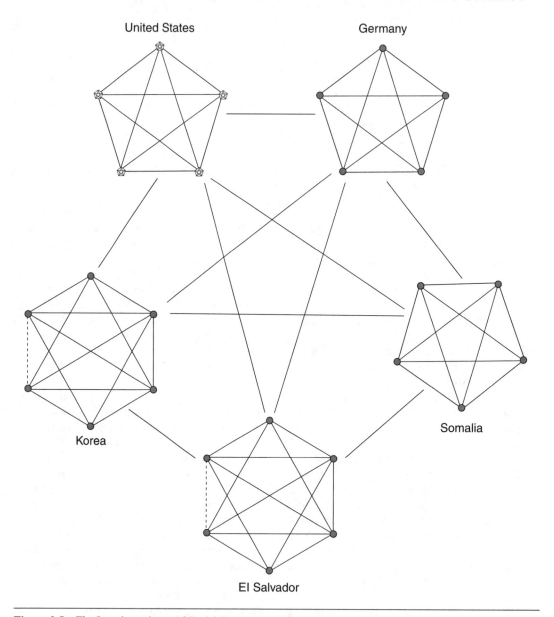

Figure 3.5 The Interdependence of Social Systems

Similar to structural functional theory, systems models stress the interrelatedness of society. Unlike functionalist approaches, there is no assumption of harmony and consensus within the system. Indeed, systems often experience conflict and tension as various units pursue their own self-interests. Further, those with power in the larger society are often able to ensure that various institutions (such as religion) serve and legitimize their own positions of privilege. Also contrary to functional theories, systems models view societies as dynamic and changing. While there may be some forces pushing toward coherence and stability, the model is anything but stagnant. The ties between units may be stressful rather than supportive, and that stress is itself not necessarily viewed as negative. Stress may bring change and enhance the health of the organization in the long run. However, at any given period of history, a system may indeed be characterized by fairly high levels of consensus and relative stability. Rather than assuming either harmony or discord, systems models leave that question open to investigation for the specific system and period of history.

> *Critical Thinking:* In what sense is open systems theory a synthesis of functional and conflict theories?

Summary

Irrespective of the ultimate truth or fiction of a religious ideology, religion has certain kinds of social effects or consequences. These consequences are what most interest sociologists. Functional theorists tend to focus on the beneficial role religion plays. For example, at an individual level, religion may offer a sense of meaning in life, sacralize and give certainty to a system of moral values, and establish a sense of belonging. Religion thereby affects an individual's sense of identity. Furthermore, religion can also serve societal functions. By providing a sense of values and beliefs around which a social consensus is formed, religion may contribute to social coherence and harmony. This is especially true for societies in which there is only one religious tradition.

On the other hand, religion is sometimes characterized by conflict—experiencing internal discord and/or contributing to conflict in the larger society. Dissension itself may be either beneficial

or disruptive, healthy or unhealthy. Conflict theorists tend to focus on dissension, disruption, and change. Conflict theorists also tend to emphasize the role of self-interests in human behavior, including supposedly religious behavior.

Conflict and functional theories need not be mutually exclusive. They each offer important insights, and each have blind spots in their views of society. We will be incorporating insights and analysis from both perspectives in this book. No analysis that relies on only one perspective can offer a rounded understanding of the relationship between religion and society.

The central fact to remember is that the balances of harmony versus dissonance and of value consensus versus interest-group coercion will vary from one society to another. Furthermore, the amount of integration or conflict within religious groups varies significantly.

The systems model of social organization attempts to integrate many of the key ideas of both structural functional and conflict theories. Systems approaches stress the fact that the structures of society tend to be rather integrated and stable, while processes of societies are dynamic and action-oriented. Systems are viewed as composed of subsystems, which are composed of subsystems, with the largest systems being global in nature. This view of organizations stresses the interrelatedness of social units, but it does not presume harmony within the system and does not predict that any one element of the system is indispensable. Religion is viewed from this approach as a complex system, linked to other elements of society in intricate ways and often protecting its own self-interests in relationship to the larger society. Open systems theory will be used to understand the dynamics of religion throughout this book.

In the next chapter, we will turn to microsociological analysis, exploring such elements as belief, ritual, symbol, nonrational transformative experiences, and worldview and examining the complex relationship among them.

4

THE CULTURAL CONSTRUCTION OF RELIGION

Experience, Myth, Ritual, Symbols, and Worldview

Here are some questions to ponder as you read this chapter:

- What does it mean to say that religion is culturally constructed?
- What are the characteristics of the sacred *experience* of the holy?
- How do myths, rituals, and symbols interact with one another?
- How do myths, rituals, and symbols support a religious worldview?
- What is the importance of a religious ethos, and how does it relate to one's worldview?
- How does a social constructionist perspective help us understand religion?

Having investigated the ways in which religion operates in society and in individuals' lives, it is appropriate next to examine religion as a cultural system and to investigate the internal interrelationships of its elements. According to sociologist Andrew

Greeley (1995), if we want to understand religion, we need to focus not on its prose or "cognitive superstructure" (doctrine) but on its poetry or "imaginative and narrative infrastructure" (experience, symbol, story, community, ritual). For Greeley (2000), "the origins and raw power of religion are at the imaginative (that is, experiential and narrative) level both for the individual and for the tradition" (p. 4).

Nonrational religious experience, myths, rituals, symbol systems, worldviews, and ethos are all part of this complex phenomenon we call religion. Scholars are interested in how each of these facets relates to the others, and how they sometimes create integrity and sometimes dissonance in religious groups.

THE SOCIAL CONSTRUCTION OF REALITY

The idea that reality is socially constructed comes from a classic work in sociology by Peter Berger and Thomas Luckmann (1966). Berger and Luckmann wanted to explain how people's conceptions of reality are created and become institutionalized in society. Berger (1967) later applied this perspective specifically to religion.

According to the social constructionist view, "Social order is a human product, or more precisely, an ongoing human production" (Berger & Luckmann, 1966, p. 52), one which "continuously acts back upon its producer" (Berger, 1967, p. 3). The social order that humans create comes about in a dialectical process of (1) externalization, (2) objectivation, and (3) internalization.

Unlike many animals, humans are "unfinished" at birth due to our underspecialized and undirected instinctual structure. Consequently, humans must make a "world" for themselves, externalizing a world we call culture. In particular, we create symbols (and symbolic acts) that help us to interpret the world around us and communicate with each other. Religious symbols, as we will see, are crucial to this. Examples of symbols include the Star of David, the cross, the star and crescent, eight-spoked wheel, and the Aum. Greetings are also a good example of externalization, from traditional handshakes or bows to newer forms like "daps" and fist bumps. In creating this cultural "world," humans not only provide order for their society but also for themselves.

The second step in the process of social construction is objectivation: "The transformation of man's products into a world that not only derives from man, but that comes to confront him as a facticity outside himself . . . The humanly produced world becomes something 'out there'" (Berger, 1967, pp. 8–9). Consider social roles, for example—what it means to be a "man" or a "woman," a "father" or "mother," an "atheist" or "believer." These roles are socially constructed, but they very much shape and constrain the actions of those playing them.

Similarly, religious symbols have a reality beyond their creators. Who was it that designed the Star of David and gave it meaning? Most people do not know, but they do know what the symbol stands for. They do share with others the meaning of this object. In fact, they may feel intense loyalty to the object. An entire history of a people may be recalled and a set of values rekindled when the symbol is displayed.

This is because other people have internalized the objectified symbol. In this final step, the objectified world is reabsorbed into the individual's subjective consciousness through the process of socialization. For example, when we internalize social roles, we do not simply see ourselves playing a role like an actor (e.g., "I'm not a professor, but I play one on TV"), but we actually identify ourselves in terms of that role. We inhabit the role; we are the role: "I am a professor."

Having briefly introduced the social constructionist perspective, let us begin exploring the roles of religious experience, myths, rituals, and symbol systems and investigating the relationships among them. We also discuss the way in which these specific manifestations of religion are related to the more intangible elements of worldview and ethos.

This Greek Orthodox painting that appears in the Stavronkita Monastery is typical of the icons through-out Greece. All Orthodox icons show a rather unusual hand configuration with the thumb and the ring finger crossed, the little finger and the pointer finger pointed almost straight up in the air, and the middle finger slightly bent. When Keith Roberts—one of the authors of this book—asked a priest about it, he was told that this is the symbol of Christianity. He was told this with considerable impatience—as if *anyone* who was Christian would most certainly know that. It is interesting that *all* Greek Orthodox icons of Jesus depict his hand in this posture. At this point, no one knows who created this symbol, but it is so objectified and internalized in the Orthodox Church that it is inconceivable to Greeks that this symbol of Christ has not always existed. Christ is even thought to have used it as a tiny infant, always with the right hand and often with both hands.

EXPERIENCE OF THE HOLY

The social construction of reality begins with human experience and the externalizations that grow out of it. It is appropriate, therefore, to begin with a consideration of the experiential dimension—some would say the experiential *core*—of religion.

Many social scientists define religion in terms of the sacred–profane distinction. For these scholars, the essence of religion has to do with a unique and extraordinary experience—an experience that has a sacred dimension and is unlike everyday life. According to some scholars, all religious phenomena evolves out of this seminal experience—the experience of the holy (James, 1958; O'Dea, 1966; Otto, 1923). Such an experience is often called nonrational, for it is neither rational nor irrational. These nonrational experiences are described by those who have them as being outside the usual categories of logical, systematic reasoning. They are not illogical; they are simply nonlogical. These experiences seem to defy the normal categories of language. Whether a nonrational experience is the essence of all religious behavior may be debatable; that such mystical experience is one important aspect of the complex experience we call religion is not. Emile Durkheim in 1912 was one of the early scholars to discuss characteristics of the sacred. Using a broad description of religious experience that he believed would be applicable in all cultures, Durkheim (1912/1995) defined the sacred realm as one that both attracts and repels individuals. The sacred is not only attractive but also repugnant; it is capable not only of being helpful but also of being dangerous. This ambiguity rests in part in the attribution of great power to the sacred; because of the overwhelming power that it possesses, it holds the potential for being either beneficial or harmful. Nevertheless, the attraction of the sacred is not based primarily on utilitarian considerations. The sacred is understood as a *nonempirical force* that is considered intrinsically valuable. As such, it places a moral obligation on the worshiper and imposes certain ethical imperatives.

Durkheim noted that mere participation in the secular world could cause one to become tainted. Ritual purification before reentering the presence of the sacred was necessary, and this often involved a rather elaborate process. For Durkheim, this emphasized the fact that the sacred was radically different from the profane world. Certain places and times were set aside as special and as belonging to another dimension of reality.

Rudolf Otto (1923) explored the nature of the religious experience in his classic book *The Idea of the Holy*. In that work, Otto insisted that there is a tendency of people in Western culture to reduce the holy to rational concepts about God. Otto's contention was that holiness cannot be reduced to intellectual concepts; hence, he attempted to describe solely the human experience of the sacred. He insisted that the experience of the holy is so unique that one can never fully understand his description unless one has experienced it; the experience seems to bring forth a "creative consciousness." A person is profoundly humbled as he or she senses an utter dependency and unworthiness before the holy. (See the "Illustrating Sociological Concepts" feature.) Rational and moral conceptions about religion come only much later, as an outgrowth of the experience itself.

Otto elaborated on the quality of this experience, which he called the *mysterium tremendum et fascinosum*. He identified five qualities of the mysterium tremendum. First, the individual is filled with a sense of awe and fear. The word *tremendum* itself expresses the tremor or terror that Otto felt was part of the experience. The story of Moses awed and frightened by the burning bush provides an example of feelings a religious experience generates. Second, one feels overwhelmed by the absolute unapproachability of the holy. Durkheim's concept of the sacred as a dangerous force is shared by Otto. The ancient Hebrew prohibition against even mentioning the name of Yahweh because of God's absolute power and unapproachability provides an example. One can also see elements of unapproachability in many "high churches" that allow only the clergy to enter the chancel area. Some Roman Catholic cathedrals and

most Greek Orthodox churches even have a screen that hides the altar from the congregation. To go behind the chancel without proper ritual purification would be to risk death. This leads us to the third characteristic: power, energy, or urgency. Otto (1923) claimed that in describing the experience of the holy, people use such symbolic language as "vitality, passion, emotional temper, will, force, movement, excitement, activity, impetus" (p. 23). Fourth, the experience of mysterium tremendum causes an awareness of the "wholly otherness" of the holy. The mystery of the experience lies in its unfamiliar and nonempirical nature. The holy is utterly unlike the profane. Fifth, one feels a sense of fascination with and attraction to the holy. Although it is potentially terrifying, it also elicits a sense of wonder and a feeling of ultimate goodness. Hence, it commands a sense of ethical imperative. According to Otto, this experience of mysterium tremendum is the universal foundation and source of all religious behavior. Thomas O'Dea has followed Otto's lead by suggesting that all other forms of religiosity are generated from this nonrational religious experience.

Religiosity is multidimensional. The kind of experience Otto described is vital to religious life for many people, but it is not the only source of religious conviction. Yet he makes the important point that religious conviction is more than ideas. To influence lives in a fundamental way, religion must have an emotional component, something that makes the ideas or belief systems "seem uniquely realistic" (recalling Geertz's definition of religion). A nonrational experience often provides such an impetus to belief.

Andrew Greeley offered clarity to this discussion of the role of sacred experiences. In his analysis of "sacredness" and the extraordinary quality of religious experience, he granted that the sacred is usually removed from everyday existence. Yet, he pointed out that sense of sacredness is not always the same thing as religious commitment. Greeley (1972)

ILLUSTRATING SOCIOLOGICAL CONCEPTS

Personal Accounts of Religious Experiences

I remember the night, and almost the very spot on the hilltop, where my soul opened out, as it were, into the Infinite, and there was a rushing together of the two worlds, the inner and the outer. It was deep calling unto deep—the deep that my own struggle had opened up within being answered by the unfathomable deep without, reaching beyond the stars. I stood alone with Him who had made me, and all the beauty of the world, and love, and sorrow, and even temptation. I did not seek Him, but felt the perfect unison of my spirit with His. The ordinary sense of things around me faded. For the moment nothing but an ineffable joy and exaltation remained. It is impossible fully to describe the experience. It was like the effect of some great orchestra when all the separate notes have melted into one swelling harmony that leaves the listener conscious of nothing save that his soul is being wafted upwards, and almost bursting with its own emotion. The perfect stillness of the night was thrilled by a more solemn silence. The darkness held a presence that was all the more felt because it was not seen. I could not any more have doubted that *He* was there than that I was. Indeed, I felt myself to be, if possible, the less real of the two. (p. 66)

I have on a number of occasions felt that I had enjoyed a period of intimate communion with the divine. These meetings came unasked and unexpected, and seemed to consist merely in the temporary obliteration of the conventionalities which usually surround and cover my life. . . . Once it was when from the summit of a high mountain I looked over a gashed and corrugated landscape extending to a long convex of ocean that ascended to the horizon, and again from the same point when I could see nothing beneath me but a boundless expanse of white cloud, on the blown surface of which a few high peaks, including the one I was on, seemed plunging about as if they were dragging their anchors. What I felt on these occasions was a temporary loss of my own identity, accompanied by an illumination which revealed to me a deeper significance than I had been wont to attach to life. It is in this that I find my justification for saying that I have enjoyed communication with God. Of course the absence of such a being as this would be chaos. I cannot conceive of life without its presence. (p. 70)

In that time the consciousness of God's nearness came to me sometimes. I say God, to describe what is indescribable. A presence, I might say, yet that is too suggestive of personality, and the moments of which I speak did not hold the consciousness of a personality, but something in myself made me feel myself a part of something bigger than I, that was controlling. I felt myself one with the grass, the trees, birds, insects, everything in Nature. I exulted in the mere fact of existence, of being a part of it all— the drizzling rain, the shadows of the clouds, the tree-trunks, and so on. In the years following, such moments continued to come, but I wanted them constantly. I knew so well the satisfaction of losing self in a perception of supreme power and love, that I was unhappy because that perception was not constant. (p. 303)

SOURCE: James, William. (1958). *The varieties of religious experience*. New York: New American Library. (Original work published 1902).

summarized his argument with an important twist:

> [People have] a tendency to sacralize [their] ultimate systems of value. Even if one excludes the possibility of a transcendent or a supernatural, one nonetheless is very likely to treat one's system of ultimate explanation with a great deal of jealous reverence and respect and to be highly incensed when someone else calls the system of explanation to question or behaves contrary to it. It is precisely this tendency to sacralize one's ultimate concern that might well explain the many quasi-religious phenomena to be observed in organizations which officially proclaim their non- or even antireligiousness. The communist, for example, may vehemently deny the existence of a "totally other" and yet treat communism and its prophets, its dogmas, its code, and its ritual with as much respect as does the devout Christian approach [Christianity]. (p. 9)

Rather than citing an experience of the sacred as the source of all religious behavior, Greeley suggested that whatever we value very highly, we tend to sacralize (to make sacred). Sacredness, then, may be the *result* of a valuing process rather than the primal *cause* of all other religious activity. Greeley did not suggest that experiences of the holy are always secondary; he merely pointed out that the relationship among values, beliefs, and the sacred may be more complex than Otto, Eliade, and O'Dea suggested.

Because not all forms of extraordinary experience are religious, Greeley offered a clarification. There are two kinds of extraordinariness that traditionally have been called religious. One has to do with the need to make sense out of life, the need for a meaningful interpretation of events. The second has to do with the need for belonging, the need to feel connected to other persons, to humanity, and to the universe. "Sacredness, then, relates to [humanity's] experience of the need for meaning and the need for belonging, as well as the fulfillment of these needs" (Greeley, 1972, p. 16). The sacred experience, then, has appeal insofar as it can help persons make sense of the world and feel a sense of belonging to something larger than themselves.

Several conclusions can be drawn regarding nonrational religious experiences. First, they vary greatly in intensity from one person to another. Some people have a "strange warming of the heart" while others have an ecstatic, orgiastic type of experience (Neitz, 1990). Second, nonrational experiences vary in frequency. For some, mystic experience becomes a goal in itself and may even take on the form of a full-time occupation. This attitude is common among certain religious groups: Buddhist monasteries in Asia, in Hare Krishna temples in the United States, and some Roman Catholic monasteries in which the pursuit of religious experience becomes the principal goal.

Third, nonrational experiences vary in context. For some people and in some religious traditions, intense experiences of holiness occur to individuals in isolation, as was the case for Plains Indian men who went off by themselves in search of a vision quest. For other people, profound nonrational experiences are primarily social, occurring in a charismatic prayer meeting or during a worship service (Neitz, 1990). In either case, social norms usually prescribe how to achieve such an experience, the value of such an experience, what a sacred encounter "looks like," and how to interpret it when it does occur.

Fourth, they vary in content. The Eskimo shaman makes a spiritual journey to the bottom of the sea to visit Sedna and to appease her for violations of taboos (Barnouw, 1982). The Lakota (Sioux) holy man experiences a visit by 48 horses that approach in groups of 12 from the four cardinal directions (Neihardt, 1961). The Christian mystic[1]

---

[1]Some writers reserve the word *mystic* solely for Eastern religions with their search for spiritual unity with the universe. Others, following Troeltsch, use the term to refer to any nonrational religious orientation that denies the importance or reality of the social order. However, most writers use mysticism in a broader sense to refer to any nonrational experience that the actors define as sacred.

may see the Holy Virgin, experience the love of Jesus, or hear the voice of God. The Buddhist mystic may experience "nonbeing" or "utter unity with the universe." The content, or at least the interpretation of the experience, is defined in culturally familiar terms.

Fifth, the people who value and expect a religious experience are those who report having had one. Abraham Maslow (1964) insisted that probably everyone has had at least one nonrational "peak experience" but that some people do not value such experiences and therefore dismiss them as insignificant or bizarre or even forget them altogether. Other researchers (Hood, Morris, & Watson, 1989; Nelson, 2005; Straus 1979; Yamane & Polzer, 1994) insisted that people who desire such experiences may cultivate behaviors and attitudes that make it likely that they will have them. This view suggests that social expectation influences such experiences, a fact that has been validated in different cultures and different religious traditions (Anthony, Hermans, & Sterkens, 2010).

Finally, a social group's definition of what is normal or desirable influences both the way individual members interpret their own experiences and the kinds of experiences they try to bring about. Most religious traditions provide symbolic resources for the construction of alternative realities and promote actions directed at breaking through to those realities. Therefore, involvement in religious traditions—especially in prayer and religious services—is conducive to religious experience (Nelson, 2005; Yamane & Polzer, 1994). This is a basic premise of social constructionist

theory. The theory is useful in understanding and predicting variations in religious experiences between groups. Roman Catholics are surrounded by much visual stimulation in churches (stations of the cross, statues, and stained glass windows). It should not be surprising that Catholics are much more likely than Protestants to have religious experiences that involve visions, while Protestants, whose worship experience is heavily auditory, are more likely to hear voices.

Regardless of variations, some form of nonrational religious experience seems to be at the root of religious behavior for many people. Such an experience gives impetus and emotion to belief systems. However, the assumption (by Otto, Durkheim, O'Dea, and Maslow) that a mystical or nonrational experience is the only source of religious conviction may be overdrawn. As Greeley pointed out, people have a tendency to sacralize the things that they value most highly and that give meaning and purpose to their lives. Note, for example, the sense of sacredness that accompanies ceremonial handling of the American flag. The directions that accompany the newly purchased flag emphasize a sense of reverence and awe that should be maintained when caring for the flag. See the "Illustrating Sociological Concepts" feature as an example. Such sacredness is not caused by a mystical experience but is created in the presence of a valued symbol. The sacralization of objects or beliefs places them above question; it ensures their absoluteness. Hence, that which we value highly tends to be perceived in reverent or sacred terms.

ILLUSTRATING SOCIOLOGICAL CONCEPTS

The American Flag as a Sacred Symbol

The American flag is often treated with a profound respect and reverence, as indicated by the directions for flag etiquette. The mere fact that detailed rules are spelled out so explicitly for the treatment of this symbol and the reality that violation of these rules will infuriate some

Americans is testimony to the awe with which this symbol is held. The extreme anger expressed by many Americans over the burning of a U.S. flag is also evidence of its sacredness, though the behavior itself is not the cause of anger. One of the rules of flag etiquette is that a tattered flag should be burned or buried as a sign of respect. The U.S. Flag Code has an extensive list of rules about sacred treatment and what constitutes desecration of the flag. A few of those rules are listed here.

- Display the flag only from sunrise to sunset on buildings and on stationary flagstaffs. The flag may be displayed at night only upon special occasions when it is desired to produce a patriotic effect.
- The flag should not be left outside when it is raining.
- The flag should be hoisted briskly and lowered ceremoniously.
- No other flag or pennant should be placed above or, if on the same level, to the right of the flag of the United States of America, except during church services conducted by naval chaplains at sea, when the church pennant may be flown above the flag during church services.
- The flag, when flown at half staff, should be first hoisted to the peak for an instant and then lowered to the half-staff position. The flag should be again raised to the peak before it is lowered for the day.
- No disrespect should be shown to the flag of the United States of America; the flag should not be dipped to any person or thing.
- The flag should never be displayed with the union down save as a signal of dire distress.
- The flag should never touch anything beneath it, such as the ground, the floor, or water.
- The flag should never be carried flat or horizontally but always aloft and free.
- The flag should never be used as a drapery of any sort whatsoever, never festooned, drawn back, nor up, in folds, but always allowed to fall free.
- The flag should under no circumstances be used as a ceiling covering.
- During the ceremony of hoisting or lowering the flag or when the flag is passing in a parade or in a review, all persons present should face the flag, stand at attention, and salute. Those present in uniform should render the military salute. When not in uniform, men should remove the headdress with the right hand holding it at the left shoulder, the hand being over the heart. Men without hats should salute in the same manner. Aliens should stand at attention. Women should salute by placing the right hand over the heart. The salute to the flag in the moving column should be rendered at the moment the flag passes.
- The flag should *never* be used for advertising in any manner whatsoever. It should not be embroidered on cushions, handkerchiefs, or scarves, nor reproduced on paper napkins, carryout bags, wrappers, or anything else that will soon be thrown away.
- No *part* of the flag—depictions of stars and stripes that are in *any form other than that approved for the flag design itself*—should ever be used as a costume, a clothing item, or an athletic uniform. In other words, wearing hats and shirts that involve a combination of red, white, and blue stripes and stars in a design that is not a complete flag is officially an offense to the flag.

It is interesting that much of the American public seems unaware of many of the flag code rules regarding sacredness and desecration (Ballantine & Roberts, 2011; Billig, 1995; U.S. Flag Code, 2007).

Anyone who doubts the fact that a firsthand religious experience can be very important need only observe the major surge of born-again and Pentecostal movements in Christianity in the past two decades. The emphasis of each of these movements is the assurance and sense of certainty provided by a personal experience of the holy. Furthermore, the New Age movement and many of the Eastern religions that grew rapidly in the 1970s placed heavy emphasis on nonrational religious experience (Wuthnow, 1976b). Indeed, the recent emphasis on "spirituality" can be seen at root as focusing on personal experience (Yamane, 1998). Although an experience of the mysterium tremendum may not be the sole source of religious behavior, a sense of sacredness is clearly one important aspect of the complex phenomenon we call "religion."

> **Critical Thinking:** Is nonrational experience of the holy *necessary* for a vibrant religious faith, or is it mostly a contributing factor? Is experience of the holy *sufficient* to mobilize a religious movement? Why or why not?

MYTH AND RITUAL

When Americans speak of religion, they usually think of a belief system. Indeed, many social scientists have even attempted to measure religiosity by questioning subjects on their agreement or disagreement with certain orthodox religious beliefs. Even the practice of referring to faithful members of a religious group as believers is indicative of this focus. As we will see, however, belief and ritual are interdependent, and in the case of some religious groups, ritual is the more important of the two.

Myth

Religious beliefs are usually expressed in the form of myths. By myth, the social scientist does not mean untrue or foolish beliefs; myths have little to do with legends, fairy tales, or folktales (Kluckhohn, 1972). Myths are stories or belief systems that help people understand the nature of the cosmos, the purpose and meaning of life, or the role and origin of evil and suffering. Myths explain and justify specific cultural values and social rules. They are more than stories that lack empirical validation; they serve as symbolic statements about the meaning and purpose of life in this world. One sociologist of religion has gone so far as to suggest that all religious symbols (including religious myths) are in a fundamental sense true (Bellah, 1970a). He did not argue their literal veracity, but he insisted that symbolic systems of meaning are true insofar as they speak to the fundamental human condition. Hence, they need to be taken seriously. Myths have a powerful impact on the subjective or mental orientation of persons because they communicate and reinforce a particular worldview or a particular outlook on life. (See the next "Illustrating Sociological Concepts" feature.)

After undertaking an in-depth study of the Hare Krishna, Stillson Judah (1974) insisted that "'myth' is actually the highest subjective reality to the devotee. It is the vehicle [which carries one] to inner integration. [Humans] can only live without 'myth' at [their] peril" (p. 196). What is important here is a recognition of the emotional power of which myth is capable. For many persons, logical, systematic, and scientific statements do not capture one's imagination so as to mobilize one's emotional resources. The capacity of myths to do this is precisely the reason they have such power in the lives of individuals.

A Central Eskimo myth may serve to illustrate. Historically, the Central Eskimo believed in the existence of a woman, Sedna, who lived at the bottom of the sea. Sedna was once an Eskimo girl. However, contrary to the wishes of her father, she married a very large bird and went to an island to live with her new husband. When her father returned from a hunting trip

ILLUSTRATING SOCIOLOGICAL CONCEPTS

Myths Can Create Powerful Moods and Motivations in People

The mythology of heaven and hell and the belief that one who is not saved by God will go to hell are vividly expanded by Jonathan Edwards in this passage. By emphasizing this myth and elaborating on it, he sought to create powerful, pervasive, and long-lasting *moods* (awe of God's power, humility, fear of God) and *motivations* (repentance, change of attitudes, and behavior).

> The God that holds you over the pit of hell, much as one holds a spider, or some loathsome insect over the fire, abhors you, and is dreadfully provoked: his wrath towards you burns like fire; he looks upon you as worthy of nothing else, but to be cast into the fire; he is of purer eyes than to bear to have you in his sight; you are ten thousand times more abominable in his eyes, than the most hateful venomous serpent is in ours.

> You have offended him infinitely more than ever a stubborn rebel did his prince; and yet it is nothing but his hand that holds you from falling into the fire every moment. It is to be ascribed to nothing else, that you did not go to hell the last night; that you was suffered to awake again in this world, after you closed your eyes to sleep.

> And there is no other reason to be given, why you have not dropped into hell since you arose in the morning, but that God's hand has held you up. There is no other reason to be given why you have not gone to hell, since you have sat here in the house of God, provoking his pure eyes by your sinful wicked manner of attending his solemn worship. Yea, there is nothing else that is to be given as a reason why you do not this very moment drop down into hell.

> O sinner! Consider the fearful danger you are in: it is a great furnace of wrath, a wide and bottomless pit, full of the fire of wrath, that you are held over in the hand of that God, whose wrath is provoked and incensed as much against you, as against many of the damned in hell. You hang by a slender thread, with the flames of divine wrath flashing about it, and ready every moment to singe it, and burn it asunder; and you have . . . nothing to lay hold of to save yourself, nothing to keep off the flames of wrath, nothing of your own, nothing that you ever have done, nothing that you can do, to induce God to spare you one moment. (Edwards, 1966; preached in 1741)

Although there is no question that this "hellfire and brimstone" style of sermonizing still exists, sociologist Alan Wolfe (2003) has argued that Edwards' approach has fallen out of favor in American culture today. The "old-time religion" has been trumped by an increasingly mass mediated American culture that is dominated by mutually reinforcing therapeutic and utilitarian forms of individualism. Wolfe highlighted contemporary worship, where the emphasis is increasingly on meeting the personal needs and interests of members (he later calls this "liturgy for dummies"), and doctrine, where feelings have triumphed over ideas. For Wolfe, sin is an idea in retreat, being replaced by an attitude of "nonjudgmentalism" that articulates with a one dimensional view of God not as "angry" (Edwards' view) but as understanding and empathetic. This creates moods and motivations of a different sort.

* * * * * * *

(Continued)

(Continued)

Critical Thinking: Readers might find it instructive to attend several different religious services in their communities and listen to the symbolism of the language. What myths are being played upon? What moods do the clergy intend to create with these myths? What motivations are sought? What is the symbolic role of emotionally laden language in these services? Another exercise would be to read sermons from clergy of various theological persuasions to identify the symbolic power of language in creating moods and motivations.

and found her missing, he came to get her. While the father was returning the girl to her home, Sedna's husband caused a violent storm to arise. Sedna fell overboard, but she hung on desperately to the gunnels of the canoe. To prevent capsizing, the father chopped her fingers off with his machete, and she sank to the bottom of the sea. The pieces of fingers turned into the various sea mammals—whales, seals, and walruses.

Eskimo women, in particular, were constrained by many taboos or moral prohibitions. For example, when seals were killed and brought home, certain behaviors were prohibited until the seal had been cut up; the skins from the sleeping platform could not be shaken out, women could not comb their hair, and young girls could not take off their boots. After the cutting was complete, products from land and sea had to be cooked in different pots. Many other taboos were related to food preparation, to giving birth, and to menstruation. If any of these taboos were broken, a vapor was believed to emanate from the body of the violator. That vapor sank through the ice, snow, and water, and settled in Sedna's hair as dirt and maggots. Because she had no fingers, Sedna could not comb out this debris. In revenge, she would call the walruses and seals (which were once her fingers) to the bottom of the sea, and the people would face starvation. Salvation came only if a shaman (Eskimo holy man) entered into a trance, traveled spiritually to the bottom of the sea, combed Sedna's hair, appeased her, and discovered who the offenders were. When he returned, all offenders were required to confess their taboo violations openly. The seals then returned (Barnouw, 1982, pp. 231–234).

Sedna was not a benevolent deity. She was vengeful and had to be obeyed. She was the cause of much anxiety, fear, and even hostility. The Sedna myth served to create these moods of fear and anxiety in people, which then motivated them toward certain types of behavior (obeying of taboos and prohibitions). This myth reinforced an overall worldview that the world is a hostile environment, that the future is fraught with danger and may be jeopardized by human acts, that survival depends on conforming behavior and obedience to rules by everyone, and that negative behavior will always be reflected back to the actor in some devastating form. The Sedna myth served to solidify and sacralize the general Central Eskimo outlook on life. Insofar as this view was consistent with the harsh environmental conditions in which the Central Eskimos live, the myths may have contributed to adaptation and survival. Further, many of the taboos helped to sustain an environmental ethic, which was important to the survival of the people.

Ritual

Although Americans tend to think of belief as the central component of religion, ritual appears to be equally important. In fact, several scholars have pointed out that *orthopraxy*, not orthodoxy, is central to Islam (Schneider, 1970; Tamadonfar, 2001; Watt, 1996). That is, precise conformity in ritual behavior (e.g., prayers five times a day facing Mecca) is what is mandated for the faithful, not total conformity in theological interpretations. Many scholars have also observed that Judaism—especially Orthodox Judaism—and the Eastern Orthodox tradition within Christianity focus much more on concrete behaviors (orthopraxy) than on theological tenets, beliefs, or attitudes, though these characteristics should not be seen as monolithic (Cohen, 1983; Moberg, 1984; Sklare & Greenblum, 1979, Vrame, 2008).

A careful observation of human behavior is enough to make one aware of the great attraction of humans to ritual experiences. Consider, for example, the elaborate pageantry and ritual of a Shriner's convention, a scouting induction, or the opening and closing ceremonies of the Olympic Games. Football games always begin with the playing of the national anthem, and colorfully uniformed marching bands perform. Many meetings of secular civic groups begin with a ritual pattern: the pledge to the flag and a prayer. Moreover, for many people, marriage is not legitimate unless the couple has been through a ceremony, however brief. Although common-law marriage (marriage without benefit of ceremony or marriage license) is perfectly legal in nine states and is recognized by seven others in certain circumstances, many people still feel the ceremony is what makes marriage legitimate (National Conference of State Legislatures, 2011).

The preceding examples do not necessarily involve a sense of sacredness or ultimacy, and we do not suggest that they are particularly religious phenomena. Our point is merely that there is something about human beings to which ritual and pageantry appeal. The myth systems associated with religious rituals need to capture the imagination of the attendees; otherwise, the formal ecclesiastical rituals risk becoming hollow rather than awe-inspiring. Still, the apparent need for elaborate ritual and ceremony at the outset of all types of human events and meetings is quite interesting.

Religious ritual usually involves affirmation of the myths and gives emotional impulse

to the belief system. Judah (1974) pointed to this when he wrote, "The enthusiasm of the devotees leaping in ecstasy with upraised arms . . . can be contagious for many" (p. 95). Not only is the mood contagious; acceptance and understanding of the belief system may be attained through continual practice of a ritual. Judah (1974) cited a number of devotees of the Hare Krishna movement who made comments such as the following: "Although we may not understand something when it is given to us, it comes to us through faith. It's revealed to us through our continuing [ritual performance]" (p. 169).

Ritual may involve the enactment of a story or myth, or it may symbolically remind one of the mythology of the faith by moving participants through a series of moods. Perhaps a brief analysis of a ritual familiar to many readers will help illustrate the point.

Biblical theology was based on the idea that God had a covenant (or contract) with the chosen people. If they obeyed the commandments

Muslims pray to God five times a day at set times, regardless of where they are. They arealso required to face Mecca. Conformity tothis and other ritual behavior is considered a more important measure of faith than doctrinal orthodoxy.

and worked to establish a kingdom of justice and righteousness, then God would protect them and provide for them. The scriptures maintain that the Hebrew people got into trouble whenever they broke the covenant, forgot the demands of justice, and ignored the sovereignty of God. In these circumstances, the prophets called the people back to the covenant. The prophets assured them that if they would repent and renew their covenant, Yahweh would forgive them. The New Testament renews this theme, with Jesus calling the wayward to repent and promising God's forgiveness. The most important sacrament in the Christian church is Communion (alternatively referred to as the Lord's Supper, the Eucharist, or the Mass). In instituting this practice, Jesus claimed to be inaugurating a new covenant. Covenantal theology, then, is a basic Christian mythology or belief system.

Many Protestant Christians are not consciously aware that the liturgy (or ritual) in which they participate is based on this theology. In fact, many laypeople believe that the order of a worship service is rather arbitrary, that the minister randomly intermixes hymns, prayers, a confession, scripture, anthems, and other liturgical devices. Let us look, however, at a consistent pattern that prevails in many American Protestant liturgies. The sample provided in the next "Illustrating Sociological Concepts" feature will serve as an example.[2] While there is some variation in the order of Protestant worship services, the majority of mainline American churches tend to move worshipers through successive movements or moods as shown in this liturgy (Hesser & Weigert, 1980; Johnson, 1959). Let us examine the relationship between these liturgies and the mythology of the divine covenant.

ILLUSTRATING SOCIOLOGICAL CONCEPTS

Order of Worship
First United Methodist Church

Service of Praise

Prelude: Walther

Our worship of God begins with the music of the prelude. Let us listen to the music. Let us read the words of the hymns we will sing today. Let us center our thoughts before God. Let us be silent, opening our lives to God.

---

[2]Many Protestant services do not have the subheadings printed in the order of worship to show the change of mood. Regardless of whether subheadings are provided in the bulletin, the general sequence described here is typical for the mainline Protestant churches. The model described here is based on patterns that prevail in several hundred church bulletins from various denominations that Roberts has collected over several decades.

Introit   Senior Choir

Call to Worship

Hymn of Praise: # 432 "Fill Us With Your Love"

Service of Confession

Prayer of Confession (in unison)

Lord, we gather on the Sabbath to worship You and to dedicate our lives to Your service. Yet in the strain and hurry of everyday living, we often forget You. We say the unkind word, or we fail to do the loving deed. We betray the values which we profess on Sunday morning, and we do not remember that You are the Center of Life. Forgive us, O Lord, and renew in us a resolve to live Christ-like lives.

Personal Silent Confession

Assurance of Pardon

Service of Proclamation

Apostles Creed

The Gloria Patri

Children's Chat

Anthem: "Different Is Beautiful" Avery/Marsh

Scripture Reading: Ruth 1:15-18, 22; 4:13-17

Galations 3:27-28

Luke 4: 16-22; 25-30

Hymn of Proclamation: # 437 "This Is My Song"

The Message: Is there a "god" in your genes?

The Morning Prayer

Service of Commitment

Concerns of the Congregation

Prayer of Intercession and Lord's Prayer

Tithes and Offerings   N. Lebegue

Doxology and Offertory Prayer

Hymn of Dedication: # 593 "Here I Am Lord"

Charge to the Congregation

Benediction

Postlude: J. Stanley

At the outset of the service, the liturgy is designed to create a mood of awe and praise. The architecture of the building may also enhance this sense. Many church bulletins request that worshipers sit in silence and focus their attention on a rose window, on some other symbol, or on "the presence of God." The prelude is frequently a piece of music that will lift one's spirits. The call to worship draws one's attention to the reason for gathering: to worship and praise God. The congregation then joins together in a hymn of praise, which is frequently a joyful, uplifting song of adoration.

Shortly after the congregation is made aware that it is in the presence of God, the mood shifts. Although most of the worshipers were in the same place dedicating their lives to God just a week before, the liturgy attempts to make them aware of the fact that they have not always lived in a way consistent with Christian values. They are reminded that in the push and pull of daily living, they have said the unkind word or failed to do a loving deed (sins of commission and sins of omission). The values they professed on Sunday they may have betrayed by Wednesday (if not by Sunday afternoon). Hence, the second mood or theme of this worship liturgy is a service of confession. No one can be loving, just, self-sacrificing, or righteous all the time, and the liturgy leads participants to acknowledge their lack of consistent faithfulness. For this reason, confession is sometimes called an "act of honesty" (between oneself and God). However, this liturgical movement does not end on a note of guilt. Confession is followed by words of assurance, assurance of pardon, or word of new possibility. At this point, the congregation is assured by the minister (often through quotation of a biblical passage) that people who sincerely confess their sins and who renew their commitment to God are forgiven. The covenantal theme of repentance prior to renewal of the covenant is enacted.

The liturgy then moves to a third phase, an affirmation of faith, or service of proclamation. This phase is frequently the major part of Protestant services. The congregation may repeat a creed or covenant, listen to an anthem or other special music, sing a hymn of proclamation, listen to scripture, and hear a message (sermon) delivered by the minister. In this phase of the ritual, the emphasis is on celebrating God's love, remembering and rehearing the Word of God, and remembering the demands of the covenant. Infant baptisms are usually a part of this movement, although churches that practice only adult baptism and view it as an act of commitment may include it as part of the final movement.

The final movement of most Protestant liturgies is a service of dedication. This part of worship calls for a response to God's word by the congregation. The movement is characterized by a monetary offering (a symbolic act of the giving of one's self), concerns of the congregation or announcements from the pulpit, a hymn of dedication, a charge to the congregation, and a benediction. In some churches, the third and fourth phases of the liturgy (proclamation and dedication) may be merged into one. In this case, some acts of dedication and commitment (such as the offering) may actually precede the sermon. However, the hymn of dedication always follows the sermon. In the more evangelical churches, this hymn may be followed by an "altar call"—a request for members to make a public commitment by coming forward and standing before the altar to commit their lives to Christ.

It should not be inferred that all Protestant worship through history has followed this mood sequence. The pattern described here was initiated by John Calvin, who articulated a rationale for an order of worship that corresponds roughly with the pattern and the theology identified here. However, Huldrych Zwingli, another reformer of Calvin's day, felt that the climax of the service should be confession. The rest of the service was to build a sense of guilt in the worshipers and culminate in the final act of repentance. This influence can be seen in Puritan liturgies of colonial America. Some denominations continue to be influenced by the Zwingli tradition in their liturgical formats. Furthermore, some Pentecostal churches and Christian sects do not have a consistent pattern of worship. The rationale for the order of worship is insignificant, for emotion is judged far more important than thought patterns.

It is also interesting that much of contemporary Protestant worship emphasizes proclamation and commitment. The Catholic Mass seems to place more stress on the service of praise. This reflects a fundamental difference in Protestant and Catholic views of worship (Pratt, 1964). The Roman Catholic Church has historically taken an objective approach to the Mass. The emphasis is on glorification of God, and the liturgy is designed with that in mind. It is best to have a congregation present at the celebration of Mass, but if no one came, the Mass would go on. On the other hand, a Protestant minister would scarcely think of conducting a full service of worship if no congregation gathered. The more subjective emphasis of the Protestant denominations—especially the ones that have a more informal, low-church tone—is on how the worship affects the worshipers themselves. Hence, the beliefs about worship itself significantly affect the order of a liturgy and the themes it includes.

From the example provided, it may be seen that ritual and belief are often closely intertwined and tend to be mutually reinforcing. That is, they tend to provide an interrelated system. It is noteworthy that sample surveys have consistently found a high correlation between regular attendance at rituals and a high level of acceptance of the belief system of the denomination. However, it is also true that many people are unaware of the logical progression of the worship liturgy and of the theological basis for its order. Hence, the liturgy is viewed as just so many hymns, prayers, and scripture readings in random order (Roberts, 1992). Given this fact and given the fact that some people attend worship services for purely status reasons (as a demonstration that they are good citizens), it is not surprising that the correlation between ritual attendance and doctrinal orthodoxy is far from perfect.

Relationship Between Myth and Ritual

Ritual often precedes myth. Judah (1974), for example, insisted that "chanting is the primary prerequisite" (p. 170) in conversion to the Hare Krishna movement. Some scholars have gone even further by suggesting that, in terms of chronological development, ritual emerges first and myth develops later to justify the existence of the ritual. Franz Boaz particularly emphasized the primacy of ritual and the secondary explanatory role of myth (Kluckhohn, 1972).

However, the process may develop the other way as well (Kluckhohn, 1972). The Mass is clearly an example of a ritual based on a sacred story. Another example is the Ghost Dance, a religious ritual based on a dream or vision by Wovoka, the Paiute holy man whose trance in 1889 regenerated a powerful religious movement among Native Americans. In this case, the interpreted dream provided both a mythology and a command to perform a ritual (LaBarre, 1972). Clyde Kluckhohn (1972) emphasized the complex relationship between these elements of religion:

> [T]he whole question of the primacy of ceremonial or mythology is as meaningless as all questions of "the hen or the egg" form. What really is important is the intricate interdependence of myth with ritual and many other forms of behavior. (p. 96)

In some cases, both myth and ritual may be viewed as factors dependent on a third component: mystical or nonrational experience (O'Dea, 1966; Otto, 1923; Van der Leeuw, 1963, p. 49). The experience of awe may be so fascinating and attractive that a ritual is established to try to elicit or recreate that experience. Furthermore, a mythology is generated to try to explain or make sense of the experience (O'Dea, 1966, pp. 24, 40–41). At this point, it seems fair to conclude that no generalization can be made about which of these three factors is primary, for there is wide variation between cultures and even within cultures. For that matter, there may even be variation within the same religion, as each denomination or sect of a faith emerges in accord with its own internal dynamic.

The image presented here of highly integrated ritual and myth needs a word of caution. The discussion of the Protestant liturgy suggested a high degree of integration and interdependence between ritual and belief system. However, that integration is largely a matter of interpretation— that is, this integration is to a considerable extent in the eye of the beholder. The participant interprets the ritual and myth as mutually supportive— hence it is mutually supportive for that person. However, many cases have been found in which diverse tribal people practice the same ritual but interpret that ritual as expressing very different myths. Likewise, people holding the same belief system may celebrate those beliefs with very different ritual patterns. In summary, we can say that beliefs, ritual, and religious experience are important components of religion, that they are usually interrelated and mutually supportive, and that the integration of the three is itself largely a matter of interpretation by the believer and the community of which the believer is a part (Batson, Schoenrade, & Ventis, 1993).

Critical Thinking: As readers attend religious services in their own communities, they may be interested in asking themselves, What is the rationale for the order of this liturgy? How are myth and ritual intertwined and mutually supportive, or is there no apparent relationship? What does the ritual mean to the *members* of this group?

THE IMPORTANCE OF SYMBOLS

The reason why myths and rituals normally have a close relationship is that they are both manifestations of a larger phenomenon—a system of symbols. Elizabeth Nottingham (1971) emphasized the unifying function of symbols:

> [It] is not hard to understand that the sharing of common symbols is a particularly effective way of cementing the unity of a group of worshipers. It is precisely because the referents of symbols elude overprecise intellectual definitions that their

unifying force is the more potent; for intellectual definitions make for hairsplitting and divisiveness. Symbols may be shared on the basis of not-too-closely-defined feeling. (p. 19)

Certainly, this is one reason that symbols are important. Clifford Geertz (1958) emphasized a slightly different reason for symbols' critical importance when he wrote, "*Meanings can only be stored in symbols:* a cross, a crescent, or a feather. Such religious symbols, dramatized in rituals or related in myths, are felt somehow to sum up what is known about the way the world is" (p. 422, italics added). Edmund Leach's (1972) research on the symbolic power of rituals supports this view. Leach has studied rituals as "storage systems" that encapsulate knowledge. He maintained that ritual provides an important form of economical thinking among many tribal people. Meaningful information and important knowledge are encapsulated in ritual in a way analogous to the loading of computer chips with information in our culture. Rituals are viewed by Leach as vessels that carry powerful symbols and that authoritatively transmit a worldview and an ethos.

A consideration of certain elements in the Catholic Mass illustrates the power of symbols. As the celebrants come before the altar, they genuflect and cross themselves. Although the act may sometimes be perfunctorily executed, making the sign of the cross acts to remind believers of a particular event. The cross has meaning because it reminds the participant of a sacred story, a particular life, and a divine event. The theological interpretation of the event may vary somewhat from one celebrant to another, but with the regularized pattern of crossing oneself, the centrality of Christ on the cross will not be forgotten. The stained glass windows may have symbols meaningful to the early church or perhaps depictions of Jesus in a well-known scene. These symbolic representations also bring to mind a whole series of events and stories that are part of the sacred myth. When the priest says, "Take, eat, this is my body," another central event is recalled. At certain points in the Mass, the congregation kneels. This act is a gesture of humility and is to remind one of an utter dependency and humility before God. In each case, the entire story does not have to be repeated in detail. If the myth is well known and if the symbolic meaning of the ritual action is understood by the celebrant, then all that is necessary to elicit certain moods and motivations is to introduce the symbol itself. The symbol stores meaning and can call forth certain attitudes or dispositions.

In the Jewish tradition, one can see the same power of symbols in a religious festival such as the seder meal. Even a Gentile cannot help but be moved by the symbolic reenactment of the escape from Egypt. The eating of bitter herbs, which actually bring tears to one's eyes, reminds one of the suffering of the ancestors. The unleavened bread, the parsley, and the haroseth (a sweet condiment) have symbolic value in recalling a sacred story. This symbolic reenactment confirms in the minds of the Jews where they have been, who they are, and what task lies before them. A sense of identity and a sense of holy mission is powerfully communicated through the ritual (Fredman, 1981).

Geertz's (1966) definition of religion articulates very well the important role of symbols: "Religion is a system of symbols which acts to establish powerful, pervasive, and long-lasting moods and motivations in [people] by formulating conceptions of a general order of existence" (p. 4). The symbol systems are important in that they act in people's lives. Ritual and myth, then, are important as symbol systems. In this text, we have stressed worldview (conceptions of a general order of existence) and ethos (powerful, pervasive, and long-lasting moods and motivations) as central components of religion. Hence, the next issue before us is the relationships among worldview, ethos, and symbols.

Judaism is rich with symbolism, and the symbols often serve on an everyday basis to remind people who they are. The man on the left wears a *yarmulke* like those worn by Jewish men since roughly the second century CE. It symbolizes respect for and fear of God. These caps are also a sign of belonging and commitment to the Jewish community. At the right is a *mezuzah*, a daily reminder of identity on doors jambs of Jews. In a parchment inside it has the "Shema Israel" (Deuteronomy 6:4–9) written on it, a passage that commands the faithful to keep God in their minds at all times. Whenever Jews pass through a door with a mezuzah on it, they kiss their fingers and touch them to the mezuzah, expressing love and respect for God and God's commandments. These symbols carry meaning and reinforce identity every day.

WORLDVIEW, ETHOS, AND SYMBOLS

Worldview refers to the intellectual framework within which one explains the meaning of life. Myths are specific stories or beliefs, the net effect of which is to reinforce a worldview (Wuthnow, 1981). A single story may not be sufficient to convince someone that God is in charge of the universe and all is well. However, a series of many such myths may serve to reinforce such an outlook on life. Worldview is a more abstract concept than myth; it refers to one's mode of *perceiving* the world and to one's general overview of life. In this sense, a worldview is more taken for granted and less questioned. Many individuals may not be fully conscious of the alternative types of worldviews, and many never question the fact that their perception is influenced by intellectual constructs. Whether one is optimistic or pessimistic in outlook is strongly influenced by one's worldview. (See the passages from Jonathan Edwards in the "Illustrating Sociological Concepts" feature on page 75.)

A religious worldview is also closely related to a group's ethos, as demonstrated in the next "Illustrating Sociological Concepts" feature. "A people's *ethos* is the tone, character, and quality of their life, its moral and aesthetic style and mood; it is the underlying attitude toward themselves and their world" (Geertz, 1958, p. 421, italics added). Ethos refers to attitudes about life (moods, motivations), whereas worldview refers to an *intellectual* process—thoughts about life (concepts of a general order of existence).[3] Both attitudes and concepts are essential to the establishment of a sense of meaning in life. The worldview is confirmed and made to seem objective by the ethos. The set of concepts is placed beyond question and is made absolute by the sacred mood in which it is transmitted. Furthermore, this basic attitude (ethos) is justified and made reasonable by the worldview. So in a well-integrated religious system, the ethos and worldview are mutually reinforcing.

ILLUSTRATING SOCIOLOGICAL CONCEPTS

Native American Ethos: Moral and Esthetic Style and Mood

Sam Gill insists that the moods and motivations of Native American religion are both critically important for understanding Native American religion and often beyond the capacity of rational discourse to explain or describe. In a discussion of the oral tradition of storytelling (myths), Gill (1982) commented on the ethos.

> Ordinarily underlying questions of meaning is the assumption that these oral traditions carry messages and that we need to translate these speech acts into their messages so that we too will know what they mean. After spending much time asking Native American people questions like What does this story mean? and feeling by their lack of response that it must have been a stupid question—or having gained answers completely incompatible with the story—I have had to seek ways of understanding how these stories bear meaning and how we can appreciate them. . . .

---

[3]It may be helpful for readers to review the material on Geertz's definition of religion in Chapter 1. The distinction between moods (which he described as having to do with depth of feeling) and motivations (which he described as directional) is elaborated. Both moods and motivations are part of the ethos.

Perhaps we can approach an understanding by certain olfactory experiences. I cannot smell the odor of juniper smoke without experiencing a series of peculiar images and feelings related to experiences I had while living among Navaho people. If you were to ask me what the smell of juniper smoke means to me, I would at first be confounded, for such a question seems inappropriate to ask. The smell bears no translatable message, although it has an emotional impact upon me; the experience is meaningful but has no meaning at all in the sense of bearing a message. Listening to music evokes similar sorts of meaning by awakening a certain emotion, often a series of images or memories connected with the music through one's personal and cultural history.

Speech acts in Native American cultures . . . have an emotional impact, a significance much more far-reaching. In their performance, they are not simply streams of words whose full significance lies in the information they convey. They are complex symbols, networks of sounds, odors, forms, colors, temperatures, and rhythms. . . . Consequently any story, any song, any prayer is a stimulus that frees strings of associated images, emotions, and patterns.

To ask what they mean and expect a translatable message is often to ask an inappropriate question. Their significance is inseparable from the whole field of symbols which they evoke.

SOURCE: Gill, S. D. (1982). *Native America religions* (2nd ed.). Belmont, CA: Wadsworth.

The inability to articulate the meaning of a particular myth may be twofold. First, part of why it is meaningful is because of the emotionally laden context in which it is told—the ethos. Second, the myth may reinforce a worldview that is only partially explicit or conscious to the subject. It is taken for granted. The worldview may, for example, assume certain categories of time and certain meanings of the concept of time that are not shared by the interviewer. Without that shared worldview, the informer hardly knows where to begin to explain the "meaning" of the story.

* * * * * * * *

Critical Thinking: Identify a smell or song (perhaps a pop music favorite from your elementary or middle school years) that elicits a strong reminiscence. What does that smell or that tune mean to you? Why is it so difficult to explain? Does the fact that it is difficult to explain indicate that it is insignificant to you? Why or why not?

Symbols, according to Geertz, play the central role of relating the worldview to the ethos. Symbols transform fact into value. The function of sacred symbols is to encapsulate, or summarize, the system of meaning and to deliver that meaning system with power and authority at appropriate times. Geertz (1958) discussed the symbolic significance of the circle among the Oglala Sioux as an example:

For most Oglala the circle is but an unexamined luminous symbol whose meaning is intuitively sensed, not consciously interpreted. But the power of the symbol, analyzed or not, clearly rests on its comprehensiveness, on its fruitfulness in ordering experience. Again and again the idea of a sacred circle . . . yields, when applied to the world within which the Oglala lives, new meanings; continually it connects together elements within their experience which would otherwise seem wholly disparate and, wholly disparate, incomprehensible. (p. 423)

Thus, symbols have power to bind worldview and ethos into a unified system of meaning.

The lack of a consistent philosophy to explain events and to justify one's values and lifestyle can be disrupting and disorienting. Peter Berger asserted that society is the guardian of order and meaning, not only at the level of social structure but also at the level of individual consciousness. Sacred images help preserve order in the structures of society and in the structuring of the individual mind. "It is for this reason that radical separation from the social world, or anomie, constitutes such a powerful threat to the individual. The individual loses his orientation in experience. In extreme cases, he loses his sense of reality and identity. He becomes anomic in the sense of becoming worldless" (Berger, 1967, p. 21).

Durkheim suggested long ago that the experience of anomie—social rootlessness or a lack of identity and purpose—can be so unsettling that it can result in suicidal behavior. The feeling of a firmly rooted worldview, with certain and definite moral rules and regulations, is a compelling need for many humans. Geertz (1966) insisted that his tribal-society respondents were quite willing to abandon their cosmology for a more plausible one (p. 16). What they were *not* willing to do was abandon it for no other hypothesis at all, leaving events to themselves.

In line with this emphasis, Mary Douglas has offered an insightful interpretation of biblical taboos (such as the taboo against eating pork or against eating a milk product and a meat product at the same meal). Not all taboos are health-related as was once maintained. Some taboos have to do with protecting the distinction between sacred and profane. Still others have to do with those things that are anomalies (unexplainable phenomena) to the accepted worldview.

In the process of socialization, a child learns to think in terms of categories of language and in terms of theories or explanations extant in a culture. When new information is received, it is interpreted in terms of those accepted categories of thought. New experiences and new information are assimilated into the present worldview. When an experience does not seem to fit into this mold, the person experiences dissonance (internal cognitive conflict). According to cognitive psychologist Jean Piaget, one is likely to revise and reorder one's worldview when too much dissonance occurs. However, a good deal of dissonant information can be tolerated by certain techniques. This is precisely where Douglas claimed that taboos enter the picture. Many taboos have to do with avoidance of those things that challenge one's worldview (Douglas, 1966, 1968).

The food taboos of the Old Testament provide an excellent example. To understand the abominations of Leviticus, we must go back to the creation story in Genesis. Here, a threefold classification is presented, with the earth, the waters, and the firmament as distinct realms. Leviticus develops this scheme by allotting to each element its proper kind of animal life. Douglas (1966) summarized this from Leviticus 11:10–12:

In the firmament two-legged fowls fly with wings. In the water scaly fish swim with fins. On the earth four-legged animals hop, jump, or walk. Any class of creatures which is not equipped for the right kind of locomotion in its element is contrary to holiness. Contact with it disqualifies a person from approaching the Temple. Thus anything in the water which has not fins and scales is unclean. (p. 55)

Contact with anything that is "contrary to nature" causes a person to be unclean and unfit to enter the temple—for example, anything in the water that does not have fins and scales. Hence, catfish are unclean because they have whiskers and no scales; they violate the principles of creation as spelled out in this worldview. Creatures that have two legs and two hands but that walk on all fours are unclean. Thus, animals that have hand-like front paws (the weasel, the mouse, and the mole) are explicitly taboo because they violate the law of creation. An animal with hands but that perversely uses its hands for walking does not fit into the proper scheme of things. Anything that creeps, crawls, or swarms on the earth is also defined in Leviticus as an abomination. "Eels and worms inhabit water, though not as fish; reptiles go on dry land, though not as quadrupeds; some insects fly, though not as birds. *There is no order in them*" (Douglas, 1966, p. 56, italics added).

Among these pastoral people, cloven-hoofed and cud-chewing animals are the proper sort of food. Wild game is edible only if it seems to conform to these characteristics, but animals that do not conform in *both* respects are unfit as food. The pig and the camel are both cloven-hoofed, but they do not chew cud. Douglas points out that the failure to conform to these two necessary criteria for defining animals as acceptable food is the only reason given in the Old Testament for not eating pork; no mention is made of pigs' scavenging habits.

Douglas offered a fascinating and well-documented thesis: Many taboos function to protect the need for order and sensibleness. Anomalies that don't fit one's worldview are simply abominations that are dirty and to be avoided. Other aspects of social order may also be protected or reinforced by taboos. Beliefs about women being unclean during their menstruation and in need of purification functioned to remind them of their inferior social position. The insistence in the New Testament that women cover their head in church served a similar function. Men and women were viewed as fundamentally different. This was symbolically emphasized by men keeping their hair short and women keeping

theirs long. Anything that threatened this essential distinction (essential to that worldview) was an abomination (it threatened concepts of order). Douglas demonstrated vividly that protection and preservation of the worldview is very important in understanding much religious behavior.

Sociologist Peter Berger insisted that one's worldview must seem authoritative, certain, and compelling. In the midst of alternative paradigms and cosmologies, one's worldview may become fragile and vulnerable. Hence, the vulnerability of a worldview is concealed in an aura of sacredness (Berger, 1967, p. 33). Without a basis in a convincing worldview, social values would not seem compelling and social stability itself may be threatened. Geertz (1958) came to the same conclusion through cross-cultural analysis of religion:

> . . . mere conventionalism satisfies few people in any culture. Religion, by fusing ethos and world view, gives to a set of social values what they perhaps most need to be coercive: an appearance of objectivity. In sacred rituals and myths values are portrayed not as subjective human preferences but as the imposed conditions for life. (pp. 426–427)

In conclusion, we might simply point to the complexity of the relationships among the various components of religion. In terms of the integration of these elements into a coherent religious system, we might summarize with the following four points:

1. Ritual and myth tend to be mutually reinforcing as symbol systems.

2. Symbols (including rituals, myths, and artifacts) encapsulate the worldview and ethos of a people. Hence they can elicit powerful emotional responses and (by repetition) help reinforce a general worldview.

3. The ethos and the worldview are themselves mutually reinforcing.

4. Together, all of these elements provide a compelling basis for social values (see Figure 4.1).

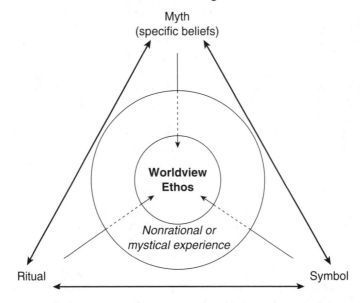

The Meaning System:
The Interrelationship Between the
Elements of Religion

Myth
(specific beliefs)

Worldview
Ethos

Nonrational or
mystical experience

Ritual   Symbol

Worldview and ethos are at the heart of what we mean by "religion." Yet these concepts are abstractions from experience; an individual's worldview and religious ethos (moods and motivations) are so pervasive and so taken for granted that the subject may not be fully conscious of all their dimensions. These abstractions are made concrete and are reinforced by more concrete expressions of religiosity: ritual, myth, and symbols. Each of these is interrelated and acts to confirm the ethos and worldview. Frequently, the entire religious realm seems uniquely compelling because of a nonrational religious experience. Some groups may emphasize some of these elements more than others (e.g., religious experience may be more important than unquestioned belief in the myth). Variations in interpretation of any of these may also lead to considerable conflict within the group. In actual practice, religion always has a good deal more internal inconsistency than this diagram would suggest. Hence, it is important to bear in mind that this model is idealized. It does represent a strain toward integrity that is a very real part of all religion. It is also true that when glaring incoherence in this system and conflict in interpretation of the elements becomes severe, some form of religious change is likely to take place.

Figure 4.1   The Meaning System

A problem may arise when a religious system is not so well integrated and mutually reinforcing. Geertz (1957) has himself described the social and cultural disintegration that can happen when ritual, myth, social values, and social structure do not harmonize (pp. 531–543). Some measure of conflict between these elements can be tolerated and is normal. This is the advantage of symbols: They can unify individuals who each adhere to the symbol but who attach alternative meanings to that symbol. However, some scholars believe that the lack of a common, shared worldview can eventually become a problem for a pluralistic society. Their argument is that lack of agreement on the big picture, the unifying ideology or outlook, leaves the culture without a unifying core. Barbara Hargrove went so far as to suggest that public school teachers may be left with no common framework for interpreting data. If this occurs, teachers will only be able to impart "disjointed facts" (Hargrove, 1979, p. 186).

It is a maxim of the social sciences that "facts do not speak for themselves; they must be interpreted." The difficulty in a heterogeneous culture is that there is no agreement on which big theory really makes sense and explains the meaning of life. The overriding, integrating worldview—which religion provides for many cultures—is not a uniting and integrating factor in a pluralistic one. For most science teachers in the past several decades, evolutionary theory has provided the big picture, the overriding theory that explains the relationships among data. With the creationist movement, this big picture is being challenged in the courts. Some functionalists view the lack of an integrated culture—with a core of common assumptions about the way things are—as devastating to social life.

> *Critical Thinking:* Is a lack of a single uniting religious outlook a serious problem for a pluralistic society like Kenya, India, or the United States? Is the result anomie, or can societies be bound together by common interests and interdependencies? Why do you think as you do? What evidence is persuasive? What kind of evidence might cause you to change your mind?

SUMMARY

Social constructionism focuses on the processes by which individuals create and internalize the cultural worlds which they inhabit. By watching how others act in certain circumstances, we define situations—ascertaining what is going on and is appropriate for that situation. Because of its emphasis on the construction of meaning in ambiguous situations, this theoretical perspective has much to offer the study of religion. Clarification of the meaning of life, death, and suffering are fraught with ambiguity and require interpretation of meaning.

In terms of religion, those worlds include experiences, myths, rituals, and symbol systems. These cultural elements are also related to the ethos and worldview of a religion. Frequently, there is a perceived integrity between these elements, and they are mutually reinforcing. Myths are embodied in rituals and vice versa; rituals facilitate religious experiences, which are shaped by the symbol system; religious experiences reinforce the ethos and worldview of a religion; and so on.

Because of this, it is nearly impossible to identify one as primary—at least when one is referring to religion as a general phenomenon. However, specific groups may emphasize one or more of these and de-emphasize others. Pentecostals emphasize direct, embodied experience of the Holy Spirit more than Presbyterians (Poloma, 1989), and African American congregations do so more than white congregations (Nelson, 2005). When a religious group begins to undergo change, the relationships between these elements of religion may be less integrated. This may be so disconcerting to believers that the conflict may lead to schisms, a topic to be discussed in later chapters.

PART III

RELIGION IN THE LIVES OF INDIVIDUALS

In this section, we explore religion in the lives of individuals. The various ways that religion manifests itself in the lives of individuals, how people understand themselves religiously, and how they practice religion in everyday life are of intense interest to sociologists. Also, the process by which people acquire faith, change religions, and become committed to religious groups is of great importance to whether a religious group will survive and thrive in the long run. In Chapter 5, we look at some of the recent research on how individuals acquire, develop, and commit to religion in their lives and how that shapes their identity and everyday lives. In Chapter 6, we examine the social psychology of religious conversion and the processes of religious "switching."

5

BECOMING AND BEING RELIGIOUS

Here are some questions to ponder as you read this chapter:

- How is religion reproduced and transmitted from one generation to the next?
- How are young people affected by religion, and how are they influencing the trajectory of religious change?
- In what ways does religious involvement change over the life course?
- How might differences in age groups (generations) cause change in religious sensibility or religious affiliation and expression?
- Why do people become committed to a faith perspective and to a religious community?
- How have the bases of individual religious identity changed over history?
- What are official, folk, and unofficial versions of a religion? How might they conflict?

In his *Invitation to the Sociology of Religion*, Phil Zuckerman (2003) made the point that people become religious through a social learning, or socialization, process. He did so by using two fictitious characters as examples: a Christian believer "Tom" living in the contemporary United States and a devout Muslim "Mustafa" living in Saudi Arabia. Zuckerman (2003) observed the following:

> had Tom been born and raised in Yemen three hundred years ago he would most likely be a devout Muslim . . . Conversely, if "Mustafa" . . . had been born and raised in northern Mississippi two hundred years ago, he would most likely be a Baptist or Methodist Christian. (p. 51)

From this, Zuckerman (2003) concluded, "Ultimately, religious identity and conviction aren't generally so much a matter of choice or faith or soul-searching as a matter of who and what one's parents, friends, neighbors, and community practice and profess" (p. 51). Put simply, religion is an accident of birth (Wuthnow, 1999, p. 5).

In this chapter, in examining the process of being and becoming religious, we will see that Zuckerman is three quarters right. Individuals are socialized into their religious identities and orientations, beliefs and practices by those close to them (Sherkat, 2003). This is a fundamental insight of sociology. At the same time, we must be careful to avoid an "oversocialized" conception of human beings that leaves no room for individual interpretation of or deviance from social norms (Wrong, 1961). As we will see throughout this chapter, the transmission of religion from a community to its residents, from parents to children, and even from a religious group to its members is not perfect. Religiosity also changes over the course of an individual's life, and this change can be magnified or diminished by broader changes taking place in the individual's sociocultural environment. Some even argue that it is the individual's lived religion, in all its uniqueness, which ultimately matters. Though this approach may lead us to an equally problematic "undersocialized" conception of human beings, it is nonetheless an important position to consider here.

RELIGIOUS SOCIALIZATION AND THE INTERGENERATIONAL TRANSMISSION OF RELIGION

Socialization is defined as the process by which individuals are taught the beliefs, values, and norms—that is, the culture—of their community and society. Religious socialization is no different. It is how religion is reproduced and transmitted from one generation to the next (Guest, 2009). The same processes by which individuals are socialized into a culture generally can be observed quite specifically in religious socialization.

For most of human history and in most parts of the world, people have become religious by inheriting their beliefs and practices from their ancestors. In the early stages of social development, religion is inherited from the clan into which the person is born. Here, the clan is the equivalent of an extended family unit, so there is no difference between one's society and one's family (Bellah, 1970c). In these early societies, religious beliefs were passed from generation to generation through stories, folktales, ballads, and chants, which were told and enacted in a ritual context. The group's children learned the beliefs and understandings as well as religious practices through this informal form of socialization (Johnstone, 2007). Under these conditions, general socialization and religious socialization cannot be distinguished, and the transmission of religion takes place more or less mechanically.

As societies grew in size and became more internally differentiated—during the industrial revolution in the West (17th and 18th centuries)—the nuclear family emerged as the core unit of "family" (Burgess, 1916). Under these social conditions, religious socialization is more distinct from general socialization and becomes primarily the job of parents, in conjunction with other distinct organizations like congregations, youth programs, and perhaps schools. As Nancy Ammerman (2009) put it, "Congregations, in tandem with families, are the primary agents of religious socialization in places where religion has been institutionally separated from an integrated culture that carries religious meaning and practices in the warp and woof of everyone's everyday life" (p. 567). Although informal socialization continues, the intergenerational transmission of religion outside the home becomes more formalized. In congregations, Sunday schools and children's ministries proliferate, increasingly supervised by professionally trained ministers and religious educators. Religious primary and secondary schools also grow in number as parents seek to reinforce their religious teachings in the home, especially as the public schools have become more secular over time (Sikkink & Hill, 2006).

Thus, as with scholars studying socialization generally, those studying religious socialization focus on the major agents of socialization in modern society: parents and the family, congregations and religious education, peers and schools (Erickson, 1992; King, Furrow, & Roth, 2002). The dominant process by which these agents facilitate religious socialization is by serving as models for children to observe and imitate and by offering positive reinforcement for the desired and negative reinforcement for unwanted religious beliefs and behaviors. This perspective on the socialization process is known as social learning theory (Bandura, 1977).

Family is one place for religious socialization. Parents read scriptures, tell stories, emphasize moral codes of behavior, and celebrate holidays bearing religious significance that become imbedded in memory and meaning for youngsters.

Social Learning Theory

Social learning theory suggests that parents are the primary models for and agents of religious socialization, and most research shows that family religiousness is the most powerful predictor of adulthood religiousness (Dillon, 2007). The first systematic study showing children's religiosity being shaped by their parents' (especially their mother's) attitude toward religion dates to 1937 (Newcomb & Svehla, 1937). In general, if parents have a strong religious identity and engage in religious education and practice at home, children will develop and maintain the religious practices to which they were exposed (Myers, 1996; Wilcox, 2006). There is also evidence that grandparents can influence their grandchildren's religious socialization by taking active roles in the process (Bengtson, Copen, Putney, & Silverstein, 2009).

Summarizing the research, Regnerus, Smith, and Smith (2004) concluded, "Parent-child transmission of religiosity and religious identity is indeed quite powerful. But it's not inevitable" (p. 28). As we noted at the outset, there is considerable truth to Zuckerman's view of the reproduction of religion from generation to generation, but it is not the whole story. Several characteristics of parenting and family life have been shown to affect religious socialization outcomes. We highlight three: (1) the quality of relationships, (2) unity of tradition, and (3) stability of family structure.

Quality

Religious socialization is stronger in families characterized by considerable warmth and closeness—where parents enjoy marital happiness and show affection toward their children (Myers, 1996; Ozorak, 1989). Having parents who are supportive and accepting of their children, rather than judgmental, also appears to affect how well religiosity is transmitted to children (Bao, Whitbeck, Hoyt, & Conger, 1999). It is likely that in these cases children are generally more receptive to all of their parents' beliefs and preferences, religion included. Research in this area has also shown that higher levels of marital conflict are found in religiously heterogamous couples (Petts & Knoester, 2007), which leads to our second point.

Unity

Religious intermarriage—marriage between people of different religious backgrounds or levels of religious commitment—disrupts the religion–family linkage and thereby weakens the socialization of children into a faith. Several studies have suggested reasons for this. In the first place, married couples who share the same faith are more religious than couples with heterogamous marriages and therefore practice religion more frequently and are more likely to want to socialize their children into a faith tradition (Petts & Knoester, 2007). Also, families in which the parents share the same religious beliefs give more reinforcing religious cues to children. Parents who come from different religious backgrounds, by contrast, send mixed messages about religion to their children (Wilcox, 2006). Religious intermarriage may also lead children to withdraw from religious commitments altogether instead of choosing one parent's religion over the other (Hunt, 2007). The research in this area confirms that parents are more likely to transmit religious affiliation and the faith perspective to their children when they have common religious commitments. An illustration is the Navajo and Hopi reservations, where intermarriage means that the parents often speak to each other in English, and the children may not learn the native language—which is essential in order to participate in a Hopi kachina dance or a Navajo healing ceremony. The mixed marriage across native tribal lines may leave the children without full socialization into either one.

Stability

Intact families are more likely to practice religious rituals, attend services, and affiliate with a religious tradition. This translates into stronger religious socialization of children. Studies have shown that children who grow up in intact families are more likely to continue with religious practice as adults than those raised in single-parent households or in step-families (Myers, 1996). This is due in part to the fact that single parents are less involved in faith communities than parents in intact families. Also, divorced fathers are less involved in religious socialization than married fathers (Regnerus & Uecker, 2006; Zhai, Ellison, Glenn, & Marquardt, 2007).

The importance of unity and stability of family life in the intergenerational transmission of religion has significant implications for the future. First, religious intermarriage has dramatically increased in recent decades as social and cultural differences between people from different religious backgrounds have diminished and as social stigmas against religious intermarriage have faded. The rate of religious intermarriage from 1955 to 2005 increased from 12% to 25% for Protestant women, from 33% to 46% for Catholic women, and 16% to 38% for Jewish women (Rosenfeld, 2008). There is no reason to expect a reversal in this trend as religious identification becomes a less and less salient factor in mate selection.

Second, rates of marital dissolution have increased significantly in the last century. Just as the rise of the nuclear family had implications for the process of religious socialization, so too will the advent of the "postnuclear" family—divorced or never married single parents, blended families, cohabiting couples with children, gay and lesbian couples who cannot legally marry (Cherlin, 1999; Popenoe, 1993). The traditional family structure in the United States for much of the 20th century has been the two-parent household in which the husband is employed and the wife is not. Only a minority of households today fit that bill, yet this type of family seems to be correlated to high levels of faith transmission. As it is unlikely that rates of cohabitation and divorce will decline, or that women will move out of the labor force in large numbers, the future intergenerational transmission of religion may be adversely affected.

Channeling Through Peers and Schools

In addition to direct religious socialization in the family through social learning processes, parents also have some indirect influence on their children's subsequent religiousness. Parents can facilitate the religious socialization of their children by "channeling" them into groups, settings, and experiences (such as friendships and schools) that reinforce their efforts in the home (Martin, White, & Perlman, 2003).

As many of us have experienced ourselves, from childhood to later adolescence friends have an increasing influence on children's lives. This includes their religious beliefs and practices. Two studies using high quality longitudinal data both found that youth religiosity was significantly predicted by peers' religiosity. When peers' attendance is low, youth attendance is low, and vice versa (Gunnoe & Moore, 2002; Regnerus et al., 2004). This does not mean that parents no longer influence their children but highlights the fact that parents can influence their children's religious beliefs and commitments both directly and indirectly by channeling their other social relationships (Cornwall, 1987).

Another extrafamilial agent of socialization over which families have some discretion is schools. The desire to align family socialization with school socialization is a major reason the number of conservative Protestant primary and secondary schools grew from 2,500 to 9,000 today (Sikkink & Hill, 2006) and also explains the choice of many to home school their children (Stevens, 2003). Although the effect is not as strong as for parents or peers, schoolmates' religious attendance patterns do matter. Mark Regnerus and his colleagues (2004) found that the more one's schoolmates attend services, the more one is likely to attend. Interestingly, they found that it is not being in a religious school that

matters but the level of religiosity within the school (regardless of type).

This is just a cursory overview of the many factors that go into childhood religious socialization. We have not discussed the role of individuals like teachers, mentors, coaches, ministers, or pastors, or groups like youth ministries, Sunday schools, the Boy and Girl Scout programs, or sports teams. Surely these agents of socialization have an influence as well on how people grow up and understand their faith (Wuthnow, 1999). We have also for simplicity sake kept our focus on childhood socialization, even though sociologists understand that socialization is a lifelong process. Obviously, most children grow up, most get married, most have children, most grow old, and all of them die. Religion continues to have a presence in many people's lives through all of these stages of the life course.

Critical Thinking: How does this discussion of religious socialization apply to your own life? If you have no religious affiliation or faith tradition or a strong one, is the pattern consistent with the explanations presented here? Why or why not?

RELIGION OVER THE LIFE COURSE

Religiosity does not proceed in a straight line over the course of life for most people, either upward or downward or straight across. Sociologists have recognized this and have attempted to find the systematic patterns of variation in religiosity over the life course. As modern society has become more pluralistic, differentiated, and even fragmented, it has become harder to find those patterns (Hunt, 2007, p. 611). We will nonetheless try to bring some order to the chaos and answer the question: How does religion evolve and change as people age and as they experience major "life course" events like high school graduation, marriage and child rearing, and retirement?

Adolescence

In discussing religious socialization, we already paid some attention to the period of adolescence. It is hard to define strictly in terms of age, but it is typically seen as the transitional period from childhood to adulthood. A broad estimate would be from 13 to 19 years of age. This is the time of life when children are maturing physically and seeking independence from their parents socially and therefore is expected to be a period of declining religious involvement.

In part because of the practical difficulties of studying children and in part because many sociologists have not been interested in the religion of youth, sociologists have not historically had good data on religion and youth. That has changed considerably lately thanks to the availability of datasets such as the National Longitudinal Study of Adolescent Health (Add Health; www.cpc.unc.edu/addhealth), Monitoring the Future (www.monitoringthefuture.org), and most recently, the National Study of Youth and Religion (NSYR; www.youthandreligion.org).

These different data sets tell a similar story about religion and adolescence: that adolescence is a major period of instability. Frequently that instability manifests itself as religious decline, for example in attendance from middle through high school. One study using Monitoring the Future's 1997 data showed that 44% of 8th graders reported attending services weekly, compared to 38% of 10th graders and 31% of 12th graders—a 13% drop (Regnerus & Uecker, 2006). The NSYR data showed that in the 3 years between Wave 1 and Wave 2 of their data collection, weekly religious service attendance declined by 13% and nonattendance increased by 10% (Denton, Pearce, & Smith, 2008). The authors highlighted some interesting caveats, though. First, most adolescents only changed their level of religious attendance slightly over this time period, from weekly to two to three times per month, for example. Second, very few adolescents (7%) went from regular attendance (once/week or more) to nonattendance. Third, 15% of adolescents reported *increased* levels of religious service attendance.

This last finding draws attention to a fact that others have previously noted: that the religious story of adolescence is not simply one of uniform decline. Rather, some scholars have called it a period of religious "polarization" (Ozorak, 1989). In line with this interpretation, Regnerus and Uecker (2006) examined the Add Health data and found that 20% to 22% of adolescents showed a decline in some form of religiosity during adolescence, but 15% to 18% showed an increase. Those whose religiosity increased during adolescence tended to come from highly religious families and from conservative Protestant traditions, suggesting the continuing importance of religious socialization in the later period of childhood.

Emerging Adulthood

In every society, adolescents make a transition into adulthood. In some societies, this involves elaborate (and sometimes dangerous) rites of passage like the Maasai lion hunt in Africa, the Australian Aborigine walkabout, and the Native American vision quest among the Plains "indians." Historically, many of these rites of passages were conducted in the context of religious celebrations. Even in the modern West, many religious groups mark the transition into religious adulthood with rituals such as baptism (in Protestant traditions), confirmation (Roman Catholicism), and the bar/bat mitzvah (Judaism). However, as these religious transitions typically take place during adolescence (the bat mitzvah coincides with physical puberty, for example), they cannot really be seen as transitions into adulthood in the secular world.

For most American youth today, the passage to adulthood is not clear (Hunt, 2007). The age of 18 used to be fairly significant, as it entailed a graduation from high school and entrance into college or the world of work, as well as the right to vote, drink alcohol, and (for men) be drafted into the military. Also, the average age of marriage and childbearing was lower, then, so the years of 18 to 22 were a real transition into adulthood. How the times have changed.[1] Today, young people stay connected financially to their parents, delay marriage (or have "starter marriages"), and postpone childbearing longer than ever. A term has even been coined to describe young people who leave home (often for college) and then return home to live with their parents— "boomerang kids" (Mitchell, 2006). Where once high school graduation was a passage to adulthood, now college graduation does not even serve that function for all young people. Jeffrey Arnett (2004) has described this new cultural life stage from roughly ages 18 to 29 as "emerging adulthood"—a sort of gray area between adolescence and adulthood.

Not surprisingly, like adolescence, the period of emerging adulthood is characterized by considerable instability, but compounded by the flux created by the process of leaving home. The effect of this on religion is clear: "Emerging adults are . . . the least religious adults in the United States today" (Smith, 2009, p. 102). On measures of service attendance, prayer, and religious importance, emerging adults consistently registered lower levels of religiosity than older adults. Smith (2009) made an important comparative point in putting these emerging adult religious practices in a broader context. Drawing on the work of political scientist Robert Putnam (*Bowling Alone*), he noted,

> Emerging adults are not only less *religiously* committed and involved than older adults but also tend to be less involved in and committed toward a wide variety of other, nonreligious social and institutional connections, associations and activities. Emerging adults, for instance, belong to fewer voluntary associations, give less money in charitable donations, volunteer less, and read newspapers less than do older adults. (p. 92)

Emerging adults who attend college are no less likely to religious involvements than those who do not seek further education (Smith, 2009, p. 250). Thus, it is not the secularizing effect of college that lowers emerging adults' religiosity but their general detachment from broader social associations, activities, and concerns (Smith, 2009, p. 93). This makes sense because it is exactly the lack of stability in work and personal relationships that is definitive of emerging adulthood as a developmental stage.

Mature Adulthood

Mature adulthood is characterized for most people by two of the most significant life course events: (1) marriage and (2) childbearing. Both of these family formation events are related to higher levels of religious practice. Of course, these are not the only family cycle events that can take place during this time. Both divorce and cohabitation are increasingly common, and generally reduce religious activity (Myers, 1996; Stolzenberg, Blair-Loy, & Waite, 1995).

Some interesting caveats apply in this stage of the life course. The exact timing of these family formation events mediates their influence on religious behaviors. For example, research shows that people who have children in mature adulthood (their late 20s or early 30s) increase their religious participation while those who rear their children earlier do not (Stolzenberg et al., 1995). This may be due to the fact that when people marry and have children at the later ages associated with mature adulthood they benefit more from the social support provided in religious organizations.

Also, unlike emerging adulthood, mature adulthood is often fraught with responsibilities and other stressors that can negatively affect religious involvement (Hunt, 2007). For middle-class mature adults, especially those who buy into the ideology of "competitive parenting," weekends may no longer constitute a Sabbath (day of rest and time of worship) but a time to

---

[1]Here, we are speaking of the broad middle class of American society. The very rich and very poor have always lived different lives, lives of much greater choice for the rich and much less choice for the poor. See Sean McCloud (2007) and Beverly Skeggs (2004).

take their children to swim meets, soccer games, dance recitals, art exhibits, tennis tournaments, photography classes, college campus visits, and the like. Stress may be work related, like increasing responsibility on the job, job related relocation, the threat of "downsizing," or the possibility of being overtaken by younger workers. Alternatively, stress may be family related, like concern for aging parents, saving for retirement, and children's education. That said, some stressful experiences in mature adulthood, like personal illnesses, may be inducements to seek consolation in religion (Ferraro & Kelley-Moore, 2002).

Later Life

Studies consistently show that religion increases in old age. For example, in his summary of the research on religion and aging, Krause (2006) noted that 79% of people between the ages of 65 and 74 claim that religion is very important, 52% attend a worship service, and 89% of older people pray during the typical week. Smith (2009, pp. 88–89), in his study of religion and youth, contrasted emerging adults' low levels of religiosity with older adults' high levels: 42% of young adults pray daily or more often compared to 76% of those over 75 years old, 22% of emerging adults characterize their religion as "strong" compare to 53% of older adults, and 15% of emerging adults attend religious services at least weekly compared to 40% of older adults.

One explanation for the rise in religiosity in later life is that religious participation provides a system of social, emotional, and spiritual support (Krause, 2006). As people age, they lose social ties and their role in society changes. They retire and lose friends. The social ties of religious congregations help to fill this void. Many religious communities also still believe in the concept of respecting one's elders. As shown by prevalent ageism in contemporary society, this is not necessarily true in the secular world. Older adults are also encouraged to perform volunteer work through religious organizations. This provides

social ties and an opportunity to participate in important productive activities.

In addition to the benefits of involvement within a religious organization stated previously, there is evidence that being involved in a religious community as an older adult has health benefits. A study by Krause (2001) suggests that people in active social networks have better mental and physical health. People with stronger support systems like those found in churches, temples, and mosques also live longer (Hummer, Rogers, Nam, & Ellison, 1999). The difference that occurs in the religious setting as opposed to secular social ties is the component of spiritual support. Unlike other social ties, religious ties are uniquely suited to relieving the fear and uncertainty of death (Siegel & Shrimshaw, 2002).

Critical Thinking: Does this description of change over the life course, including modifications in religious sensibility, seem consistent with older members of your family or with other people you know? If not, what might be different to have altered the outcomes?

AGE, COHORT, AND GENERATIONAL EFFECTS ON RELIGION

In our review of religiosity at various stages of the life course, we focused our attention only on different age groups (13- to 18-year-olds, 18- to 20-year-olds, and so on). We specified how religiosity fluctuated according to age or age-based life course events (e.g., leaving home, marriage). These are called "age effects" on religiosity. Although this offers an important developmental perspective on variations in religiosity, it assumes that the effects of aging on religiosity are constant over time. For example, emerging adults have low levels of religious involvement *now*, but when they are 40 and married with kids their religiosity will *increase*, and when they are 70 and facing death it will be *even higher*.

This assumption would be sound if age effects were all that mattered in explaining religiosity over time. It seems unlikely, however, that someone experiencing adolescence in the 1920s or 1930s (the Great Depression era) would experience religiosity over the life course in the same way as someone coming of age in the 1940s (WWII era) or 1960s ("The Sixties"). The age effect in all of these cases combines with the effect of living through a specific period in history (called, not surprisingly, a "period effect") to produce what sociologists call a "cohort effect." A cohort effect is the combined effect of being a certain age at a certain period of time.

Cohort analysis reminds us that cultural context shapes the expectations of age-related behavior. In fact, although we discussed it as an age-based developmental stage, "emerging adulthood" may in fact be more properly characterized as a cohort phenomenon: It is the particular way that early adulthood is experienced by those who came of age in the last 10 to 15 years. "Emerging adulthood" as a part of the life course did not exist 50 years ago, and it may not exist 50 years from now. This highlights the importance of looking at cohorts but also the difficulty of separating age and cohort effects in any particular analysis.

At this point, you might be saying to yourself that this sounds like an extremely complex way of describing a commonsense idea: "generation." To some extent, you would be right. Karl Mannheim (1927) long ago stressed the importance of understanding how shared formative experiences forge in a collection of individuals particular worldviews that they carry with them the rest of their lives. He called his paper "The Problem of Generations." So, generations and cohorts are closely related. A generation is typically seen as encompassing several cohorts (Ryder, 1965).

A good example of how age and cohort/generation can independently affect religious participation is Michele Dillon's (2007) summary of her analysis of data collected by UC Berkeley's Institute of Human Development (IHD) in the 1920s (see also Dillon & Wink,

2007). The study contains longitudinal data over a 60-year time span, including two cohorts born 8 years apart (the older in 1920 and 1921 and the younger in 1928 and 1929). The older subjects were children of the Great Depression, old enough to serve in the military during WWII, and experienced the social upheaval of The Sixties when they were in their 50s. The younger subjects did not experience the deprivation of the Great Depression, were not old enough to serve in the military during the war, and were almost to middle age during the 1960s. In Dillon's view, the drastic differences in experiences of the subjects represent not just different cohorts but also different generations.

Results of the IHD study showed that religiousness did not follow a linear trend across the life course but was described as a U-curve. Religiousness was high in adolescence and in later life and generally lower in middle adulthood. That the pattern of high religiosity at the beginning and end was the same for both of the generations suggests that age effects might be at work rather than cohort effects. On the other hand, the timing of the postadolescence decrease differed between the generational groups. The younger generation declined in religiousness in their 30s, while the older generation declined much later, in their late 40s. The different timing of these drops suggests a cohort or generational explanation. Given their different experience—the Depression and WWII compared to the 1950s and 60s—members of the older group *as a generation* remained more committed to social institutions, including religion, for a longer period of time.

Using more contemporary data, generational analysis in the sociology of religion has grown increasingly popular in the past 25 years as the "baby boom generation" has worked its way through the life course, spawning generation X and generation Y (also known as "millennials" or "echo boomers"). Table 5.1 illustrates how age and period come together to forge these cohorts into recognizable generations for sociological analysis.

Table 5.1 Recent Generations of Interest to Sociologists of Religion

Generation	Birth Years	High School Class	Age in 2010	Some Period Influences
Baby boomers	1943–1960	1960–1978	50–67	The Sixties (Vietnam War, civil rights movement, sexual revolution, drugs, rock and roll, Kennedy–King–Kennedy assassinations, Stonewall riots), *Roe v. Wade*, OPEC oil crisis, Watergate, Nixon resignation, drugs–disco, stagflation
Generation X (baby bust)	1961–1981	1978–1999	29–49	disco dies, MTV born, beginning of cable TV, Iran hostage crisis, Ronald Reagan/George H. W. Bush, energy crisis and recession followed by unprecedented stretch of economic growth, HIV/AIDS, end of Cold War/fall of Berlin wall, hip-hop, PC revolution, creation of the World Wide Web, cell phones
Generation Y (post-boomers, echo boomers, millennials)	1982–2001	1999–2019	9–28	ubiquity of computers, cell phones, iPods and MP3 players, e-mail/IM/texting, Facebook and YouTube ("Digital Natives"—see Chapter 14); globalization and pluralism; Harry Potter; hooking up; sex scandals; 9/11 and war on terror; dot-com bubble burst (2000); U.S. housing bubble burst (2008); global financial crisis (2008–)
Homeland generation (new silent generation)	2001–	2018–	under 10	war on terror, global financial crisis

NOTE: There is no definitive dating of these generations. For example, the U.S. Census Bureau dates the actual demographic boom in births from 1946 to 1964, but Neil Howe and William Strauss (1993, 2000) categorize baby boomers as those born from 1943 to 1960, and hence coming of age from the late 1950s to the late 1970s. We follow Howe and Strauss's periodization here for the sake of convenience.

Critical Thinking: Examine Table 5.1, and determine the generation with which you would be identified. Do you feel a part of that generation? Do you think of yourself in those terms? Look at the period influences that have shaped your generation. Do those influences shape the way you approach religion, or do you feel your religion is pretty typical for a person your age, regardless of the time period?

The key to a generational analysis of religion is to understand how the period influences fundamentally shape the outlook of individuals who live through them at a key developmental moment (typically adolescence). That unique outlook then shapes the specific way the generation understands and practices religion. Wade Clark Roof was among the first to do this for baby boomers. In *A Generation of Seekers: The Spiritual Journeys of the Baby Boom Generation* (1993) and *Spiritual Marketplace: Baby Boomers and the Remaking of American Religion* (1999), Roof argued that the particular generational experiences of baby boomers led them to fundamentally reorient themselves to religion by adopting the posture of spiritual seekers. Having come of age largely in the 1960s and 1970s—fully exposed to the cultural ferment of the time as well as the authority damaging revelations of the Vietnam War and Watergate— the boomers Roof studies feel authorized to choose their own religious path, emphasize the experiential aspect of religion over the doctrinal, and are skeptical of traditional religious organizations. They are, in a word, "spiritual not religious."

This brings us back to cohorts. According to Ryder (1965), "cohort replacement"—the replacement of older cohorts (through death) by younger cohorts—is a fundamental engine of social change. Roof's argument that baby boomers are remaking American religion is based in part on this idea: cohorts of boomers and their children—who they will socialize into this new seeker orientation toward religion— will gradually replace older cohorts who were oriented toward more traditional expressions of religion.

Of course, for this transformation of American religion to take place, boomers need to effectively socialize their children into their spiritual seeking orientation. Have they done so? No comprehensive study of generation Y—sometimes called echo boomers because they are generally the children of boomers and hence also a larger generation in number—yet exists. However, some scholars have begun to grapple with the religion of those born after the baby boomers.[2] Richard Flory and Donald Miller (2008), for example, highlighted the spiritual quest of the post-boomer generation. They argued that, like the baby boomers, post-boomers have formative experiences that are specific to their time and place and which influence their orientation to religion. Among the formative experiences is that as children of boomers, they inherit an ethos of questioning institutional authority and focusing on the personal journey. This has been reinforced by unethical and hypocritical actions of some major institutions, including business scandals like Enron and religious malfeasance like the Catholic Church's handling of pedophilia by priests. Another major influence on post-boomers is the increasing pluralism of the world, whether they experience it directly in their neighborhoods and schools or indirectly through the ubiquitous "new media" these "digital natives" employ (Palfrey & Gasser, 2008). This leads to an attitude of tolerance and acceptance of difference.

How do these cultural influences and generational responses affect the practice of religion? Flory and Miller (2008) argued that post-boomers approach religious institutions with some suspicion and so focus on the quality of relationships within the institution instead of the institution itself (p. 10). Post-boomers also emphasize experience and embodiment in their religious practices. Although they are in the spiritual marketplace created in part by their parents' generation, they are not content simply to browse and buy (consume). They want to have a hand in the production of the religious goods to be consumed. The authors call this combination of these elements "expressive communalism."

---

[2]Although he does not use generational terminology, the sample in Christian Smith's NSYR are millennials, so that study is arguably the most comprehensive study of the religion of the millenial generation available.

Cohort groups may well influence religious loyalties and understandings of a faith tradition, and religious communities sometimes group people by age—perhaps reinforcing those differences. Still, religious communities are often one place in the society—in prayer groups, religious retreats, mission programs, and religious education programs—where there is often a cross section of generations. It is one place in our society where older and younger generations interact intensely and learn to understand other points of view.

According to Wade Clark Roof (2009), generational patterns not only have significance in terms of individual-level beliefs and practices but also have implications for the religious organizations that have to adapt to generational shifts. Faced with children of baby boomers who have no strong attachment to particular religious institutions but who, as "digital natives" (who have never lived in a world without digital technology), are strongly attached to their iPhones, BlackBerrys, Droids, and other smartphones, religious organizations must respond. They are scrambling to figure out ways to harness the power of new media in order to connect with this millennial generation. As we will discuss in Chapter 14, use of the Internet is one of the areas of fastest growth for congregations recently, but much of that use is simply informational, not interactive. In the "Illustrating Sociological Concepts" feature, sociologist Gerardo Marti highlighted the fact that successful congregations need to recognize which direction the cultural winds are blowing and find ways of putting up sails to harness the power of those winds rather than fighting them or ignoring them.

ILLUSTRATING SOCIOLOGICAL CONCEPTS

The Wi-Fi Church of the Future (and Present)

Gerardo Marti

Churches have climbed on the digital bandwagon in the past decade. Computers and internet connections are no longer considered luxuries, and most churches have a web presence (or at least feel like they should have one).

Today, the pervasiveness of portable computing combined with the proliferation of creative "apps" provides yet another opportunity to expand the ministry of the church. The past two years have seen a rapid acceleration in the adoption of portable computing by the average person. This will inevitably prompt changes by church leaders.

One such innovation could be closer to us than we think. I am impressed with the GPS capabilities expanding to cell phones, in particular Apple's iPhone and Motorola's new Droid powered by Google's Android platform. A basic mantra among entrepreneurs is "Location, Location, Location," and the importance of "location" is now being harnessed by the latest generation of smartphones. GPS embedded in these phones takes advantage of satellite technology so that your device always "knows" where you are. The Apple iPhone already has remarkable apps that take advantage of location specifics so that users can find local restaurants, movie theaters, and gas stations—all based on where one happens to be at that moment.

We can't be too far from inventive congregational leaders creating applications that take advantage of the GPS in these devices. For example, congregations could allow users to discover church services and other congregational events. GPS devices could allow people to connect automatically with ministry offerings in regions by city, zip code, or mile-radius. I can imagine a day soon when I visit a city for a conference and can look up services and events (perhaps indexed by religious orientation) and see what is happening religiously in the area—whether it starts in 12 minutes or in 2 days time.

This smartphone technology does not have to be tailored only for new visitors. New features can expand the liturgical experience for the faithful. Religious organizations could pre-program mobile phone events to begin as people arrive to services. Because of GPS capabilities, these can automatically begin when a person comes within "range" of the congregation.

Users could elect to connect to their congregations either by pre-set time (20 minutes before services or events) or by pre-set radius (within 1 mile of the meeting place) and have devotional music play as a "pre-event" or "pre-service" preparation. Prayers could be read aloud. Readings from scripture or church history could be included. Announcements could also be programmed to play for users either before or after events.

For study groups that require advance reading, these apps could remind members of their assignment and allow people to download the required reading right to their phones. Alternatively, the device could have selections read aloud to them using speech-to-text software as they are traveling. All of these features could be hyperlinked for more information. They could also include a "forward link" so that people could invite friends, neighbors, and co-workers to participate. People who have moved away for work or school could also keep in touch with their church, especially if such services are complemented with downloadable video streams and continued interactive features like "comment" or "chat" extras.

Inventive programmers in specific congregations could do this right now if they had the motivation and encouragement from their pastors to do it. Perhaps the initiative of larger networks of denominations or parachurch support organizations could commission programmers to create these apps. Seminaries could design apps that extend and accentuate offerings to students and alumni as a way to extend their ministry.

However it happens . . . if such apps are created, they will be implemented by thousands of people overnight—whether they are members of local churches or not. In all of this, the point is to discover innovative ways to expand land-based local experience through online global connections.

In short, the wi-fi church of the future is just around the corner. The emerging electronic capacities of portable computing and the overwhelming presence of these devices in the consumer world will soon be co-opted by religious organizations. It only remains for the first set of innovators to create the apps that take advantage of these capacities in religiously creative ways.

SOURCE: First published in the Faith & Leadership (www.faithandleadership.com) "Call and Response" Blog, Duke University Divinity School, December 1, 2009.

A MULTIDIMENSIONAL
MODEL OF COMMITMENT

The study of religious socialization and religiosity over the life course tends to focus on the patterns evidenced by people involved in the mainstream religions of society—appropriately so, because the vast majority of people in society are involved in such groups. Still, analytically we can gain some significant insights into social processes by looking at "outliers"—groups that deviate from the mainstream. The study of religious commitment is such an area. In this section, we begin with insights from Rosabeth Kanter's (1972) study of the elements of commitment in utopian communities and communes and then suggest how those same elements can be seen in more mainstream religious groups.

Utopian communities—the best known of which in her study is the Shakers—provided a good test for her model of commitment because their attempts to create ideal social orders took place in the context of a broader society which competed for followers' loyalties. Clearly, living an austere lifestyle with all things shared in common demanded a high level of commitment. The result of her research was to identify a number of mechanisms of commitment found in successful communities. These commitment mechanisms apply to a variety of different types of organizations— Kanter's work has been cited in over 1,000 other studies of businesses, nonprofits, and social movement organizations over the past four decades—but the model is particularly applicable to religious movements.

Kanter (1972) found that commitment occurs on three different levels: (1) commitment to the organization (*instrumental commitment*); (2) commitment to other persons in the group (*affective commitment*); and (3) commitment to the rules, regulations, ideas, and mores of the group (*moral commitment*). Any group or institution may elicit one or more of these types of commitment. Although they are interrelated, they are also distinct aspects of social organization that can be analyzed separately, for they are forms of commitment to the institutional system, the belonging system, and the meaning system, respectively—the three subsystems we found to make up a religion. According to Kanter, each of the three types of commitment involves two mechanisms that enhance commitment.

Instrumental Commitment

Instrumental commitment means the individual must be convinced that continued association with the group or organization is worth the time and effort it demands. Hence, the individual engages in a sort of cost–benefit analysis. At the cognitive (thinking) level, the committed individual must conclude that the profits associated with continued participation are significant and well worth the time and energy expended. Furthermore, if individuals feel that there would be substantial cost associated with leaving the group, the institution stands to gain in its *retention* of members. Two organizational mechanisms tend to enhance instrumental commitment by influencing this cost–benefit ratio. They are *sacrifice* and *investment*.

> Sacrifice means that membership becomes more costly and is therefore not lightly regarded nor likely to be given up easily. Investment is a process whereby the individual gains a stake in the group, commits current and future profits to it, so that he must continue to participate if he is going to realize those profits. Investment generally involves the giving up of control over some of the person's resources to the community. (Kanter, 1972, p. 72)

Members of religious movements are sometimes asked to make substantial sacrifices. Many times a comfortable life has been sacrificed for the rigor of an austere lifestyle, in which numerous popular recreational and social activities are denied. Devotees may be forbidden alcohol, drugs, or coffee, and they may find that card playing, social dancing, and movies are disallowed. For example, Stillson Judah (1974) wrote this of the attitude of the

Hare Krishna: "The devotee must be ready to relinquish anything material for the satisfaction of Krishna. He must be ready to give up something that he strongly desires, while accepting something he does not like" (p. 91).

Even being active in the local religious community may mean one cannot sleep late on the Sabbath morning (a sacrifice, indeed) and one may be expected to fast (go without food) on certain holy days. Attitudes of asceticism or self-denial are especially common among sectarian and cultic groups. Kanter (1972) showed that this sacrifice is quite functional for commitment:

> Once members have agreed to make the "sacrifices," their motivation to remain participants increases. Membership becomes more valuable and meaningful. Regardless of how the group induces the original concessions or manages to recruit people willing to make them, the fact is that those groups exacting sacrifices survive longer . . . The more it "costs" a person to do something, the more "valuable" he will consider it, in order to justify the psychic "expense." (p. 76)

Sacrifice causes a person not to want to leave because leaving would be to admit that the sacrifices were not worthwhile. Whether consciously or unconsciously, people do not like to admit that they have made foolish sacrifices. Hence, once a part of the group, demands for self-denying behavior act to increase and sustain members' commitment.

The other mechanism that enhances instrumental commitment is investment. Although people sacrifice by not using their time, energy, and resources in activities they might otherwise find enjoyable, they are expected to invest that time, energy, and money in the group. Many vibrant religious movements require at least a tithe (usually 10% of earnings), and some new religious movements (NRMs) require that for full membership the devotee turn over *all* of his or her earnings to the organization. If the donation of all one's worldly goods is irrevocable, the investment becomes even more important, for it deters one from leaving the group later. Most

people will continue their association with a group rather than to admit to themselves that they have been foolish.

The devotee may also be required to invest substantial amounts of *time* in the group. The expectation of the Church of Jesus Christ of Latter-day Saints (the Mormons) that young men devote 2 years proselytizing or missionizing is a good example. Likewise, hours of public chanting required of new members of the Hare Krishna enhanced their commitment (Judah, 1974). Members of many religious communities will spend untold hours in meetings, service activities, and other time-expensive activities.

Research by social psychologists has indicated that the act of making a public statement on any issue is one of the most important factors in solidifying one's commitment. If a person is asked to serve on a panel to discuss the benefits her university offers to the community, she would be much more likely to make a financial donation at a later time to that university. She would have invested time and energy in defending the school's importance. Furthermore, after she had taken a stand publicly, she is likely to perceive any attack on that organization as an attack on her. She does not want to lose face, so she would likely become defensive on behalf of the institution. Likewise, the public door-to-door proselytizing of the Mormons and the Jehovah's Witnesses or a Roman Catholic signing a published statement on abortion serve to reinforce commitment.

The important factor in understanding retention of members is that a group that requires little investment and little or no sacrifice is likely to elicit little instrumental commitment. Persons may be committed at other levels, but commitment to the organization itself will be low. Because religious movements bent on radical transformation of the world usually exact a high cost (alienation from family, rejection of opportunities of an affluent life, etc.), they must emphasize the benefits and de-emphasize the costs. Moreover, their major source of recruitment will be among those who do not yet have much invested in the status quo. It is for this reason that youth are the major target of many NRMs. In conventional religious denominations, the recruitment process need not be so intense and need not be limited to a particular clientele; the *costs* of joining a conventional group are much less severe. Regardless, some measure of instrumental commitment is important for any organization, for the organization needs the time, the talents, and the financial resources of members if it is to thrive.

Affective Commitment

Affective commitment refers to an emotional dependence on the group. The group's members become one's primary set of relations. Many sects and cults work to become the *only* reference group of the members, but affective ties are important in conventional religious groups as well. A number of studies have shown that the vitality of any local religious community is profoundly affected by the number of small, intimate groups in which people share their lives with one another (Miller, 1997; Wilson, Keyton, Johnson, Geiger, & Clark, 1993; Wuthnow, 1994b). Affective commitment is of central importance in religious behavior. Andrew Greeley suggested that while meaning functions may be *primary*, the belonging functions are often chronologically *prior*. We see this clearly as we explore the commitment and conversion processes in the NRMs. For example, "warmth and friendship among the devotees" was one of the reasons most frequently cited by members for original attraction to the Hare Krishna movement (Judah, 1974, pp. 153–154). It is also noteworthy that the Moonies consciously looked for signs of isolation or symbols of transiency—like backpacks—to identify prospective converts (Bromley & Shupe, 1979). Such persons are less likely to have other affective commitments and are more likely to be in need of a group that can offer emotional support. Helen Berger (1999) has found that commitment to Wicca is usually initiated through friendship networks as well.

Prayer groups build an intense sense of intimacy and belonging. In prayer groups, people develop trust and display vulnerability—sharing weaknesses, confessing faults or failings, and asking for prayer on these matters. The group is likely to be supportive rather than judgmental, and other members reciprocate with confessions of matters where they need support. Very intense sense of belonging develops—thus enhancing affective commitment.

The affective commitment process involves two mechanisms: (1) detachment or *renunciation* of former ties and (2) *communion* with the new group. Many of the NRMs have demanded that new recruits not contact their families, especially for the first few weeks or months. Among the Hare Krishna, for example, "progress depends on willingness to give up the company of anyone who is not a devotee" (Judah, 1974, p. 91). Renunciation of one's former friends, family, and social groups typically involves defining outsiders as evil or as insignificant. For example, a Hare Krishna devotee responded in the following way when asked whether he corresponded with his parents:

No, not really. Sometimes they write me a telegram. They want to know where I am so I tell them. . . . Your parents are so temporary. When I speak to my mother on the phone, it's like a stranger . . . What is a father? A father is one who . . . gave me this material body. It's a relation of bone and stool and blood. That's all it is . . . But Prabhupad . . . told me what to do—how to live like a human being, how to elevate myself . . . My father is Prabhupad and my mother is the scripture. (Judah 1974, p. 179)

Among the Amish, the Bruderhof, the early Christian church, the early Mormon church, and many holiness groups, we find high boundaries and a tendency to define nonmembers as degenerate, evil, or confused. Hence, contacts with outsiders are limited and are controlled by norms that make those interactions rather formal, ritualized affairs. The experience of being treated as a stranger—and as a rather suspect degenerate stranger—may be confusing and disconcerting to the family member, who can only make sense of the behavior by assuming that the devotee is under some sort of trance. Certainly, confusion, hurt, and anger are understandable responses. Still, the process of becoming committed to NRMs has to do with a change of reference groups, not some sort of hypnotic trance or "brainwashing." Further, many groups enhance commitment by heightening the boundaries between "us" and "them:" think about Greek houses on campus.

The second affective mechanism is communion—emotional solidarity with others. While new recruits are removed from former reference groups, they are warmly embraced and provided with a high degree of emotional support in the new reference group. The Moonies have been very intentional about this process, which they refer to as "love bombing." Particularly for isolated persons, transient people, or loners, this intensity of concern and emotional warmth may be the most significant experience of family that the individual has known. Many communes and religious groups refer to themselves as family, for the group actually becomes a surrogate family for the members.

It is not necessary to focus only on cults to see the importance of this process. Most of the more conservative churches and temples with high retention rates have extensive meetings and social events. If a person attends morning worship on the Sabbath, has another worship experience on Sunday evening (perhaps preceded by a potluck dinner), attends the men's prayer breakfast or the women's prayer group on Tuesday morning, goes to Bible study on Wednesday evening, and serves on a committee of the congregation, he or she has little time or energy for other social commitments. The other persons attending those functions become one's primary social relations.

As mentioned in the photo of the young people praying, prayer groups often share deeply within the group and become tightly bonded. Still, many congregations of mainline denominations do not seem to elicit high levels of commitment; many are lacking programs that develop close and supportive interpersonal relationships among members. In fact, Hare Krishna members frequently cited a lack of meaningful friendships in their original religious group as a reason for leaving that faith (Judah, 1974, p. 151). Feeling of communion—of truly belonging, being cared for, and caring for others—is an extremely powerful commitment mechanism.

Once recruits become a part of the group and begin to identify its members as their best friends, they may come to realize that the group has an internal stratification system. Some people are more highly respected than others. Because one wants to be liked and respected by one's new friends and colleagues, following the group's rules becomes important. As one receives responsibilities and is recognized for contributions to the group, one has begun to climb the stratification ladder. Having taken a step or two up the ladder, the individual has made an initial investment in the group. Hence, affective commitment leads naturally to instrumental commitment. Commitment to the group members and commitment to the organization are intertwined—as indeed, the belonging and institutional subsystems of religion are linked.

Retention of members also depends on an ongoing set of primary relationships with believers. Bromley and Shupe (1979) reported that when Moonie devotees are not harbored in supportive environments, defections are very high (p. 184). The insistence on endogamy (marriage only to individuals within the group) is one way to ensure interlocking relationships, and the Reverend Moon was quick to recognize this. If one's spouse and eventually one's children are within the faith, then defection from the faith community will also involve separation from one's family of procreation. By ensuring endogamous marriages, as Reverend Moon did by arranging marriages, a set of interlocking and mutually reinforcing relationships is created. Cross-religion marriage is, after all, one of the strongest predictors of low religious practice. It all has to do with in-group affective ties.

Moral Commitment

Moral commitment refers to commitment to the norms and values of the group, indeed, to the meaning system of the religion. If the group is to develop in a coherent way, members must accept the mandates of the ideology—the worldview and ethos—as it is formulated by the leaders. To put it another way, the leadership must be able to control the group in order to direct its development. Because religious groups do not have military powers or total economic control over the entire populous, they are limited in the extent to which they can coerce members to obey the norms of the group. They must depend on voluntary compliance. The control issue is more problematic for groups that deviate from the dominant society, for compliance with the religious group means deviation from the dominant culture. Hence, the cause of the group must seem compellingly true, eternal, and just. There are two mechanisms that can enhance the moral commitment of devotees: *mortification* and *transcendence*.

The mortification process places heavy emphasis on the willfulness, egotism, selfishness,

and conceit of people, and this generates a sense of profound humility. This is done by emphasizing that without this group or the faith perspective of this charismatic leader, the members would be worthless degenerates. For example, Oneida was a religious commune in upstate New York that survived for 33 years in the mid-1800s. It is a fascinating community but is especially remarkable because it is the only free-love commune to survive for such a long period of time.[3] One reason Oneida proved so viable is because of a practice called mutual criticism. This involved a public confession of all one's faults, weaknesses, temptations, and areas of needed growth. Furthermore, the assembled members would add to the confession if the confessor failed to include everything. The founder of Oneida, John Humphrey Noyes, was especially forthright and aggressive in probing a person's inner feelings and motives.

The journals and diaries of members offer a fascinating account of how members of Oneida perceived this process. They uniformly reported that after such a session they felt utterly humble and worthless. However, the more important emotion (and perhaps the more surprising one) was a feeling of utter exhilaration and joy. Kanter (1972) wrote that

> the use of mortification is a sign that the group cares about the individual, about his thoughts and feelings, about the content of his inner world. The group cares enough to pay great attention to the person's behavior, and to promise him warmth, intimacy, and love . . . if he indicates he can accept these gifts without abuse. (p. 105)

The act of being emotionally naked before others whom one trusts tends to engender ecstatic feelings of intimacy and union with one's associates. Beyond that, the feeling of worthlessness is

---

[3]Oneida's system of complex marriage meant that any adult man and any adult woman might engage in sexual relations on a given night. However, there were many norms that guided these practices. In this sense, the free-love term, which is commonly applied to Oneida, is somewhat of a misnomer. Oneida defined itself as a biblically based and profoundly religious community. The members did not view their sexual practices as promiscuous (Carden, 1969; Parker, 1935).

also important in understanding why people follow the leadership of the group with such unquestioning loyalty. Any personal initiative, critical thinking, or alternative explanation of things will be defined as egoistic, self-centered, and self-aggrandizing behavior. Any challenge to the doctrine of the group is a sign of "the Devil at work" in the individual or an indication of a lack of humility. In either case, the person may be demoted in the stratification hierarchy of the group. This demotion may induce humility if the other techniques did not.

Many other groups could be cited for similar mechanisms to induce feelings of humility or individual worthlessness. Regardless of specific variations in the way different religious groups engender the attitude, a sense of personal worthlessness is very functional for the group. When people are convinced that they are truly insignificant as individuals, they are more humble and more willing to obey their superiors. They are also less likely to assert themselves in conflict with peers, and deference to others enhances harmony and cooperation.

Note that if Roberts and Yamane (your authors) have huge egos and they disagree about something, each may dig in his heels and insist that the other is just plain wrong. It could get nasty if it were in a public setting. If, on the other hand, they have some measure of humility, they may each figure that the other is a bit smarter than himself. In such a case, it is amazing how well they can get along. Humility helps groups cohere, and mortification mechanisms foster that humility.

This leads to the second process: transcendence—the sense of ultimate purpose and meaning. Although the person may feel worthless as an individual, his or her life has ultimate and eternal meaning as a member of the group. The group offers a hope of final victory that obliterates the meaninglessness of mundane existence. The group offers "The Truth" and claims an exclusive hold on that Truth. This generates a sense of mystery and a feeling of awe for the leaders and/or for the myths, symbols, and rituals of the group. These processes of de-emphasizing the individual, glorifying the group, and creating a sense of awe about the group's ideology are the foundation stones of moral commitment. Table 5.2 summarizes the types of commitment and the processes that enhance each type.

Table 5.2. Kanter's Commitment Theory

Type of Commitment	Processes That Enhance This Commitment
Instrumental commitment (Commitment to the *organization*)	1. Sacrifice 2. Investment
Affective commitment (Commitment to the *members*)	1. Renunciation 2. Communion
Moral commitment (Commitment to the *ideas* as spelled out by the leaders)	1. Mortification 2. Transcendence

Some of the mechanisms that contribute to commitment in sects and cults are antithetical to many mainline denominations. First, the doctrine of the "priesthood of all believers" in some religious communities is contrary to the "spiritual hierarchy" of certain other groups. Second, the emphasis on open-mindedness and tolerance toward others in most mainline denominations counters the sectarian emphasis that one's group has the whole truth, the only truth, and the exclusively held truth. The high boundaries of "us" and "them" have been falling away in the more moderate and liberal denominations as they have stressed cooperation and ecumenism. Third, and perhaps most important, clergy in mainline denominations have tended to emphasize the importance of self-esteem, thereby reversing the sectarian denigration of one's self. Liberal clergy, especially, see a positive self-image as essential to mental health and well-being and maintain that religions encouraging dependency are not allowing individuals to nurture their God-given creativity (Clinebell, 1965). People who respect themselves and their own ideas are more likely to engage in independent and critical thinking, but this does not enhance commitment to the group's ideology. Hence, liberal religious groups may have lower levels of commitment because they emphasize individuality and personal self-esteem.

Many different types of groups and organizations use one or more of the commitment mechanisms discussed by Kanter. The unique quality of nonconventional religious groups is that they must elicit and sustain a very high level of commitment, for they need many resources from their limited membership. The NRMs and sectarian groups that survive are those that use all or most of the commitment processes discussed here.

Research does find that affective commitment usually comes first, followed by instrumental commitment to the organization and then moral commitment to the ideas and beliefs of the group. The sequence is not absolute, but when Greeley (1972) said "instead of Americans belonging to churches because they believe . . . there may be a strong tendency for them to believe . . . because they belong" (p. 115), he was describing a typical scenario. Note that we are born into groups and feel belonging from infancy. No wonder that most people stay in the religious group of their birth and come to believe (accept) the meaning system that it affirms.

Commitment, then, can be commitment to the organization, to the group members with whom one has strong emotional ties, or to the ideology and the moral rules of the group. Often, commitment at one level will lead to commitment at another. Although it is possible for people to be committed at only one level (the old man who gives thousands of dollars to the church or temple but never attends), religious groups usually need commitment at multiple levels, for the subsystems that compose a religion are linked together in ways that mutually affect one another.

Critical Thinking: In your own religious tradition, which type of commitment do you think *usually* comes first as people become committed members? Why do you think so? Can you think of people in your congregation or people you know for whom the sequence is different than the one you think is typical?

RELIGION AND IDENTITY

Identities are a person's conceptions of who they are and where they belong in the world. Some of these are very individually indiosyncratic (e.g., I am a sensitive, new age guy), while others are connected to various group identities (e.g., I am an Asian American professor) (Postmes & Branscombe, 2010). Not surprisingly, given the belonging dimension of religion that we stress in this textbook, religious groups have historically been among the most important identity groups for people. A person's religious identity, then, can be understood (at least initially) as the religious group with which a person identifies (including behaviorally and in terms of attitudes).[4] For example, "I am a Methodist," "I am a member of First Baptist Church in Mocksville," or "I am a druid."

Most sociologists of religion, usually implicitly, have adopted this approach to identity, perhaps because it is particularly conducive to the closed-ended survey method of data collection which dominates the field. In these quantitative studies, religious identification or "affiliation" is typically measured using a two-step question. For example, since 1972, the National Opinion Research Center's General Social Survey (GSS) has been asking Americans the following questions: What is your religious preference? Is it Protestant, Catholic, Jewish, some other religion, or no religion? Those who respond "Protestant" are then asked this: What specific denomination is that, if any? From this we know, for example, that 51.1% of the American population in 2008 identified as Protestant, and 4.3% of Protestants identify as United Methodist.

In 1990, the American Religion Identification Survey (ARIS) began and took a radically different approach to religious identity. Rather than offering people predetermined response categories, ARIS simply asked, What is your religion, if any? Although interviewers did have a list of many different religions that they could mark as responses, they were specifically instructed not to read the list to respondents. As with other surveys, ARIS also had a second follow-up question: If "Christian" or "Protestant," "Which denomination is that? The first ARIS survey has a sample size of (an astounding) 113,723 cases (Kosmin & Lachman, 1993)[5]. The survey was replicated in 2001 with 50,281 respondents and again in 2008 with 54,461 respondents (Kosmin & Keysar, 2009). The benefit of having these very large sample sizes is that very small religious groups can be identified and analyzed. For example, in the 2001 ARIS, 33 respondents identified themselves as "druid" (Kosmin & Keysar, 2006). Table 5.3 summarizes the ARIS findings regarding religious identity in the United States in the nearly 20 years from 1990 to 2008.

Although there are many interesting developments represented in this table—for example, the rising number of Buddhists, Muslims, and other Eastern Religions due to immigration (see Chapter 15)—the most significant in terms of changing patterns of religious identification are as follows: (1) The proportion of the American population that self-identifies as Christian in one form or another declined from 86% in 1990 to 76% in 2008. (2) The proportion of the American population self-identifying as mainline Protestants (Methodist, Lutheran, Presbyterian, Episcopal, and UCC) declined by nearly 6% in the last 20 years. At the same time, there has been a 40-fold increase in the number identifying as nondenominational Christians, from an estimated 194,000 in 1990 to over 8,000,000 in 2008. This represents a nearly 3.5% increase in the proportion of Americans whose religious identity is nondenominational. (We discuss the former development at greater length in Chapters 6 and 8.) (3) The number of Americans who identify themselves as having no religious preference ("nones") continues to grow dramatically, as does their proportion of the American population. From 1990 to 2008, the proportion of the American population self-identifying as nones increased from 7.5% to 13.4%. Note that this explicitly excludes those who self-identify as agnostic or atheist. As we discuss in Chapter 6, about half of these estimated 34 million Americans are religious but choose not to identify themselves as having a religious affiliation—that is, their religious identity is not connected with a particular religious group identity.

---

[4]We hasten to add that this social identity theory inspired approach is just one of many approaches to the study of identity within sociology in general and even the sociology of religion in particular. For an alternative view, see Ammerman (2003).

[5]By comparison, the GSS cited just previously surveys 1,500 to 3,000 people.

Table 5.3 Religious Self-Identification of U.S. Adults, 1990 and 2008

	1990	2008	Change
Religious Identity	%	%	%
Catholic	26.2	25.1	−1.1
Baptist	19.3	15.8	−3.5
Nones/No Religion	8.2	15.0	+6.8
Christian Unspecified	4.6	7.4	+2.8
Methodist	8.0	5.0	−3.0
Lutheran	5.2	3.8	−1.4
Non-Denominational Christian	0.1	3.5	+3.1
Pentecostal Unspecified	1.8	2.4	+0.6
Protestant Unspecified	9.8	2.3	−7.5
Presbyterian	2.8	2.0	−0.7
Mormon/Latter Day Saints	1.4	1.4	—
Jewish	1.8	1.2	−0.6
Episcopalian/Anglican	1.7	1.1	−0.6
Evangelical/Born Again	0.3	0.9	+0.6
Eastern Religions	0.4	0.9	+0.5
Churches of Christ	1.0	0.8	−0.2
Jehovah's Witness	0.8	0.8	—
Muslim	0.3	0.6	+0.3
Buddhist	0.2	0.5	+0.3
Assemblies of God	0.4	0.4	—
United Church of Christ	0.2	0.3	+0.1
Church of God	0.4	0.3	−0.1
Agnostic and Atheist	0.7	1.6	+0.9

SOURCE: Kosmin, Barry A., & Keysar, Ariela. (2009, March). *Summary report. American Religion Identification Survey (2008).* Hartford, CT: Trinity College. Retrieved from www.americanreligionsurvey-aris.org/

These data can be read in light of our previous discussion of generational change. As older cohorts of Americans, who identified themselves according to traditional denominational labels, are replaced in the population by baby boomers, genXers, and millenials, who have more fluid religious identities, the population who identify as nones will likely continue to grow. In the spiritual marketplace, which has been developing for the past 50 years, religious consumers are increasingly shopping for religious identities that are not closely tied to historic religious groups (Roof, 1999). They are increasingly individualized in their tastes and preferences for religious identity (Cimino & Lattin, 2002).

These data can also be read in a larger historical context as reflecting the flowering of what Philip Hammond (1992) called the "third disestablishment" in American religion. The formal disestablishment of religion in the First Amendment to the Constitution of the United States encoded an open market for religion into the DNA of the country (Warner, 1993). That open market allowed for—and perhaps even promoted—considerable religious vigor in the early decades of the new republic, including revivalism and even the birth of new, distinctively American religions like Mormonism (Finke & Stark, 2005). Still, American culture remained steadfastly Protestant during this time. It was not for another century that cultural disestablishment of American (Protestant) religion took place, aided in particular by internal divisions between fundamentalists and modernists within Protestantism and by waves of Catholic immigrants who challenged Protestant hegemony (cultural dominance) in politics, education, and elsewhere. Hammond (1992) called this the "second disestablishment" of American religion (pp. 9–10). Consequently, Will Herberg (1955) famously argued in the mid-1950s that a moderate pluralism existed that allowed for Protestants, Catholics, and Jews to be seen as equal contributors to the "American Way of Life."

In the decades following WWII, the moderate denominational pluralism of Protestant–Catholic–Jew would give way to the seemingly unlimited diversity of religious options in present-day America. Religion as an integrative force and source of collective identity gave way to a more individualized approach to faith which centered on "'personal autonomy,' meaning both an enlarged arena of voluntary choice and an enhanced freedom from structural restraint" (Hammond, 1992, pp. 10–11). This is the "third disestablishment" of American religion (see also Roof & McKinney, 1987). It is driven by structural and cultural changes in American society. Increasing geographic mobility (especially suburbanization), social mobility (driven by rapidly expanding higher education), and familial mobility (rising rates of divorce and blended families) all acted to loosen the connection between place, family, and inherited faith—all of which had previously sustained a "collective-expressive" sense of the faith community or "spirituality of dwelling" (Hammond, 1992; Wuthnow, 1998).

Alongside and related to these structural changes was a profound cultural change that significantly increased the centrality of individualism and individual choice in religion. Religious identity went from being something "ascribed"—a characteristic someone is born with—to being "achieved"—something a person earns or chooses. This was the social and cultural environment (the "period") in which the baby boom generation came of age. The danger of this for traditional religious identification was obvious to scholars from the start. In their analysis of mainline religion in the United States in the 1980s, Wade Clark Roof and William McKinney (1987) concluded that "the enemy of church life in this country is not so much 'secularity' as 'do-it-yourself religiosity'" (p. 56).

For many Americans, religious authority lies in the individual rather than in the Bible, the historical tradition of the faith, or the hierarchy of clergy. Each person is expected—even required—to pick and choose what they believe to be true. This has two consequences. First, some religious leaders begin to "market religion" (explored in more detail in Chapter 8). A second consequence is that religion comes to be viewed as a matter of "opinion," which can be easily modified or discarded. The idea of religious conviction or commitment along the lines that Rosabeth Kanter described comes to be replaced with the more noncommittal notion of "religious preference." The salience of social identity connected to religious groups is diminished.

Reginald Bibby (1987a) suggested that the same process is very much at work in Canada. Canadians do not switch denominations very often but have begun to treat their religion as casually as a menu choice in a restaurant:

> Canadians are still eating in the restaurants [i.e., mainline denominations]. But their menu choices have changed. . . . [Many] are opting only for appetizers, salads, or desserts [funerals, weddings, and baptisms], rather than full course meals [of full commitment and weekly Sunday worship]. The minimum charge has been lifted— it is now possible to skip the entree page altogether. (pp. 133–134)

Bibby also believed that such privatized religion is very vulnerable because it lacks the sanction and support of a group. The ideas set forth by the community may have less plausibility—less staying power—without the context.

At the same time that Americans have nearly unlimited discretion to choose and an unprecedented number of religious identity options to choose from (including no religious identity), they frequently opt for *convention*. Indeed, the main chronicler of the spiritual quest of the baby boom generation, Wade Clark Roof (1999), documented that 59% of American baby boomers are "born-again Christians" or "mainstream believers" and only 14% are "metaphysical believers" or "spiritual seekers" (p. 178). Christian Smith has extensively documented this conventionality in his work on American religion. In his recent studies of American youth, for example, Smith (2005) found the following:

> the vast majority of American teenagers are *exceedingly conventional* in their religious identities and practices. Very few are restless, alienated, or rebellious; rather, the majority of U.S. teenagers seems

> basically content to follow the faith of their families with little questioning. (p. 120)

In explaining the vitality of American evangelicalism, Smith (1998) went so far as to argue that choice in fact is key to a stronger religious identity: "Moderns authenticate themselves through personal choice. Therefore, modern religious believers are capable of establishing stronger religious identities and commitments on the basis of individual choice than through ascription" (p. 104).

In terms of religious identity, we may be seeing a dividing up of the American population into two great camps: (1) those whose religious identities are more social and (2) those whose religious identities are very personal. Like the individual mentioned earlier whose self-identity is "the sensitive new age guy," more and more people may craft identities that do not exactly fit the established group categories. The baby boom generation's mantra, "I am spiritual not religious," may be foremost among these personal religious identities and appears to be spreading. How does this boomer influence echo through the culture into the outlook of the millennial generation? One answer to this question that is readily accessible to many readers of this text can be found on Facebook. A cursory scan of the expressed religious views on the profile pages of the junior author's millennial generation Facebook friends finds identities such as "I play dreidel for fun," "'love is my religion'—ziggy marley," and "everybody love everybody." Even those who seem to be expressing a conventional religious identity do it in an unconventional, or at least nongroup orientated, way: "follower of Jesus Christ," "he's beside me every step of the way," "Appropriating Grace through Existential Faith." Very few people use Facebook's option for expressing conventional, group-based religious identities like "Christian–Catholic" or "Christian–Baptist." If you have access to Facebook, try collecting your own systematic data on Facebook religious identity according to the guidelines in the next "Illustrating Sociological Concepts" feature.

ILLUSTRATING SOCIOLOGICAL CONCEPTS

Exploring Religious Identity on Facebook

Is religious identity becoming increasingly personalized? Although Facebook has nothing formally to do with religion, it does allow users to specify their "religious views." Looking at individuals' presentation of their religious views on Facebook is a way to explore this. In the spirit of sociology's methodological empiricism, systematically examine your Facebook friends' religious views, as follows (you obviously need to have a Facebook account to do this):

1. When you are looking at your Facebook profile, on the left-hand pane six of your "friends" appear. Click on the uppermost left friend.

2. If the person's religious views do not appear on the left-hand side of their profile, click on the "Info" tab to see them.

3. Write down the individual's religious views, or note with an *X* if the person does not list religious views (an interesting finding in itself).

4. Keep clicking and recording until you get 10 friends who list some religious views.

What does this exercise reveal about religious identity today (especially among young people)? To analyze what you have just observed, think about issues of personal vs. group identities, being "spiritual not religious," and the like. For example, how many friends did you have to look at to find 10 who even listed religious views? Did those who listed religious views tend to use "conventional" labels (e.g., "Roman Catholic," "LDS"), other theological characterizations ("follower of Jesus Christ"), or perhaps more unconventional, even silly, ones ("I play dreidel for fun")? Are your Facebook friends' religious identifications similar to or different from your own (if you even have one)?

The possibility of a growing gap between individuals' religious identities and religious group identities raises a related issue: the gap between the official teachings of religious groups and the nonofficial, more individualized understandings of "the people."

OFFICIAL AND NONOFFICIAL RELIGION

The conflict between Christian modernism and more conservative Christianity has frequently been a conflict with theology professors, professional clergy, and church bureaucrats on one side and with common folks on the other. Jeffrey Hadden (1969) maintained four decades ago that this conflict was inevitable because of an allegiance to a common symbol system by two groups of people—people who attributed entirely different meanings to the same symbol system. The two groups were the ecclesiastical officials and the laity of the various Christian denominations.

A number of scholars have made distinctions between the folk religion of tribal peoples and the world religions that attempt to be universalistic (Menschung, 1964). Folk religions preserve the local culture and customs, but world religions tend to evolve a complicated rational theology, a system of ethics based on that theology, a formal cultic ritual, and a professional clergy that elaborates the theology and ethics in a way that has universal (not culture-specific) appeal. However, world religions also tend to develop folk versions of the faith. The masses are seldom moved by complicated rationalized theologies, and a localized version of the meaning system evolves. Hence, most world religions have within them an official and a folk version of the faith that blends with local norms, rituals, and myths. Sometimes ethnic groups within the nation have their own unique interpretation of a faith. This is another form of folk religion, which may also diverge from the official form of that faith. *Official religion* is the orthodox faith as is presented by ecclesiastical officials, and it usually involves a more systematic theology and a more universalistic application of the faith.

Two central dimensions of the conflict between official and folk religion concern us here. The first has to do with the relative importance of rationality: official religion tends to develop a systematic theology. Theology professors, publications' editors, directors of boards of social concerns, and other clergy in the bureaucratic structure may struggle with the relationship between science and mythos. They may develop rather secularized theologies, or they may develop orthodox theologies that may require mental gymnastics to understand. In either case, the theology is founded on a principle of logical consistency and coherence. (The theology is to be consistent within itself even if it does not relate to everyday experiences and perplexities of laypeople.)

Sometimes the theological elite engages in the demythologizing or remythologizing of biblical stories—using modern metaphors involving the Internet, for example, rather than employing agrarian stories about sheep and shepherds.[6] All the while, the common layperson may not be troubled at all by the lack of logical congruence in the theology. Many people seem to be quite comfortable with a highly incoherent assortment of beliefs and practices. Furthermore, they may be quite satisfied with traditional myths—and find the demythologized or revised versions rather sterile. In fact, the laity of some faith communities are hostile to the idea of exegesis and biblical criticism. The rise in this century of snake-handling cults and other mystical beliefs among the laity may suggest that secularization is more a phenomenon of the clergy—the elite of trained religious professionals—than of common folks. Rationalized theology may be less capable of providing a powerful, pervasive, and long-lasting system of meaning for the average person. Greeley (1972) argued that conflict over the worldview of science and that of religion is not a concern of most Americans:

> To some extent American believers have been able to avoid the conflict between science and religion by simply denying that it exists. Whether this is intellectually honest or not may be questioned, but the point remains that it has been successful; religion and science can go on their merry ways, not conflicting with each other very much, despite the arguments of some elite religionists and elite scientists that they should. (p. 106)

The struggle for logical coherence and consistency in theology is only one aspect of the divergence between folk religion and official religion. A second and equally important factor is the desire on the part of the elite to make the faith relevant to all cultures and all peoples. Hence, the theology and ethical principles of the faith are articulated in such a way that the faith

---

[6]Demythologizing is reducing myths to logical principles or philosophical statements. Remythologizing is a process of recasting or reinterpreting myths in a way that is compatible with secular scientific thinking or contemporary realities. Berger insisted that the demythologizing strategy of dealing with secularization is theologically bankrupt and will ultimately lead to a denial of the existence of any supernatural realm (Berger, 1979).

does not appear culture-bound. The desire is to emphasize those principles of the faith that would have universal appeal. The values and attitudes that are specific to one particular culture are downplayed. The folk religion, on the other hand, involves a synthesis of the historic faith with local customs, values, beliefs, and traditions. The myths and symbols of the religion may be interpreted in such a way that they confirm and justify the local concepts of morality. The ethnocentric biases of the community may be so strong that this localized version of the religion may seem to be the only true understanding of the faith. Due to the modification of a religion so that it is compatible with a particular culture, some writers distinguish folk religion from "true religion" (Southwold, 1982). However, the terms *true Buddhism* or *true Christianity* involve unwarranted value judgments. Frequently, official religion is also a modified version of the worldview that was set forth by the founder of the faith. Each usually represents consistency with the original tenets in certain respects and deviation in other ways. Neither official nor folk religion is entirely static.

What we see in some forms of evangelical Christianity is powerful folk religion that blends Christianity and Americanism. It endorses the American way of life, the free enterprise system of economics, and middle-class American values and lifestyles as central to Christianity. It is not uncommon to have a pledge to the American flag during morning worship in these churches. Sometimes the leaders of this movement embrace American militarism and bless a war, a position that certain other Christian theologians find inconceivable. These critics claim that this is directly contrary to the historic position of the Christian church, which adamantly forbid participation in the military for the first four centuries of its existence (Bainton, 1960).

The posture of most official religions is that ethical positions are expected to emanate almost entirely from theological beliefs. To some extent, any religion must bend and adapt if it is to have wide appeal in a given culture. The values of the faith must be, at least in some basic respects, compatible with the values of the culture in which it hopes to have adherents. In this sense, all American Christianity is influenced by American values and culture. The official group of trained professionals who seek to articulate a universalistic Christianity is not exempt from this influence. Their adaptations tend to tilt toward principles of individualism and a notion of universal human rights.

Several scholars have treated the conflict between ecclesiastical officials and the laity as a struggle over power (Berger, 1981; McGuire, 2002). The trained professionals who work in the national headquarters of the denomination may exercise their authority and come to understand "the church" as being the national organization. The laity in the local community may comprehend "the church" as being the local congregation. The parish minister serves as a buffer between these two views of what comprises the faith community (Hargrove, 1979). The conflict is over the ultimate source of authority.

Critical Thinking: Who has the right to speak for a religious group? Who has the right to act on behalf of the entire denomination? How is conflict best resolved over authority to speak on behalf of the faith or for a particular denomination?

Meredith McGuire went a step further in pointing to the controls the professionals in the hierarchy establish to ensure the continuance of their power and privilege. (Refusal to ordain women is one such method of control.) She demonstrated that women have been systematically excluded from significant roles in the official religion and have frequently been required to wear head coverings and/or veils as a means of setting women apart and reminding them of their inferior status (but see our discussion of veiling in Chapter 11). On the other hand, women have often had important roles in the emergence of nonofficial religion. Hence, nonofficial religion has served an important function in allowing for religious leadership by very able women (Baer, 1984; McGuire, 2002). This raises an important point: Not all nonofficial religion is necessarily "folk religion" as Menschung defined it. Hence, we may define nonofficial religion as any "set of religious and quasi-religious beliefs and practices that is not accepted, recognized, or controlled by official religious groups" (McGuire, 2002, p. 113).

McGuire's insight is important; she insisted that there is more than one process that can lead to the emergence of nonofficial religious groups. For our purposes here, we can distinguish two. First, common laypeople may reject the emphasis on rationality and on abstract universalistic concepts. In place of this systematic, logical theology, they may affirm a localized version of the faith that incorporates many of the local attitudes, values, and customs. Second, nonofficial religion may arise through disenfranchised groups seeking to exert their own leadership skills and express their own religiosity. In this case, emergence of nonofficial religion is largely a result of exclusivity on the part of the official elite. McGuire (2002) also pointed out that official religion has traditionally affirmed masculine values and concepts and the substantial involvement of women in nonofficial religion represents a search for alternative expressions of spirituality.

This line of investigation suggests that more work is needed to understand fully the relationship between official religion and its various nonofficial forms. Although folk religion may be one type of modification of official religion, other nonofficial variations of Christianity include faith healing, spiritualism (attempts to communicate with the dead), astrology, and stichomancy (the method of receiving a divine message by randomly opening the Bible and pointing blindly to a passage). Hence, nonofficial religion often coexists with official religion. Many people who participate in nonofficial religion or who hold folk religious beliefs are also lay leaders in mainline denominations. In any case, the worldview of nonofficial religion and a given individual's experience of the faith is often quite different from the official religion and the personal faith journeys of the trained specialist. Increasingly, social scientists have sought to understand both the religious orientation of the elite and the religion of the common folk. Both forms are important parts of the religiosity of a society.

> *Critical Thinking:* From the discussion in this chapter, what insights have you gained about religion in the lives of individuals? What continuities do you see over time in individual religion, and what are the major differences?

SUMMARY

In this chapter, we have examined some of the dimensions of individual religiosity: belief, practice, commitment, and identity. We have also considered how religion is lived out over the life course, with attention to differences in generational experiences. As we analyze religion in the lives of individuals, we find remarkable diversity and complexity. Efforts by sociologists to understand the complex social configurations require simplification, and that always means risk of oversimplification. Still, some patterns of religious behavior and affiliation are clear.

We see this in our studies of religious socialization and of the role of faith over the life course. Religion is transmitted from one generation to the next via certain recognizable social patterns, even though there are variations on and exceptions to those relationships. Family is the main influence on religious socialization of children. Children exposed to active and devoted parents tend to continue that tradition and remain religious as young adults and beyond. Mature adulthood is usually a time of heightened religiosity as family formation events drive connections to religious values and institutions. At some points, however, the demands of adult life make religious responsibilities difficult to fulfill. Adults may leave their religious community with the possibility of a return when life becomes less demanding. They may also reengage with religious communities as a result of aging. For adults in late life, the benefits of religious engagement include social, emotional, and spiritual support.

Religious commitment is both a property of individual religiosity and an important factor in the survival and prosperity of religious organizations. Interestingly, the three types of commitment also relate to the three interrelated subsystems that make up a religion: (1) meaning, (2) belonging, and (3) institutional/ instrumental. In any case, successful religious commitment is related, in large part, to the formation of a particular religious identity. When one says "I am a Muslim" or "I am a Catholic," that says something about their loyalty to that religious organization or tradition. However, in modern society religious identity has become more personal and less group oriented, resulting

in multifaceted and idiosyncratic identities for some modern people. This is compounded by the fact that within any particular religious group there can be coexisting official and nonofficial versions of the group's beliefs. Who is to say what defines a Catholic? From the perspective of official religion, for example, the Pope is the final authority on Catholic theology, but an unofficial version of the faith may spawn new syntheses with other traditions or with national and ethnic loyalties.

We have focused on individuals becoming and being religious—how one grows into a faith perspective and how one comes to have particular religious commitments, practices, and networks. The next chapter also looks at religion in the lives of individuals but focuses on the processes of religious transformation or change, which we know as "conversion" and "switching."

6

CONVERSION AND SWITCHING

Here are some questions to ponder as you read this chapter:

- Are many of the new religious movements (NRMs) involved in *brainwashing*? (What *is* brainwashing, and how do we know whether some religious movements do it?)
- Is there any typical order or sequencing of contributing factors in bringing about intense religious commitment?
- Is a decision about religious commitment driven by the same forces as secular economic decisions: a cost–benefit analysis in which people seek that which is in their own self-interest?
- What social forces are at work today in people switching from one religious community or denomination to another?
- What is the story and the trend-line regarding religious "nones"—those with no religious preference?

In Chapter 5, we explored the process by which individuals are socialized into and practice religion throughout their lives. Yet most religions give considerable attention to the process of personal change or transformation. In this chapter, we investigate the change of worldviews that is expected and even anticipated by religious practitioners: the conversion process. We also explore the patterns of religious change in individuals' lives when they switch from one religious group to another ("switching") or from religion to no religion ("apostasy"). The increasing diversity and voluntarism in American religion discussed in the last chapter creates a context in which religious choices like conversion, switching, and apostasy are more likely. Although we focus our attention on the American context, a recent analysis of 40 countries finds that countries with more religious pluralism have higher overall levels of conversion (Barro, Hwang, & McCleary, 2010).

CONVERSION, BRAINWASHING, AND THE NEW RELIGIOUS MOVEMENTS

The term *conversion* refers to a process of "turning around" or changing direction in life. Specifically, it refers to a change of worldview. It often is viewed as a sudden crisis event, but the process can also be a gradual one. In any case, conversion represents a transformation in a person's identity or self-image. The change is often symbolized by a change of name (e.g., the Hebrew Saul became Paul when he converted to Christianity, and, in the 20th century, the boxer Cassius Clay became Muhammad Ali when he converted to Islam).

Part of the problem in understanding conversion in mainline groups is that the rhetoric of religious groups can be misleading. What is sometimes called a conversion is often a ritualized reaffirmation of a worldview that is already held. Many adolescent "converts" at revivals and crusades do not really change their worldviews. They merely go through an experience by which they publicly affirm the faith in which they have been socialized since infancy. This is a significant experience, but it is not necessarily conversion. In fact, the evangelist Billy Graham defined conversion as a "change in the direction of one's life to a totally new direction" (quoted by Wimberley, Hood, Lipsey, Clelland, & Hay, 1975, p. 162). According to the research of Ronald Wimberley and his colleagues, most of those who respond to the crusades are not really converts by Graham's definition. What they really experience is an *intensification* of their faith.

> **Critical Thinking:** What is the meaning of the term *conversion*? Does it make sense to say that at many evangelical revival meetings people experience an intensification of faith rather than a conversion? Why or why not?

Research on conversion in the sociology of religion flourished from the 1960s through the 1980s, coinciding with the rise in prominence of nonconventional religious groups known as "New Religious Movements" (see Chapter 7). Perhaps this is to be expected. Conversion to a group that affirms the basic values of the dominant society does not seem as mysterious and puzzling to most people as conversion to a cult.[1] People who convert from Methodism to Catholicism have not engaged in terribly unconventional or "abnormal" behavior. The individual who departs from his or

---

[1]The term *cult* will be treated in more detail in Chapter 7. For our purposes in this chapter, we define a cult as the nascent organization of a new religion. It involves a break from the traditional religions in a society and the creation of a nonconventional form of religion (Stark & Bainbridge, 1979).

her Episcopal heritage to join a congregation of Reform Jews is somewhat more of a curiosity because the change involves adherence to a different religion and an acceptance of minority group status. However, the person who gives up all of his or her possessions to join a religious commune is even more of an enigma to the average American. Indeed, this behavior can be terribly puzzling and somewhat frightening to many middle-class Americans. The charge of brainwashing has sometimes been a ready explanation for this otherwise inexplicable behavior.

The public is often attracted to the brainwashing thesis, but most people have little idea what brainwashing actually is. The term has a specific meaning in the language of the social psychologist. Brainwashing is usually used to refer to a process by which persons are *involuntarily* caused to adopt a belief system, a set of behaviors, or a worldview. To force a person to make such a change, one must have total physical control over the individual. The captors must control all the necessities of life and be able to control life and death itself. The captive must be in a circumstance in which no alternatives and no other choices seem available. Even in such a total control situation, only small numbers of American soldiers yielded to the brainwashing techniques of the North Koreans in the 1950s. Furthermore, most of the acquiescence to the North Korean and Chinese torture procedures were merely verbal. The Americans were eager to return to their previous culture as soon as the total control situation was alleviated. Only about a dozen men were permanently converted by North Korean thought reform out of thousands that experienced the severe treatment (Farber, Harlow, & West, 1951). New religious movements (NRMs)—sometimes called "cults"—have much movement in and out of membership and clearly do not hold the same sort of physical control over members' lives that prisoners of war experienced in the North Korean or the Chinese camps in the mid-20th century.[2]

When most Americans use the word *brainwashing,* they have in mind some form of hypnotic trance or mysterious mind control. The implication is that the new religious groups manipulate the minds of potential recruits so that the latter are unwitting and somewhat passive victims of the process. The actual studies of conversion and commitment suggest otherwise. For example, Roger Straus (1979) insisted that the recruit is usually actively involved in choosing to be converted.

> The act of conversion, we find, is not a terminal act. Rather, guided by the principle that the way to be changed is to act changed, the new convert works to make conversion behaviorally and experientially real to self and others . . . It is not so much the initial action that enables the convert to experience a transformed life but the day to day actions of living it. (p. 163)

Rather than a passive victim, researchers find that recruits are active seekers who want the conversion experience and go to considerable effort to cause it to happen (Balch 1980; Judah, 1974; Staples & Mauss, 1987; Straus 1976, 1979). In short, the "new religions" have not induced people into some sort of hypnotic trance (Barker, 1984; Batson, Schoenrade, & Ventis, 1993; Beckford, 1985; Bromley & Shupe, 1981; Levine, 1984; Stark & Bainbridge, 1985). Thirty-nine members of the Heaven's Gate movement committed suicide in 1997 as a way to be transported via a flying saucer to the "next level," but recent studies conclude that while some odd group dynamics were at work and some strange beliefs were held by the members, "brainwashing" was by no means part of the process (Balch & Taylor, 2002).

---

[2]There is some evidence of a near total control environment, with inducement of utter exhaustion and threats to the lives of individuals and their families, in Jonestown (Moberg, 1980). Hence, there may be a few isolated cases in which converts experience something similar to Korean brainwashing procedures. This seems to be the exception, however.

Why, then, has there been so much talk of brainwashing, or "mind control"? Essentially, the conflict is over resources (Bromley & Shupe, 1981). NRMs are recruiting members who will expend their time and energy on behalf of the new religion. The conventional religious organizations lose those resources. Further, many of the new religions demand such total commitment that recruits have little time or energy to devote to their families. In fact, the religious group comes to be an alternative family unit, with emotional commitments to the group replacing family ties.

Thomas Robbins and Dick Anthony (1978) pointed out that the brainwashing term has been used as a weapon to suppress these nonconventional groups. Anson Shupe and David Bromley (1978) even insisted that the anticult movements have distorted information to the point that the attacks are actually very akin to the witch-hunts of an earlier era. The same might be said regarding the sensationalism surrounding so-called "Satanic cults" (Bromley & Melton, 2002; Richardson, Best, & Bromley, 1991).

Stigmatization of new and growing religious movements by established forces are common throughout history. When Christianity was still an NRM, pagans claimed that the Christian movement was a dangerous one because Christians worshipped an ass's head, murdered children as sacrifices, and committed other atrocities (Baroja, 1964, p. 41). As recently as 60 years ago in the United States, Roman Catholics were characterized as subversives who committed all sorts of abominations (Bromley & Shupe, 1981). The charge of brainwashing is primarily a way of discrediting NRMs and making them appear illegitimate and dangerous.

Stigmatizing unconventional groups makes them appear more dissimilar from conventional religions than they really are. As we will see in

this chapter, the processes by which individuals convert to NRMs are fundamentally social and do not involve any forms of recruitment or socialization that we would not also expect to find among more mainstream religious groups. Moreover, the conception of large numbers of NRMs in the latter part of the 20th century created a unique opportunity to study conversion and recruitment at the critical period when a religious group is just getting started.

PROCESS MODELS OF CONVERSION

For the first six decades of the 20th century, conversion was viewed by social scientists as a single event that brought radical change in the orientation of an individual. More recent analyses have tried to identify a process or series of steps involved in conversion (Richardson, 1985). In the first real process model—based on a study of conversion among the Moonies—John Lofland (1977) identified a sequence of factors that operate to move a person from nonmember status to committed devotee.[3] Lofland's theory is based on the "value-added" model of Neil Smelser (1962), which maintains that a social movement or process may best be understood by identifying "successively accumulating factors." According to this perspective, it is the cumulative effect of many different factors that gives rise to specific types of behavior.

According to Lofland, only persons who experience all of these conditions are likely to convert to a new religious group. Hence, these conditions help determine who will convert and who will remain uninvolved. "The sequential arrangement of the conditions may be conceived as a funnel"; that is, they "systematically reduce the number of persons who can be considered available for recruitment" (Lofland, 1977, p. 31). A large

---

[3]Sometimes this model is referred to as the Lofland–Stark model because the ideas were first set forth in skeletal form in a jointly authored article with Rodney Stark (Lofland & Stark, 1965). Lofland's more elaborated version appeared the following year in book form. Several other sequential theories of conversion have also been set forth by social scientists, but most follow a schema quite similar to Lofland's (Downton, 1980).

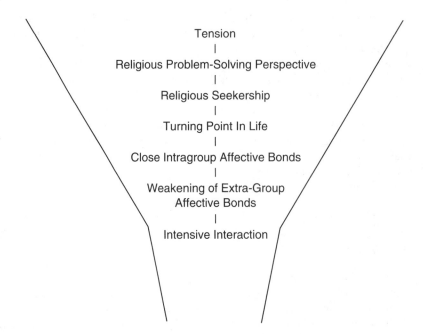

Tension
|
Religious Problem-Solving Perspective
|
Religious Seekership
|
Turning Point In Life
|
Close Intragroup Affective Bonds
|
Weakening of Extra-Group
Affective Bonds
|
Intensive Interaction

One might view Lofland's model as a "filtering" of people. Each step involves a filtering out of some people and a filtering or even funneling in of others. Those who have experienced the first six filters become "verbal converts." They must go through the seventh step as well to become "total converts." Total converts really believe the theology or ideology of the group. Verbal converts feel committed to the members of the group, and they verbally assent to belief, but they are not yet really committed at a "moral level."

Figure 6.1   Lofland's Process Model of Conversion

portion of the population may experience stress or tension, the first condition. A somewhat smaller portion of the population may adopt a religious problem-solving perspective, the second condition. The third condition may apply only to a segment of *that* population, and only a few who experience the first three conditions may confront the fourth. Hence, the number of converts, and who those converts are, is largely a function of the proper combination of events in the lives of individuals (see Figure 6.1). Lofland maintained that there are three *predisposing conditions* (dispositions, attitudes, outlooks) and four situational factors that must be present for the individual to be susceptible to influence by a new and innovative religious group. The predisposing conditions

are attributes of individuals that exist *prior* to contact with the religious group.

Predisposing Conditions

Tension

It is a commonly accepted maxim of the social sciences that personal change is generally the result of some felt need for change—a dissatisfaction with the current situation. In societies such as our own that undergo a great deal of change, persons often experience anomie or normlessness. They feel that they are without roots. This anomie creates a great deal of stress in persons. Anomie is, of course, only one source of stress that can cause discontent. Further, while Lofland viewed tension or dissatisfaction as necessary, it is certainly not sufficient by itself to cause a religious conversion. Dissatisfaction may be worked out in a psychiatrist's office, in a political campaign designed to change social conditions, or in any one of a dozen different ways.

Religious Problem-Solving Perspective

The second personal characteristic is an inclination to solve problems by turning to religious leaders or methods rather than to political or psychiatric ones. Only when a person begins to attach religious or spiritual meaning to events does the individual become amenable to the message of religious groups. Otherwise, one will seek political, psychiatric, or other resolutions to problems. James Downton (1979) maintained that young people were not favorably inclined toward the Divine Light Mission and the Guru Maharaji until they first sensed that religion was a viable problem-solving route. Hence, Lofland posited that a spiritual outlook is necessary before a potential recruit will take seriously the message of a religious group.

Religious Seekership

Lofland insisted that among the Moonie converts there was a dissatisfaction with the conventional religious groups and a feeling of being a religious seeker. Virtually all converts had drifted from one religious group to another. They had already concluded that the worldview of the religion in which they were raised was inadequate. The new converts had identified themselves as seekers of truth. However, not all seekers were converted. There must be a certain amount of "cultural conduciveness" between the worldview of the group and that of the recruit. To illustrate, in the 19th century, the Ghost Dance religion spread rapidly among Native American nations. However, it failed to attract the Navajo. The Ghost Dance religion predicted the rise of the dead, but the Navajo were afraid of ghosts. Hence, rather than celebrating such a future event, they refused to have anything to do with

this movement. Likewise, Lofland suggested that for a person to be a likely candidate for Moonie recruitment, he or she must already have held some attitudes that are consistent with Moonie theology. Uniform congruence of beliefs is not required, but a general compatibility of outlook seemed to be necessary for a potential recruit to become an actual convert.

Situational Contingencies

If these predisposing conditions are operative, the *situational contingencies* become relevant. Situational contingencies are circumstances that influence the social interaction between a potential convert and a recruiter to a religious group.

Turning Point in Life

The converts to the Unification Church had reached important turning points in their own lives. Lofland (1977) put it this way:

> Each had come to a moment when old lines of action were complete, had failed, or . . . were about to be disrupted, and when they were faced with the opportunity or necessity for doing something different with their lives. (p. 50)

Potential converts had recently migrated, had lost or quit a job, or had completed or dropped out of college. On the other hand, Lofland found that marital dissolution and illness seldom served this turning-point function. Apparently, some types of turning points stimulate the meaning and belonging needs more profoundly than do others. Nonetheless, persons who are not at turning points in their lives are less likely to be responsive to proselytization—especially to proselytization of unconventional groups (Moonies, Hare Krishna, etc.). Turning points are times of new beginnings when investment in the status quo is minimal.

Close Intragroup Affective Bonds

In Lofland's study of the Moonies, almost all recruits were gained through preexisting friendship networks. The Moonies learned that cognitive appeals to ideology (with a focus on the moral commitment level) did not win converts. As the Moonies became aware of this, they gradually began to modify their recruitment strategy. Lofland (1977) emphasized that they "learned to start conversion at the emotional rather than the cognitive level" (p. 308). In other words, they learned that affective levels of commitment precede moral levels of commitment.

Weakening of Extra-Group Affective Bonds

Because of migration away from one's family or because of disaffection from it, an individual may experience isolation, alienation, or loneliness. Such a person is more in need of emotional support as he or she faces a turning point than is one who has a close network of friends and family. As we have mentioned previously, Moonies consciously looked for signs of transiency (backpacks, individuals who were alone). The person who has no other immediate reference group is less likely to have someone oppose or intervene in the conversion process. Without another individual or group to offer an alternative interpretation of the group, the individual is more likely to be drawn into that group. Furthermore, without another significant reference group, the process of renouncing "outsiders" is much less complicated.

Intensive Interaction

Lofland maintained that some Moonies were only verbally committed—that is, they were not yet totally committed. This seventh condition solidifies commitment to the group through intensive interaction among members. In this way, "moral-level commitment" is enhanced by providing "communion" with the group. Once the intensive interaction provides a sense of unity and oneness, the devotees actively *want* to believe the ideology. By consciously working on

strengthening their faith, they come to believe the ideology, to feel a sense of awe about the leader, and to uphold the group's values. The conversion becomes total, and engagement becomes intense in all three modes of commitment.

Even members of some of the recent groups who have gone to their deaths for the sake of the group, an act that represents extreme commitment, seem to have been motivated by these processes. Recent studies of Heaven's Gate, the California group that involved the suicide of 39 members, and the Branch Davidians, who went to fiery deaths in Waco, Texas, indicate that these same processes of intense bonding were at the core of the movements (Balch & Taylor, 2002; Hall, 2002).

Critique and Evolution of Lofland's Process Model

Several scholars have voiced reservations about the applicability of this particular "process model" to all groups. One study of an American Buddhist movement indicated that several of the stages may not be universal even among NRMs (Snow & Phillips, 1980). First, there is no hard evidence that personal tension is necessarily higher for converts than for the general population. Tension is reportedly high among the entire population. Further, much of the reported tension of religious converts may be a retrospective interpretation based on their new ideology—reading into an earlier situation something that was not felt to be present or important at that time. This can be seen in the comments of converts reported by Snow and Phillips in the "Doing Research on Religion" feature. It is also possible that religious conversion actually *creates* the increased tension.[4] For example, belief that the world will soon end or that all nonmembers are evil—including one's family members who refuse to join—may cause great stress.

---

[4]We discussed in Chapter 3 the view of A. R. Radcliffe-Brown that religion sometimes creates tension and anxiety in order to provide social solidarity.

DOING RESEARCH ON RELIGION

Tensions and Personal Problems as Factors Leading to Conversion

Interviews with people who had converted to new religious movements resulted in some of the following kinds of responses:

* * * * *

Male, Caucasian, single, under 30:
When I joined I didn't think I was burdened by any problems, but as I discovered, I just wasn't aware of them until I joined and they were solved.

* * * * *

Female, Caucasian, single, under 30:
After I attended these meetings and began chanting, I really began to see that my personal life was a mess.

* * * * *

Male, Caucasian, single, under 30:
Now as I look back I feel that I was a total loser. At that time, however, I thought I was pretty cool. But after chanting for a while, I found out that my life was just a dead thing. The more I chanted, the more clearly I came to see myself and the more I realized just how many problems I had had.

* * * * *

Male, Caucasian, married, over 30:
After you chant for a while you'll look back and say, "Gee, I was sure a rotten, unhappy person." I know I thought I was a saint before I chanted, but shortly after I discovered what a rotten person I was and how many problems I had.

Conducting research on religious groups often calls for extremely careful interpretation of the data. The reports of new religious movement members cited here would seem to support the thesis that tension and unhappiness are factors that contribute to cult conversion. However, these reports may represent substantial revisions or reinterpretations of pre-cult experiences. This involves assignment of motives and use of an interpretive schema that was learned from the cult. Hence, the perception of prior strains and problems may be a *result* of conversion rather than a cause of it. The statements quoted here are generally typical of those made by devotees to cult researchers. However, these statements were chosen because their wording illustrates the retrospective nature of the analysis: "as I discovered, . . . I wasn't aware of them until . . ."; "after I attended these meetings . . . I really began to see"; and "as I looked back. . . ."

SOURCE: Snow, David A., & Phillips, Cynthia L. (1980, April). The Lofland-Stark Conversion Model: A critical reassessment. *Social Problems*, 430–447.

Having a religious problem-solving perspective was not a prerequisite to persons becoming members of some religious communities. Likewise, the question about whether one was at a turning point in one's life at the time one encountered a religious group is a matter of interpretation. Snow and Phillips (1980) wrote the following:

> Whether a particular situation or point in one's life constitutes a turning point is not a given, but is largely a matter of definition and attitude. There are few, if any, consistently reliable benchmarks for ascertaining when or whether one is at a turning point in one's life. As a consequence, just about any moment could be defined as a turning point . . . We again face the problem of retrospective reporting. (p. 439)

The importance of weak extra-group affective bonds may depend on the extent to which the group's values represent a break from the values and perspectives of one's family. If group membership does not involve a transformation of the values one shares with one's family, then alienation from family is unnecessary. If the group is not stigmatized, then the strategy of alienating members from nonmembers is less important.

Several empirical studies have found that while many of the factors identified by Lofland were present, they seemed to vary independently. They did not appear to be cumulative in all cases as is suggested by the funnel concept (Kox, Meeus, & Harm't, 1991). Still, most other researchers have found affective bonds and intensive interaction to be central to conversion (Cornwall, 1987; Greil & Rudy, 1984; Kox et al., 1991; Roof & McKinney, 1987; Stark & Bainbridge, 1985), and the *lack* of intensive interaction has been linked to deconversion or apostasy (Jacobs, 1987). While the idea of a sequential cumulative process seems to be a fruitful way to think about conversion, with

various factors contributing, at this point there is no consensus about which factors might be *necessary* ones or what precise sequence might be most typical.

Actually, Lofland and a colleague have offered a modification of his original theory. They now suggest that there appear to be several conversion processes and that different types of groups tend to employ different processes. Lofland and Norman Skonovd (1981) discussed six different conversion motifs, each involving a slightly different series of factors. Some conversions are induced by nonrational experiences, others involve a more intellectual process of study, and still others stress affective ties and belonging functions. Lofland and Skonovd also conjectured that different modes of conversion may be more common in different epochs of history. Hence, the heavy emphasis on nonrational experience by William James and Rudolf Otto early in the 20th century may have been due to the fact that mystical modes of conversion were more common then, while affective modes have been more widely employed by groups in the 1960s, 1970s, and 1980s.[5] It is interesting to note, however, that of the six conversion motifs outlined by Lofland and Skonovd, four emphasize belonging and group participation *prior to* belief. Intense affective involvement normally precedes total conversion. In the case of contemporary sects and cults, this sequence seems to be nearly universal.

RELIGIOUS CHOICES AND COMMITMENTS: A RATIONAL CHOICE MODEL

In the past two decades, a new paradigm has emerged in the sociology of religion: rational choice theory. Rational choice theory in much of sociology evolved from exchange theory, a long-standing micro theory that focuses on individual

---

[5]A few scholars think that what has changed has not been the types of conversion, but the perspectives and paradigms through which scholars have viewed conversions in different eras. See, for example, Kilbourne and Richardson (1989).

decision making and interaction between individuals and within small groups. However, rational choice theory in the sociology of religion has drawn heavily on macro and micro economic theories as well. We will begin by looking at the early critique and contribution of rational choice theory to analysis of conversion, then turn to broader issues of commitment.

The Convert and Active Choice

Historically, many social scientists have explained conversion using a rather passive model of human behavior. Conversion was an event that happened to the individual because of unconscious psychological processes or compelling social tensions (Richardson, 1985). Certainly, the "mind control" hypothesis is consistent with this perspective. The process models of Lofland and others represent a break from this determinism. A series of events are believed to be at work—events in which the participant has some choice and makes some decisions. However, several researchers believe that even these models have too much of a passive view of the convert, that these models depict conversion as the result of various social pressures *outside* of the individual.

Several social scientists have set forth a view of conversion in which the individual is an active agent purposefully making choices and seeking conversion (Balch, 1980; Balch & Taylor 1976, 1977; Dawson, 1990; Finke & Stark, 1992; Kilbourne & Richardson, 1989; Richardson, 1985; Straus 1976, 1979; Warner, 1993). This activist perspective stresses that individuals are seeking meaning in life, and they consciously join groups that they believe may fulfill their needs. To give the faith a fair chance, they thrust themselves into the roles and behaviors of the group.

One major version of this perspective focuses on role theory. As people play the role of convert, sometimes they find the role rewarding. They make an investment in the group, they gain certain ego gratifications for playing the role well, and they may come to believe in the ideas that justify and explain those roles. In essence, the new recruits convert themselves. However, some who join do not find rewarding roles or find that the faith does not meet their meaning needs, and they drop out. Still others may find the roles fulfilling for a while, then find that as role partners change and the organization evolves, the roles become unsatisfying. These people then fall away from the group.

Critical Thinking: In what ways does the "activist" model differ from other conceptions of conversion? Do you think people always make deliberate and conscious choices, or do social factors often incline people toward identification with a certain group or toward a particular outlook on life and death?

Religious Choices: Costs and Benefits in a Supply and Demand "Market"

The rational choice model stresses economic principles of behavior. The basic idea is that the kind of economic decision making that people use in all sorts of choices is also at work with religious choices: What will the benefit be, and what will it cost me? Do the benefits outweigh the costs? The benefits, of course, are nonmaterial when it comes to religious choices—sense of meaning, assurance of an afterlife, sense of communion with God, and so forth. This approach views religiously engaged people as consumers who are out to meet their needs or obtain a "product"; it depicts churches, mosques, and temples as entrepreneurial establishments trying to compete in a supply and demand market. Converts are thus regarded as active and rational agents pursuing self-interests, and growing religious communities are those that meet "consumer demand" (Finke & Stark, 1992; Iannaccone, 1994). In one sense, we might acknowledge that rational choice theorists focus on instrumental commitment as most essential, for they see all religious decision making as governed by cost/benefit calculations, either fully conscious or semiconscious in character. However, these scholars apply this perspective to congregations and denominations as well as to the individual convert.

Like other commodities, religious "merchandise" is produced, chosen, and enjoyed by consumers. In the weighing of benefits and costs, many benefits are supernatural compensators for things that cannot be satisfied by other economic processes. The advantage of supernatural compensators—promise of an afterlife, hope for rewards in this life through extraordinary interventions by nonempirical forces and beings, sense of meaning—is that these hoped-for benefits cannot be disproved with empirical methods. Yet if there is uncertainty about the benefits, there may be a lessening of the worth of the compensators. Thus, groups with very high commitment and greater certainty about their claims to Truth would offer more benefits. Such groups would be expected to grow more quickly and to lose fewer members.

There are essentially two major approaches to rational choice theory, especially at the level that stresses macro analysis: supply-side and demand-side explanations (Sherkat & Ellison, 2001). *Supply-side* theories stress the way in which religious communities produce religious "commodities" (rituals, meaning systems, sense of belonging, symbols, and so forth) to meet the "demand." They believe that preferences for religious "goods" remain fairly stable, so the interesting issue is how firms (denominations or macro religious organizations) operate through their franchises (local congregations) led by local "entrepreneurs" (ministers) (Finke, 1997; Stark & Iannaccone, 1994; Stark et al., 1995). Firms generally develop a specific kind of "product"—a specific type of theology and a particular style of worship, for example—and these "products" usually appeal to people who have a specific socioeconomic status, ethnicity, geographic identity, or other traits. If one grows up in a tradition with a specific style of music or with particular interpretations of the meaning of life, death, and spirituality, that person will be drawn to some "firms" over others. If an individual has "religious human capital"—that is, knowledge and familiarity with hymns, myths, memorized creeds, and insider language—then she is in a position both to produce

and to benefit from collective actions that foster this outlook (Iannaccone, 1990). Since demand for religion is believed to be stable, it is the production of religious communities that interests these theorists. Moreover, if people do switch "firms," it is likely to be to other "companies" that are quite similar, so that their capital is still relevant and useful (moving from Methodist to Presbyterian, for example). In short, people are committed because they benefit by the production of a religious enterprise. They use their own resources to collectively meet their own needs.

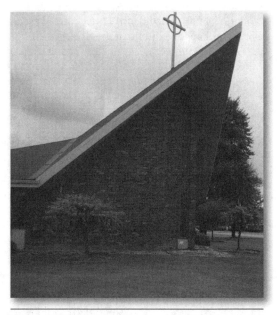

Rational choice theorists would call this local church a "franchise" that is part of a larger "firm" that produces a particular "product"; in this case, the "firm" is the Evangelical Lutheran Church in America.

Demand-side theorists argue that prior socialization and early experiences with religion help to create a need—a demand—for religious "products." However, the "consumers" remain open to various styles and types of "products." Preferences are shifting and variable. Thus, it is innovators—aggressive entrepreneurs—who reap the benefits of large constituencies and loyal customers. The focus is on organizations (both local congregations and national denominations) as competitive enterprises, and it is these organizations that must make the investments worth the effort, time, and resources. Supply is stable in that there are many religious entrepreneurs out there seeking to meet the need; the challenge is for the various firms and franchises to beat the competition by meeting the demand of the current market place (Finke & Stark, 1992; Sherkat, 1997; Stark & Bainbridge, 1996). In other words, commitment is variable, and congregations and denominations must earn it each week and each year or lose their place in the marketplace. Commitment becomes the issue: how to elicit it and how to sustain it.

The rational choice theorists insist that when there are more choices in the marketplace, activity is stimulated. Similarly, when there are more religious groups competing for the hearts and minds of members, concern about spiritual matters is invigorated, and commitment is heightened. Religious pluralism and spiritual diversity should increase rates of conversion as each group seeks its "market share" and as more types of individual needs are met in the society (Finke & Stark, 1992; Iannaccone 1994, 1995; Stark & Iannaccone, 1994; Stark et al., 1995; Stark & Bainbridge, 1996; Warner, 1993). When there is a religious monopoly—that is, less diversity of religious options in the society because an established religious organization is supported with tax dollars—religion may become less vigorous. When there is a state-supported religion, the clergy do not have to "sell" their product in order to earn a living. Religious competition, in this view, creates vigor. The spiritual "business climate" of the society influences the religious marketplace and the level of participation in faith communities.

One can readily see that this perspective is quite compatible with the open systems model we have used in this text. The outputs of the religion can influence the society. The "supply" of religious options or "products" can influence how highly religious a society becomes. On the other hand, characteristics of the society, such as how pluralistic the society is and how competitive the faith marketplace is within the society can shape the nature of religious "firms" or the behavior of ministers and local congregations.

Interestingly, rational choice models began largely because theorists felt that religious people and religious commitments were not being taken seriously. At least one rational choice scholar has taken other sociologists of religion to task for not treating religious people and their decision making with respect; religious behavior cannot be reduced to materialistic concerns. Decisions of religious people should be respected on their own grounds (Stark, 2000a). Yet this strictly utilitarian/economic analysis of religion seems to be counter to the way most religious people understand their own behavior. Many would argue that analysis of a *covenantal relationship* with a purely *contractual* set of concepts confuses very different kinds of relationships (Bromley, 1991, 1997; Bromley & Busching, 1988). Covenantal relationships stress commitment to one another, willingness to sacrifice one's own interests for the sake of others or for a higher principle, and deeply personal caring. The key to social interaction is bonding with others. The family and religious groups are the best examples. Contractual relationships are characterized by utilitarian relationships, people seeking their interest or advantage, and exchange-based impersonal calculations. The key to interaction is negotiation as people pursue their own self-interests. Governmental, bureaucratic, and managerial relationships are examples (Bromley, 1991; Bromley & Busching, 1988). Many economists and exchange theorists would argue that there simply are no altruistic or self-sacrificing human behaviors; they regard the distinction between covenantal and contractual relationships as illusory. However, religious people themselves often view their commitments as covenantal; the attribution of ultimate self-concern to their behavior is not an interpretation that is perceived as particularly sympathetic to their view of themselves or their behavior. As religious scholar Meredith McGuire (2002) put it,

It is one thing to describe a person's decision to join a church *as if* the person were trying to maximize his or her benefits; it is something different to claim that the person's *actual* thought in joining measured his or her potential gains against potential losses. It is impossible to demonstrate that people actually think this way (and few people would admit to doing so). Empirical motivations are much more complex than the rational-choice approach allows. (pp. 298–299)

Rational choice theory is very controversial in the field right now (Bruce, 1999; Hamilton, 2009; Lechner, 2007; Lehman, 2010; Spickard, 1998; Young, 1997). Some of those controversies will be explored later in this text. However, it has clearly generated a new approach and an active research agenda as scholars seek to explore religious commitment as a form of consciously made choices based on costs and benefits. Unlike many theories, it also makes contributions at both the micro and macro levels of analysis.

> **Critical Thinking:** In what way does the rational choice approach seem to you to make sense of religious behavior? If you find this approach appealing, how do you respond to the charge that it shows disrespect for the self-definitions of religiously motivated people? If you find it unconvincing, what reasoning or what evidence seems to you compelling in rejecting this approach?

The activist perspective that rational choice models offer need not be entirely incompatible with the reference group models discussed previously; it simply serves as a corrective. It is a mistake to view the convert as a passive participant, who is the unwitting "victim" of social processes beyond his or her control (Dawson, 1990). The recruit is a participant in the definition of his or her situation. Most sociologists do not view humans as robots who are totally controlled by outside stimuli. Persons are active agents who help shape their environments. However, it is also true that reference groups are

extremely powerful forces in any person's interpretation of experience. The individual actively chooses his or her reference group, but the reference group in turn helps the individual define norms and make sense of experience. If theorists lose sight of either side of this dual character of humans, their theories will be simplistic and distorted. When more choices are available, it is probably true that people feel an increased commitment to the choices they have made.

"SWITCHING" AMONG DENOMINATIONS

Changes in religious affiliation have become fairly common in the United States. According to the Pew Research Center Forum on Religion & Public Life (2008b),

> More than one-quarter of American adults (28%) have left the faith in which they were raised in favor of another religion–or no religion at all. If change in affiliation from one type of Protestantism to another is included, 44% of adults have either switched religious affiliation, moved from being unaffiliated with any religion to being affiliated with a particular faith, or dropped any connection to a specific religious tradition altogether. (p. 5)

As expected by rational choice theorists, the American religious marketplace is very open and therefore very dynamic. Some scholars call this change in religious affiliation "everyday conversion" or "mundane conversion." Although the terms are sometimes used interchangeably, denominational switching does not necessarily involve a conversion. It may not necessarily involve any change of worldview, but it does involve a change in organizational membership. Hence, the term *reaffiliation* is more accurate than conversion.

The Pew Forum's U.S. Religious Landscape Survey, based on responses from a representative sample of 35,000 Americans, charts these patterns of religious switching. Every religious group has gained some members and lost some members due to religious switching. Table 6.1 shows the percentage of the U.S. adult population that reports growing up in a particular denomination and the percentage currently a member of those same denominations. The table shows that the biggest "winners" in the American switching game are nondenominational Protestants and the religiously unaffiliated (more on them later). The biggest "losers" are Catholics, Baptists, Methodists, and Presbyterians. Many groups, however, essentially break even, with roughly equal numbers of individuals entering and leaving the religion.

Table 6.1   Changes in Religious Affiliation from Childhood to Present Among Adults in the United States

Religious Group (Listed in Order of Increase to Decrease as Percentage of Population)	*Childhood Family Religious Affiliation (percentage)*	*Current Religious Affiliation (percentage)*	*Total Change (percentage)*
Unaffiliated ("nones")	6.6	12.1	+5.5
Nondenominational Christian	1.5	4.5	+3.0
Agnostic	<0.3	2.4	+2.1
Protestant (nonspecific)	3.4	4.9	+1.5
Lutheran	5.5	4.6	+1.1
Atheist	0.5	1.6	+1.1
Other faiths	0.3	1.2	+0.9
Pentecostal	3.9	4.4	+0.5
Holiness	0.8	1.2	+0.4
Buddhist	0.4	0.7	+0.3
Jehovah's Witness	0.6	0.7	+0.1
Adventist	0.4	0.5	+0.1
Muslim	0.3	0.4	+0.1
Congregationalist (United Church of Christ)	0.8	0.8	0.0
Orthodox	0.6	0.6	0.0
Hindu	0.4	0.4	0.0
Reformed	0.3	0.3	0.0
Other Evangelical/Fundamentalist	<0.3	0.3	0.0

(Continued)

(Continued)

Religious Group (Listed in Order of Increase to Decrease as Percentage of Population)	Childhood Family Religious Affiliation (percentage)	Current Religious Affiliation (percentage)	Total Change (percentage)
Anabaptist	0.3	<0.3	0.0
Pietist	<0.3	<0.3	0.0
Friends (Quakers)	<0.3	<0.3	0.0
Mormon	1.8	1.7	−0.1
Restorationist	2.3	2.1	−0.2
Jewish	1.9	1.7	−0.2
Anglican/Episcopal	1.8	1.5	−0.3
Presbyterian	3.4	2.7	−0.7
Methodist	8.3	6.2	−2.1
Baptist	20.9	17.2	−3.7
Catholic	31.4	23.9	-7.5

SOURCE: Pew Research Center Forum on Religion & Public Life. (2008b). *U.S. religious landscape survey*. Retrieved from http://religions.pewforum.org

Critical Thinking: As you study Table 6.1, what patterns do you see in members switching between religious groups? Does the pattern seem to suggest a move toward more conservative and evangelical groups, toward disaffiliation with denominations, or toward dissatisfaction with religion entirely? Why?

Table 6.1 shows the overall pattern of people's switching from the faith communities of their youth to new ones or to no churches. Table 6.2 elaborates that pattern by showing the original religious tradition of respondents and their current religious tradition. Among the interesting findings in this table is that Hindus are the least likely to switch faith traditions, and Protestants leave that faith tradition to become, among other things, Jehovah's Witnesses (one third of whom were raised in Protestant households), Buddhists (32% former Protestants), or another Protestant tradition. It is interesting that 44% of all unaffiliated adults were raised Protestant, double the number of unaffiliated who were raised without religious identification (21%). Note also that being unaffiliated is a much less stable religious identity than most others.

Table 6.2   Retention and Switching: From What Groups Are People Exiting?

Current religion . . . (listed by highest retention)	*Unchanged from Childhood*	Percent Who Were Raised:			
		Protestant	*Catholic*	*All other Faiths*	*Unaffiliated*
Hindu	90	2	4	4	2
Catholic	89	8	NA	1	2
Jewish	85	5	2	3	5
Orthodox	77	12	5	1	4
Mormon	74	13	7	1	5
Muslim	60	24	4	5	8
Protestant	54	29	9	2	6
Within Protestant		*To other* Protestant			
Historic black Prot.	69	21	4	1	4
Mainline Prot.	54	30	9	2	5
Evangelical Prot.	51	31	1	2	6
Jehovah's Witness	33	33	26	1	8
Buddhist	27	32	22	6	12
Unaffiliated	21	44	27	8	NA
Other Faiths	9	50	23	7	11

SOURCE: Pew Research Center Forum on Religion & Public Life. (2008b). *U.S. religious landscape survey*. Retrieved from http://religions.pewforum.org

Switching by Individuals and Families

What accounts for switching among the 44% of Americans who have changed their religious affiliation? The Pew Forum has not done complex statistical analyses necessary to make causal inferences about religious switching, but earlier sociologists have done so on other data sets. The two most consistent predictors of religious switching in these studies are intermarriage and education.

Individuals who marry outside their religious tradition are more likely to switch to bring their affiliation in line with their spouse's. This comprises a significant proportion of religious switchers (Loveland, 2003; Musick & Wilson, 1995; Suchman, 1992). Hence, the "friendship network" factor is extremely important in this case as well (one's spouse being the "friend").[6] One reason that Roman Catholics, Orthodox Jews, and conservative Protestants tend to have higher retention rates than do liberal Protestants is that the former are much more likely to marry within their own denominations (McCutcheon, 1988).

Education level is also a significant factor in predicting religious switching. What happens is that people tend to worship with others of a similar socioeconomic class. As people are upwardly mobile, they often change their denominational affiliation (Sherkat & Wilson, 1995). It is not clear whether the move to a denomination of higher status is usually a planned "image-enhancing" strategy or merely a function of joining a religious group where one has a friendship network (coworkers and colleagues). In any case, the move to higher-status churches is commonly a move to a more modern or liberal theology, for the high-status churches tend to be more modern in theology. For someone who is in a professional occupation or who has moved up the social ladder, the more modern, more secular theology may be appealing. A worldview that accepts the advances of science may provide more personal coherence and meaning to a highly educated scientist than does a worldview

that rejects Darwin. Further, religions of the have-nots tend to condemn materialism and the accumulation of possessions as signs of depravity. This may be comforting to the dispossessed, but it can be most uncomfortable to those who are affluent. Liberal theology tends to embrace *this* world more completely than do orthodox traditions that stress rewards in an afterlife. Regardless of reasons, changes in socioeconomic status do seem to be a significant source of denominational switching (Hoge & Carroll, 1978; Newport, 1979).

In addition to those sociodemographic factors, there are also aspects of an individual's personal background that increase their likelihood of switching. In particular, instability in an individual's connection to religion makes them more likely to switch. Loveland (2003) found that not formally joining a faith community and having a lapse in religious practice are both significant predictors of religious switching. Along these same lines, having one's parents divorce in childhood increases the likelihood of religious reaffiliation (Lawton & Bures, 2001). The next "Doing Research on Religion" feature explores in more detail one rational choice perspective on religious switching that puts some of these observations in a broader theoretical context and generates propositions that can be tested.

DOING RESEARCH ON RELIGION

Rational Choice Theory and Religious Switchers

Some scholars regard rational choice theory as the "new paradigm," offering a new "big picture" and fresh analysis. Others think it is intellectually bankrupt and reductionist. The "demand" side of rational choice theory emphasizes the behavior of individuals choosing on the basis of a cost/benefit analysis. With regard to "switching," this approach points to some factors that might influence a decision to change one's religious affiliation.

---

[6]Hoge and Roozen (1979) reported that marriages in which partners continue to belong to different denominations are highly correlated with religious inactivity.

Using a rational choice approach, Sherkat and Wilson (1995) set forth some predictions about which people in the society are more or less likely to become "switchers." Since the effectiveness of any theory lies largely in its ability to allow us to predict human behavior, these hypotheses make for an interesting test of rational choice theory. The following summarizes some of the predictions that these scholars suggest based on rational choice assumptions.

1. We know that social position governs preferences—ranging from art and music to sports and children's names. Many studies for more than a century have also found a close correlation between socioeconomic status and religious affiliation. We know that lower socioeconomic status groups usually have stricter ideas about child rearing and obedience to authority. Likewise, their religion is usually more authoritarian and absolute about answers to questions about the meaning of life or about the authority of scripture. By contrast, in the upper or affluent groups, the emphasis is on intellectual curiosity and on individual autonomy over obedience to authority, whether in the workplace or in theology. Commentators over the years—as far back as John Wesley in the 18th century—have noted that as people move into a higher social class, they often modify their religious ideas and even affiliations. The switch has sometimes been interpreted as a symbol of change in the person's social position. Alternatively, it may be that the person experiences a higher comfort level when surrounded by others with similar values and self-interest. So switching might be predicted to be high among those whose social status has changed, and we would predict that they would adopt a tradition more compatible with their own social status.

2. Another aspect of socioeconomic status may also be a factor. People normally develop a preference for the "religious goods" in their denomination of origin. Yet if the denomination is upwardly mobile, as the United Methodists have been over the past two centuries in North America, some people are bound to be left behind. While many affluent people are comforted by changes within the religious organization toward more liberal stances on creation of the earth, on biblical authority, or on acceptance of homosexuality, those who have not moved up in socioeconomic status may find this threatening. The theology is no longer compatible with their own social position. Since more conservative "firms" may offer a worship experience and a theology more in keeping with the person's social standing, we would predict that these would be the strongest candidates to "switch" to a more conservative denomination.

3. Cultural preferences or choices are also usually associated with the familiar, whether it is the familiarity of music or art or aromas. We might expect people to make choices that are compatible with what is familiar to them. Frequency of childhood participation is likely to instill familiarity, so we might predict that there would be greater benefits for not switching for those who were intensely socialized and were highly active as children. Those who were not active with religious groups as children would be more likely to switch and would be more likely to switch across styles of worship and theologies, since they had less familiarity with a single tradition.

(Continued)

(Continued)

4. Isolation, of course, means that people have less exposure to alternatives, and lack of exposure will influence choices and how one assesses potential benefits. Isolation means that other options would be less familiar and might seem more threatening or bizarre. So isolation breeds loyalty by making a switch to another tradition more costly. One might predict from this that groups which isolate themselves (and conservative groups of *any* religious tradition are more likely to do so) would have fewer "switchers." This includes quasi-ethnic groups—in-groups like Jews or Irish Catholics—where individual identity is closely bound to the group. Switching away from the group would be highly costly.

5. We may also make choices because those choices make our friends happy, not because there are intrinsic satisfactions or benefits. We may choose something not because it makes sense to us, but because it keeps our social bonds in order. This is another type of rational choice factor. Theologically liberal parents are less likely to be offended by a child's apostasy or by a switch to another tradition, but children of theologically conservative parents may feel that they must remain within the conservative family of religious "firms" or they will pay a high cost—the breech of an important relationship. Thus, conservatives would be less likely to switch; if they do switch, it would be in a manner that is not too distant from the original family tradition.

6. All voluntary organizations provide a number of collateral incentives or benefits that are not related to the product offered by the organization: friendships, access to mates, confirmation of social legitimacy, self-esteem, social interaction opportunities, and the like. Conservative religious "firms" are likely to insist that one invest all of one's resources in the firm in order to reap these benefits; thus, they limit crosscutting ties and affiliations in order to fully "belong." Liberal "firms" are more likely to encourage crosscutting memberships and networks within the community. Thus, benefits of members are not restricted to those with total loyalty. Because of this, conservative firms are less likely to lose members due to switching, for they have had to make total commitment if they wanted the collateral benefits. Leaving the group will require a much higher cost, since they have fewer sources of these benefits.

7. Choices are influenced by geographic mobility, since moving to a new community breaks old ties, introduces movers to new options, and may actually make previous choices inaccessible. Thus, one could expect geographically mobile people to be more likely to be "switchers."

Although the research is only in the early stages, early findings suggest that this demand-side rational choice theory is effective in making predictions, a key step in conducting research. The hypotheses in this case are generally supported. While critics of rational choice theory complain that people do not make religious decisions based on the same sort of calculations that govern everyday decisions, Sherkat and Wilson indicated that rational choices about faith communities by individuals may include a wide range of types of benefits and potential costs. Continuing tests of these hypotheses will advance our knowledge.

It is clear that denominational switching is not the same as an internal conversion experience. Much of the change from one denomination to another is a matter of merely changing organizational membership. Yet the feeling of belonging—because of friendship networks or because of socioeconomic homogeneity—seems to be the primary influence in this decision to switch. Reginald Bibby's (1987a) summary of the research on recruitment among Canadian evangelicals seems to apply equally well to American faith communities of all kinds: "If they are serious about recruiting 'real live sinners,' the best approach is either to befriend them or marry them. For this is how the majority of outsiders are actually recruited" (p. 30).

Mainline denominations—including Presbyterian Church (USA)—have generally been declining in membership at a higher rate than the national average. Others that have been declining are Roman Catholic, Baptist, and United Methodist denominations.

> *Critical Thinking:* How or why is "switching" between denominations *similar* to "conversion"? How is it *different* from conversion experience?

Switching and the Growth or Decline of Religious Bodies

Beyond the individual causes of religious switching, some scholars have been interested in its potential social consequences. They want to know why some religious groups are growing and others are declining and whether switching is a primary factor in the change of fortunes of various religious groups.

Dean Kelley's (1972) book *Why Conservative Churches Are Growing* was one of the first to explore this phenomenon. Kelley's thesis was that people are attracted to churches that have strict standards of membership and that expect members to invest much of their time and resources in the group. He pointed out that conservative (fundamentalist, charismatic, and evangelical) churches like the Churches of God, Nazarenes, Seventh-day Adventists, Jehovah's Witnesses, Southern Baptists, and Mormons are growing rapidly while liberal denominations like the United Methodist, United Presbyterian, and Episcopal churches are declining. See Table 6.3 for data on this pattern.

Data on denominational growth and decline 40 years later show the same pattern as Kelley sought to explain (see Table 6.3).

With the exception of Catholicism, all of the denominations in Table 6.3 that are growing in membership are conservative varieties of Protestant Christianity, and all of the denominations that are shrinking in membership (with the exception of the conservative Lutheran Church-Missouri Synod) are mainline Protestant. In America today, there are now more Jehovah's Witnesses than there are Congregationalists (United Church of Christ) and more members of the Pentecostal Assemblies of God than mainline Presbyterians. Mormons today have higher membership numbers than every mainline Protestant denomination except the United Methodists.

Kelley explained this trend by arguing that people are leaving more liberal denominations and switching to more conservative denominations. However, Kelley's interpretation of the data has been challenged. Most of the converts to the conservative churches are rejoining after a period of absence or they are coming from other

Table 6.3 Membership in 13 of the Largest U.S. Denominations in 2008, Compared to 1998

Denomination	Characterization	1998	2008	% Change
Jehovah's Witnesses	Millenarian restorationist Christianity	825,570	1,114,009	+34.9
Church of God (Cleveland, TN)	Holiness Pentecostal denomination, conservative theology	870,039	1,072,169	+23.2
Church of Jesus Christ of Latter-day Saints	American upstart denomination, conservative theology	4,923,100	5,974,041	+21.3
Roman Catholic Church	Historically White, increasingly Latino	62,018,436	68,115,001	+9.8
Assemblies of God	Pentecostals, conservative theology	2,687,366	2,899,702	+7.9
Southern Baptist Convention	Historically Southern & White Baptists, conservative theology	15,729,356	16,228,438	+3.2
United Methodist Church	Mainline Protestant	8,400,000	7,853,987	−6.5
Lutheran Church-Missouri Synod	Theologically conservative Lutherans	2,594,404	2,337,349	−9.9
Evangelical Lutheran Church in America	Mainline Protestant	5,178,225	4,633,887	−10.5
Episcopal Church	Part of Worldwide Anglican Communion, Mainline Protestantism	2,300,461	2,057,292	−10.6
American Baptist Churches in the USA	Mainline Protestant	1,507,400	1,331,127	−11.7
Presbyterian Church (USA)	Mainline Protestant	3,595,259	2,844,952	−20.9
United Church of Christ	Mainline Protestant	1,421,088	1,111,691	−21.8

SOURCES: 1998 data collected in 1999 and published in Eileen Lindner, ed., *Yearbook of American and Canadian Churches, 2000* (Nashville, TN: Abingdon Press). 2008 data collected in 2009 and published in Eileen Lindner, ed., *Yearbook of American and Canadian Churches, 2010* (Nashville, TN: Abingdon Press).

*Three historically black churches—Church of God in Christ, National Baptist Convention USA, and National Baptist Convention of America—are reputed to be among the largest denominations in the country, but they have not reported new membership figures to the *Yearbook* in many years, and the numbers previously reported were unsubstantiated self-reports. Therefore, we have excluded them from this list.

conservative congregations. The conservative churches actually seem to be growing because of two factors: (1) recruits from other evangelical or fundamentalist groups and (2) a high fertility rate in conservative churches.

In one series of studies, from 10% to 15% of the new members to conservative churches were converts from outside the evangelical community (Bibby & Brinkerhoff, 1973, 1983, 1994). In terms of numbers, this averaged 2.3 new converts per evangelical church who had grown up in mainline churches. On the other hand, liberal churches seem to be drawing a larger percentage of their new members from the more conservative churches. Many scholars believe the movement of denominational switching is not primarily from liberal churches to conservative ones but vice versa (Bibby, 1978; Newport, 1979; Roof & McKinney, 1987).[7] A similar recruitment pattern holds in Canada as well, where conservative churches have been slightly less successful than the more liberal mainline churches in reaching the unchurched. Most recruits to conservative churches in that country are from other conservative groups (Bibby, 1987a).

Despite the overall switching trend from conservative groups to liberal ones, many conservative churches in the United States are growing in membership while most liberal denominations are either declining or remaining stable. One reason is that most liberal and moderate churches do not *retain* members as well as most conservative churches. We must be wary not to generalize too broadly here. Conservative churches typically have more commitment mechanisms (discussed in Chapter 5), and these tend to enhance retention. The early research indicated that Jehovah's Witnesses, Seventh-day Adventists, Southern Baptists, Pentecostals and holiness groups, and Mormons had some of the highest retention rates. However, conservative

fundamentalist and evangelical sects, whose members are largely of lower socioeconomic class, have had some of the worst retention rates (Roof & McKinney, 1987). Interestingly, the recent Pew Forum research is not showing strong retention among Jehovah's Witnesses or the Baptists (Pew Research Center Forum on Religion & Public Life, 2008b). Although these groups have many commitment mechanisms, they foster values and outlooks—especially on a range of issues related to gender—that are in conflict with the larger societal trends.

Still, the poor retention rates of conservative churches are countered by very high birth rates (Newport, 1979; Perrin, 1989; Roof & McKinney, 1987). Over 30 years ago, Dean Hoge and David Roozen (1979, p. 322) reported that, over time, birth rates have been the most consistent correlate with growth and decline of church membership. They insisted that "contextual factors" (aspects of the larger society that the churches cannot control) have usually had a greater effect on church membership levels than have "institutional factors" (policies or actions within the church itself). This finding was supported by a more recent analysis that the main explanation for the changing shape of American Protestantism— the growth of conservative denominations and the decline of liberal ones—is demographic not theological. Conservative denominations "grow on their own" because of the higher fertility rates and earlier childbearing among women in conservative denominations (Hout, Greeley, & Wilde, 2001).

Hence, the growth of conservative religious groups makes sense. The conservative faith communities in the United States employ more commitment mechanisms, have higher fertility rates, and have more religious training in the home, accounting for better overall membership trends (Bibby & Brinkerhoff, 1973; Hoge & Roozen,

---

[7]Switching seems to be primarily to the high status (generally most liberal) and the very low status (generally most conservative) churches and away from middle status congregations. Hence, the movement is not entirely in one direction (Newport, 1979; Roof & McKinney, 1987).

1979; Newport, 1979). On the other hand, liberal mainline churches in Canada have better retention rates than their counterparts in the United States and are not experiencing as much decline in numbers. One recent study of a mainline congregation in the United States that *is* growing at a very healthy rate revealed that the key to their success lays largely in a strategy of implementing affective and instrumental commitment mechanisms—small support groups for emotional bonding, eliciting a commitment of time and energy in the organization by new members, and so forth (Wilson, Keyton, Johnson, Geiger, & Clark, 1993).

This entire field of church growth and decline has been enormously controversial for some time in the sociology of religion. Many books and articles have been devoted to exploring the trends. Some scholars side with Kelley, asserting switching to evangelical churches is occurring and that strictness, absoluteness, and otherworldliness are powerful attractions that cause growth (Finke & Stark, 1992; Perrin & Mauss, 1991). The rational choice theorists have been especially strong advocates of Kelley's thesis. One of their arguments is that religion is a "collectively produced commodity" in which the participants share in creation of an experience (such as a worship service) and they each benefit from that collective experience. However, some people "give" a great deal to the creation of the experience and support of the collective enterprise, while others may be "free-riders"—people who benefit from the experience but are not as committed. Rational choice theorists claim that this results in resentment or demoralization by those who give a great deal, and it reduces commitment by everyone. Strict churches force high levels of commitment by everyone, so the "free-riders" are forced to leave; thus, the argument goes, strict churches remain more vigorous and vibrant (Iannaccone, 1994).

Other scholars counter the argument with alternative data and interpretations (e.g., Bibby & Brinkerhoff, 1992; Smith, 1992). One team of researchers has found that church members like ministers who are *authoritative*—that is, they are confident and firm in their faith. However, people do not necessarily prefer a minister who is *strict*—that is, they demand obedience (Tamney & Johnson, 1998). Some sociologists—using information from sample surveys rather than from denominational reports—have even argued that as an absolute proportion of the population, conservative churches as a whole are *not* growing (Smith, 1992). On the other hand, research on "New Paradigm churches" that combine evangelical theology with very modern music, upscale but nonchurchy environments, and nonauthoritarian pastors are drawing members from mainline Christian denominations, especially former Catholics (Miller 1997; Perrin, Kennedy, & Miller, 1997). (New Paradigm churches will be discussed in more detail in Chapter 14.) There are many trends within trends, but it does appear that in the United States certain conservative groups and congregations are gaining a larger overall proportion of the *active, religiously affiliated population*. On the other hand, the category that has grown the most is believers who have left religious communities entirely and those who have moved to nondenominational congregations (Pew Research Center Forum on Religion & Public Life, 2008b).

Newport (1979) concluded that "the present evidence argues against the notion that Americans pick and choose their religious affiliation on the basis of some well thought-out and *theologically* based criteria" (p. 550). He went on to say that this does not imply a total absence of theological considerations but that such concerns appear, from current evidence, to be secondary.

Insofar as some very conservative churches have grown, this is clearly the result of a complex mix of factors: higher birth rates, higher percentages of endogamous marriages, more commitment mechanisms, higher overall retention of members, and less ambiguity about authority. Their numerical success is not due primarily to switching of members from liberal faith communities to conservative ones or to recruitment of large numbers of people from among the unaffiliated. The old mainline denominations in the United States, on the other hand, seem to be suffering from apathy and dropping out more than from massive out-migration to other denominations. Indeed, those who leave mainline churches are increasingly likely to answer the "religious preference" question on a survey form with "none."

> ***Critical Thinking:*** Why do you think people are leaving mainline churches and disaffiliating? What social factors might explain the trend?

Religious "Nones"

All of the attention paid to denominational winners and losers has obscured to some extent the social group that has grown the most in recent decades: those with no religious preference, or religious "nones." In 1968, Glenn Vernon characterized these individuals as "a neglected category." This was perhaps understandable at the time, because in the 1950s and 1960s, the proportion of Americans claiming no religious preference was just 3% to 4% (Condran & Tamney, 1985). Today the situation is much different. As previously suggested, the largest area of growth in American religion is among religious nones. The percentage of Americans claiming no religious preference in survey studies has grown from 3% to 4% to 16.1% in the Pew Forum survey.

An important part of Vernon's analysis—replicated many times in subsequent research—was that religious nones are not synonymous with "nonreligious." Many nones believe in God, pray, participate in religious services, and report having religious experiences. The Pew Forum,

This young lawyer and most of his friends grew up in churches—in his case, a Methodist church. However, he and his friends have disaffiliated in their 20s and 30s. They seek spirituality, but the main thing alienating them from childhood religious groups is the stance of so many Christian churches with highly conservative social positions—positions they find morally unconscionable. One of these is the stand of many religious bodies against same sex marriages.

therefore, distinguishes among four types of religious nones among the 16%:

Atheist	1.6%
Agnostic	2.4%
Secular unaffiliated	6.3%
Religious unaffiliated	5.8%

Only one tenth of religious nones do not believe in God, along with another 15% of that group who are unsure. The "secular unaffiliated" are those who say that religion is not important in their lives (though do not call themselves atheists or agnostics) and the "religious unaffiliated" are those who say that religion is either somewhat important or very important in their lives, but they are not part of an existing religious community. Religion, therefore, is important in the lives of one third of those who claim no religious preference.

How are we to understand the rising number of Americans who claim no religious preference? According to sociologists of religion, religious nones have not only been increasing over the years but the reasons for nonaffiliation also seem to be changing. Previously, "structural" reasons for non-affiliation were dominant. For example, as Condran and Tamney (1985) argued, some people were geographically or socially isolated and so had little connection with religious institutions (e.g., people living in rural areas). Today, "cultural" or ideological reasons appear more important.

Michael Hout and Claude Fischer (2002) made this point when they revisited this issue 20 years after Condran and Tamney. By the early 2000s, the percentage of nones had doubled, gaining considerable attention in the mass media. Hout and Fischer argued that the doubling of those with no religious preference from 1991 to 2000 was influenced by "cohort replacement"—individuals born in the second half of the 20th century replacing the most religious generation in American history, those born in the first third of the 20th century (see also Schwadel, 2010). They also observed that individuals who were raised with no religion were less likely to affiliate in adulthood than had been the case historically, perhaps because they find themselves enmeshed in social networks with others—spouses, friends—who are also nones (Baker & Smith, 2009). Finally, they found that the growth in religious nones was heavily influenced by individuals with liberal political identities becoming alienated from organized religion due to the alignment of religion and conservative politics by the "religious right." In an updating of their study using data through 2008, Hout and Fischer (2009) again found "unaffiliated believers" to be a significant segment of religious nones, and symbolic protest against the cooptation of religion by conservative Protestants to be an important explanation of their unwillingness to identify themselves as having a religious preference.

As the percentage of individuals claiming no religious preference continues to grow, we might expect that this trend will continue into the future since one of the strongest predictors of being a religious none is having parents who are religious nones. Although some of these individuals will be atheists or agnostics, many of them will be unaffiliated believers. They may be the Americans who are most likely to describe themselves as "spiritual not religious" (recall Chapter 2). In any event, what the continuing increase in Americans with no religious preference means for the future of American religion is an open question that sociologists of religion will be grappling with for some time.

Critical Thinking: Some well-educated young adults say they are disaffected with their denomination's conservative stance on homosexuality or other issues that they see as matters of social justice. How important do you think social positions of denominations have been in the exodus of younger adults from formal religious communities? What other factors might be at work?

SUMMARY

Conversion and switching are complex processes, and there is much we do not know. What does appear entirely clear at this point is that the NRMs are not engaged in brainwashing as was once charged by the anticult movement and by the media.

While the idea of a sequential cumulative process first advanced by Lofland and Stark seems to be a fruitful way to think about conversion, with various factors *contributing*, at this point there is no consensus about which factors might be *necessary* ones or what precise sequence might be most typical. Most conversions probably involve a complex interplay between factors.

Rational choice theorists stress the activist role of actors in the process of making choices. Whether the focus is on converts or on the action of religious "firms" acting through local "franchises" and "entrepreneurs," this form of analysis stresses the role of decisions that are based on a religious market. The "winners" are those who are aware of market forces of supply and demand and the interplay of costs and benefits in individual religious commitments.

Switching denominations also seems to be a result largely of affective factors. Friendship networks and marriages across denominations appear to be the most powerful factors, but other factors are at work as well. Interestingly, the conservative Christian churches and nones are the fastest-growing. The growth patterns among conservative groups are because birth rates are very high; strict membership rules in those organizations may also contribute to higher levels of retention. The largest growth and arguably the most important development in American religious affiliation is among those with no religious preference. Although they have long been a part of the American religious scene, religious nones are now one of the largest religious categories in America. The connection between politics and alienation from established religions promises to be an important issue into the future, especially since the demographic underpinnings of disaffiliation make continued growth of religious nones very likely in the near future.

In this and the previous chapter, we have sought to understand socialization into and commitment to religion, as well as religious conversion and denominational switching. In the next section, we will explore other factors that are necessary if a religious group is to survive over time, examining religion at the organizational level.

PART IV

FORMATION AND MAINTENANCE OF RELIGIOUS ORGANIZATIONS

I n the previous chapters, we have explored two of the three interrelated subsystems that comprise the larger system we call religion. We have investigated the meaning system (e.g., the network of myths, symbols, rituals, ethos, and worldview), and we have pointed to the importance of the belonging system (e.g., the network of friendships and affective ties). We have learned something about the processes by which individuals become and remain religious, as well as patterns of changing religious affiliations.

The question of the organizational bases and forms of religion will be a central focus of this unit. In Chapter 7, we examine the major typologies that sociologists have used to categorize religious organizations: church, sect, denomination, and cult (or new religious movement [NRM]). As part of that consideration, we examine the process of institutionalization that is necessary for any religious group to survive and grow. In Chapter 8, we turn our attention more narrowly to American religion and examine two distinguishing characteristics of the social organization of U.S. religion: denominationalism and congregationalism. Included in our consideration of congregations is the rise of megachurches, new paradigm and seeker churches, and the marketing of religion.

7

ORGANIZED RELIGION

Churches, Sects, Denominations, and "Cults"

Here are some questions to ponder as you read this chapter:

- What is a charismatic leader, and how does this kind of leadership influence a group?
- How does institutionalization of religion figure into a group's survival?
- What dilemmas does institutionalization create for a group?
- What are the core characteristics of a church, denomination, sect, or cult?
- What kinds of social conditions help to create a new religious movement (NRM) or a sect?
- What forces cause a fledgling religious movement to evolve into denominations of an existing religion or a new religious tradition?
- Are the ideas of ecclesia, denomination, sect, and NRM the best way to think about religious groups and movements?

Adequately socializing young members, gaining new members, and maintaining a high level of commitment from existing members is necessary for a religious group to survive. Such commitment, however, is not sufficient to ensure the group's ongoing viability. In this chapter, we look at the organizational development of religious groups from their beginnings as fledgling "cults" to their status as established and stable religious organizations.

Ever since Max Weber first developed the proposition that NRMs generally start out as cults headed by a charismatic leader, much of the sociological work in religion has evolved around the concept of charisma. However, Weber insisted that charisma is inherently unstable; if the group does not institutionalize, it will die shortly after the founder dies. Many NRMs never make the transition and therefore never develop into viable religions. Even if a group does institutionalize, the process of institutionalization is itself fraught with dilemmas, usually causing changes in the religious movement. In this chapter, we explore the role of charismatic leaders in the emergence of NRMs, discuss the importance of institutionalization in the survival of these movements, and examine problems that this institutionalization creates for the group.

Having examined the changes religious organizations undergo in the process of institutionalization, we turn our attention to the various ways that sociologists of religion have classified religious organizations. "Church," "sect," "denomination," and "cult" (or "NRM") are the most common categories used, though the precise definitions of each and their relation to one another have been the source of considerable debate, as we will see. (We will be considering denominationalism more fully in Chapter 8.)

CHARISMA AND THE CHARISMATIC LEADER

Weber was the first sociologist to write extensively on the importance of charismatic leadership. He maintained that new religions generally get their impetus from the attraction of a charismatic leader, a dynamic person who is perceived as extraordinary and set apart from the rest of humanity. Weber (1947) wrote,

> The term "charisma" will be applied to a certain quality of an individual personality, by virtue of which (s)he is set apart from ordinary [people] and treated as endowed with supernatural, superhuman, or at least specifically exceptional powers or qualities. These are such as are not accessible to the ordinary person, but are regarded as of divine origin or as exemplary, and on the basis of them the individual concerned is treated as a leader. (pp. 358–359)

The charismatic leader is able to use this power to mobilize followers and to create within them a sense of mission.

Weber insisted that charismatic leadership is not to be judged as intrinsically good or evil, for it could be either. The point is that some individuals come to be regarded as exceptional and perhaps even divine by followers or disciples. Although the followers do not necessarily experience the mysterium tremendum in quite the way it is described by Rudolf Otto (see Chapter 4), they do become convinced that their charismatic leader is a direct agent of God, or perhaps God incarnate. Believers may feel a sense of mystery and awe in the presence of such a leader. The feelings of members of the Unification Church for the Reverend Moon provide one example. He is believed to be the new Christ, God incarnate. Whatever he says is believed to be true simply because he said it.[1] Likewise, whatever

---

[1]This is a different sort of religious experience than Otto described. Nonetheless, some elements of this phenomenon are quite similar to those described by Otto and by Emile Durkheim (see Chapter 4). Although the absoluteness of power and unapproachability are not as pronounced as with the mysterium tremendum experience, the charismatic leader's authority is nonempirical, total, and unquestionable. The charismatic person offers the only source of truth or salvation and in this sense has a good deal of power. Hence, the leader has a considerable amount of personal space—that is, he or she may be approached but not too closely. A second important characteristic of the charismatic authority is that it is specifically outside the realm of everyday routine. Like Durkheim's concept of the sacred, charisma is foreign to the profane world. The experience is based on something beyond the ordinary empirical world. Hence, we might identify this awe in the presence of a charismatic leader as an alternative form of nonrational religious experience.

Jesus Christ said is true for Christians because he is, in their eyes, God incarnate (God in the flesh). Jesus is a classic illustration of a charismatic leader who began a new movement.

Later writers, building on Weber's concept, have elaborated the difference between charismatic and ideological leaders. Margrit Eichler (1972) suggested that the charismatic leader's wisdom is considered to be truth simply because he or she utters it. The ideological leader, on the other hand, is one whose leadership comes from an ability to interpret an existing belief system in a compelling manner. Final authority rests outside the person, but that leader is still the primary interpreter. This distinction is consistent with much of Weber's writing about charisma, although at other times he used the term *charisma* in a broader context and seems to accept many types of leadership as charismatic. Hereafter, we will reserve the concept of charismatic leader for those whose authority resides in their very personhood or in their utterly unique relationship with the deity. What they say does not need to be legitimated or confirmed by some other source. What they say is viewed as truth itself.

One element of charismatic leadership that Weber thought to be especially important was its antiestablishment and revolutionary tendencies. Weber (1947) believed that charismatic authority is intrinsically unstable and is antithetical to social order:

> Charismatic authority is specifically outside the realm of everyday routine and the profane sphere. In this respect, it is sharply opposed both to rational, and particularly bureaucratic, authority and to traditional authority. . . . Both rational and traditional authority are specifically forms of everyday routine control of action; while the charismatic type is the direct antithesis of this. Bureaucratic authority is specifically rational in the sense of being bound to intellectually analyzable rules, while charismatic authority is specifically irrational in the sense of being foreign to all rules. Traditional authority is bound to the precedents handed down from the past and to this extent is also oriented to rules. Within the sphere of its claims, charismatic authority repudiates the past and is in this sense a specifically revolutionary force. (pp. 361–362).

Weber was emphatic in asserting that the Old Testament prophets were not spokesmen for some economic group. Their interests were explicitly religious, although there were social and economic implications to their messages. Weber was interested in the power of ideas in bringing social change. Research in the latter half of the 20th century focused on why people attribute superhuman powers to others. This line of research stressed the fact that adherence to a charismatic leader and the development of an NRM is likely when certain social conditions prevail, and it occurs in stages by which the leader comes to be elevated to sacred status (Wallace, 1972).

Nonetheless, once the charismatic leader has emerged and developed a following, it is necessary for the group to undergo a transformation. Charismatic leadership is not only revolutionary but it is inherently unstable. Hence, the movement undergoes a process that Weber referred to as the "routinization of charisma."

Critical Thinking: How does the experience of awe before a charismatic leader differ from the experience of awe in a nonrational mysterium tremendum encounter? How are these experiences similar? How might these two types of intense religious encounter result in different social consequences?

THE ROUTINIZATION OF CHARISMA

If the religious movement is to survive for any significant period of time, a stable set of roles and statuses must be established and a consistent pattern of norms generated and practiced. Hence, the nature of the charismatic authority is transformed. Weber (1947) wrote,

> In its pure form charismatic authority has a character specifically foreign to everyday routine structures. If this [religious movement] is not to remain a purely transitory phenomenon, but to take on the character of a permanent relationship . . . it is necessary for the

character of charismatic authority to become radically changed. Indeed, in its pure form charismatic authority may be said to exist only in the process of originating. (pp. 363–364)

The community gathered around a dynamic leader must evolve into one with a stable matrix of norms, roles, and statuses. This process of *routinization* (developing stable routines) is commonly referred to by sociologists as institutionalization. Any group that fails to institutionalize its collective life simply will not survive.

Institutionalization serves both ideal and material interests of followers and leaders. The ideals of the group can be furthered only if it survives and only if it mobilizes its resources. Hence, institutionalization serves the ideological interests of adherents. Yet the followers and the leaders also have a material stake in the survival of the group. Insofar as they have invested time, energy, and financial resources in the group, they are likely to feel that they have a vested economic interest in its survival and success. Therefore, the inherent forces working for routinization are strong.

Perhaps the most critical test of a group's routinization is the way it handles the issue of succession. When the charismatic leader dies, the group may quickly disperse. However, many people have a vested interest in the survival of the group and will seek to ensure its viability. The problem is, who will provide the group with leadership? Equally contentious is how that decision will be made.

The transfer of power to the next designated leader has important implications for the subsequent evolution of the group. First, the charisma that was once identified with a personality must be associated with the religious ideology and with the religious organization. The group, the body of beliefs, and perhaps a written record (a scripture) become sources of veneration. This more stable source of authority in itself changes the character of the group. Second, the decision making process itself becomes sacralized as the divinely appointed method of choosing the successor. This method may involve the designation of a successor by the original leader, some form of divinely sanctioned and controlled election, the drawing of lots, a

hereditary succession, or any one of a number of other procedures. In any event, the followers must recognize the new leader(s) as the legitimate heir(s) to leadership. Otherwise, the group may be torn by schisms as various splinter groups identify different persons as the rightful leader.

The new leader or group of leaders is not likely to possess the same sort of unquestioned authority that was vested in the personhood of

Joseph Smith was the founder of the Church of Jesus Christ of Latter-day Saints (also called Mormons), but when he was murdered in Illinois in 1844, Brigham Young (above) became the president and leader of the group. He was already the president of the church's leadership core—the Quorum of the Twelve Apostles—and second in command after Smith. Young led the migration west, established church settlements throughout the western states, founded Salt Lake City, and served as territorial governor of Utah. For 33 years, Brigham Young was the leader of the church, further stabilizing the resource base and leadership of the group. Without this smooth transition of leadership—part of routinization—the church may have split or simply died without its charismatic founder, Joseph Smith.

the original leader. Some of the awe and respect will have been transferred to the teachings and to the continuing organization itself. The rules and values of the group must be attributed with transcendent importance in and of themselves. Commitment is now to the organization and to the ideology of the movement, and the authority of the *new* leader(s) may be restrained by these stabilizing forces. No longer are the sayings of the leader taken as true simply because that person said them. They must be evaluated in light of what the original leader said and did. In short, the new leader is usually an ideological leader rather than a charismatic one.

Another issue of routinization has to do with provision of a stable economic base. If some members are to devote full time to the service of the leader and the movement, then a continuing and consistent source of income must be provided. This may come from some obligatory payment to the organization by members who are employed in secular positions (a tithe), from members begging or soliciting (as some NRMs in the 1970s and 1980s did), or from the establishment of some industry sponsored by the organization (as in the Bruderhof production of *Community Playthings* or, in the mid-19th century, Oneida's production of steel traps). However accomplished, this provision of a stable source of income is a critical part of the routinization process. Without this financial base, there can be no full-time clergy, administrative staff, or other regular employees. If there are no career opportunities within the organizational structure, instrumental commitment may begin to wane. Furthermore, some type of full-time administrative staff may be necessary if the group is growing and expects to continue expanding.

Sometimes routinization is a slow process that occurs largely after the death of the leader. After the death of Jesus, his disciples were a band of frightened and discouraged individuals. Through a series of experiences, they became convinced that Jesus had risen from the dead and their faith was renewed. However,

it was the new convert Paul (previously known as Saul of Tarsus) who was responsible for the expansion of the group, for he universalized the faith and recruited Gentiles on an equal basis with Jews. Paul also established organizational links between the various congregations. He collected money from his Gentile converts, which was given to the parent organization in Jerusalem, and he traveled from one congregation to another forging a bond between them and creating a loyalty to himself. In fact, the parent Christian group in Jerusalem often sought to undermine his authority, and Paul constantly had to work at maintaining his position of leadership. (By most accounts, Peter and James, leaders of the Jerusalem church, were not adept at organizational matters. Moreover, in the early period they felt that Paul was wrong to allow Jew and Gentile converts to eat together and to be treated as equals.)

Paul established norms and expectations that were spelled out in his teachings and in his epistles. He often "corrected" local congregations because they had begun to follow norms that had emerged spontaneously out of the group. He assumed authority and was sought out as the person to resolve conflicts and to define customs and mores (e.g., whether women could worship without a head covering, and what format the love feasts should follow). He also authorized teachers at various churches. In fact, in his later epistles, he even referred to bishops and deacons. These officials appear to have been local ones at the time, but eventually the roles expanded to be more centralized, with broad supervisory functions. The routinization process was by no means complete by the time of Paul's death, but it was well under way. It is clear that Paul was the organizational genius who initiated the institutionalization process; it was his organizational initiative that eventually resulted in that bastion of European religious life, the Roman Catholic Church.

So in some cases, a later organizer initiates institutionalization of the faith after the death of the charismatic founder. In other cases, the charismatic leader also happens to be an excellent organizer. In any event, the routinizing function must be performed.

Perhaps the most important test of the routinization of charisma is that of determining leadership succession. However, normalization of roles and statuses, establishment of relatively stable norms, and provision of a stable economic base usually have to be addressed long before succession becomes an issue. If routinization does not occur, the viability of the group for any substantive period after the death of the original leader is unlikely.

There have been thousands of NRMs that are spawned but do not survive. One example is the California group known as Heaven's Gate. Studies indicate that when one of the leaders died in 1985, there was a major crisis. Because there had been two leaders, the community rallied behind the remaining leader, known to them as "Do." (The leaders were a woman and a man called "Ti" and "Do"—as in the musical notes.) Ti had been the more charismatic figure, but they worked together and when she died, followers came to believe that Do had direct communication with Ti on a regular basis. As Do began to enter his 60s and he experienced some health problems, it was clear that the group would not survive his death intact. The group had failed to fully institutionalize and was still dependent upon primary charisma. Ultimately, of course, the group did not survive the leader since they all committed suicide with him. This was the only way they could join "Ti" in a spaceship on the other side of comet Hale-Bopp, thus ascending to the "Next Level" (Balch & Taylor, 2002). There are many other examples of collapsed religious movements that failed to institutionalize, but few have as dramatic an ending as Heaven's Gate, whose group suicide made international news in March 1997. Indeed, the reason most people cannot name many groups that did not institutionalize is precisely because they failed to survive as a group. They are relegated to the history of religious movements.

Many scholars believe that most, if not all, religions begin as charismatic cultic movments. However, if a new group is to prosper and survive,

it must undergo a process of institutionalization. The extent and form of institutionalization will vary from group to group, but some routinization must take place. Beyond this, institutionalization itself creates new problems or dilemmas.

> *Critical Thinking:* Refute or support the following statement, providing your reasoning for your position: "When religion becomes 'organized religion' it ceases to be real religion."

DILEMMAS OF INSTITUTIONALIZATION

As Thomas O'Dea (1961) put it, "religion both needs most and suffers most from institutionalization" (p. 32). Although institutionalization is necessary, it tends to change the character of the movement and to create certain dilemmas for the religious organization. O'Dea elaborated five such dilemmas.

The Dilemma of Mixed Motivation

When the enthusiastic band of disciples is gathered around the charismatic leader, there is a single-minded and unqualified devotion to the leader and to his or her teachings. The followers are willing to make great sacrifices to further the cause, and they willingly subordinate their own needs and desires for the sake of group goals. However, with the development of a stable institutional structure, the desire to occupy the more creative, responsible, and prestigious positions can stimulate jealousies and personality conflicts. Concerns about personal security within the organization may cause members to lose sight of the group's primary goals. Mixed motivation occurs when a secondary concern or motivation comes to overshadow the original goals and teachings of the leader. Conventions of clergy sometimes debate pension plans and insurance programs more heatedly than statements of mission (O'Dea, 1961).

It is important to recognize the dilemma in this process. Religious institutions do need to provide for the economic security and well-being of their full-time professionals if they expect them to maintain high morale and commitment. Likewise, if those professionals are to be satisfied and fulfilled by their work, they need to feel that they can really use their creative talents and abilities. These secondary concerns are important to the individuals in question. The problem for the organization is that these secondary matters can take on primary importance for some members and subvert the original sense of mission.

Such secondary concerns may come to the surface while the charismatic leader is still alive. For example, the New Testament reports that at the time of the Last Supper, Jesus' disciples got into an argument over who was the most important and who would have the most exalted position in the kingdom of God (Matthew 18:1; Mark 10:37). However, such self-oriented motivations can be rather easily overcome by a simple command of the charismatic personality. Mixed motivation is more likely to develop when the group's sense of security is institutionally based rather than charismatically based. Later generations are much more likely to belong to the group for reasons that are unrelated to the teachings of the charismatic leader.

Mixed motivation may also involve a more subtle form of goal displacement. Max Assimeng (1987) studied six millennial and Pentecostal Christian mission groups in Africa, which began with a goal of converting people to their own faith. Their focus was otherworldly and theological. To attract converts and to demonstrate their concern for people, some groups started schools; others established hospitals. While the religious message itself was not always well received, the hospitals and schools were enthusiastically embraced. Indeed, promises of offering such this-worldly services were often necessary before the government authorities would allow the group to proselytize in their country. It is natural, one might suggest, for people to repeat those behaviors that are successful and well received. The enthusiastic and warm response to

this-worldly benevolent aid was an enticement for missionaries to focus more and more of their energies on provision of those services. Goal displacement began to occur in the allocation of time and resources because the areas in which success did occur provided ego gratifications for the individuals involved. Such goal displacement did not occur equally in all groups or in all regions of Africa; for a variety of reasons, some missionaries were more effective than others in resisting goal displacement. Still, the original goals in many mission stations were displaced. Success can be intoxicating and alluring and can cause one, consciously or otherwise, to redefine one's goals.

The Symbolic Dilemma: Objectification Versus Alienation

In Chapter 4, we explored the importance of symbols in religion. For a community to worship together, a common set of symbols must be generated that meaningfully expresses the worldview and the ethos of the group. However, this process of projecting subjective feelings on to objective artifacts or behaviors can proceed to the point that the symbols no longer have power for the members. O'Dea (1961) wrote,

> Symbolic and ritual elements may become cut off from the subjective experience of the participants. A system of religious liturgy may come to lose its resonance with the interior dispositions of the members. . . . In such a case the forms of worship become alienated from personal religiosity. (p. 34)

Variant interpretations of a symbol in a religious community can clearly have divisive effects. Yet according to O'Dea, symbols may also become utterly meaningless to members of the religious community. Many Christian churches have stained glass windows that depict a fish. This was a powerful symbol to the early Christians. In Greek, the first letters of each word in the phrase, "Jesus Christ, Son of God, Savior" spelled the word fish. The fish came to be used as a coded symbol of insiders: a drawing of a fish identified one as a

Christian to other believers without giving one's Christian identity away to Roman soldiers. Hence, a representation of a fish became an important symbol of solidarity and conviction for Christians, a highly persecuted group. Most modern-day worshipers do not know the origins of the fish as a symbol of faithfulness, and the symbol does not particularly enhance the worship experience or bond the members together. In this case, the symbol is treated with indifference rather than with hostility. In any event, the fish is no longer a powerful symbol that acts to create powerful moods and motivations and reinforce the worldview. (See the "Illustrating Sociological Concepts" feature for an example of forgotten meanings of symbols in the Greek Orthodox tradition.)

ILLUSTRATING SOCIOLOGICAL CONCEPTS

Official Symbolism and Actual Behavior in the Greek Orthodox Church

In the Greek Orthodox church, when people cross themselves they are suppose to use the thumb, the middle finger and the index finger pressed together to represent the Trinity. The other two fingers are to touch the palm of the hand and stand for the two characteristics of Christ: (1) human and (2) divine. The interesting thing is that when Orthodox Greeks cross themselves, they are supposed to touch their heads (representing the fact that the faith affects their intellect), then they touch their hearts (representing the fact that the faith is to affect their emotions), and then they are to touch their right shoulder and then the left shoulder (to indicate that the faith is to also influence their limbs, their activities). However, when Greeks cross themselves in Greek churches, few of them touch their head and then their heart; most touch their throat and then about 2 inches below that and then move a couple of inches to the right and a couple of inches to the left. In addition, their hands were not usually held in the way prescribed by orthodox manuals. So the intended meaning attributed to this symbol—faith in Christ influencing intellect, emotions, and behavior—appears to be lost on the people. The "official" meaning of the symbols is less powerful than just the kinesthetic activity of making a sign of the cross.

*****

Review the hand symbol of Christianity in Greek Orthodox icons as illustrated on page 68. This symbol is always made with the right hand and is used by priests for blessings. The index finger is supposed to be straight forming an *I*; the middle finger is slightly bent forming a *C*. The little finger is slightly bent (also forming a *C*), and the thumb and ring finger are slightly crossed forming an *X*. The *I* and *C* of the index finger and the middle finger are the first and last letters of the Greek name of Jesus. The *X* and the *C* of the remaining fingers stand for Christ.

Especially interesting in this symbolism is the fact that in many of the "later edition" icons, the index finger is bent in the shape of a *C* instead of being straight to form an *I*. Furthermore, in most of the icons of Jesus the thumb and ring finger—which are supposed to cross—often simply meet at the tips to form a circle. So again, there is symbolism that has obviously gotten sloppy over the years and is very much lost on most of the people, including some of the priests and the religious artists.

In other cases, the symbols may come to be seen by some as a barrier to communication with the transcendent. The antisymbolism of the Puritans, who rejected stained glass windows and identified Roman Catholic stations of the cross as "idols," provides an example. In the minds of these people, visual symbols had come to symbolize an overbearing bureaucratic organization rather than a transcendent experience. Such alienation from established forms, symbols, and rituals can lead to either apathy toward religion or radical revolt and reformation of the faith.

The process of developing objective, observable symbols that express a deeper subjective experience is part of the process of institutionalization. Symbols are necessary to bind a group together and to remind them of a common faith, but that very objectification of experience into symbols may eventually create problems. If the symbols lose their power and meaning for members, the group must either create new symbols or it will face internal problems of meaning and belonging.

The Dilemma of Administrative Order: Elaboration of Policy Versus Flexibility

As a religious group grows and institutionalizes, it may develop national offices and a bureaucratic structure. In so doing, a set of rational policies and regulations may be established to clarify the relationships between various statuses and offices in the organization. Like the federal government or any other bureaucratic structure, these rules and regulations and the plethora of interrelated departments and divisions may create an unwieldy and overcomplicated structure. Ecclesiastical hierarchies are no less susceptible to red tape than any other bureaucratic structure. As O'Dea pointed out, attempts to modify or reform the structure may run into severe resistance by those whose status and security in the hierarchy may be threatened. Resolution of the dilemma of administrative order may be impeded by the existence of mixed

motivation. Many persons within the hierarchy may view reorganization as a threat to their own security or positions of power and prestige.

Again, the dilemma must be recognized as such. Students sometimes see unwieldy organizational complexity and excessive rules and regulations as silly and unnecessary. However, people in complex organizations frequently feel a profound need for guidelines for decision making. The search for concrete policy in the face of some new problem is common in any large, formal organization. If no policy has been clearly established, people look to the resolution of similar cases in the past. The precedent often then becomes an unwritten rule. The development of elaborate and sometimes overly complex policy is a natural outgrowth of the desire of people to know how to solve problems and deal with unusual cases. People at lower levels in the organization do not feel comfortable making decisions without the sanction of those with more authority. Yet, this felt need for clearly articulated policy can mean that an organization is run entirely by rigid rules and regulations. Not only is this uncomfortable and frustrating, but it also greatly reduces flexibility.

A group of clergy appointed to a committee to screen candidates for ordination and ministry may not feel authorized to establish their own standards for ordination. They want to know what guidelines have been set by the denomination as a whole. If none are available, they would likely seek to have some guidelines formally approved to help them make decisions. (In the absence of formally approved policy, their own decisions may take on the authority of official policy.) The guidelines that are adopted may specify a college degree and a seminary education from an accredited university and theological school, respectively. However, years later an ordination-screening committee may encounter a case in which an individual brings impressive experiential credentials and enthusiastic letters of reference but holds degrees from nonaccredited institutions. Does the committee have the authority to waive the national guidelines in this case? Whatever the members decide, their decision may be seen as a binding precedent for future cases.

Policy continues to be elaborated, but it may also restrict a group in its ability to be flexible with new circumstances.

The Dilemma of Delimitation: Concrete Definition Versus Substitution of the Letter for the Spirit

In the process of routinization, the religious message is translated into specific guidelines of behavior for everyday life. The general teachings

The substitution of the letter for the spirit of a set of beliefs is illustrated in this linocut by Robert 0. Hodgell. Some religious people seem so intent on literal interpretation of the written word that the spirit of the faith seems to be lost; there seems to be little love, compassion, or joy in such religion.

about the unity of the universe or about the love of God must be translated into concrete rules of ethical behavior. If this is not done, the religious belief system may remain at such an abstract level that the ordinary person does not grasp its meaning or its importance for everyday living. The meaning of many of Jesus' teachings were not understood even by his disciples until he gave specific illustrations or elaborated through parables. In the ancient Hebrew tradition, the covenant that Abraham had made with God was comforting, but to understand the human obligations in it, specific ethical codes were spelled out. The Torah, the first five books of the Hebrew Bible, is a record of this religious law; it includes the Ten Commandments as the best known such code in the Western world.

In the process of translating the outlook of the faith into specific rules, however, something may be lost. Eventually, the members may focus so intently on the rules that they lose sight of the original spirit or outlook of the faith. The religion may then degenerate into legalistic formulas for salvation and/or may become moralistic and judgmental in a way inconsistent with the original intent of the founder (O'Dea, 1961). The legalism and ritualism of the late classical period of Pharisaic Judaism provides one example. The fundamentalistic and rule-bound interpretations of Islam by some of the leading Shi'ite ayatollas of Iran provide another example; their posture is clearly inconsistent with the central thrust of Islam—love, compassion, and peace. The linocut print by Robert O. Hodgell expresses more eloquently than words the problem created when the letter of the religious ethical code becomes more important than the spirit of that code. The image could apply equally well to some types of Protestant Christians, to Islamic fundamentalists, or to Hasidic Jews (at least as the Hasids are seen by Reformed Jews).

Hence, the dilemma is created: the abstract moods, motivations, and concepts must be made concrete so that common laypeople can comprehend their meaning and implication for everyday life. However, by translating the spirit of the faith into specific moral rules and ritual requirements, the possibility is established that later generations may become literalistic and legalistic about those rules and may miss the central message.

The Dilemma of Power: Conversion Versus Coercion

If a religious group is to stay together and sustain its common faith, conformity to the values and norms of the group must be ensured. Although occasional deviation from established norms may not threaten the group, most of the beliefs, values, and norms of the faith must be adhered to most of the time. In its early stages, the religious group is composed of members who have personally converted to the group. They feel a personal loyalty to the charismatic leader, they have had a nonrational conversion experience, or they have been motivated to internalize the faith through some other process. By contrast, later generations who have grown up within the religious organization may never have personally experienced anything that compelled them to accept the absolute authority of the faith in their own lives, or to accept the authority of the religious hierarchy to interpret the faith. They may be inclined to challenge official interpretations. Although such challenges can also come in the first generation, internalization of the authority structure is usually more complete in the first generation of believers than in later ones (O'Dea, 1961).

To maintain the integrity of the organization and ensure consensus in their basic worldview, religious organizations may resort to coercive methods of social control. Excommunication is one procedure. When Sonia Johnson, a Mormon, took a public stand in 1979 in favor of the Equal Rights Amendment for women, the Mormon hierarchy informed her that her stance was heretical. When she refused voluntarily to accept the position of the Mormon Church and the authority of its leaders, she was excommunicated. Likewise, other religious bodies have excommunicated members who fail to conform to basic doctrinal or ritual standards. Of course, in some periods of history, the unorthodox were subject to torture and death. Regardless of the specific methods used, the maintenance of conformity through coercion rather than conversion is significant. Conformity due to internalization of norms is much more powerful and lasting than the use of coercion, but voluntary internalization of the norms by all members is difficult to sustain. Therefore, an institutionalized religious organization may come to rely on coercive methods as a last resort to maintain conformity and consensus.

Edward Lehman (1985) pointed out that in democratic, pluralistic societies such as the United States, religious organizations have little coercive power over their members. In fact, it is the members who have some control over their churches; they may threaten to withhold financial support. Thus, churches in the modern world have much less clout to enforce and maintain their interpretations of the faith.

Institutional Dilemmas and Social Context

Milton Yinger (1961) pointed out that O'Dea's discussion of institutional dilemmas failed to specify the conditions under which the dilemmas are most and least likely to occur. Indeed, O'Dea suggested that the dilemmas are inherent, unavoidable, and inevitable. If they *are* inevitable, it seems clear that in some cases it may take several generations, and perhaps several centuries, to occur.

James Mathisen (1987) analyzed the Moody Bible Institute in Chicago, tracing its 100-year history, with attention to how it has dealt with these institutional dilemmas. Several of the dilemmas appear not to have occurred, and the movement seems to have coped fairly effectively with most of the others. Mathisen's analysis highlights the fact that some social organizations and some social conditions may be more conducive than others to the occurrence of certain dilemmas.

It appears from Mathisen's study that the symbolic dilemma is less likely to occur in groups that stress intense emotional conversion experiences and that have strong mechanisms to sustain the plausibility of the worldview and the symbol systems. Alienated or marginal groups in the society may also be less susceptible to this dilemma because the religious symbols may serve to enhance group boundaries. Conflict with outsiders provides a context in which symbols gain emotional force and relevancy for the members.

Likewise, the dilemma of delimitation (concrete definition versus substitution of the letter for the spirit) may be less of a problem in circumstances where the group perceives itself to be a persecuted or vulnerable minority group. Having a clearly defined set of "fundamentals" by which to measure faithfulness may become a source of strength for the group and may be defined as a virtue. Not only was this the case for fundamentalists running the Moody Bible Institute but also appears to be operative among Orthodox Jews in North America as well. Both groups are minority movements that are substantially concerned with maintaining clear boundaries between themselves and the larger society. If concrete definition sometimes does lead to a problem for the group, it is clear that this is not always the case.

The dilemmas of mixed motivation and of power may also be lessened in circumstances in which many commitment mechanisms are used. As we saw in Chapter 5, commitment mechanisms heighten one's commitment to the needs and interests of the group by causing one to identify self-interest with group interests. This would lessen the likelihood of mixed motivation and would reduce the need for coercion as a means of social control. These generalizations are at this point reasonable hypotheses that need to be tested empirically. Continued research is needed regarding the conditions under which the dilemmas are most and least likely to occur.

Dilemmas of institutionalization do plague organized religion, but the five dilemmas outlined in this chapter are not necessarily "inevitable," and it is clear that they need not be fatal. The "Doing Research on Religion" feature, for example, discusses a researcher's work on a group that still deals with these issues. It does appear that the dilemmas are more pronounced when the organization is a majority religious movement than when it is a minority protest movement preoccupied with tensions with the larger society.

DOING RESEARCH ON RELIGION

Researching a Religious Group that Still Celebrates Charisma

Margaret Poloma, a sociologist who practices glossolalia or speaking in tongues (such Christians are sometimes called "charismatics"), conducted research on a religious organization that celebrates the unexpected and the spontaneous messages of God through nonrational experiences. Unlike Weber's findings, she maintains that the Assemblies of God has insisted that charisma is disperse among all of the people, not just a charismatic leader. She has conducted a study to find ways that the group has institutionalized over time but has fought against lessening of spontaneous nonrational euphoria and inspiration by members of the group. She finds that the dilemmas of institutionalization have played a role in the institutional life of the Assemblies of God, but that they have found ways to affirm charisma and to slow the eroding effect of rationalization (bureaucratization) and rationalism. The following passage describes Margaret Poloma's research and how it evolved into two major books.

* * * * * * *

The development of most research projects is as fascinating as the findings such projects generate. From the day I first became acquainted with the Assemblies of God while gathering data on the larger charismatic movement, planning the project, collecting and analyzing the data, and writing up the findings have all provided adventure.

Until 1979 my knowledge about the Assemblies of God was limited to recognizing it as a sect-like Pentecostal group best known for its puritanical proscriptions and its practice of glossolalia. While doing preliminary research for my book *The Charismatic Movement: Is There a New Pentecost,* however, I met two ordained Assemblies of God ministers who challenged my stereotype. One was Edward E. Decker, then a vice-president of Emerge Ministries (a local counseling center founded by another Assemblies minister); the second was E. Eugene Meador, pastor of Akron's First Assembly of God (one of the denomination's oldest congregations in the United States). Both men defied the common stereotype of a Pentecostal preacher as a pre-modern fundamentalist who unthinkingly accepts outmoded theology and rejects the development of the modern world. They were invaluable informants for my research and astute critics of drafts of my manuscript on the charismatic movement.

By 1982 and the publication of *The Charismatic Movement,* my view of Pentecostalism in general and the Assemblies of God in particular was significantly altered. Sociologically, I was intrigued by the significant growth and institutionalization of the Assemblies as well as by the struggle to maintain the charisma that had launched the sect earlier in the century. There was ample evidence of what sociologist Max Weber had termed the "routinization of charisma," but there were also many signs that the Assemblies were being revitalized, partially through influences from the larger charismatic movement. A healthy tension seemed to exist between the flow of charismata and the development of a bureaucratic institution.

My original research design called for limiting data collection to the use of focused interviews and participant observation. For each congregation selected, interviews were to be conducted with the pastor(s) and a few other key persons to facilitate a congregational portrait. These interviews were to be supplemented by my own participant-observation of the congregations, including attending services, meetings, and other church activities.

These plans to limit my research to qualitative methods were altered when a bright undergraduate student, James Hibbard, expressed an interest in writing his honor project on the Assemblies of God. His one reservation was a preference for a quantitative study rather than the qualitative methods which we had proposed. Demonstrating that professors can indeed be challenged by their students—and can respond positively to the challenge—I soon found myself working with Jim in putting together a survey instrument that became an indispensable part of this research project.

What was originally proposed as a qualitative study developed into an example of triangulated research that employed both qualitative and quantitative techniques. The different approaches worked well together, supplementing and complementing findings and enabling better interpretations.

Although I thought my research on the A of G was finished when the book was published in 1982, I found myself completing another monograph at the turn of the new millennium. My interest and participant involvement in the larger Pentecostal/Charismatic movement (of which the Assemblies of God is but one component) was rekindled with the unexpected revivals that broke

(Continued)

(Continued)

out at the Toronto Airport Christian Fellowship (1994) and at Brownsville Assembly of God in Pensacola, Florida (1995). In what Carl Jung would describe as "synchronicity," I began to research the revival at the Toronto Airport Christian Fellowship and was later invited by the Jim Lehrer News Hour to accompany a news team to Pensacola as PBS reported on the new revival in America.

Shortly after the "outpouring" (a common term for a "revival") began in Pensacola, I was privileged to participate in a Lilly Endowment sponsored project on Protestant denominations conducted through Hartford Seminary. Once again we combined our participant observation of the revival with a survey of Assemblies of God pastors. The observations and the survey results constituted an unplanned longitudinal dimension to my original research and permitted me to "revisit" the thesis we developed and tested in my earlier research. The findings appear in a scholarly article (Poloma, 2005).

The outcome of the late 1990's study was a reaffirmation of the earlier analysis. Rekindled by the revival fires of the 1990s, charisma and structure remain in a healthy dialectical tension within the Assemblies of God.

SOURCE: Revised by Margaret Poloma from Poloma, Margaret M. (1989). *The assemblies of God at the crossroads: Charisma and institutional dilemmas.* Knoxville: University of Tennessee Press.

Once created, institutions take on a life of their own and may focus more on their own survival than on anything else. Anson Shupe found that denominations would often try to cover up clergy malfeasance (including sexual misconduct by ministers) because of a concern for how publicity might hurt the church as an organization. Survival and health of the religious organization became more important than ethical misconduct and injury to members of the church (Shupe, 1995, 1998; Shupe, Stacey, & Darnell, 2000). Thus, the process of institutionalization is a mixed blessing—or perhaps a necessary curse—for religion. If an NRM does not institutionalize, it is not likely to last long. The absence of routinization will mean that the group will expire with the death of the charismatic leader. The particular pattern of routinization will vary from group to group, and some organizations become more bureaucratic and centralized than do others. For example, some religious bodies have successfully routinized without establishing a professional clergy, while others have highly trained professionals as ministers and a hierarchy of area ministers, bishops, and archbishops.

The student of religious organizations will find considerable variation in the extent and style of institutionalization, but some form of routinization is essential.

Critical Thinking: Make two lists: one enumerating the benefits of institutionalization for a religion and the other identifying problems that institutionalization creates for a religion. How do you evaluate the net impact on a religion?

CLASSIFYING RELIGIOUS ORGANIZATIONS

We have already seen in this chapter that religious groups vary a great deal and undergo significant changes as they develop. To conduct research on religion and make generalizations about religious behavior, social scientists have categorized groups with significant similarities into types, comparing and contrasting characteristics of churches, sects, denominations, and cults (the latter now usually called NRMs). Max Weber called such groupings *ideal types*. By this, he did not mean that they were "the best" or somehow preferable, but only that these classifications exist entirely in the world of ideas. For example, if there are seven typical characteristics of a dysfunctional family, the ideal type would list all seven of these. However, most troubled families may only have some combination of four or five of the features.

Generalizations about any social behavior can be made only by comparing and contrasting phenomena that are in some respects similar. However, it is essential that the conception of various types be clear and accurate as possible. Otherwise, the concepts themselves may cause faulty generalizations. In fact, the concepts of church, sect, and denomination have been used in a variety of ways by various researchers, and this causes some confusion. Nonetheless, many scholars continue to try to work with these concepts because they feel that they continue to be valuable as analytical tools. As we explore the ways in which terms such as church, sect, denomination, and cult have been used, it is important to keep in mind that they provide only one of many possible ways to organize the data and to think about religious groups. Readers must always be sure they understand how a given study is using the terms (Dawson, 2009).

THE CHURCH–SECT TYPOLOGY

Early Formulations and Process Approaches

Most discussions of the church–sect typology begin with the work of Max Weber and his student, Ernst Troeltsch. Weber emphasized that the sect is an exclusive group. To be a member, one must meet certain conditions such as adherence to a particular doctrine or conformity to particular practices (like adult baptism or abstinence from alcohol). For Weber, the sect involved three characteristics: (1) membership is voluntary, (2) it is limited to those who "qualify" for membership, and (3) it involves a substantial commitment by

the members. The church, on the other hand, is depicted as: (1) a group that one is typically born into rather than choosing, (2) inclusive—that is, encouraging all members of the larger society to join, and (3) minimal commitment is required to remain a member (Weber, 1922/1963).

Troeltsch and American theologian H. Richard Neibuhr, who each wrote much more extensively on this topic, included Weber's defining criteria but expanded the concept to include many other factors. For them, the central characteristic of the church is its acceptance of the secular order—including even reproduction of the larger society's hierarchy of prestige within the church itself. In short, the church "compromises" Christian values and makes accommodations to the secular society. The sect, on the other hand, tends to reject the social order and to maintain a prophetic ministry. While this was their central emphasis, Troeltsch and Niebuhr generated multi-variable typologies, with *many* distinguishing characteristics of church and sect (see Table 7.1).

H. Richard Niebuhr (1929/1957) elaborated Troeltsch's suggestion that the sect and the church are *stages* in the evolution of a religious group, adding a new type: the denomination. He also moved beyond Troeltsch by identifying factors, including internal structural characteristics, that cause a group to move from one end of the continuum to the other.

These two scholars believed the division of churches into different (and often exclusive) groups was rooted in socioeconomic inequalities and ethnic prejudices and was therefore contrary to core religious teachings. As Niebuhr (1929/1957) put it, "The inequality of privilege in the economic order appears to contain a fundamental denial of the [religious] principle of brotherhood" (pp. 8–9). For Niebuhr the emergence of the "church" is caused by a deterioration of religious ethics and adoption of secular definitions of human value. The reemergence of a sect is an attempt to recapture and reassert the faith commitment to social justice and the equality of all people under God.

Table 7.1 Characteristics of Sects and Churches as Delineated by H. Richard Niebuhr

The Sect	The Church
1. Volitional membership (emphasis on adult conversion and commitment).	1. Volitional membership (emphasis on adult conversion and commitment).
2. Exclusive membership policy.	2. Inclusive membership—may coincide with national citizenship or geographic boundaries.
3. Particularism—judgmental attitude toward those who do not accept the one true path; self-image that of the "faithful remnant" or the "elect."	3. Universalism—acceptance of diversity and emphasis on the brotherhood and sisterhood of all humanity.
4. Small faithful group.	4. Large, bureaucratic organization.
5. Salvation achieved through moral purity, including ethical austerity or asceticism.	5. Salvation granted by the grace of God—as administered by church sacrament and church hierarchy.
6. Priesthood of all believers; clergy de-emphasized or nonexistent; lay participation high.	6. Leadership and control by highly trained professional clergy.
7. Hostile or indifferent to secular society and to the state.	7. Tendency to adjust to, compromise with, and support existing social values and social structures.
8. Fundamentalistic theology—only the original revelation is an authentic expression of the faith.	8. Either orthodox or modernist theology— formulations and interpretation of the faith in later periods of history are legitimate in their own rights.
9. Predominantly a group of lower-class persons or those otherwise socially disfranchised. (Worldly prestige is rejected.)	9. Membership composed of upper- and middle-class people, but with professional classes controlling most leadership positions.
10. Informal, spontaneous worship.	10. Formal, orderly worship.
11. Radical social ethic—emphasizing the equality of all persons and the necessity of economic equality.	11. Conservative social ethic—justifying the current socioeconomic relationships.

Sectarian reaction against "compromises" of the faith results in renewal and revitalization. The first generation of sect members stresses adult conversion and commitment, but they also establish religious education programs for their children. Eventually, the children are accepted into full membership on the basis of their knowledge of the faith (memorized scriptural passages and so on) rather than personal conversion and dramatic life changes. Often the later generations also experience upward social mobility and are no longer disfranchised. The sect gradually institutionalizes at the same time it assimilates secular outlooks. In emphasizing the process of evolution, Niebuhr developed the concept of denomination, another type of religious group that represents a midpoint in the continuum between sect and church (Figure 7.1). Formality and orderliness (lack of spontaneity) were also marks of the trend to denominationalism— a reflection of institutionalization. This was even reflected in the tendency to a more sober, literate, intellectual, and orderly style of worship as opposed to the emotional expressiveness of sectarian worship.

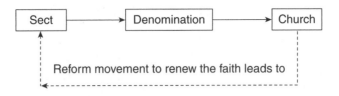

Figure 7.1 The Continuum Between Sect and Church

John Wesley began a renewal movement in the Church of England. He wanted to call people back to the basics of the faith. His evangelical revival called followers to lead Christian lives through a method of strict discipline—prayer, worship, study, and mutual support groups. Hence, the name *Methodists* arose. Today, Methodists have become one of the largest Christian bodies and a mainstream denomination in the United States. The photo on the left is a United Methodist Church in a medium-sized town in Ohio. Eventually, as the Methodist church became more prosperous, more open, and more institutionalized, there were reform schisms—such as the Wesleyan Church (right photo), which is now institutionalizing.

Niebuhr especially stressed that the existence of different denominations is not due to mere ideological differences, as is often believed. He asserted that religious ideology is often used to justify economic self-interests and pointed out that groups from different social classes tend to develop different theological outlooks. The stratification of society affects both the social organization and the theology of a religious group. The real source of the schism that eventually creates new denominations, then, is *social* rather than theological. This was a radical idea at the time.

The formulations of Troeltsch and Niebuhr assumed a high level of correlation among many diverse factors. Although their descriptions may have been accurate for many religious groups of their days, we can point to many groups now that have some features of the sect and some of the denomination. For example, many of the more emotional, conversion-oriented Christian groups of today (born-again Christians) have many traits of a sect, but they do not object to the stratification system in the society, nor do they speak forcefully on behalf of the poor. Many of them support the current economic system (the status quo) and defend the national values of the larger society; they are well acculturated. Niebuhr and

Troeltsch would probably assert that they are not manifestations of early Christianity but further forms of compromise: They fail to restore the original social ethic of Christianity. So the problem of the church–sect model is the problem of all ideal types: A significant number of groups do not seem to fit *all* the characteristics of any one type. The question becomes this: Which characteristics are most important in classifying a group?

Although many religious groups do fit Niebuhr's sequential model—with new groups returning to the purity of the original faith and then gradually assimilating to the dominant culture—some groups do not. Several scholars (Cohen, 1983; Steinberg, 1965) have analyzed the emergence of Reform Judaism from Orthodox Judaism as a reform group resembling the church model. Reform Judaism did not represent a protest against the "perversion of true religion" and did not call for a return of the "old ways." Rather, it represented a fuller acculturation or modernization of the faith. This new group was rejecting what its adherents considered an excessive tie to the past on the part of the parent body. The reforming body sought greater assimilation—in effect, more openness as a system. Likewise, Mauss and Barlow (1991) documented a shift in Mormonism *away* from assimilationism and toward a deliberate religious traditionalism, an evolution opposite that predicted by Niebuhr. Finke and Stark (2001) pointed toward evidence of movement toward higher levels of tension with the larger culture in the United Methodist Church during the 1990s as well. Therefore it is important to remember that Niebuhr described the predominant trend, but

the sequential evolution of sect-denomination-church and revolution back to sect is not universal.[2]

> ***Critical Thinking:*** Why is Niebuhr's diachronic (or historical) model of analyzing sects and denominations an improvement over simple (nonhistorical) typologies? That is, what are some advantages of seeing sect, denomination, and church as evolving patterns rather than stagnant, unchanging "types"?

Single-Variable Models

Various sociologists have continued to simplify the concepts of sect and church and then to identify factors that contribute to the formation of one type of group as opposed to another. Some have suggested that tension or conflict with the dominant society should be the central criterion that distinguishes the sect (Johnson, 1963; Stark, 1967; Stark & Bainbridge, 1985). Simply stated, "A church is a religious group that accepts the social environment in which it exists. A sect is a religious group that rejects the social environment in which it exists" (Johnson, 1963, p. 542). Using this single criterion, the classification of a group will not depend on its intrinsic qualities (extent of institutionalization, emphasis on conversion and voluntary adult commitment, and so forth.) but on its relationship to the larger culture in which it resides. For a discussion of how tension with the larger society relates to open systems theory, see Figure 7.2.

---

[2]Troeltsch also described another type of religious group that received less attention by scholars. He described *mystics* as that cadre of loosely associated individuals who emphasize nonrational personal experience as the cornerstone of religion. This is a matter we will return to later. At this point, it is interesting merely to note that Troeltsch was aware that some religious expressions did not seem to fit on his continuum from sect to church.

While most of the scholars discussed in this chapter do not talk about open and closed systems in explaining the difference between church and sect, the model can be useful. Most church/sect typologies essentially depict the church as an open system, with fairly free flow of influence *into* the larger society (outputs such as symbols, benevolent actions, definitions of morality, and so forth) and *from* the larger society (inputs such as new members; ideas from education, science, and business fields; predominant economic attitudes and assumptions; and even reproduction of the secular system of social stratification—with high-income professionals more respected as church members and more likely to be elected to church offices). The danger of openness for the church is that "the salt may lose its flavor"—the uniqueness of the religious perspective may be endangered as the church becomes just another social club. No religious group can allow this to happen, so all religious groups do *some* filtering of inputs. Nonetheless, the filter used by churches allows much to pass through it into the faith system and the life of the congregation.

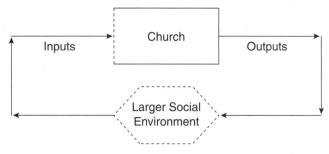

The sect chooses to protect its uniqueness at all costs, barricading itself against the secular culture, carefully screening any inputs. It is a much more closed system, though few systems can be totally closed and still survive. They need members and resources and even technology to enhance efficiency, provided by the larger society. The sect allows inputs, but these inputs must pass through a dense filter. The danger of this choice is that the group may risk becoming so isolated that its message may seem anachronistic and irrelevant to the larger society. The closed system also risks stagnation as little that is new or innovative can enter the system.

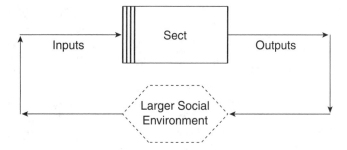

It is useful to think of open and closed systems not as categorically different types but as opposite ends on a continuum. Any given religious group can be closer to one end than the other. As you read various ideas about sects, denominations, and churches, it may be helpful to keep in mind the idea of more open and more closed social systems. Further, each group may make different choices as to what kinds of things are threatening: Some will find certain economic ideas incompatible with their worldview; others will reject modern technologies; still others may find definitions of sexual morality or scientific theories especially threatening. Finally, keep in mind that open and closed systems have their own advantages . . . and their own problems.

Figure 7.2   The Sect and the Church as Closed and Open Systems

One problem with this tension with society criterion has to do with determining which value conflicts are so significant as to warrant classification as a sect. The "dominant value system of the United States" includes many different values, attitudes, and beliefs. These various values may themselves be found independent of one another. For example, no one who studies the Church of Jesus Christ of Latter-day Saints (Mormons) can fail to notice the emphasis on hard work, the importance of the family, individualism, and national loyalty. Yet, there are significant ways in which the Mormons have deviated from the cultural values of this country. Although most of the aforementioned values were part of the value orientation of this group in the 19th century, Mormons also endorsed the practice of polygyny (one man with two or more wives). Non-Mormon neighbors and the federal government were firmly opposed to this policy. The conflict between the Mormons and Gentiles at one point became violent. In 1857, President James Buchanan commissioned federal troops to move into Utah, and the resultant skirmishes are referred to as the Utah Mormon War (O'Dea, 1957).

Is conflict over one value perspective (such as polygyny) enough to classify a group as sectarian, or must there be conflict in several major areas? Alternatively, is it merely the *intensity* of the conflict that is the defining factor? Of course, intense conflicts may also arise over issues other than value differences. In fact, much of the prejudice against Mormons was due to conflict over scarce resources and valued lands. The problem was not strictly a conflict of value orientations, but one over who would possess resources that *both* groups valued. Johnson, and more recently Rodney Stark and William Bainbridge, would recognize a group as a sect if it is in *intense conflict* with its larger social environment on one or more issues.

Still, Johnson (1963) pointed out the long-term *acculturating* role of many sectarian groups:

It may well be that one of the most important functions of the conversionist bodies in the United States, both now and historically, has been to socialize potentially dissident elements—particularly the lower classes—in the dominant values which are our basic point of reference. (p. 547)

The use of value conflicts with the dominant culture is one way to distinguish sects from churches, but this defining criterion is not without its shortcomings.

Andrew Greeley (1972) and Geoffrey Nelson (1968) both argued that the most salient—and most measurable—characteristic of sects is the complexity of organization: the degree of routinization. In any event, the trend is toward a simplified definition of church and sect (Knudsen, Earle, & Schriver, 1978). It may be useful to visit the webpage and look at the document on "Defining Sect and Church—Divergent Views" to see which variables some key scholars have used. That document is at **www.pineforge.com/rsp5e.**

Yinger's Multilinear Evolution Model

Like many contemporary scholars, Yinger held that Troeltsch's concept of church and sect includes too many variables, many of which are not highly correlated in actual groups. Unlike Johnson, he preferred to use several factors in defining sects, denominations, and churches. Yinger suggested three social factors that seem to him to be central, and he established a model that lends insight into types of groups and their evolution. The three defining criteria suggested by Yinger (1970) are as follows:

1. The degree to which the membership policy of the group is exclusive and selective or open and inclusive.

2. The extent to which the group accepts or rejects the secular values and structures of society.

3. The extent to which, as an organization, the group integrates a number of local units into one national structure, develops professional staffs, and creates a bureaucracy. (p. 257)

Yinger pointed out that in actual cases, the first two variables are very closely correlated. Those groups that reject secular values are likely to be exclusive and selective in their membership policies—indications of a closed social system; those groups that accept secular values are likely to be inclusive and open in their membership—indications of an open social system.

By using the first two factors as one axis of variation and the extent of institutionalization as the other, Yinger developed a model that suggests several different types of groups. (See Figure 7.3.) This two-dimensional model is capable of showing a progression from sect to church, but it also demonstrates that increased institutionalization may occur somewhat independently from membership policy and from the acceptance of secular values. There are pressures that cause groups to evolve generally from the types in the lower left side of the grid toward those in the upper right side. Nonetheless, the rate, the direction, and the extent of this evolution is not inevitable. Whether a group stabilizes as an "established sect" or continues to acculturate to the stage of "institutional denomination" or "ecclesia" has to do with the group's belief system. A group that focuses on the sin and salvation of individuals will acculturate rather easily. Sects whose primary concern is social evils and injustices—reformative and transformative movements—are more likely to become established sects and may never become ecclesiastical or denominational bodies (Wilson, 1970; Yinger, 1970). Ronald Lawson (1995, 1996, 1998) found that a number of other elements of a sect's ideology may shape its trajectory toward becoming established. Passionate commitment to pacifism, and thus an internationalism in the face of warfare, was one factor that

has kept Jehovah's Witnesses more sectarian. Meanwhile, the Seventh-day Adventists (which came from the same Adventist movement in the 19th century) has embraced patriotism and become more mainstream in the society. The refusal to embrace jingoistic values during war has also led to persecution for the Witnesses, further alienating them from the dominant society in a way that Seventh-day Adventists never experienced.

Two types of religious organizations do not fit this schema. The first is the shamanistic religion of many nonindustrialized societies. The religion is universally held by all members of that society. It does not pose values that are antithetical to the secular values, for it serves as a sort of glue that helps unify the culture. On the other hand, religion in such a culture is not highly institutionalized. Yinger called such a religion a "universal diffused church," for it is diffused through the culture rather than being specialized and maintaining its own autonomous organization. The second case that does not fit is the "universal institutionalized church," which most perfectly fits Troeltsch's description of the church. This is best exemplified by the highly institutionalized Roman Catholic Church of the Middle Ages. It was universal in that no other religion had a significant influence in Europe; national boundaries and religious loyalties were therefore coextensive.

These two exceptions highlight an important fact: The existence of sects and denominations is a phenomenon of pluralistic societies. In fact, scholars have maintained that the pure form of "Church" described by Troeltsch is impossible in a democratic and pluralistic society. Likewise, the Universal Diffused Church is a phenomenon of only very simple and homogeneous societies; it is not found in complex and heterogeneous ones. Moreover, it may be the case that in a very open, pluralistic society, the distinction between sect and denomination also becomes less important as all groups converge on a single denominational model. We explore the possibility of

Inclusiveness of Membership
Extent of Acceptance of Dominant Society Values

	Low $\longrightarrow$		High
High ↑	Null No known examples	Rare (Seventh Day Adventists)	Institutional Denomination (Church of God)	Institutional Ecclesia (Episcopal Church)
*Extent of Organizational Complexity**	Null No known examples	Established Institutionalized Sect (Old Order Amish)	Diffused Denomination (Beachy Amish; Mennonite)	Diffused Ecclesia (Christian Scientist)
	Sect Movement (The Bruderhof)	Established Lay Sect (Amana Church Society)	Rare	Rare
Low	Charismatic Sect (The Way)	Null. No known examples	Null. No known examples	Null. No known examples

* The criteria for determining placement on the vertical axis are (1) whether local religious units are integrated into a national organization, (2) whether there are religious professionals, and (3) whether there is a bureaucratic structure with rules and procedures that govern group decisions. A yes to all three questions places the group at the highest level.

Figure 7.3 Yinger's Schema for Types of Religious Organizations (Adapted)

SOURCE: Adapted from Yinger, J. M. (1970). *The scientific study of religion*. New York: Macmillan.

such a "denominational society" in the next chapter.

We have found Yinger's conceptualization to be particularly helpful, for he reduced the number of variables to be considered while he shows how several different types of groups can emerge. The primary value of his schema is in identifying several possible patterns in the evolution of groups.

SOCIAL CONDITIONS THAT GENERATE EACH TYPE OF GROUP

We might summarize this discussion of churches, denominations, institutionalized sects, and sects by reviewing the factors that seem to cause the emergence of new groups. Niebuhr pointed to four factors. First, when

Christian denominations begin to ignore the original concern of the faith for poverty and inequality, sectarian groups are likely to arise. As members of those sects (who are disproportionately from the lower classes) begin to achieve some affluence, the sect begins to accommodate secular values and becomes comfortable with the status quo. The group begins to denominationalize. Those persons who are still economically disfranchised are likely to break again from the group and reject social inequality as contrary to the original ethic of the faith. To put it in other terms, the poor may reject worldly affluence—defining it as depraved—and seek otherworldly compensation. By affirming asceticism and condemning the comforts of this world, the sect tends to alienate persons who are affluent (Stark & Bainbridge, 1985). One cause of sectarianism, then, is the existence of social inequality.

Second, Niebuhr pointed out that some groups are expressions of ethnic values and national loyalties. In such cases, the belonging function operates to separate German Methodists from Welsh Methodists and Italian Catholics from Irish Catholics. One analysis of sectarian movements concludes that approximately 21% of American sects were founded almost entirely on the basis of racial or ethnic identity (Stark & Bainbridge, 1985, p. 132). On the other hand, the fact that ethnic and religious groups are mutually reinforcing in North America may explain why religious participation is so much higher and religious groups so much stronger than in Europe (Roof & McKinney, 1987), and why Italian Protestant churches assimilate and dissolve more quickly than Italian Catholic ones (Form, 2000). The sense of ethnic belonging, then, has been a source of organizational diversity within a larger religious "family" (Greeley, 1972; Niebuhr, 1929/1957; Roof & McKinney, 1987).

Third, Niebuhr pointed out that churches and denominations become bogged down in bureaucratic structures. Some sect movements

are expressions of a desire for religious groups that are smaller, more informal, and less under the control of a professional clergy. Individuals may become alienated by the institutional constraints and formalized structures of large denominations. Sectarianism then is sometimes an expression of rebellion against complex organizational structure.

Fourth, he believed that sectarian movements are sometimes spawned by a desire for more spontaneity and more emotional expression in worship. Worship in the denominations tends to become formal, orderly, and highly intellectual. Some people find that stifling and inauthentic as a way to worship God.

Of course, sectarian schisms can also occur over seemingly trivial issues. Max Assimeng (1987) pointed out that Seventh-day Adventists split from other Adventists largely over the issue of whether the Sabbath should be on Saturday or Sunday.

Whether and how a sect develops into a denomination also depends largely on how the larger society responds to this new group. If the sect challenges values that are central to the dominant society, if the host culture does not have a tradition of religious freedom and tolerance, and if the strategy of the sect is aggressive militancy (rather than avoidance and withdrawal), the group will either be crushed or will reinforce its antiestablishment posture. If the group does manage to survive, its acculturation is likely to be very slow. In fact, such a group may remain forever an established sect. On the other hand, if the sect rejects values that are tangential to the dominant society and if there is a tradition of religious tolerance in the host culture, the group may accommodate and become acculturated rather quickly. The movement from sect to denomination is determined in large degree by outsiders' response to the group (Redekop, 1974).

In a comparison of Protestantism in the United States and Canada, Harold Fallding (1980) suggested that separation of church and state also contributes to the emergence of sects. Canada, with state-established religion, has generated far fewer religious sects than has the United States and has historically shown a greater trend toward denominational mergers. The struggle for religious liberty and disestablishment from government structures may unwittingly result in intensification of boundaries between groups and an environment conducive to religious innovation and independence. Rational choice theorists point out that in an environment where the state does not control religion, there will be more religious vitality simply because there will be more religious "firms" started. They see the increased numbers of sects as a sign that more religious niches in the religious market will be filled.

The internal belief system of the group can be another factor important in the evolution of a group. If a sect's definition of evil and corruption is individualistic (matters of personal decision making and individual determination), then acculturation and accommodation is likely. If evil is considered to be social in nature, if the structures of the society are defined as incompatible with religious values, then the group is less likely to denominationalize. The Quakers (Society of Friends) have a very strong social ethic, which includes a rejection of any participation in warfare. This position often puts them at odds with the larger society when international conflicts arise. Although they are not currently persecuted in American society, Quakers remain an institutionalized sect. In this case, the resistance to denominationalizing is internal (based in their own social ethic) rather than external.

Brian Wilson (1959) argued that some types of sects are more likely to denominationalize than are others. Groups that have a theological orientation encouraging a simple ascetic lifestyle are more likely to generate an affluent membership, for their lifestyle leads to savings and accumulation of resources.

Affluence often leads to accommodation. Wilson believed that those sects that lack this emphasis on asceticism are less likely to accommodate. Wilson also demonstrated that the movement toward denominationalism is strongly influenced by an expanding economy. As members become more affluent, they tend to acculturate to the values of the dominant society and lose their fervor for revolution or reform. Because opportunities for upward mobility are most likely in a climate in which the economy is growing, the state of the economy may affect the likelihood of a group becoming a denomination or stabilizing as an institutionalized sect. Stagnant economic conditions are more likely to generate permanent sects. Lawson (1995, 1996) agreed that both internal and external factors are at work, showing that upward social mobility, organizational openness, ideological rigidity, urgency of expectation of the end of the world, the intensity of the indoctrination of converts, relations with the state, and the extent to which the group was stigmatized and persecuted all affect a sect's trajectory.

The formation and evolution of any group is the result of many interacting processes: the social standing of members, ethnic factors, survival forces that impel a group to institutionalize, responses of outsiders to the group and its message, the belief system of the group, and the state of the economy in the host society. Both internal and external forces determine its evolution.

Critical Thinking: Using sociological perspectives and insights, *refute* or *support* the following statement: "Denominations are functional to society and to the groups' members, but sects are disruptive and dysfunctional."

Thus far, we have discussed only the development of sects, institutionalized sects, denominations, and churches. What is meant by a religious "cult?"

NEW RELIGIOUS MOVEMENTS, AKA "CULTS"

The term *cult* is used by sociologists in two distinct ways—and by the popular media in yet a third way. Sociologists feel that the popular usage has terribly misconstrued and distorted a useful sociological concept. The popular media and many anticult movements define a cult as a religious group that holds esoteric or occult ideas, is led by a charismatic leader, and uses intense and highly unethical conversion techniques. (See the discussion in Chapter 6 of the charges of "mind control" or "brainwashing" in NRMs.) The group is almost always depicted as totalitarian, capable of bizarre actions, destructive of the mental health of members, and a threat to conventional society. The tag "cult" is often used in this context as a stigmatizing label intended to discredit a group rather than as a nonjudgmental technical term that describes a social unit (Zuckerman, 2003). This view of cults is explicitly rejected by sociologists of religion, but this media use has managed to cause confusion about NRMs and about the very concept of "cult."

One way that sociologists use cult has been to describe a group without much internal discipline and with a loose-knit structure. This is an attempt to elaborate Troeltsch's third category of mysticism and to rename it as a cult. Following Howard Becker (1932), the cult is seen as an urban, nonexclusive, loosely associated group of people who hold some esoteric beliefs. Members are kindred spirits who have some common views relative to one particular aspect of reality, but such persons may also belong to other, more conventional, church groups. The presence of a charismatic leader is common but certainly not necessary. The Spiritual Frontiers Fellowship provides an

example. It is a group of people who believe firmly in life after death and in various forms of parapsychology. They emphasize spiritual healing, the power of mind over matter, and the possibility of communication with the dead. They hold occasional workshops and lecture/seminar programs to expand their understandings and to inspire one another to deeper belief in the power of the spirit. Much of the membership consists of professional people who are well educated, belong to mainline denominations, and do not think of themselves as esoteric "kooks." The group has no charismatic leader; in fact, authority is at a minimum.

Commitment to such groups is nondemanding, and membership is likely to be transient. The chief defining characteristics of the cult are the loose structure and the lack of application of the worldview to all aspects of life. A number of sociologists use this approach in understanding a cultic movement (Becker, 1932; Hargrove, 1979; McGuire, 2002).

The second sociological approach is to define the cult as the beginning phase of an entirely new religion. The group may be loosely structured or it may demand tremendous commitment, but it must provide a radical break from existing religious traditions (Nelson, 1968; Stark & Bainbridge, 1985; Yinger, 1970). Although both of these definitions have their advocates, the latter approach seems to be the more common one and seems to us to provide a more helpful analytical tool. Note, however, that many sociologists now recommend that the meaning of the term has become so muddled and tainted by media misuse that it should be abandoned entirely. Because of the misunderstanding, most sociologists now use NRM (new religious movement) in place of "cult" (Bromley, 2009; Robbins & Lucas, 2007). We have tended to use the phrase *NRM* through much of this book, but since the word *cult* is still frequently used, students should be aware of its proper sociological meanings.

Perhaps some new term is also needed to identify the quasi-religious movements that have little or no sense of group identification or cohesion—the groups Becker described. Stark and Bainbridge (1985) referred to these groups as either *audience cults* or *client cults*, depending on the type of appeal. Audience cults involve use of mass media appeals—advertisements, direct mail, and other publicity to promote a lecture circuit, a series of workshops, or sale of books or tapes on esoteric or occult topics. Supporters of such movements are essentially consumers and not members. Spiritual Frontier Fellowship, astrology magazines, *Fate* magazine, and *Science of Mind* are examples of audience cults and their media. Client cults tend to involve relationships based more on a patient–therapist model. Adherents to client cults seek help in specific areas: psychological adjustment, contact with the dead, forecasts of the future, medical miracles, and so forth. Scientology, Transcendental Meditation (TM), Silva Mind Control, and palm readers are examples of client cults.

Both audience and client cults involve very loose bonds between members and little sense of group identity. Stark and Bainbridge pointed out that these groups are concerned primarily with manipulating nonempirical forces in the service of specific this-worldly needs—akin to magic (discussed in Chapter 1). Occasionally, audience or client cults may evolve into NRMs. Since client and audience cults are, at best, only quasi-religious phenomena, they are not a primary focus of this chapter.

In an effort to distinguish cults from sects, Stark and Bainbridge (1985) wrote,

> Because sects are schismatic groups, they present themselves to the world as something old. They left the parent body not to form a new faith, but to reestablish the old one, from which the parent body had "drifted." Sects claim to be the authentic, purged, refurbished version of the faith. . . . Whether domestic or imported, the cult

is something new vis-à-vis the other religious bodies of the society in question. If domestic—regardless of how much of the common religious culture it retains—the cult adds to that culture a new revelation or insight justifying the claim that it is different, new, "more advanced." Imported cults often have little common culture with existing faiths; they may be old in some other society, but they are new and different in the importing society. (pp. 25–26)

This emphasis distinguishes a cult from a sect in that the latter attempts to renew or purify the prevailing religion of the society, whereas the cult or NRM introduces a new and different religion. Sometimes it is difficult to determine whether a religious movement is attempting to renew or replace the traditional religion, for a new religion often tries to gain legitimacy and acceptability by exaggerating its continuity with existing faiths. Nonetheless, the issue of whether a group is trying to purify or to replace the traditional religion of a society is central to this concept of cult. Yinger (1970), after his careful development of the sect-denomination-ecclesia grid, turned to a discussion of the cult:

> [Some religious groups] do not appeal to the classical, the primitive, the true interpretation of the dominant religion, as the sect does, but claim to build *de novo*. The term *cult* is often used to refer to such new and syncretist movements in their early stages. It often carries the connotations of small size, search for a mystical experience, lack of structure, and presence of a charismatic leader. They are similar to sects, but represent a sharper break, in religious terms, from the prevailing tradition of a society. (p. 279)

NRMs often place researchers in awkward situations when the latter seek access to the group and the group struggles for legitimation in the society. This is discussed in the next "Doing Research on Religion" feature.

DOING RESEARCH ON RELIGION

Researching the New Religions

Doing research on NRMs raises logistic problems of how to establish rapport and gain access to the groups; but perhaps the ethical dilemmas involved in studying these groups are even more troubling for the researcher. Two sociologists who have experienced these problems firsthand share their perspectives with us.

* * * * * * * * *

Most Americans have learned about new religious movements (NRMs), popularly called "cults," not through direct personal experience but through media reports. The media often has painted these groups as sinister and secretive. Recall the numerous stories by journalists who "infiltrated a cult compound" and "escaped" with an expose of "life in a cult." Most recently, we were informed that the only way government officials were able to obtain information about the Branch Davidians was to infiltrate undercover agents into the community. In most cases the reality is far less sensational. NRMs do vary considerably in their openness to nonmembers, and like corporations and government agencies, are likely to limit access by outsiders wanting inside or sensitive information.

There are various reasons for this reticence. One is that NRMs have granted the media access only to conclude that journalists betrayed their trust. Another is that groups think that ousiders will not understand them. They feel that revealing the revolutionary truths they possess without a proper foundation simply opens them up to ridicule. Finally, these groups all perceive themselves to be engaged in a world-saving mission; taking time out to give interviews to social scientists has very low priority unless it serves some function for their mission.

We have studied the three groups—Unification Church ("Moonies"), International Society for Krishna Consciousness (Hare Krishnas), and The Family (formerly called the Children of God), which precipitated the cult conflict that began in the 1970s and has waxed and waned ever since. Why did these NRMs allow us to study them firsthand? What did they expect in return?

NRMs grant researchers access for several common reasons. One is that members often believe that if you get to know them and listen to their message you will convert. (Anthropologists call this "going native.") This, of course, is very rare, but in a few cases it has happened. Since we did not convert, what we had to offer was an understanding of their beliefs and ability to speak their language, which they interpreted as a second-best alternative of sincere interest and acceptance. Another reason is that groups sometimes use sociologists as a status symbol; they boast to potential converts that the movement must be important—after all, they are being studied by social scientists! Most commonly in recent years, groups have opened themselves up to sociologists because they hope for a more "objective" hearing than they have received from

(Continued)

(Continued)

journalists. These NRMs seek reciprocity when they become embroiled in public controversy. Each has come to us asking us to take their side in a dispute, serve as an information source to whom they can refer journalists or governmental officials, and even to testify in court cases. When this happens, we move from walking a tightrope between professional responsibilities and fairness to research subjects to an even more complex task of balancing professional norms, NRM expectations, and the public interest.

David G. Bromley and Sydney Newton

Virginia Commonwealth University

The Unification Church provides an interesting example of an NRM. The members of this group believe that Sun Myung Moon is the Messiah. They accept his doctrine that Jesus was supposed to have been the Son of God but that he failed in his mission and got himself killed. Jesus provided a partial salvation (purely spiritual), but he failed to redeem the social, economic, and political structures of this world. Because Jesus failed in the total task that was assigned to him, God has now blessed and empowered Reverend Moon to fulfill this divine role. Obviously, such a doctrine is a sharp break from traditional Christianity. While the organizational structure and the conflict with the predominant societal values cause the Moonies to appear to be a sect, the development of a new and unique religious doctrine distinguishes them from the sects. To gain legitimacy and acceptance, however, the Unification Church has downplayed these radical doctrines and has stressed its endorsement of traditional American values (pro-capitalism, pro-family, and so forth).

The early period of the Church of Jesus Christ of Latter-day Saints provides another example. Although they believe that Jesus was the Messiah and they believe in the Bible, Latter-day Saints also have a second book that they hold as sacred scripture. The new scripture (*The Book of Mormon*) came from Joseph Smith's "translation" of some golden plates, which he found in upstate New York. Smith taught that prophets in the early Americas had made a written record of messages from God and of the visit that Jesus made to this continent. The last of these prophets buried the written record on golden plates, and Smith was told in several visions where to find them and how to translate them.

The Mormons emphasize that they are a branch of Christianity, but many of their theological innovations are not accepted as Christian doctrine by other Christian groups (such as the belief that unmarried individuals do not attain the celestial kingdom, the highest of their levels of heaven). It seems fair to say that at least the early Mormons were an example of an NRM rather than a case of sectarian reform (Johnson & Mullins, 1992; Shipps, 1985; Stark, 1984). Whether it is driven by a desire for legitimacy or is a reaction to gross misrepresentations of their faith, many Mormons reject the NRM designation (Barlow, 1991). The same was true of the initial decades of Christianity; the first Christians tended to think of their group as a faction of Judaism rather than the beginnings of a new religion. The difference is that Christianity eventually made a clean break and developed on its own path quite independent of its parent group; Mormons have in some ways moved closer to traditional Christianity in the 20th century rather than asserting independence and stressing differences. Is the modern-day church of Latter-day Saints just another denomination of Christianity, or a new religion that evolved out of Christianity? The dominant scholarly interpretation seems to affirm the latter pattern (Shipps 1985; Stark, 1984; Warner, 1993), but the issue remains controversial, especially among Mormon scholars.

Although the matter of whether a group is initiating a new religion is the central issue, there are several other characteristics that are common in NRMs. While sects often place a strong emphasis on the authority (perhaps even literal authority) of scripture, NRMs frequently stress mystical, psychic, or ecstatic experiences. There is not a categorical difference between the two groups on this, but NRMs seldom use previously existing scriptures as a sole source of truth. In fact, it is not uncommon in the United States for NRMs to generate their own scriptures. The Mormons are one example. The Unification Church also generated a written record that seems to have the aura of sacred scripture. This penchant for scriptural basis is probably caused by the scriptural (written record) orientation of American culture. NRMs in other cultures do not ordinarily generate new scriptures. Hence, if the traditional religion emphasizes the role of scripture, the NRM is likely to develop its own alternative form of "written word."

Another pattern that prevails in many NRMs is the centrality of a charismatic leader (Bromley, 2009). As discussed earlier in this chapter, a charismatic leader is a person who is believed to have extraordinary insights and powers. Such a person is attributed with certain divine qualities and is believed to have direct and unique contact with the supernatural. It is the unique insights of these individuals that are the basis for the alternative faith. Although most NRMs are founded by such charismatic leaders, some sociologists believe that they can also develop in a more spontaneous way—through "spontaneous subcultural evolution" (Bainbridge & Stark, 1979). Such movements, like the groups that focus their

faith on the saving power of unidentified flying objects (UFOs), are usually more loosely structured and more democratic. They have no identifiable charismatic figure, but stigmatization and hostility by outsiders may force adherents to draw more closely together and raise boundaries against outsiders. Such groups usually start out as nonreligious subcultures but may eventually emerge into esoteric cultic groups with a worldview and ethos that contradicts that of the dominant society (Nelson, 1968; Bainbridge & Stark, 1979). Hence, Bainbridge and Stark believed that there are really three different ways in which NRMs may emerge: through the process of spontaneous subcultural evolution; through the dynamic leadership of a charismatic leader who genuinely believes in the veracity of his or her teachings; or through the leadership of a charismatic "entrepreneur" who sees religion as a money-making scheme. Social scientists are therefore not in agreement on the necessity of a charismatic leader in the formation of an NRM.

Externally, the sect and the NRM look much alike, although NRMs tend to appeal to the middle and upper classes, while sects generally appeal to the lower classes (Stark, 1986). However, both rebel against the predominant cultural values, both lack trained professional leaders and a bureaucratic structure, and both insist on a stringent membership policy that requires significant commitment (although the NRM may not require this in its very earliest stage of development). The NRM, like the sect, is also capable of institutionalization. In fact, some institutionalization is necessary if the group is to outlive its founder.

The evolution of sects and NRMs is comparable as indicated in Figure 7.4. This diagram, while oversimplifying the process, does suggest the parallel manner in which these two types of groups may evolve. If we allowed for the two-dimensional analysis Yinger proposed (Figure 7.3), we would have a more accurate picture: Institutionalization and accommodation to secular values do not necessarily occur

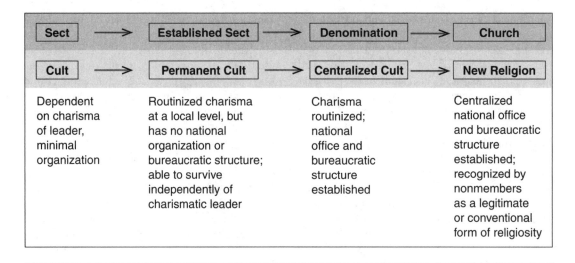

Sect →	Established Sect →	Denomination →	Church
Cult →	Permanent Cult →	Centralized Cult →	New Religion
Dependent on charisma of leader, minimal organization	Routinized charisma at a local level, but has no national organization or bureaucratic structure; able to survive independently of charismatic leader	Charisma routinized; national office and bureaucratic structure established	Centralized national office and bureaucratic structure established; recognized by nonmembers as a legitimate or conventional form of religiosity

Figure 7.4   Evolution of Sects and NRMs

SOURCE: Nelson, G. K. (1968, November). The concept of cult. *Sociological Review,* 360. Used by permission of Routledge and Kegan Paul Ltd.

at the same rate. Sects and NRMs, then, tend to evolve along parallel lines.

To understand fully the evolution of a sect or NRM, however, we must give attention to the hostility toward the group by traditional religious groups in the society and to the strength of the desire for legitimacy within the new group. In other words, we must take into account the pressures at work on a group as it attempts to mobilize its resources and to counter opposition to its existence. Bainbridge and Stark (1979) insisted that hostile forces may cause a sectarian group to evolve into an NRM; as its beliefs are modified to define its separateness and to cope with being made "alien," it develops a unique worldview. Jim Jones's People's Temple of Jonestown is an example of this sect-to-NRM movement. With this added variable that affects the development and self-image of a group, our view of sect and cult emergence becomes more complex. (See Figure 7.5.)

A central point here is that both sects *and* NRMs lack organizational complexity and

both reject at least some of the values of the secular society. However, the sect views its role as one of purifying the traditional faith by calling members back to what are believed to be core principles. NRMs (or cults) represent the nascent stages of the development of a new or a syncretistic religion. Other factors may affect the evolution of a sect or an NRM, but the institutionalizing and accommodating tendencies frequently cause sects to become denominational in form. The cult usually either dies out or institutionalizes into a new religion. The debate previously discussed regarding whether Mormonism is an NRM or another denomination of Christianity is an argument about how that group has evolved.

No one diagram is capable of demonstrating the many factors that affect the evolution and development of a particular religious group. This one looks at sects and NRMs from a slightly different lens than does the one developed by Yinger. It is not possible graphically to combine all the complexity of both models, but Figure 7.6 on page 182 is an attempt to synthesize

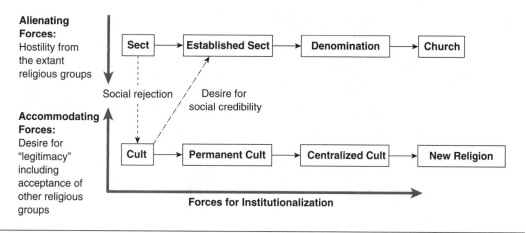

Figure 7.5   Social Forces and the Sect–Cult Transformation

these two perspectives into a three-dimensional or cube model. It emphasizes the fact that the force field on any religious movement involves three different types of pressures, and as each group chooses strategies to cope with each set of dynamics, the evolution of the group is affected. If we were to name each stage or position within this model, we could potentially have 64 "types" of religious groups. However, our real purpose here is to focus on the field of forces at work on groups rather than creating another typology.

If all of this seems terribly complex, it is because the evolution of religious groups *is* complex. Various scholars have focused on different aspects of religious group evolution and therefore have defined such terms as sect, cult, and church in alternative ways. This is an important point for the new student of religion to bear in mind when reading a study that discusses sects and cults. Readers should always be sure they understand how a given author defines his or her terms and how those terms were operationalized or measured.

Stark and Bainbridge added another important dimension to our understanding of religious group emergence. Using demographic data on church affiliation in various regions of North America and comparing these data with information on location of NRM headquarters and on highest concentrations of NRM memberships, Stark and Bainbridge (1985) concluded that where traditional religious organizations are weak, NRM formation is highest. Where religious tradition is strong and church membership is high, NRM formation is less common. The West Coast region, from Southern California to northern British Columbia, has the lowest levels of church membership and the greatest incidence of NRM success on the continent. These scholars believe that secularization breeds NRM development; they insist that secularization involves a vacuum in which certain human needs are left unmet. If existing religious organizations are not satisfying those needs, NRMs will arise to fill the vacuum. Their thesis will be explored in more detail in Chapter 12, but we note here that sects are more likely to form than are NRMs if traditional forms of religion are well established.

Theories about sect-to-church transformations are generalizations about how this change normally occurs. It may or may not be precisely descriptive of the process as it occurs in a specific group one chooses to study. Furthermore, such terms as *sect, cult, established*

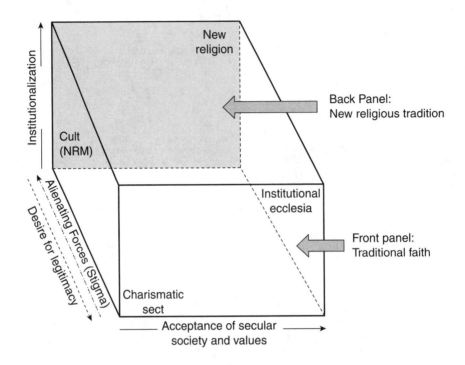

It may help to understand this diagram if you again familiarize yourself with Yinger's model in Figure 7.3. Imagine the front panel of this cube as Yinger's schema, depicting evolution from Charismatic sect to Institutional Ecclesia. The back panel would be similar but would represent evolution from the charismatic cult to a new religious tradition. The alienating and accommodating forces from the larger society, represented in Figure 7.5, are here depicted as forces which could move a group from sect toward cult (front panel to back panel) or vice versa. Thus, the forces in the evolution of the group represent three types of choices regarding organization and environment that groups must make. This accounts for the diversity of types of groups that exist at any one time. This must be seen as a dynamic model, however, with groups moving over time from one position to another.

Figure 7.6 The Force Field of Religious Groups: A Three-Dimensional Model

sect, and *denomination* are normally not ones that can be permanently assigned to a group. They simply describe organizational characteristics of a group at a particular moment in its history. With this in mind, church–sect theory can enable us to describe the normal process of organizational change in religious collectivities. References to churchlike or sectlike groups also allow us to identify similarities of religious style among certain religious groups and differences in character with others at any given time. One interesting application of church–sect theory is the use of the adjectives *sectlike* and *churchlike* to describe the religiosity of individuals—regardless of the groups to which those individuals belong (Demerath, 1965). This has yielded fruitful insights, but students encountering such a study should be sure they understand how *churchlike* and *sectlike* were defined and measured.

In short, if one seeks to generalize about how most groups evolve and the stages they go through, the concepts of church, sect, and denomination can be helpful. However, if one seeks to understand a specific group, one must learn that group's characteristics on each issue. One must discover various dimensions of the group, including theological orientation, relationship with the secular society, amount of formality of the ritual ("high church" being very formal, elegant, and decorous and "low church" being informal, unpretentious, and spontaneous), appeal of the ritual (with a "Dionysian" appeal to emotions or an "Apollonian" appeal to reason), and the extent of organization (the complexity of the structure; whether there are official clergy and whether the clergy are formally trained; and the relative restrictiveness of membership policy). Other elements may also be included in this analysis (Greeley, 1972). Looking at each issue separately rather than constructing a typology is sometimes called "choice point analysis" (O'Dea, 1968; Winter, 1977). Because these elements do vary somewhat independently of each other, analyzing these dimensions without reference to sectlike or churchlike categories can sometimes be very useful.

> *Critical Thinking:* Using sociological perspectives and insights, respond to the following statement: "Cults are inherently evil and are led by dangerous, dictatorial leaders."

SUMMARY

Many religious movements appear to get started through the dynamism of a charismatic leader (Jesus, Mohammed, Sun Myung Moon, Ann Lee). The individual is viewed as an agent of God or perhaps as God incarnate (in the flesh). What that person says is held as true simply because he or she said it. However, Weber held that charisma is inherently unstable and if a group is to survive it must routinize—that is, it must develop norms, roles, and statuses; it must transfer the sense of awe from the individual personality to the teachings and the organization; and it must make provisions for succession of the leader when he or she dies. As these policies are spelled out and various statuses gain specialized job descriptions, the movement is taking its first steps toward institutionalization. Many decades later the group may be highly bureaucratized.

Some new religious groups are started by ideological leaders rather than charismatic ones. They are the interpreters of some scripture or other ideological source that is inherently authoritative. Yet even in cases where an ideological leader initiates a new religious group, routinization is essential if the group is to survive.

Regardless of the cause of institutionalization, it brings many changes in the religious movement itself. Bureaucracies take on a life of their own, somewhat independent of the lay members, and tend to influence the workings of the whole organization. Over time, the institutionalized religion is likely to face the dilemmas that routinization creates. Institutionalization is a necessity, but it has mixed consequences for the group.

In attempting to classify different styles of religious groups, various scholars have described different types of religious groups—sect, denomination, and ecclesia (church)—and have examined the typical mode of religious group evolution.

The lack of precise correlation of the many variables in multidimensional formulations has led to efforts to simplify the number of factors that are used to define each type of religious group. Some scholars suggest using a single variable (Johnson; Greeley); others (such as Yinger) use two or three variables and identify several subtypes of religious groups (charismatic sect, established sect, diffused denomination, institutionalized denomination, etc.). This procedure has contributed to the identification of alternative paths of religious group evolution, depending on the extent to which a group institutionalizes and accepts the values of the dominant society.

The concept of cult or NRM has also been added to the types of religious groups. Although the concept of cult has been used in various ways by different scholars, we have (with Nelson, Yinger, Bainbridge & Stark, and others) distinguished the cult as a new (or imported) religious movement in a society (also frequently called an NRM to avoid the stigma of popular usage given to "cult"). This type of group is differentiated from the sect, which represents the regeneration of a traditional religion in that society. The NRM, like the sect, goes through processes of institutionalization and modification. However, if it survives and eventually gains legitimacy in the society, the cult becomes a new religion rather than a denomination of an existing religion. As in the case of the sect, many social forces affect the development of the NRM and shape its style and character. The extent of institutionalization, pressures to be "legitimate" in the society, experiences of rejection and stigmatization by the traditional religions, and the extent to which the group seeks to be an "open social system" are just a few of the key factors shaping the evolution of a religious group.

8

ORGANIZED RELIGION

Denominationalism and Congregationalism

Here are some questions to ponder as you read this chapter:

- What factors—social and theological—have caused the fission (schisms) and fusion (mergers) of denominations?
- Why have denominationalism and congregationalism become the central features of organized religion in the United States?
- What are the ways in which denominations organize themselves?
- How are religious special purpose groups and paradenominational groups supplanting denominations?
- How do congregations work, and how do megachurches differ from the typical local congregation?
- Is marketing religion—selling it like any other product—a perversion of the faith, or is it a normal and healthy process?
- What are "new paradigm" churches, and how are they using marketing ideas to grow at phenomenal rates?

In the previous chapter, we looked at some important aspects of organized religion, such as the processes by which religions institutionalize and the various ways that sociologists have tried to categorize different types of religious organizations (church, sect, denomination, and new religious movement [NRM]). In this chapter, we continue to think about the way religion is organized, taking a distinctly American focus as we consider denominationalism and congregationalism as central organizing principles of religion.

Recall that church–sect theory was initially developed by Max Weber and Ernst Troeltsch, both Germans and well aware of the tradition of state established religions, which was a central part of German (and European) history in the early modern period. In Chapter 7, we suggested that the concept of "church" as Weber and Troetsch imagined it was not possible in a democratic and pluralistic society like the United States. Not surprisingly, it was an American, H. Richard Niebuhr, who introduced the concept of "denomination" into the church–sect equation, and another American, Milton Yinger, who elaborated the concepts of sect and denomination even further (identifying various types of sects and denominations). This expansion of the original model has led some to conclude that church–sect theories have become such a hopeless hodgepodge of definitions and variables they have no real meaning or utility (Goode, 1967; Greeley, 1972; Murvar, 1975).

At least in the modern American context, it may be wiser to begin understanding the organization of religion from different starting points. The unique history of religious development in the United States has led one observer to describe America as a "denominational society" and to specifically "reject the notion that denomination is a compromise or halfway house between sect and church" (Greeley, 1972, p. 71). Another focuses his organizational lens even closer to the ground by looking at congregations (local churches, synagogues, mosques, and temples) and arguing that the central organizing principle of American religion is "de facto congregationalism" (Warner, 1993, 2005). To understand organized religion, from this point of view, we have to understand denominations/denominationalism, congregations/congregationalism, and the relationship between the two.

Although religious establishments existed in the American colonies, the United States as a nation is constitutionally diverse in two senses: (1) multiple religious groups have been woven into the fabric of American life from its founding and (2) this pluralism became embodied in the disestablishment of religion in the U.S. Constitution. Disestablishment allowed for free exercise of religion—religious voluntarism—which has taken organizational form in the proliferation of religious groups in America. In the terms of the "religious economies" perspective (Stark & Finke, 2000), American religion is a free market in which religious organizations from corporate giants like the Roman Catholic Church to entrepreneurial start-ups in storefront churches compete for the attention of religious consumers. As we will see, there are winners and losers in this marketplace. This religious pluralism in a free religious marketplace raises the issue of religious marketing, which we discuss in this chapter.

As Nancy Ammerman (2006, p. 355) has argued, both denominations and congregations are part of larger "organizational fields," and as we would expect from the open systems perspective, they are susceptible to influence from this larger field. Hence, both denominations and congregations from all religious traditions have a tendency to become like each other. Organizational theorists call this propensity to morph into similar forms *institutional isomorphism* (DiMaggio & Powell, 1983).

One major development in the organizational field of religion that has the potential to influence both denominations and congregations through this process of isomorphism is the rise of very large congregations known as *megachurches*. In addition to understanding the rise, structure, and spread of megachurches, we will also consider their implications for the marketing of religion, especially to the unchurched and religious "seekers."

THE DENOMINATIONAL SOCIETY

Denominationalism is a unique and recent way of organizing religion in the long history of human society (Ammerman, 2006, pp. 361–362). In his famous work on the evolution of religion, Robert Bellah (1970c) maintained that the early stages of religious development are characterized by an undifferentiated religious worldview, so that individuals experienced the world as a single cosmos in which religion was diffused through all of life. Under these conditions, religious community and society are one and the same.

In the course of societal development, religion becomes symbolically differentiated from other social institutions; in particular, the political and distinct religious organizations arise. Initially, a single religious organization dominates a particular geographic area. This is evident in the domination of Europe by the Roman Catholic Church. Over time, the religious sphere itself comes to be internally differentiated. In the Western world, the Reformation era (1517–1648) brought about the emergence of different "confessions" which were doctrinally and organizationally distinct: Catholicism, Lutheranism, and Calvinism (Gorski, 2000).[1]

Although this gave rise to some religious pluralism between societies, each confession attempted to align itself with the political powers to become the officially sanctioned church in its territory. Thus, Lutheranism became the official religion in Scandanavia and parts of Germany (e.g., Saxony); Calvinism predominated in Switzerland, the Netherlands, and southern Germany; and Roman Catholicism continued to have official status in France, Spain, and Italy. In each case, political citizenship and confessional identity were linked. Consequently, conflict rather than peaceful coexistence dominated the scene, and religious persecution was prevalent. As Gorski (2000) observed, "The sixteenth and seventeenth centuries witnessed mass movements of religious refugees, a sort of confessionally driven *Völkerwanderung* in which Protestants drove out Catholics, Catholics drove out Protestants, and everybody drove out the Baptists and other sectarians" (pp. 157–158).

Some of those driven out of Europe for religious reasons were followers of the Calvinist Puritan movement who were seeking to reform the Church of England. They ended up founding the New England colonies and creating Congregationalist religious establishments of their own in Connecticut, Massachusetts, and New Hampshire. In six other colonies, the Church of England was established (Georgia, Maryland, New York, North Carolina, South Carolina, and Virginia). Rhode Island, Delaware, New Jersey, and Pennsylvania had no religious establishment, the latter three being among the most religiously diverse colonies. As noted, the result of this situation was that no single confession dominated the American religious scene.

Religious disestablishment and religious freedom are key social structural preconditions for denominationalism to flourish. When Greeley referred to America as "the denominational society," he means a society that is characterized neither by an established church nor dissenting sects but religious bodies or associations of congregations that are united under a common historical and theological umbrella, that are presumed equal under the law, and that generally treat other bodies with an attitude of mutual respect. As a consequence of this "social organizational adjustment to the fact of religious pluralism" (Greeley, 1972, p. 1), there are hundreds of denominations in America (Mead, Hill, & Atwood, 2005). Indeed, the *Handbook of Denominations in the United States* (Mead et al., 2005) lists 31 Baptist denominations and

[1]These are called "confessions" because they have distinct "confessions of faith." A confession of faith is a particular religious group's doctrinal statement. Among the most significant of these historically are The Augsburg Confession (1530), written by Martin Luther and severing ties with the Roman Catholic Church, and the Geneva Confession (1536), originally credited to John Calvin (Pelikan, 2003).

17 denominations in the Calvinist/Reformed tradition alone. This denominationalism has also become a global phenomenon; the *World Christian Encyclopedia* reports 33,830 denominations within Christianity worldwide (Barrett, Kurian, & Johnson, 2001).

William Swatos (1998) observed the following:

> Although, strictly speaking, denominationalism is a Protestant dynamic, it has become fully accepted in principle by all major religious groups in the United States; in fact, one could say that the denominationalizing process represents the *Americanizing* of a religious tradition. (p. 135)

In this understanding, other Christian religious traditions like Roman Catholicism, Mormonism, and Seventh-day Adventism are also considered denominations. Beyond Christianity, some have even written about Jewish "denominationalism" in reference to the four branches of Judaism— Orthodox, Conservative, Reform, and Reconstructionist (Lazerwitz, Winter, Dashefsky, & Tabory, 1998). The Nation of Islam and American Society of Muslims are distinctively American Islamic denominations (though the former is often viewed negatively by traditional Muslims). In fact, the American Society of Muslims evolved from the Nation of Islam under the leadership of Warith Deen Muhammad and once established saw a splinter group calling itself the Nation of Islam break from it with Louis Farrakhan as head (see Chapter 12 for more details). If Swatos's observation is correct, we could in time expect to see Sunni and Shia Islamic organizations develop as distinctive Islamic denominations, perhaps with a national twist: Nigerian Sunni Muslims of America, American Federation of Pakistani Shia Mosques, the Association of Indonesian Sunni Mosques in America, and so on. Buddhism and Hinduism also have different branches or schools, though

like Islam they were brought to America in large numbers only recently so time will tell whether there will be an evolution of those branches into recognizable denominations.

Theology and Social Processes in the Proliferation of Denominations

The development of this diversity of American denominations has been the result of theological differences and disputes, as well as theological innovations by religious entrepreneurs but also of fundamental social processes such as immigration and racial conflict. (An excellent, free online resource for information on American denominations, including profiles, membership statistics, and family trees, is the Association of Religion Data Archives: http://www.thearda.com/Denoms/Families.)

Race

Methodism, for example, has been heavily influenced by theological differences and race. The denomination itself began as a revitalization movement in the Church of England under the direction of John Wesley (an ordained minister of the Church of England) whose preaching emphasized personal holiness. As the movement spread, it eventually broke from the Church of England when the Methodist Episcopal Church in the United States was organized in 1784. By 1816, the African Methodist Episcopal (AME) Church was founded by African American congregations seeking independence from white Methodists, and not long after, the AME Zion Church ordained its first bishop, symbolizing its split from the larger (white) body of the Methodist church. In 1844, the Methodist Episcopal Church, South split from the Methodist Episcopal Church over the issue of slavery[2] and later amicably split with the Colored Methodist Episcopal (CME) Church

---

[2]It is well known that John Wesley, the founder of Methodism, was vehemently opposed to American slavery, and early in its history, the Methodist Episcopal Church opposed the practice. See Wesley's "Thoughts Upon Slavery," published in 1784 (available online at http://docsouth.unc.edu/church/wesley/menu.html).

in 1870. In 1939, the two historically white Methodist Episcopal denominations, along with the Methodist Protestant Church, reunited to form the Methodist Church, which was joined in 1968 by the Evangelical United Brethren Church to become the United Methodist Church—the second largest Protestant denomination in America. All these splits and mergers were related to issues of race, not theological differences.

Other denominations were also affected by racial differences and differences over racial issues. Presbyterians were split over the issue of slavery, dividing in the mid-19th century between the United Presbyterian Church in the USA and the Presbyterian Church in the United States ("Southern"). The Southern Baptist Convention was formed in 1845 when Baptists in the South—who held to a biblical defense of slavery—split to form their own denomination.[3] Free slaves after the Civil War founded the National Baptist Convention in 1895, creating yet another race-related schism.

Immigration

Lutheranism has been a predominantly white denomination in the United States, owing to its historic establishment in parts of Germany and Scandinavia. Its history, therefore, has been shaped not by race but by immigration and theological disputes. When Lutherans immigrated, they brought their national churches with them, as well as their particular languages and cultural practices, which prevented unity among groups that shared many religious views in common. Lutheran denominations (typically called "synods" or "churches"), therefore, proliferated in the late 19th and early 20th centuries as Lutheran immigrants flowed into the country. For example, Swedes founded the Augustana Evangelical

Lutheran Church in 1860, Danes the American Evangelical Lutheran Church in 1872, Finns the Finnish Evangelical Lutheran Church of America in 1890, and Norwegians the Lutheran Free Church in 1897.

Over time, as these immigrants assimilated into American society, theological differences became more prominent than differences of nationality. This is reflected in the current state of Lutheran denomination in the United States. Currently, there are 10 Lutheran denominations organized in the United States.[4] The three largest of these are the Evangelical Lutheran Church in America (ELCA), Lutheran Church-Missouri Synod (LCMS) (est. 1847 by Germans), and Wisconsin Evangelical Lutheran Synod (WELS) (est. 1850 by Germans). (The word "Synod" in Lutheran usage can refer either to an entire religious body like the Lutheran Church-Missouri Synod or a local administrative region like the Texas Synod, Iowa Synod, etc.).

Important theological differences and religious practices exist among these three groups (e.g., over the interpretation of scripture, ordination of women and gays, and the role of women in lay authority positions), with the ELCA being the liberal or mainline Lutheran church, the LCMS more conservative, and WELS the most conservative.

The history of denominations is not only one of division and schism but also of mergers and consolidation (Chaves & Sutton, 2004). As Figure 8.1 shows, the largest Lutheran church in America, the ELCA, was formed out of a complex series of mergers beginning in the early 20th century. It was formed in 1987 with the merger of three Lutheran denominations, each of which were formed from the merger of several other Lutheran denominations, which were themselves formed from mergers of various Lutheran denominations and synods.

---

[3]In 1995, the Southern Baptist Convention for the first time acknowledged its racist origins and apologized for its defense of slavery. See *This Side of Heaven: Race, Ethnicity, and Christian Faith* (Priest & Nieves, 2007).

[4]American Association of Lutheran Churches, Apostolic Lutheran Church of America, Association of Free Lutheran Congregations, Church of the Lutheran Brethren of America, Church of the Lutheran Confession, Evangelical Lutheran Church in America (ELCA), Evangelical Lutheran Synod, Latvian Evangelical Lutheran Church, Lutheran Church-Missouri Synod (LCMS), and Wisconsin Evangelical Lutheran Synod (WELS) (Mead et al., 2005).

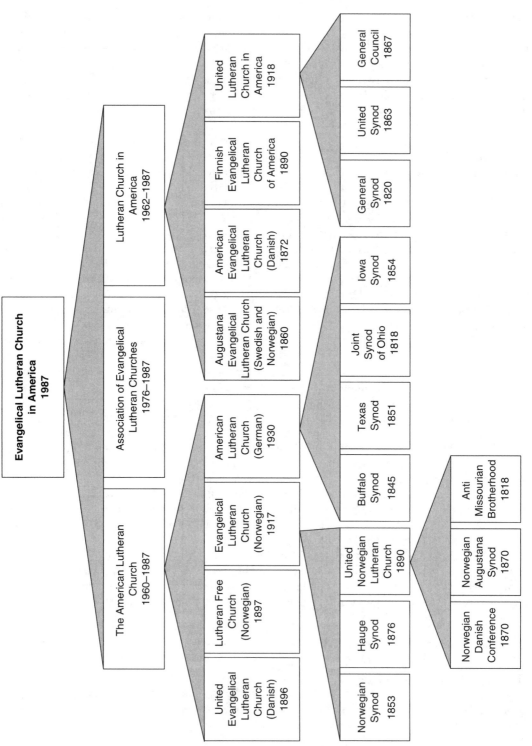

Figure 8.1 Mergers Resulting in the Evangelical Lutheran Church of America

There are 10 Lutheran denominations in the United States as of 2010, and most of the splits have been based on immigration and ethnicity patterns rather than race or doctrinal differences. This ELCA congregation is affiliated with the largest of the Lutheran groups—being the largest because of a series of consolidations. So denominationalism is characterized by fission (splits) and fusion (mergers).

These developments are not just remnants of a historical past. In the dynamism of the American "religious economy," denominations are continually developing and dying, splitting and merging. As of this writing, there may possibly be yet another split in the Lutheran church over a theological issue. As we discuss at more length in Chapter 11, the ELCA voted in 2009 to ordain noncelibate gay clergy, and in 2010, the first gay Lutheran ministers were welcomed onto the church's clergy roster. As a consequence, 185 ELCA congregations have voted to leave the denomination, and a new Lutheran denomination is in the works: the North American Lutheran Church.

Innovation

A final significant source of new denominations in America is the theological innovations of religious entrepreneurs. Joseph Smith founded the Church of Jesus Christ of Latter-day Saints

(the Mormons) in the 1820s, Charles T. Russell the Jehovah's Witnesses in 1870, and Mary Baker Eddy the Church of Christ, Scientist, aka "Christian Science" in 1879. All of these groups are innovative updatings of the Christian tradition, and although they were attacked early on as "cults" (in the negative sense sociologists reject), they have since grown to relatively large and respected denominations (Jenkins, 2000). As we saw in Chapter 6, amazing growth has particularly been among Mormons and Jehovah's Witnesses.

BEYOND DENOMINATIONS?

Although the United States is still a "denominational" society, there are some significant developments in American religious life that suggest the importance of looking beyond denominations. These include transdenominational evangelicalism, nondenominationalism, and paradenominational groups and organizations.

Transdenominational Evangelicalism

By all accounts, one of the most dynamic sectors of the American religious economy—and, indeed, the global religious economy—is evangelical Christianity. Although still in the minority, evangelical Protestantism has made significant inroads in Africa, Latin America, and Asia (Freston, 2007). In the United States, individuals belonging to evangelical traditions now constitute over half of all Protestants (Pew Research Center Forum on Religion & Public Life, 2008b).

Although for ease of identification in survey-based studies some scholars define individuals as evangelicals according to their denominational affiliation (commonly named evangelical denominations are the Southern Baptist Convention and Churches of God), most recognize that evangelicalism is a movement, not a denomination or even a collection of denominations. It is, in fact, a *transdenominational movement* that had its origins at a specific place and specific point in time: with the founding of the National Association of Evangelicals in April 1942 at the Hotel Coronado in St. Louis, Missouri (Marsden, 1991). According to sociologist Christian Smith (1998), "Evangelicalism is not primarily denominationally based but . . . 'a transdenominational movement in which many people, in various ways, feel at home.' . . . Institutionally, this transdenominational evangelicalism is built around networks of parachurch agencies" (p. 135). These agencies include seminaries like the Moody Bible Institute, publications like *Christianity Today*, colleges like Wheaton, and publishing houses like Zondervan. Some churches and denominations do contribute to the movement—Vineyard Christian Fellowship and Calvary Chapel are examples. Some observers have also noted the influence of the evangelical movement within otherwise mainline denominations (Kosmin & Keysar, 2009). Nonetheless, Smith and others are right to say that evangelicalism in the United States is not fundamentally denominational.

Evangelical Protestantism's transdenominational organization may be one of the sources of its strength today. The theoretical issue of interest with respect to evangelical strength has to do with the plausibility of traditional religious worldviews in the modern pluralistic world. One widely respected sociologist of religion, Peter Berger, has devoted much of his writing to the exploration of *plausibility structures* (Berger, 1967; Berger & Luckmann, 1966). These are social interactions and processes within a group that serve to protect and sacralize the shared meanings and outlooks of the group. Much of Chapter 4 was concerned with this issue, and the emphasis in Chapter 5 on the importance of affective commitment in maintaining moral commitment (adherence to ideology) points to the same phenomenon.

Berger's view is that belief systems, if they are to survive, must be rooted in a plausibility structure. An understanding of the religious group as a reference group is particularly critical

to any analysis of plausibility. Individuals are capable of accepting all sorts of strange beliefs if enough other people seem convinced. It is a basic maxim of social constructionism that individuals look to others for a definition of the situation if they are uncertain themselves. When dealing with issues of the meaning of life, with the supernatural, and with the ultimate cause of perplexing events, ambiguity is a given. Plausibility structures—which typically include rituals, symbols, music, architecture, reference groups, self-validating beliefs—are especially critical in making beliefs credible if they run contrary to those of the larger society, as we see in the "Illustrating Sociological Concepts" feature. Moreover, plausibility is especially problematic in a pluralistic society where people are constantly exposed to meaning systems and believers who seem to contradict their own ideas. Thus, for Berger (1967, p. 152), pluralism itself undermines the plausibility structures that support religious belief.

Some of the world's great cathedrals elicit a sense of awe and humility in the individual just through the effective use of space and the breathtaking beauty of the chancel area. Anyone entering the Notre Dame Cathedral in Montreal is likely to feel such awe. The architecture, by creating a mood, enhances the plausibility of belief in the transcendent power of God.

ILLUSTRATING SOCIOLOGICAL CONCEPTS

Plausibility Structures Support Beliefs

The plausibility or believability of the worldview is essential if a religious group is to survive. If everyday events or if scientific explanations seem to disprove the religious worldview, the survival of the group may be threatened. Belief systems, if they are to survive, must be rooted in a social base and reinforced through a sense of sacredness or absoluteness about the beliefs. Plausibility can be enhanced through a variety of mechanisms: reference groups committed to the belief system, dualistic belief systems, norms requiring that members engage in evangelism, a sense of profound respect or awe for the leadership, rituals that elicit a sense of awe, music that evokes strong emotions, emotionally laden symbols, and even especially beautiful and expansive architecture.

Individuals are capable of accepting all sorts of strange beliefs if the plausibility structures are strong enough. If a religious leader stakes his or her reputation on a specific time, date, and year when the end of the world will come, the movement may be in serious trouble the day after the predicted apocalypse. What happens to the belief system when time drags on and the end does not arrive? If they are going to survive, world-transforming movements must develop belief systems that are self-validating (Festinger, Riecken, & Schachter, 1956).

An important characteristic of the ideology of many world-transforming religious groups is *dualism*, the belief that reality is ultimately a battle between the forces of good and evil—perhaps personalized in the form of God versus Satan. Dualism is very functional in sustaining plausibility of an otherwise implausible worldview. In other words, groups that have a worldview that contradicts the scientific and "commonsense" explanations are more likely to survive if they have a worldview that is dualistic. Any outcome or event can be explained as a cosmic battle—dualistic perspective appears to be self-validating. As Lofland (1977) put it, "the believer cannot lose. He derives confirmation from any outcome" (p. 197). Lofland explained that the Moonies he studied often used the conventional mass media to confirm their worldview. When national and international events reflected unrest, deterioration, and disorganization, the devotees were jubilant. Surely this was evidence to everyone that the end was near! Surely this victory by Satan would convince people to repent and join the unification cause. If a month went by when no tragedies occurred, this was because God was restraining Satan for some special reason.

On the other hand, some doomsday prophets have set a *specific date* on which the final judgment will commence. What happens to such groups when the predicted date passes? In many cases, passage of the doomsday date results in the movement's collapse. However, the Unification Church passed several critical dates established by the founder, Reverend Moon. In these cases, the ideology was modified by simply proclaiming that the end of the world will come in several phases. The cataclysm that was expected was simply declared to have occurred—but in the unseen spiritual realm. Some Christian groups that predicted the end of the world have similarly modified their stance. In some cases, they simply set a new date. A more effective strategy is to proclaim that the new era is now in progress but that only the "saved" or the "elect" are able to see or experience the transformation. This has been the strategy of the Jehovah's Witnesses for whom the "new age" began in 1914 (Kephart & Zellner, 1998). In any case, basic dualism can continue to provide a simple and self-validating system of meaning.

(Continued)

(Continued)

Another apparently self-validating belief system is taking a body of scripture—whether it's the Quran, the Bible, or the Book of Mormon—as literally and absolutely true. Passages testifying to the veracity of the contents may serve to reinforce the believer's conviction. All other forms of evidence (scientific or otherwise) can be readily dismissed if they contradict those scriptures. The sacred scrolls become the final authority on all things, and the sense of sacredness that surrounds that scripture makes anything in it seem plausible. In this case, the utter respect for the scripture serves as a self-validating plausibility structure.

Close-knit reference groups and awe-inspiring rituals and symbols serve as important plausibility structures as well. However, for groups that reject the outlook of the dominant culture, dualism provides a worldview that not only offers to explain the meaning of events but also provides a neat tautology (circular reasoning) that allows for self-validation of the worldview. It is not surprising then that many religious movements, especially the world-transforming ones that have managed to survive, are dualistic, evangelistic, and otherworldly.

In his analysis of American evangelicals, Smith took issue with Berger's argument. He maintained that "American evangelicalism as a religious movement is thriving . . . very much because of and not in spite of its confrontation with modern pluralism" (Smith, 1998, p. 1). Smith suggested a *subcultural identity theory* of religious strength in modern society that accords well with the understanding of religion we suggest in this textbook: "Religion survives and can thrive in pluralistic, modern society by embedding itself in subcultures that offer satisfying morally orienting collective identities which provide adherents meaning and belonging" (Smith, 1998, p. 118).

Part of evangelicals' collective identity has to do with their distinguishing themselves from non-evangelicals as a negative reference group ("out-group") and thereby creating some boundaries but at the same time keeping the boundaries somewhat permeable so as to not cut themselves off entirely from the wider society (as fundamentalists do). By doing so, they create the social basis (plausibility structure) necessary to support their religious beliefs. About this Smith (1998) concluded the following:

> In the pluralistic modern world, people don't need macro-encompassing sacred cosmoses to maintain their religious beliefs. They only need "sacred umbrellas," small, portable, accessible relational worlds—religious reference groups—"under" which their beliefs can make complete sense. (p. 106)[5]

Evangelicalism as a transdenominational movement helps provide the plausibility structures that support individuals' beliefs and that help make evangelical Protestantism one of the strongest traditions in American religion.

Nondenominationalism

A second religious development that takes us beyond denominations proper is the rise of

[5]Smith (1998) elaborated the metaphor: "We would like to suggest, however, that in the modern world, religion does survive and can thrive, not in the form of 'sacred canopies,' but rather in the form of 'sacred umbrellas.' Canopies are expansive, immobile, and held up by props beyond the reach of those covered. Umbrellas, on the other hand, are small, handheld, and portable—like the faith-sustaining religious worlds that modern people construct for themselves. We suggest that, as the old, overarching sacred canopies split apart and their ripped pieces of fabric fell toward the ground, many innovative religious actors caught those falling pieces of cloth in the air and, with more than a little ingenuity, remanufactured them into umbrellas" (p. 106).

nondenominationalism, both in the form of nondenominational congregations and individual religious identities. We have already seen in Chapters 5 and 6 the increasing number of Americans who choose not to identify their religion according to a denominational label (including those who specifically say "nondenominational" but also those who are "just Christians" and the "religious but unaffiliated"). Organizationally, nondenominational congregations are the single largest category of faith community in terms of affiliation. Roughly 18% of congregations in the United States are nondenominational. Mark Chaves (2004) has observed the following: "If the unaffiliated congregations were all in one denomination, they would constitute the third-largest U.S. denomination in number of participants" (p. 25).

Nondenominationalism is actually related in part to evangelicalism, and vice versa. In one survey, 20% of evangelicals reported attending nondenominational churches (Smith, 1998). Another study—the American Religious Identification Survey (ARIS)—found that from 1990 to 2008, most of the growth in the Christian population occurred among those who would identify only as "Christian," "evangelical/born again," or "nondenominational Christian." Taken altogether, these three groups grew from 5% of the population in 1990 to 11.8% in 2008. Looking to the future, Kosmin and Keysar (2009, p. 12) observed that among all Christians, these Generic Christians have the youngest age composition.

In the press release announcing the release of ARIS 2008, Mark Silk (director of the program that sponsors ARIS) suggested that a "generic form of evangelicalism is emerging as the normative form of non-Catholic Christianity in the United States." If this is true, it may soon be possible to view nondenominational and transdenominational evangelicalism as a sort of "denominational label" (Ammerman, 2006, p. 364)—a category people can use to identify themselves as holding certain beliefs and engaging in certain practices in common with others across the United States and, perhaps, around the world.

Paradenominational Groups and Organizations

Robert Wuthnow (1987) argued in *The Restructuring of American Religion* that religion in the United States is experiencing a dramatic shift rooted in special purpose groups. Since early in the 19th century, special purpose groups have operated alongside of denominations, crossing boundaries and providing for joint efforts between various groups. The American Bible Society (founded in 1816) was followed in that century by such groups as the American Sunday School Union, the Women's Christian Temperance Movement, the Anti-Saloon League, the American Anti-Slavery Society, and the YMCA. People of diverse religious affiliations banded together to address some problem or provide some service.

The 20th century has seen a continuation of the spawning of special purpose groups, with religiously motivated people mobilizing associations across denominational lines to address public issues: war and peace, abortion legislation, world hunger, civil rights, changes in gender roles, business ethics, and the recruitment activities of cults. The Full Gospel Business Men's Fellowship, the Christian Legal Society, the Fellowship of Christian Athletes, and Clergy and Laity Concerned about War are only a few examples of special purpose groups for which religion is intrinsically tied to social and political issues.

Until the 1970s, said Wuthnow, the members of these organizations still felt primary loyalty to their denominations and it was the faith communities, as such, which defined their core values and their sense of morality. The special purpose groups were secondary in their sense of belonging and identity. However, in the last three decades intense conflicts have arisen over many issues relating to moral conduct and how to redefine reality. Wuthnow believed that the nation has polarized into conservative and liberal camps, each with their own sets of special purpose groups. Members of each camp may belong to the same congregation/denomination, so that community is split and is no longer able to claim the moral authority or to elicit the deep loyalty necessary to define meaning and to sacralize values. The deeper loyalty, he says, is going to the paradenonimational groups, and these paradenominational groups are growing remarkably in both numbers of groups and in memberships.

This could fundamentally alter the structure of religion in the United States, so that religious conservatives from various denominations who are adherents of, say, The Christian Voice, may find that the paradenominational group elicits more loyalty and does more to provide a sense of identity and belonging than do the individual denominations to which the members each belong. Wuthnow believes the face of religion in the United States is being fundamentally restructured by these paradenominatinal groups. Denominations, he believes, are no longer the central structural element of American religion.

Robert Wuthnow's assessment is based on some sweeping interpretations of history, and it remains a controversial thesis. It, along with the idea of transdenominationalism and nondenominationalism, does provide a new way of thinking about the historic shifts that appear to be occurring in American religion.

Critical Thinking: From your own experience and observations, does it seem feasible that special purpose groups (cross-denominational groups concerned with issues of abortion or peace and justice or prayer in schools) are replacing denominations in providing belonging and in defining religious meaning and values? Why do you think as you do? What kind of evidence is persuasive?

DENOMINATIONS AND DE FACTO CONGREGATIONALISM

Organizationally, denominations are a group of congregations linked through centralized bodies to which congregations grant some overriding authority and to which individuals direct some loyalty as members.[6] Although denominations are an organizational reality in American society and people often use denominational labels to describe themselves, no one actually "belongs" to a denomination. People are members of particular congregations (Chaves, 2004, p. 21). These congregations are often, though decreasingly, affiliated with denominations, but most people's experiences of denominations are wholly mediated by the local congregation to which they belong. This is the centrifugal force that R. Stephen Warner (1993) described as "an institutionalized bias of American religious life" called "de facto congregationalism" (p. 1066). Even churches that are hierarchically organized by church law become "congregationalist" in fact. Regardless of the formal structure of authority in a denomination or religious tradition, local congregations tend to have a significant measure of autonomy to do their own thing (Ammerman, 2006).

The way that denominations *officially* relate to local congregations is determined by the denominational "polity" structure. Three ideal type polities have been identified: *congregational*, *episcopal*, and *presbyterian* (Moberg, 1984).[7]

In a congregational polity, the authority of the local congregation is supreme. For example, although the thousands of Baptist congregations in the United States are typically part of various

---

[6]Of course, not every religious group conforms to this pattern. Exceptions include Native American religions and religious groups practicing Wicca (witchcraft). We are unable to treat another interesting way of looking at denominational organization—Mark Chaves' (1993) analysis of denominations as "dual structures."

[7]Interesting research that uses polity structure as an independent variable in understanding denominations includes Takayama's work (Takayama, 1975; Takayama & Cannon, 1979; or more recently, Cantrell, Krile, & Donohue, 1983; Mullen, 1994).

denominations (like the Southern Baptist Convention or American Baptist Churches), they are nonetheless formally independent. They choose their own pastors, control their own finances, own their own property, and so on. This is not to say that some denominations that are formally organized as congregational polities do not exercise influence on their member congregations. As Ammerman (1990, 2009) observed, at the height of its power, the Southern Baptist Convention "provided such comprehensive programmatic support that local congregations from Atlanta to Dallas bore strong resemblance to each other" (p. 574).

The Roman Catholic Church provides a good example of the episcopal polity. Also called "hierarchical," this type of church polity places ultimate authority over local churches in the centralized hands of bishops (the Greek word is *episcopos*). Each of the Roman Catholic Church's nearly 20,000 congregations in the United States (called "parishes") are geographically defined and clustered into a "diocese," which is under the authority of the local bishop, and ultimately, under the authority of the Bishop of Rome, that is, the Pope. This does not mean, however, that de facto congregationalism does not exist in Catholic churches. In fact, there is a great deal of diversity and difference between parishes due to their local autonomy. In his survey of Roman Catholicism in America, theologian Chester Gillis (1999) observed that

> differences in composition of the parish community, leadership, interests, preaching, programs, worship, and organization can make one parish, even in the same diocese or city, very different from one another. . . . To this extent, all Catholicism is local. (p. 32)

De facto congregationalism is evident in how local dioceses and parishes implement—or choose not to implement—universal rites promulgated by the church hierarchy (Yamane & MacMillen, 2006). It is also seen on Sunday mornings when Catholics across America "church shop," floating from their geographically defined parish to their parish-of-choice, often passing several other Catholic churches on the way (Warner, 2005). Catholics choose their parishes because Catholic parishes have their own local cultures and identities through which official church policies are interpreted (McCallion, Maines, & Wolfel, 1996). Indeed, one of the major conclusions of the Notre Dame Study of Catholic Parish Life in the 1980s was that "relative to the life of the rest of the church, parishes seem to have a life of their own" (Gremillion & Castelli, 1987, p. 47).

Other denominations that have some form of episcopal polity, with a bishop having some authority over the churches, include Eastern Orthodox, Anglican, Episcopalian, Methodist, AME, AME Zion, and some (but not all) Lutherans. Many of these denominations are also affected by congregationalism and local authority. For example, the United Methodist "General Conference," where policy is decided, has an equal number of ordained clergy and laypeople who vote representing their congregations. Likewise, the local United Methodist church building is actually *owned* by the larger denomination but maintained by the local congregation. So power and responsibility is shared.

Presbyterian is a kind of intermediate polity, between the congregational and episcopal, which is derived theologically from the Swiss reformation. Like episcopal polities, it is hierarchical; like congregational polities, it gives some authority to the local church. Some say it was a model for the U.S. Constitution's federal system of government. Like the U.S. government, there are typically levels of governance: Local churches are grouped into larger assemblies sometimes called presbyteries, which are grouped into synods, which are brought together in general assemblies. General assemblies have authority to set policy for the denomination, but decisions at that level must be filtered down through the lower levels of governance for final approval. Local congregations have authority insofar as they elect members to represent them at the higher levels of the denomination's governance structure. Not surprisingly, denominations like the Presbyterian Church USA (PCUSA) and Presbyterian Church in America (PCA) are organized in this fashion. Yet, as Warner (2005) wrote, many Presbyterian congregations use hymnals other than the ones authorized by the denomination and choose ministers trained at seminaries other than denominationally sponsored ones. This de facto congregationalism means that the "label *Presbyterian* on the door no longer conveys a great deal of information to the first-time visitor to a local church" (Warner, 2005, p. 163).

Although the idea of a congregation comes from Christianity, de facto congregationalism means that this organizational form is not limited to the Christian tradition. Scholars studying the Muslim community in America have found convergence toward the congregational model, even though most of the nearly 2,000 mosques (*masjids* in Arabic) in the United States are relatively new, with 87% having been founded since 1970 and 62% since 1980 (Bagby, 2003). According to Islamic scholar Ihsan Bagby (2003), "Most of the world's mosques are simply a place to pray. . . . A Muslim cannot be a member of a particular mosque" (p. 115) because mosques belong to God, not to the people. The role of the imam—the minister—is simply to lead prayers five times a day and to run the services on the Sabbath, including delivery of a sermon. The imam does not operate an organization and does not need formal training at a seminary, unlike most Christian and Jewish clergy. Mosques were historically government supported in other countries, so they had to make major changes in how they operate in North America. Because they could not depend on government funding, Islamic mosques needed to adapt to the congregational model: Members who were loyal to a particular mosque would support it. They also began to put more emphasis on religious education (which had been done largely by extended families in the "old country"), religious holidays at the mosque rather than with families, socializing (potlucks and coffee hours), and life cycle celebrations (birth and marriages solemnized, conducting funerals) (Waugh, 1994). This is a major change in the role of the mosque and the imam for many Muslims.

This same pattern of transformation from the religious structure of the homeland to a congregational structure in America has been seen in other religious communities founded by non-Christian Asian immigrant groups in recent years, including Indian Hindus (Kurien, 1998) and Japanese, Korean, and Laotian Buddhists (Warner, 2005). For example, Carl Bankston and Min Zhou (2000) reported on a fieldwork study of a Laotian Community in New Iberia, Louisiana, in the mid-1990s. The Theravada Buddhist temple constructed there reproduces the festivals and rites of the home country, but its organizational functioning has also been transformed by the American context. In the home country, the *sangha* is defined as the community of monks who have authority over the temple, and all men at some point in time are expected to become monks at least for a period of time. In America, the sangha is dissolved into the local congregation and being a monk becomes a specialized, professional status. Authority becomes vested in the lay members of the temple who "call" the monk to service in their community (a very "Baptist" or "Congregational" idea). The Korean Buddhist Kwan Um Sa temple in Los Angeles grew in membership when it began offering the types of social services often found in American congregations: "marriage and youth counseling, hospital arrangement, hospital visits, arrangement for Social Security benefits, . . . transportation for elderly members" (Yu, 1988, p. 90). The Japanese American Jodo Shinshu Buddhists call their place of worship a "church" and sit in church pews rather than traditional Japanese mats (Warner, 2005, p. 167).

De facto congregationalism is a structural reality of American religious life. It reflects principles of the open systems model: Organizations are influenced by inputs from the larger social environment. In this case, the key aspect of the environment is the pattern of organization constituted by other religious congregations. That environment exerts influence particularly under conditions of uncertainty for new organizations, like those founded by immigrant religious communities. As Warner (2005) argued that "organizations that copy other organizations" (p. 168) have a "competitive advantage" and "reliance on established legitimated procedures enhances organizational legitimacy and survival characteristics" (p. 155; see also DiMaggio and Powell 1983).

CONGREGATIONAL DEMOGRAPHY

No one knows how many congregations exist in America today, but most educated observers estimate that there are over 300,000 (Ammerman, 2009). Because there is no single list or directory of all congregations, it has been very difficult to study a representative sample of them so as to draw generalizable conclusions. Most "congregational studies" have been qualitative case studies (Ammerman, Carroll, Dudley, & McKinney, 1998). Indeed, sociology of religion students are often asked to do case studies of congregations—visiting a congregation and writing or presenting about what they observed—as one of their class assignments. The "Doing Research on Religion" feature in Chapter 2 offers students some suggestions from a veteran sociologist on how to observe congregations sociologically.

The most important recent contribution to the study of congregations in America is the National Congregations Study (NCS). The NCS uses a technique pioneered in the study of other organizational fields to draw a representative sample of congregations in America: it asks individuals participating in a nationally representative study of Americans to name their congregation (Chaves, Konieczny, Beyerlein, & Barman, 1999). To date, there have been two waves of data collection, the first in 1998 had a sample of 1,236 congregations, and the second in 2006 and 2007 had a sample of 1,506 congregations. The longitudinal design of this ongoing study allows scholars to track continuity and change in congregational life (Chaves & Anderson, 2008).[8] Analyzing a representative sample of congregations at two points in time has yielded many insights. Unless otherwise noted, the findings we discuss are from the 1998 NCS data, since published research from the second wave of the NCS is limited at this time.

The first insight from the NCS has to do with the demography of congregations, and the curious relationship between looking at congregations from the perspective of the organization as compared to congregational experiences of individuals. As Chaves (2004) wrote, "although most *congregations* are small, most *people* are associated with large congregations" (p. 18), in both 1998 and 2006 and 2007, the median *congregation* had 75 regular participants (including children), but the median *person* was in a congregation with four hundred regular participants.[9]

One way to understand this "double aspect" of congregations is to think about lining all congregations up from largest to smallest. The

---

[8]Of the congregations surveyed in 2006 to 2007, 262 were also surveyed in 1998. This is known as a "panel" design, and adds even more precision to the study of congregational stability and change because comparisons can be made not between one sample of congregations in 1998 to another sample of congregations in 2006 and 2007 but between the same congregations at two points in time. This allows researchers to make stronger causal claims in explaining change (Halaby, 2004; Hannan & Tuma, 1979).

[9]In statistics, *median* is one way of understanding the *average* value in a collection of values. It refers to the value that is in the middle of a distribution, where half of the values are above it and half of the values are below it. Like the more commonly used *mean* (or arithmetic mean or simple mean), it is a measure of central tendency, but has the benefit over the mean of eliminating the influence of outliers (very large or very small cases). For example, median income is often reported because the income of the superrich like Bill Gates would skew the mean. Similarly, Chaves looks at the median size of congregations to eliminate the excessive influence of the few extremely large congregations that exist. We consider these megachurches more fully later in this chapter.

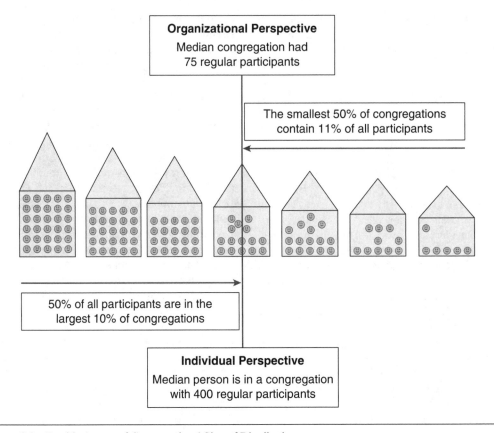

Figure 8.2   Double Aspect of Congregations' Size of Distribution

median congregation is the one in the very middle (see Figure 8.2). If you walked from the smallest congregation toward the largest and stopped in the middle, you would be at a church or temple with 75 regular participants. The proportion of the total population who attend the congregations you passed would only be 11% (Chaves, 2004, p. 18). Eighty-nine percent of regular attendees are in the 50% of congregations that are above the median.

Looking at it from the individual perspective, if you walked from the largest religious communities toward the smallest, once you walked past half of all the people who attend religious services regularly, you would be at a congregation with 400 members. The proportion of all of the congregations you have passed, however, is just 10%.

There are significant denominational differences in congregation size also between the organizational and the individual perspectives. Chaves (2004, p. 24) reported that only 6.2% of all congregations in America are Roman Catholic, but 28.6% of people attend Catholic churches. By contrast, over twice as many congregations are Southern Baptist (16.9%) than Catholic, but half as many people attend Southern Baptist churches (11.2%) compared to Catholic churches.

One consequence of this is that when you ask people to estimate the average size of a congregation, they almost always estimate well over 75. As we can see, this overestimate is based in their individual experiences with congregations. From the individual perspective, the average size of a congregation *is* larger than it is from the organizational perspective. Also, if you ask a Catholic to estimate average congregation size, she is likely to give a higher estimate than a Baptist (or most any other Protestant for that matter).

WHAT DO CONGREGATIONS DO?

As one of the most prevalent voluntary organizations in American society, congregations do many things. As we have already seen (Chapter 5), they are central to the socialization—especially the religious socialization—of children. They are the organizational center for the belonging function that we have identified as central to religion. As such, they play a key role in community formation (Ammerman, 2009).

Because communities can be exclusive, there is a dark side to this, as we will see in Part V when we consider religion and social inequality. The segregation by people according to economic class (Chapter 9) and race (Chapter 10) are two organizational realities of religious life.

The flip side of this same segregation coin is that congregations can also be the main community-based organization involved in the preservation of ethnic or racial minority cultures and therefore important organizational bases for survival and empowerment. This is why religious communities have historically been so important to immigrant populations (as discussed in this chapter and in Chapter 15 on globalization). We discuss at length the reality of churches as the core institution of African American communities in Chapter 12.

This segregation notwithstanding, when the "imagined community" is drawn more widely, congregations perform many valuable social services that benefit individuals well beyond their memberships (Chaves & Tsitsos, 2001; Trinitapoli, 2005). They feed the hungry, clothe the naked, welcome the stranger, care for the

sick, visit those in prison, give comfort to the abused and elderly, tutor the young, and so on. For all of their difficulties in terms of recent membership declines, mainline Protestant churches are the most active congregations in providing social services to their local communities (Ammerman, 1997, 2005).

Finally, congregations are very active in political life. Using the 1998 NCS data to look at seven different types of political involvement—from distribution of voter guides to demonstrations and marches to voter registration drives—Kraig Beyerlein and Mark Chaves (2003) found that 41% of congregations had engaged in at least one political activity in the previous 12 months. Congregations from different religious traditions tend to be drawn to different types of activity, though. More than others, Catholic congregations are involved in demonstrations, marches, and lobbying. Mainline Protestant congregations more often organize discussion groups. Black Protestant congregations register voters. Mainline and Black Protestant congregations host political candidates, while evangelical and Black Protestant congregations distribute voter guides more than other congregations (Beyerlein & Chaves, 2003).

These various and important activities of congregations notwithstanding, at the heart of congregational life is worship (Ammerman, 2005). As Chaves (2004) put it, "Congregations' central purpose is of course the expression and transmission of religious meaning, and corporate worship is the primary way in which that purpose is pursued" (p. 127). There have been many fascinating case studies of congregational worship, especially changes in worship style to more contemporary forms (see Wolfe, 2003, chap. 1). As we will see later in this chapter, much recent attention has been paid to the nontraditional worship services held at megachurches, especially those that are seeker-oriented.

A particularly insightful recent study is Timothy Nelson's (2005) analysis of Eastside Chapel, an African Methodist Episcopal (AME) church of some 300 members in Charleston, South Carolina. More than some other AME churches, Eastside Chapel is Pentecostal in orientation, emphasizing the "gifts of the spirit." Eastside Chapel members believe they can and should have relationships with God, Satan, and other spiritual beings; they are an objective reality for them, as consequential as human beings (Nelson, 2005, p. 49). Nelson showed how the Pentecostal beliefs and the ritual structure in Eastside Chapel's service work together to organize and direct what otherwise seem to be spontaneous, enthusiastic outbursts from the members in worship. Jumping up and down, raising hands, shouting out, and even moving out into the aisles to dance are all governed by social norms prescribing what can be done, by whom, and when.

Table 8.1 Frequency of Worship Elements, From Perspective of Congregations and of Individuals, 1998 and 2006 to 2007

Elements	% of *Congregations* with Element (1998)	% of *Congregations* with Element (2006–2007)	% of *Attenders* at Services with Element (1998)	% of *Attenders* at Services with Element (2006–2007)
Singing by congregation	96	97	98	97
Sermon/speech	95	95	97	98
Silent prayer/meditation	74	74	81	82
Written program	71	68	84	75
People speak/read/recite together	63	60	75	71
People call out "amen"	63	71	53	60
Applause	55	61	59	59
Singing by choir	52	44	72	58
People other than leader raise hands in praise	45	57	48	55
People speak in tongues	24	27	19	21
Electric guitar	22	34	30	36
Drums	20	34	25	37
Adults jump/shout/dance spontaneously	19	26	13	17
Visual project equipment	12	27	15	32

SOURCE: National Congregations Study Wave 1 (Chaves, 2004, Table 5.1, p. 132) and NCS Wave 2 (Chaves & Anderson, 2008, Appendix B, pp. 433–434).

NOTE: All elements are found in most recent main service, except for speaking in tongues, which was any time in the past year.

As with any case study, Nelson's offers great insight into the substance and process of worship. However, the question of representativeness also arises: How typical is Eastside Chapel? However, because it is a national sample of congregations, the NCS provides a national sample of worship. This first wave of the NCS explored worship according to a "repertoire" of worship elements (29 in total) that can be assembled by congregations in different ways (Chaves, 2004). The second wave asked about many, but not all, of those same elements. Table 8.1 shows the frequency of 14 different worship elements in 1998 and 2006 to 2007, both from the perspective of congregations and attenders.

As Table 8.1 indicates, there are two nearly universal elements in congregational worship: singing and preaching. In 1998, 96% of congregations (housing 98% of attenders) included singing in their weekly worship services, and 95% (with 97% of attenders) had a weekly sermon or speech. The average worship service was 70 minutes long, of which 20 minutes were devoted to a sermon and 20 minutes to music (Chaves, 2004, p. 133). These figures were essentially unchanged in the second wave of the NCS.

Beyond singing and preaching, other elements of worship were less ubiquitous and also evidenced more change from 1998 and 2006 to 2007. We will return to these changes in a moment. First, we observe with Chaves that there are systematic patterns in the elements congregations use and how they put them together. Congregations tend toward social homogeneity—that is, members of congregations tend to be like one another economically, racially, and ethnically—and Chaves (2004, p.135) found that worship elements vary systematically according to social class.

As Table 8.2 makes clear, poorer, less educated congregations are more spontaneous and demonstrative in their worship styles. In these congregations, one is more likely to see people jumping, shouting, dancing, raising their hands in prayer, speaking in tongues, and calling out "amen." To return to the question previously posed, it is quite clear that Timothy Nelson's case study of Eastside Chapel is generalizable. The worship style at Eastside is what Chaves (2004) would call "enthusiastic," and as we would predict based on the NCS, Eastside is located in one of Charleston's most impoverished neighborhoods.

In contrast, more affluent, well-educated congregations are more formal in style. In these congregations, one is more likely to see elements like singing by choirs, silent prayer or meditation, written programs, and incense. We discuss the reasons for this social class homogeneity in congregations at greater length in the next chapter (see also Chaves, 2004, pp. 139–143).

Although informal styles of worship are associated with lower socioeconomic status congregations, one of the most significant changes between the first and second waves of the NCS is the increasing informality of worship (Chaves & Anderson, 2008, p. 422). As Table 8.1 showed, in just 8 years, the percentage of congregations that report using visual projection equipment has increased by 15%.

Use of visual projection equipment rather than hymnals, praise bands, and praying with hands upraised in praise are on the increase in many congregations. While this type of emotionally intense worship was once associated with a particular social class, it is spreading through the culture, in part due to megachurches and televangelism.

Table 8.2   Congregational Social Class Differences in Use of Various Worship Elements

Elements	Mean % of people with bachelor's degrees	Mean % of people in poor households (<$25,000 annual income)
Spontaneous and Demonstrative		
Adults jump/shout/dance spontaneously	15	47
People speak in tongues	16	46
People other than leader raise hands in praise	18	42
Drums	18	43
People call out "amen"	19	42
Applause	23	38
Electric guitar	23	39
Visual project equipment	23	37
Common Features		
Singing by congregation	25	37
Sermon/speech	25	37
Formal		
Singing by choir	26	31
Silent prayer/meditation	27	36
Written program	28	33
People speak/read/recite together	28	36
Communion	30	34
Incense	42	29

SOURCE: National Congregations Study Wave 1 (Chaves, 2004, Table 5.2, p. 135).

NOTE: All elements are found in most recent main service, except for speaking in tongues, which was any time in the past year.

Drum use is up by 14%. Electric guitars and people raising their hands in praise increased by 12%. Indeed, all eight spontaneous and demonstrative elements listed in Table 8.2 became more common by 2006 and 2007. In contrast, the elements that become less common (written programs, people speaking/reading/reciting together, singing by choirs) are all formal worship elements.

What explains the increasing informality of worship? Analyzing the first wave of the NCS study, Chaves (2004) highlighted "a long-term trend toward informality in several cultural domains, including clothing, manners, and naming" (p. 158, citing Lieberson, 2000). The open systems perspective suggests that these broader cultural shifts cannot help but influence religious organizations.

Another explanation for the spread of informality in worship is the growing number and influence of megachurches, especially those "new paradigm" and "seeker" churches that use contemporary marketing strategies to reach bigger and bigger audiences. We turn our attention to these very large and influential religious communities next.

MEGACHURCHES

Stories about megachurches in the popular news media abound. For example, a 2005 *New York Times Magazine* cover story on "The Soul of the New Exurb" reported on Surprise, Arizona's, Radiant Church (Mahler, 2005). Although it was only founded in 1997, the church had since grown to 5,000 weekly attendees in a new 55,000 square foot church with five 50-inch plasma screen TVs, a bookstore, café (including drive-through), Xboxes for the kids, and Krispy Kreme doughnuts at every service (the doughnut budget is $16,000 per year). The story quotes the church's pastor, Lee McFarland, as saying, "We want the church to look like a mall. We want you to come in here and say, 'Dude, where's the cinema?'"

More recently, the website of *Forbes* magazine reported on megachurches using Second Baptist Church of Houston (est. 1978) as an example. Second Baptist is the sixth largest megachurch in America, claiming 53,000 members, over 22,000 weekly attendees, and an annual budget of $53 million dollars. It doesn't just have a building with a sanctuary; it has five *campuses* that house fitness centers, bookstores, coffee shops, and an auto repair clinic. Of course, it does have a building for worship,

which it is currently refurbishing to the tune of $8 million. It will seat 6,500 under a six-story high dome (Bogan, 2009).

Other well-known megachurches include the Willow Creek Community Church in South Barrington, Illinois (average attendance 23,400), and LifeChurch of Edmond, Oklahoma (26,776), not to mention the megachurches associated with famous pastors like Joel Osteen's Lakewood Church in Houston (43,500), Rick Warren's Saddleback Valley Community Church in Lake Forest, California (22,418), T. D. Jakes's The Potter's House in Dallas (17,000), and Creflo Dollar's World Changers Ministries in College Park, Georgia (15,000).

Critical Thinking: The majority of megachurches (over 60%) are located in the southern Sunbelt of the United States—with California, Texas, Georgia, and Florida having the highest concentrations. Most are also located in suburban areas of rapidly growing sprawl cities such as Los Angeles, Dallas, Atlanta, Houston, Orlando, Phoenix, and Seattle. Although megachurches are geographically concentrated, there are megachurches in almost every state. The Hartford Institute for Religion Research maintains an interactive database of megachurches in the U.S. on its website: http://hirr.hartsem.edu/megachurch/ database.html. You can sort their list by congregation, denomination, or size. You can also look at megachurches by state. What megachurches in the database are near where you live? Have you ever attended one of these megachurches? If so, how did your experience compare to what is described in this chapter?

By any measure, megachurches are a significant social phenomenon. According to Scott Thumma and Warren Bird (2008), megachurches are "Protestant congregations that draw 2,000 or more adults and children in a typical weekend (attendance not membership)" (p. 1). (Due to their geographic—parish—organization and consequently larger average size, Catholic parishes are excluded from this definition.) Thus defined, the best estimate is that the total number of megachurches in the United States has increased from 50 in 1970 to over 600 in 2000 to more than 1,200 by 2010. Even though this is a small segment of the 300,000+ congregations in America, the average megachurch has weekly attendance of 4,142 persons, compared to 75 in the average of congregation. This means that over five million people attend services weekly in megachurches, giving them a disproportionately large influence.

In line with our earlier discussion of denominationalism, which connected the movements toward nondenominationalism and transdenominational evangelical Protestantism, 37% of all megachurches are nondenominational (up from 29% in 2000) and 65% self-identify as theologically evangelical (up from 48% in 2000) (Thumma & Bird, 2008). Of denominationally affiliated megachurches, 16% are Southern Baptist, 10% are other Baptist, and 6% are Assemblies of God. Regardless of their denominational affiliation, virtually all megachurches have a conservative theology, even those within mainline denominations (Thumma, Travis, & Bird, 2005).

The ever-present singing and preaching we find in congregations generally are translated in the megachurch ritual vernacular into electric guitars (often or always used in 96% of megachurches), drums or other percussion instruments (97%), and visual projection equipment (97%). Megachurches are also leading the way in terms of informal, spontaneous, and demonstrative contemporary worship styles. As a flier commissioned by the Radiant Church's Pastor Lee McFarland and sent to the residents of Surprise, Arizona, put it: "You think church is boring and judgmental, and that all they want is your money? At Radiant you'll hear a rockin' band and a positive, relevant message. Come as you are. We won't beg for your money. Your kids will love it!" Some sociologists have argued, and many megachurch pastors would readily admit, that everything done in megachurches is done intentionally for the purpose of attracting members and attenders. In this view, most megachurches can be considered "new paradigm" or "seeker churches."

The state of Alaska may be the largest state in the Unit States (more than double the size of Texas), but the populati is only a bit more than half a million. Still, even Anchora Alaska, has its megachurch—ChangePoint—with coffee ba the largest gymnasium/recreation space in the state, Sund school classrooms that look like the kids are in Noah's A and many other features to market the faith and draw peop to the church.

"NEW PARADIGM" AND "SEEKER" CHURCHES: CONGREGATIONS WITH A MARKETING STRATEGY

Some newly founded and very rapidly growing religious communities have been extremely deliberative about marketing strategies. They have identified a "market segment" that they believe to be undersupplied or for which they believe a demand can be created. They assess the demographic characteristics of the population and seek to provide a service (pun intended) that will "sell." Not all of these congregations are huge, but they do tend to grow quickly. Donald Miller (1997) called these churches geared to marketing the "new paradigm churches." Indeed, Miller believed these religious organizations are so significant that they represent the first wave of a new reformation that will shake the foundations of religion as we know it. He believed this new movement will be as profound as the reformation led by Calvin, Luther, Zwingli, Knox, and the Anabaptists of the 16th century.

Miller called these "new paradigm churches" because they seem to break many of the rules for reform movements. These religious organizations are unlike previous renewal or reform movements within Christianity. Some of the unique aspects can be summarized as follows:

- Unlike most reform movements, these churches are less interested in revising the *message* of the dominant religious establishment (that is, a rethinking of the meaning system) and more focused on a radical transformation in *medium* of delivery (for example, breaking away from stodgy 16th-century hymns accompanied by a pipe organ and presenting the faith through the music and technology of the youth culture).
- Whereas previous renewal groups have rejected the dominant culture and presented an image of world rejection, the new paradigm churches have assimilated greatly to the larger culture, adopting most of its technology and *some* of its values. This defies the typical sect-to-church-to-sect renewal pattern.
- As these organizations have developed, they have been remarkably "postmodern" organizations, with more emphasis on energetic activity than on hierarchy—more characterized by networking than by formal structures that control people's behaviors within the organization.

The new paradigm churches have 12 common features, according to Miller (1997):

1. They were all started after the mid-1960s.
2. The majority of congregation members were born after 1945.
3. Seminary training of clergy is optional (indeed, there is fear that seminary training may taint one's pure spirituality).
4. Worship is contemporary.
5. Lay leadership is highly valued.
6. They have extensive small group ministries.
7. Clergy and congregants usually dress informally.
8. Tolerance of different personal styles is prized.
9. Pastors tend to be understated, humble, and self-revealing.
10. Bodily, rather than merely cognitive, participation in worship is the norm.
11. The "gifts of the spirit" (a phrase that normally embraces "speaking in tongues") are affirmed.
12. Bible-centered preaching predominates over topical sermonizing. (p. 20)

Kimon Howland Sargeant (2000) did a study of a similar (but not identical) phenomenon, a group of loosely connected Christian churches designed very specifically to market the faith to the unchurched. Sargeant looked at very rapidly growing churches, many of which either are or are becoming megachurches. These religious movements are referred to as "seeker churches" because they are designed to appeal to religiously alienated "seekers." The population they address, the adoption of modern media, the insistence on upbeat messages, the ideas of marketing the faith through paying attention to method of delivery, the willingness to modify aspects of the faith to meet consumer demand, and the leadership by innovative (and often theologically untrained) religious entrepreneurs makes these organizations relevant to our focus on marketing of religion. For example, Sargeant (2000) said that "seeker church experts often proclaim the shopping mall, Disney, and other customer-sensitive companies as models for the twenty-first-century church" (p. 8). We will draw from the literature on both new paradigm and seeker groups in the remainder of this chapter. With Miller, we will consider "seeker-sensitive churches" as one type of new paradigm church.

The Audience: Targeting Markets

Like any good business that pays attention to marketing, seeker and new paradigm churches have focused on specific target markets. They analyze the characteristics of their target group and make adjustments to attract that population. Indeed, they sometimes hire consultants to conduct surveys of interest and needs of the local population and to develop market strategy action plans (Sargeant, 2000). It is important to realize what a profound shift this represents in evangelism. It would never have occurred to Great Awakening preachers like Jonathan Edwards to target the market. He simply preached what he believed to be the truth and trusted it would fall on receptive ears. While later evangelists such as Charles Finney, Dwight Moody, and Billy Sunday introduced a marketing perspective (Billy Sunday calculated the "efficiency" of revivals by determining the cost per convert and advocated "scientific management in the pulpit"), they did not focus explicitly on target markets or market shares.

While there are variations on the theme, new paradigm churches have largely targeted middle- to upper-middle-class families, especially those who were part of the baby boom generation and are college educated. Often the targeted group is "the unchurched," but since white baby boomers, young adults, and the college educated are the least likely folks to be active in a church, and since baby boomers are a very large segment of the population, the target quickly becomes this same cohort. People who fit that profile generally have certain characteristics in common, and the new paradigm clergy are highly sensitive to those traits.

Thus, a key element of new paradigm churches is the desire to be *culturally relevant*. Miller (1997) wrote, "Simply put, if the message is going to communicate with an intended audience, it must be culturally appropriate" (p. 28). He further quoted a new paradigm minister as saying, "If [nonbelievers] are going to reject the message I preach, let them reject it, but let them reject the message and not all the peripheral things that are secondary" (Miller, 1997, p. 66).

Market-sensitive churches therefore try to appeal to people through themes that have cultural currency with middle-aged, well-educated Americans:

- They use the language of the therapeutic community (though they may reject the narcissistic elements of that subculture).
- They stress individualism.
- They are intensely antiestablishment (opposed to bureaucracy, but not necessarily antagonistic to the dominant culture, per se).
- They employ the metaphors of consumerist America, with the mall as a model of the good life.
- They appeal to the cultural styles that are current with the targeted cohort, such as music styles.

What this means is that direct requests for money are carefully avoided and offerings are unobtrusive (sometimes as minimalist as a box at the exit for contributions). Because "product loyalty" is very weak (including loyalty to a denomination), any affiliation to a denomination—if it does exist—is downplayed. Most new paradigm churches are independent, and those that *are* affiliated with a mainline denomination do not indicate that affiliation on their signs or their literature. This seems to be in keeping with the anti-bureaucracy attitudes of the baby boomers. Music is decidedly contemporary and upbeat, with a very strong preference for vocal and band music over the ponderous melodies of Bach played on an organ. Indeed, as was noted, guitars are everywhere, and meeting houses in the early years of a church's growth are likely to be rented space in a mall, a renovated warehouse, or a rented auditorium with no stained glass or other of the accoutrements of traditional religious settings. Baby boomers by and large have a negative image of "organized religion" so the setting and the attire are not conventional "high church." People dress very casually—including the ministers in many such congregations. Pastor Lee McFarland was pictured in the *New York Times Magazine* preaching in a Hawaiian shirt. This target group wants direct access to spirituality, not a mediated experience through ordained clergy. They are often free with bodily expression in worship, reaching levels of euphoria that would be embarrassing in mainline churches.

The historical Protestant idea of the "priesthood of all believers" resonates well with the members of new paradigm churches. Lay leadership is encouraged, pastors are understated in their leadership styles (even though they may be very much in charge behind the scenes), and members are invited to make their own interpretations of the meaning of scripture to their lives.

Modern technology has also facilitated the antiestablishment and local autonomy trends. The availability of computers, desktop publishing, and communications networks has meant that sizable local churches have sufficient resources and talent to write their own curricula for religious education and can produce other materials that congregations have normally obtained from the larger denomination (Miller, 1997). Each congregation or each program within a congregation is more capable of producing their own tailor-made materials with desktop publishing. The benefits of being affiliated with a denomination are therefore a bit muted.

The middle-aged professionals—who are the target population—value self-actualization highly. Moreover, as predominantly suburbanites, they are used to the many conveniences of suburban life, including having lots of choices. The expectation is carried over into their religious lives. Sargeant (2000) wrote, "One might describe the dominant characteristic of this cultural ethos as consumerism. People today, especially middle-class baby boomers, expect, even demand, choice in their workplace, home, shopping options, *and* their religious commitments" (p. 42). If religious individuals are the consumers in this equation, then religious communities are the producers of religious products that must be appealing in a competitive marketplace.

Creating an Appealing "Product"

Many unchurched people who are white-collar professionals are more comfortable in a business or consumer environment than in a traditional religious one. Thus, the huge Willow Creek Community Church near Chicago has developed a campus of new buildings that look more like a modern community college or a corporate training center than a faith community. Indeed, a mall is often explicitly used as a model of what these churches should be like, once they become large enough to build their own facilities. Sargeant (2000) wrote,

> Every aspect of the church's facilities emulates the best of corporate America in quality, design, and style. Willow Creek's aim is to reduce or minimize any cognitive dissonance between the religious realm and the working and shopping world of suburban middle-class Americans. (p. 19)

Many seeker-sensitive churches are even afraid of making people uncomfortable with too much religious symbolism, so a surprisingly high number of these evangelical churches do not even have a cross in the worship auditorium or in any prominent place (Sargeant, 2000, p. 61).

Tradition is often avoided as well. Sargeant (2000) reported that

> seeker churches have taken a very low view of tradition in all of its various meanings. Tradition, according to many pastors, poses an unnecessary barrier for seekers. . . . Tradition, in short, represents the old paradigm, an outdated way of doing things that is largely ineffective in the current religious environment. (p. 63)

So eager are such churches to be user-friendly to the unchurched that American holidays are often highlighted over Christian holy days. In roughly a fourth of the seeker-sensitive congregations, there is not even any communion as part of worship, despite the fact that these are led by emphatically evangelical Christians (Sargeant, 2000, pp. 70–72). In the interests of recruitment, these churches have jettisoned aspects of organized religion that alienate young adults and baby boomers (Miller, 1997).

Services are designed to appeal to the cultural preferences of the target group; the pre-worship music is often light rock or soft jazz "because rock music has played such an important role in the lives of baby boomers" (Sargeant, 2000, p. 65). The music style helps to define what kind of people a faith community wants to attract. Praise music is often written locally and is very rhythmical and upbeat. An important point is that these sounds are *love songs to God* rather than lyrics *about* God. Miller (1997) even describes the musical portion of the services as "a form of sacred lovemaking" (p. 87) to God. They are easy to sing and are emotionally evocative.

Of course this is also a population that seeks the novel and the innovative, so spontaneity prevails over ritual and contemplation; the services are often designed to surprise and intrigue attendees. Baby boomers generally do not cope with boredom patiently. Innovation is the name of the game. Drama is often employed as a major element of the morning service, as are multimedia entertaining "shorts." Everything is very carefully choreographed and rehearsed, with timing as important as in a Broadway theatrical production. The pastor, while giving the image of a down-home sort of person, is actually performing as a CEO in a major corporation (Sargeant, 2000).

Seeker-sensitive churches try to present the gospel in a way that is relevant to the lives of attendees. Unlike many new paradigm churches that have sermons based on verse by verse explication of the Bible, the explicitly "seeker-sensitive churches" present messages that are topical and pragmatic. Message titles include "Fanning the Flames of Marriage," "Authenticity," "The Art of Decision Making," and "Energy Management" (Sargeant, 2000, p. 18). Kimon Howland Sargeant (2000) pointed out that one danger of emphasis on "relevance" is that "the audience and not the messenger, determines, at the very least, the topics and tone of the message" (p. 81).

There is a danger in marketing religion, however, which brings us back to the issue of mixed motivations in a complex organization. In her study of televangelism empires, Frankl (1987) found that like any other institution, a religious corporation can face dilemmas that can compromise the original mission of the organization. The survival of a huge organization requires generation of large amounts of money to keep the organization going. Organizational maintenance of the huge financial enterprise may lead to goal displacement so that even the core beliefs of the group may be modified. According to Frankl, the content of religious programming in a market-driven religious enterprise often becomes transformed; it is no longer based on the message of the founder but on rationally calculated economic needs of the "business" to prosper. In today's media world, content must be simple, fast-paced, and entertaining—not complex and thought-provoking—so the message itself may have to be truncated to meet the market needs. The message of Christianity may undergo scrutiny for its marketability and be modified accordingly. If a message is popular with the public, it "sells" well: The money pours in. So there is a temptation to preach only on those things that are profitable and to avoid preaching on those things that do not yield a financial reward. So the marriage of Christian message with marketing can sometimes change the message itself. The marketing process—changing the message to adapt to the market—may distort the message which the founder intended. This is a worst case scenario that you may consider as you learn about marketing a faith and efforts to avoid compromising it.

There is little that is challenging or complex in these churches regarding the ethics or the environmental impact of a consumerist lifestyle. Indeed, a message focusing on personal and private issues is so pervasive that there is little attention to social justice issues or to public or global causes. The church becomes a place for therapeutic comfort or for maximizing one's full potential. God is *immanent* (a nearby source of comfort) but not a *transcendent* being who sets standards for a life lived in covenant with the divine. As Sargeant (2000) put it, "Seeker church pastors de-emphasize God's inscrutability, God's mystery, and God's ultimate judgment" (p. 85). These things just do not "sell" well with this population.

Sin is recast as preventing realization of one's potential as a human being. The measure of profound spirituality remains: "Is your soul being satisfied?" By contrast, First and Second Great Awakening evangelists believed that God had chosen the elect and God's satisfaction was all that mattered; using human subjectivity as a standard of authentic Godliness would have been offensive and profoundly un-Christian to them. However, we live in a different time. Sargeant (2000) summarized it this way: "For today's religious consumers, the search for meaning and fulfillment is as important—and initially even more important—than the search for God. . . . God becomes the means toward our fulfillment, rather than the end toward whom we owe our allegiance" (pp. 98, 121).

In addition to the worship hour being adapted, two other types of programs cater to the needs of the congregation. These seeker-sensitive churches are not necessarily different from other congregations, but they are clearly more intentional and more focused. First, they provide a wide range of entertainment and social opportunities: movie nights, marriage renewal weekends, family camping retreats, basketball and softball leagues, handball tournaments, and so forth. These are sometimes called the "side doors" of recruitment by church-growth consultants, since they are ways of attracting people to the church other than the historically preferred "front door": the Sunday morning church service (Mahler, 2005).

Second, small groups are at the core of the ministry. Opportunities for deep intimacy are rare in a competitive capitalist society, and feelings of alienation from the structures and from other people are common. People in a dog-eat-dog, rapid-paced modern society need to feel human connection. New paradigm churches provide this. Indeed, they consider small groups to be the "*real* church experience," not the Sunday morning worship hour. Virtually every member is urged to join at least one small group, for "this is a movement built on relationships" (Miller, 1997, p. 36). Sargeant (2000) cited the adage among leaders

of these churches that "the church only grows bigger by growing smaller" (p. 118). These small groups can be Bible study groups, prayer groups, exclusively men's or women's gatherings, healing or therapeutic clusters, or other topical groups. However, they remain small, they challenge people to apply faith principles to their lives, they share intimately with one another about their lives, and they become potent reference groups. Many studies have shown that in churches of all sizes and denominations, small intimate groups are at the core of local church vitality, for we live in a society that atomizes and isolates us (e.g., see Wuthnow, 1994a).

Critical Thinking: New paradigm and seeker-sensitive churches are being very self-conscious about their marketing strategies to draw members. Yet Max Weber insists that people have always been more likely to join churches where there is "elective affinity" to their social, cultural, or economic circumstances. Is this any different? Does it make a difference whether the leaders unintentionally draw people with similar cultural values or interests or intentionally target an audience and modify the message to have greater appeal? Why?

Pragmatism About Methods and an Uncompromising Message

New paradigm clergy believe that they have changed the medium or the methods of delivery, but not the message. Scholars who study these groups are less sanguine that they have been successful in not compromising the message itself. Sargeant (2000) concluded that "changing the *method* can not only change your results; it can also change your *message*" (p. 131). A basic principle of marketing, of course, is that the product *ought* to change in order to meet the demand.

We have seen that marketed Christianity tends to become infused with local or national values—embracing consumerism, shying away from asking the larger ethical questions about the consequences of one's behavior on others or on the natural environment, and even blessing a certain amount of self-absorption. Yet leaders of these new paradigm churches have tried to be self-conscious about what they believe to be the core of the faith—what is *not* open to compromise. It may be that their decisions about what is the kernel and what is chaff may differ from yours or mine or the mainline minister down the street, but there are some things on which they will not budge. The centrality of Jesus Christ is affirmed, as is the importance of a personal relationship with Jesus. The notion that human depravity is part of the human condition is affirmed, even if it is given a soft sell and is sold as a matter of self-actualization. Interestingly, a central theme in the baby boom generation is tolerance for others who are different—an openness to acknowledging that what is right for me is not necessarily right for you. This is perhaps the first generation of Americans in which the majority affirms that there are multiple paths to God, each with equal legitimacy. Indeed, a timely Gallup poll reports that of Americans who identify with a religious group, 82% said other paths to God are equally good as their own path (Gallup Organization, 2000). However, new paradigm churches are quite uniform in rejecting this notion. They proclaim that Jesus offers the exclusive access to God. This is an issue that for them is a core belief and is not to be compromised.

Another core belief is the authority of the Bible. This is a firm affirmation. Yet the new paradigm churches have avoided stridency or the appearance of intolerance on this issue. This seems to be related to the fact that they "are doctrinal minimalists. Their emphasis is on one's relationship to Jesus Christ, not on whether one believes [specific dogmas]. New paradigm Christians view doctrine as being of human origin" (Miller, 1997, p. 121).

What typically gives stridency to biblical literalism is its use in the service of doctrinal dogmatism, as the justification for an attack on other beliefs. However, the deemphasis of doctrine by new paradigm Christians puts a different spin on the way scripture is utilized. Although they view the Bible as authoritative, there seems to be a certain modesty in terms of the conclusions that one should draw. (Miller, 1997, p. 131).

While the authority of scripture is affirmed, the intolerance that middle-class baby boomers find offensive is avoided. This rather "soft" approach to interpretation is consistent with new paradigm principles of priesthood of all believers in interpreting the meaning of the faith. In this case, compatibility with target-group values is achieved without compromise of basic principles.

The new paradigm ministers try to keep clarity about the message, the model, and the market: how these interact, which are most important, and which cannot be compromised. They do not all agree on where the lines should be drawn. Indeed, while some leaders in new paradigm churches actively utilize the language of the business world, not all are comfortable with this. Sargeant (2000) quoted a minister of music in one Baptist church: "People may think because we are a church, maybe we shouldn't market. But any organization, secular or otherwise, if [it's] going to grow, [it's] got to get people to buy into the product" (p. 5). On the other hand, Miller found that most new paradigm pastors do not use the rational choice language about "selling a product." They generally avoid the crass utilitarian terminology of the marketplace. Still, the leaders of these new paradigm churches are "entrepreneurs" in the sense that they innovate freely and seek to meet demand in the marketplace, but their purpose is spiritual. Ultimately a faith community is not a business seeking economic profits. For pastors and members, the purpose is spiritual and that makes a huge difference in standards of conduct. It is mostly social scientists using marketing perspectives who are likely to refer to individual congregations as local "franchises."

Critical Thinking: Is "marketing" of religion intrinsically a problem that causes distortions of religion, or does competition in the marketplace of ideas add vitality and intensified commitment to religion? Why do you think as you do?

Summary

Denominationalism was born as a fundamentally American phenomenon, rooted in the pluralism of the new continent and the new country. Separation of church and state became central principles of the founders as they realized that religious tolerance and public policy need to be based on rational deliberation. This is the way the best interests of the country can be served in a culturally and religiously diverse society. In this environment, each religious group had to organize without state tax support. As congregations with similar outlooks or similar issues and cultures looked for ways to provide mutual support, denominations became the vehicle to do so. When religious groups that did not organize themselves into larger umbrella groups immigrated to the United States (Muslims, for example), they have tended to follow the pattern of "institutional isomorphism"—becoming like other religious communities in the environment. Moreover, this pattern of denominationalism has now spread around the globe to other places that are religiously and culturally diverse. Still, as we enter the third millennium, denominationalism seems less compelling as people organize around religious special purpose groups. Nondenominational Christian churches are now the largest "brand" of religious congregation in the United States.

Congregationalism—in which local congregations have a good deal of power and authority and no bishop controls policy—is also preeminently American. Even the more hierarchically organized churches—those with episcopal or presbyterian polity—have elements of congregationalism when they are in the United States. Congregations do vary in many ways: when we compare congregations, the median size of a congregation is 75, but because some congregations are huge, the typical American attends a congregation of about 400. Regardless of size or style of worship (which varies by social class), congregations provide a range of resources and support to their members. The most notable phenomenon in the past three decades has been the rise of megachurches: congregations of more than 2,000 and often becoming small cities with tens of thousands of members and a vast array of services. Often associated with these megachurches is a relatively new phenomenon—the explicit and intentional marketing of the faith so it will "sell" and so the organization can grow and thrive. This marketing involves finding target "markets" where there is a need and producing a "product"—a religious product—that will sell to that market. Critics claim that this includes a truncating and distortion of the message, but adaption of a faith to a new environment is hardly new or innovative. It has existed for a very long time.

One of the ways that religion is modified is that it must meet the needs for meaning of people in different socioeconomic circumstances. Thus we turn next to the issue of religion and inequality—how religion affects economic development and how it is affected by economic factors in the environment.

PART V

RELIGION AND SOCIAL INEQUALITY

In the next three chapters, we examine the interactive relationship between religion and the social structures and processes of the larger society. These chapters are designed to illustrate the sociological method of analysis; they are certainly not exhaustive treatments of the influence of religion on other social structures or of other social structures on religion.

One of the important elements of social structure is the stratification system. Four main dimensions of social stratification in contemporary society are (1) social class, (2) race, (3) gender, and (4) sexuality. Therefore, in Chapter 9, we investigate the interrelationships between social class and religion. Chapter 10 considers race and religion; and Chapter 11 examines gender, sexuality, and religion. In each case, religion is examined as both a dependent and an independent variable.

In Chapter 12, we address an important issue regarding religion and social status that arises from the considerations of inequality in Chapters 9 through 11, the relationship between religion and social activism aimed at redressing social inequality.

9

RELIGION AND CLASS STRATIFICATION

Here are some questions to ponder as you read this chapter:

- Can religion influence economic behavior and socioeconomic status? If so, how?
- Did religious beliefs have any role in the formation of capitalism as an economic system?
- What specific aspects of religious commitment might influence one's economic conduct or circumstances?
- How might socioeconomic status influence religious behavior or beliefs?
- Which is more likely to influence the other: religion or economics?

One of the most important factors in shaping the life chances and life experiences of people is economics. In this chapter, we explore how socioeconomic circumstances affect religion and how religion, in turn, impacts economic behavior. We begin by investigating the role of religion as a causal variable that affects one's economic behavior.

RELIGIOUS ETHICS AND ECONOMIC ACTION

Weber's Protestant Ethic Thesis

Among the important factors determining the social status of individuals in any society are the economic behavior of those individuals and the nature of the economic system itself. Many scholars have wondered about the extent to which religion affects economic behavior: To what extent is religion an independent variable influencing economics? The landmark study on this issue has been Max Weber's (1904–1905/1958a) seminal work, *The Protestant Ethic and the Spirit of Capitalism*, for it has generated an incredible amount of research and discussion.

Weber's study was undertaken in response to two issues. First, he was interested in the relationship between religion and economic activity. This study was one of a series of comparative studies of religion and its effects on economic development (Weber, 1920–1921/1951, 1920–1921/1952, 1920–1921/1958b, 1922/1963).[1] In the first sentence of Chapter 1 of *The Protestant Ethic*, Weber (1904–1905/1958a) pointed out that Protestants tend to be more affluent than Catholics and to occupy the higher-status positions in virtually all industrialized societies. He was interested in the relationship between religious affiliation and social stratification, including the differential effects of various faiths.

Second, Weber wished to address a larger theoretical issue. He hoped to provide a corrective to the simple economic determinism of Karl Marx. Marx had maintained that one's economic status was the principal determining factor in all behavior. He felt that it was fruitless to try to understand human behavior as an expression of values, ideals, or beliefs. Marx believed that beliefs and values are a *result* of economic forces, that one's ideas and ideals act to justify one's economic fortunes—or to compensate one for a lack thereof. Religion served to justify and sacralize the current social arrangements.

Because of this, it helped reinforce the status quo and served to retard change. For Marx, values, beliefs, and ideals do not serve as primary causal forces; they are secondary factors that result from economic forces.

Marx recognized that beliefs—including religion—could serve as *proximate* causes, even as he insisted that they are not *ultimate* causes. He acknowledged that humans are active agents and that beliefs organize and propel one's action. It was for this reason that he stressed the importance of the working class changing from "false consciousness" to "class consciousness." He also emphasized that religion often served as an opiate of the masses—again revealing his awareness of the role of beliefs and ideas on action. Without ignoring this awareness, it is also accurate to say that Marx believed ideas are powerfully conditioned by material (or economic) circumstances. Economic forces were viewed as the principal factors in shaping human behavior; ideas, values, and beliefs were only proximate influences and were themselves largely shaped by economic forces. Religion, which deals in values, ideals, and beliefs as its primary currency, was viewed as a relatively unimportant force—at least for those interested in the principal factors that cause social change or ultimately shape human behavior.

Some writers have claimed that Weber's study was intended as a direct refutation of Marx, but this clearly was not the case. Weber agreed with Marx's contention that economic self-interests have a powerful effect on the beliefs and values of people. Yet, Weber viewed this position as only a partial truth. He insisted that while economics can affect religion, religion

---

[1]Weber's interest in accounting for the distinctive pattern of development in the modern West, especially the rise of modern capitalism, was a key problematic at the founding of the discipline of sociology. *The Protestant Ethic* (Weber, 1904–1905/1958a) was one part of a planned multivolume work on the religions of the world. Weber did not ultimately complete the project, but three volumes were published in 1920 and 1921: *Ancient Judaism* (Weber, 1920–1921/1952), *The Religion of China* (Weber, 1920–1921/1951) (Confucianism and Taoism), and *The Religion of India* (Weber, 1920–1921/1958b) (Hinduism and Buddhism). His thesis in this massive comparative–historical project was that civilizations outside the West could not internally generate a "spirit of capitalism." The "inner-worldly asceticism" of Protestantism in the West thus played a key role in helping motivate and diffuse the spirit necessary for capitalism to arise.

can also affect economics. In fact, he held that Protestantism (especially Calvinism) was a significant force in the formation of capitalism as an economic system. Rather than being a refutation of Marx, Weber viewed his study as a modification or corrective to the overly simplistic analysis by Marx. Although he accepted Marx's view that people behave in ways that enhance their own self-interests, Weber felt that perceptions of self-interest were not limited to the economic realm. A religious self-interest (e.g., concern over salvation) could also motivate people.

In *The Protestant Ethic and the Spirit of Capitalism*, Weber (1904–1905/1958a) attempted to identify how religious beliefs and self-interests had affected economic behavior. In focusing on the Protestant "ethic," he was really referring to the overall perspective and sense of values of Protestantism. He felt that the breakthrough from the feudalistic to the capitalistic economic system was substantially enhanced by this particular worldview. In sum, he felt that ideas could be important factors that facilitate social change, including changes in the economic system.

For capitalism to thrive, several conditions had to be met. There had to be a pool of individuals with the characteristics necessary to serve as entrepreneurs. They had to be individualistic and had to believe in the virtues of hard work and of simplicity of lifestyle. Protestantism tended to create a supply of such people. Although economic self-interests create such people today, Weber maintained that the original pool was formed largely by people motivated by religious beliefs and self-interests.

One of the primary concerns of people in the 15th and 16th centuries was their eternal salvation. One way to be assured of salvation was to serve God directly. If one was called by God to the priesthood or to holy orders in a monastery, one seemed to have a better chance of eternal salvation. However, taking a new slant, Martin Luther stressed the "priesthood of all believers" and insisted that one could be called by God to a variety of occupations that served humanity. This new concept of a calling is referred to as his *doctrine of vocation*, for the word *vocation*

means calling. According to Luther's teachings, one may be called to many types of secular positions, as well as to the priesthood. Because secular positions were also viewed as service to God, any form of work could be a means of expressing one's faith. Hard work became a way of serving and glorifying God. Idleness or laziness, by logical extension, came to be viewed as a sin. Although Luther initiated this concept of vocation and held to it in principle, it was the Calvinists who fully implemented it. In fact, John Calvin even suggested that when a person was hard at work, he or she was most in the image of God.

Industriousness was an essential quality for the emergence of a class of entrepreneurs, and this quality was a central virtue among Calvinists. However, it was not enough for people simply to develop an ethic of hard work. The rational investment of earnings was also critical. Weber believed that the Protestant emphasis on asceticism and on *delayed gratification* served to enhance this aspect of capitalistic enterprise. Asceticism was a major theme in the preaching of a number of reformers, for the pleasures of the world (gambling, drinking, secular forms of entertainment, and luxuries) were all viewed as sinful and evil. God required a simple, even austere, lifestyle. Calvin even described self-discipline as the "nerves of religion"; hence, self-denial became a central virtue. This denial of present desires was accomplished because of the principle of delayed gratification. Believers were willing to forgo pleasures now for the promise of much greater rewards in the future. In the case of reformers, the future rewards were anticipated in the afterlife. Nonetheless, the principle of delayed gratification is an important one for the development of economic capital in *this* world. In fact, Weber defined capitalism as the systematic investment of time and resources with the hope of significant returns in the future (profits). The idea of denying one's immediate desires in the hope of a greater return in the future was basic to both Calvinism and to capitalism.

Among many of the early Protestants—and especially among Calvinists—hard work was a moral and religious duty, but because income from one's industriousness was not to be spent on luxuries or sensual pleasures, the only thing to do was to invest it. Among Methodists and certain other Protestant sects, excess income was to be given to the poor, but according to R. H. Tawney (1924/1954), the Calvinists felt that people who were impoverished were poor simply because they were lazy. Not wishing to contribute to this wickedness, Calvinists were not inclined to donate much of their income to the needy. Most of their profits were available for investment. This created a situation in which increased amounts of capital were available in the society and economic growth was enhanced.

Other characteristics of Calvinism were also significant. One of the important doctrines in Calvin's theology was that of *predestination.* According to this doctrine, God has already decided who is saved and who is damned. One's fate is predetermined, and there is really nothing a person can do about it. This doctrine might easily have resulted in a sense of fatalism, despair, and despondency, but in this case, people coped with their anxiety about whether they were saved by acting as if they were. Of course, this would not improve their chances, but any impious or un-Christian behavior would only ensure to themselves and others that they were damned. Righteous behavior would not earn a person a position in heaven, but one's behavior was believed to be an outward sign of one's eternal status. Being righteous, thrifty, hardworking, and ascetic was a way of hedging one's bets. When so much was at stake, it seemed foolish to take chances.

Furthermore, this predestined state placed one in a position of radical individualism. One was not saved because of anything others did or because of the groups one belonged to. One was strictly on one's own in this matter of salvation. Weber (1904–1905/1958a) wrote,

In what was for the man of the age of the Reformation the most important thing in his life, he was forced to follow his path alone to meet a destiny which had been decreed for him from eternity. No one could help him. . . . In spite of the necessity of membership in the true church for salvation, the Calvinist's intercourse with his God was carried on in deep spiritual isolation. (pp. 104–107)

Because an attitude of rugged individualism was important for the entrepreneur, the sense of religious individualism provided a compatible outlook. In fact, Weber suggested that the religious individualism may have partially predated and thereby contributed to the attitude of economic individualism. Economic individualism, in turn, means that individuals make rational economic decisions based on their own self-interests. Once capitalism is formed, according to Weber, it is capable of sustaining its own individualistic motivations. Still, he suggested that religious self-interests and beliefs may have contributed to the *original formation* of capitalism.

Finally, Calvin taught that regardless of one's eternal salvation, everyone is to glorify God and to work for the creation of a divine kingdom on earth. In fact, Calvin emphasized that the proper aim of humanity is not personal salvation but the glorification of God through the sanctification of this world. This focus created a strong this-worldly component to the theology, for one's labors in transforming this realm—the here and now—were the best indicators of one's devotion to God. Later Calvinists (Puritans) reinforced a this-worldly orientation even further by suggesting that one's socioeconomic status was an indicator of one's spiritual grace and eternal destination. Delayed gratification and asceticism became increasingly this-worldly and rational (i.e., based on concrete self-interest).

Due to certain beliefs, then, 15th-century Europe was supplied with an increasing supply of capital for investment and a pool of individuals who had the values and attitudes appropriate to becoming entrepreneurs. Although these characteristics were not sufficient to bring on the development of capitalism, they were necessary, and Weber believed that religion had contributed to the formation of the capitalistic system of economics by supplying these characteristics. He maintained that a religious outlook had influenced the economic behavior and financial fortunes of individuals (Protestants becoming more affluent than non-Protestants) and also had contributed to the development of a new economic system. Contrary to Marx, Weber felt that religion was capable of being a cause of economic conditions—not just a result.

Weber's thesis has been highly controversial and in the decades following its publication generated a large number of essays and research projects, some historical and some focused on efforts to measure a "Protestant ethic" in the modern world. (The "Doing Research on Religion" feature examines an attempt to operationalize the Protestant ethic in order to study its effects.)

Some scholars have argued that Calvinism had little to do with the development of capitalism. Some insist that the beginning of colonialism (with its influx of new capital resources) and changes in postmedieval technology sparked the advent of capitalism. Others emphasize the fact that capital was available for entrepreneurs through Jews and through Catholic bankers in urban areas. In these cases, the concurrent rise of capitalism and Protestantism is seen as a simple coincidence (Samuelsson, 1957/1961). The outlook described as the Protestant ethic is viewed by these scholars as insignificant—or at least its uniqueness to Protestants is denied.[2]

DOING RESEARCH ON RELIGION

How Does One "Operationalize" or Measure a "Protestant Ethic"?

Weber's thesis claims that certain values, attitudes, and dispositions influenced the economic behavior and standing of those who held them. There have been many efforts to prove or disprove Weber's thesis through gathering empirical research in the modern world, but one challenge is to figure out who has a Protestant ethic and who does not.

In one widely cited research project, Gerhard Lenski (1963) attempted to test the Weberian thesis. First, he asked people to rank the following in order of importance in a job:

1. Receiving a high income

2. Having no danger of being fired (job security)

3. Being able to work short hours and having lots of free time

4. Having chances for advancement

5. Feeling that one's work is important and provides a sense of accomplishment

(Continued)

[2]Those interested in detailed critiques of Weber's Protestant ethic thesis will want to see R. H. Tawney (1924/1954), Winthrop Hudson (1949), Amintore Fanfani (1936), H. M. Robertson (1933/1959), Robert Green (1959), and Kurt Samuelsson (1957/1961).

(Continued)

Although items 1 and 4 reflect an interest in upward mobility, Lenski pointed out that only item 5 is consistent with the original Protestant concept of vocation.

As a second means of measuring the Protestant ethic, Lenski sought to ascertain the attitudes of people toward heavy use of credit to charge things—referred to as "installment buying." Installment buying (have now, pay later) seemed to Lenski to be the utmost rejection of the principle of delayed gratification. A critical attitude toward installment buying would be consistent with the Protestant ethic; a willingness to go into debt now to have the things one wants immediately would indicate less delayed gratification and a weak "Protestant ethic."

Critical Thinking: Review the central ideas of Weber's notion of the Protestant ethic. Then consider the way Gerhard Lenski measured or operationalized this core concept. Do you think his method of determining whether someone had a strong Protestant ethic was sound? Why or why not? If you think his measures were flawed, can you suggest a better way to determine who does or does not have a strong Protestant ethic as described by Weber? If you think his method of operationalizing these ideas was valid, what other variables would you want to identify and control before you would know whether the conclusion was reasonable? Could any outside variables have explained the relationship? If so, what are some of those?

Other writers have basically agreed with Weber but have suggested modifications in his thesis. For example, R. H. Tawney (1924/1954) suggested that most contributions of Protestantism to the development of an individualistic and laissez-faire economic system were entirely latent (unconscious and unintended). He pointed out that in the Middle Ages, the Catholic Church held usury (lending money for interest) to be on a par with adultery and fornication.[3] Not only was lending money at interest considered immoral but prosperity itself was viewed as a source of spiritual corruption. Both of these attitudes were carried over in the teachings of the reformers.

Luther, who represented a rural peasant orientation, was particularly opposed to self-interested individualism and to laissez-faire policies. Furthermore, he felt that it was immoral to make money through investments, lending, or any form of speculation. He held to the idea that one should be rewarded financially only in proportion to the labor one actually performs.

Calvin, on the other hand, was much more urban and secular in background and orientation. Although he was no defender of laissez-faire capitalism, he did believe that capital investment was essential to a healthy economy. Hence, he insisted that there was nothing inherently evil about investing or lending money at moderate interest. On the other hand, he did insist that no interest should be charged the poor, and he put strict ethical guidelines on economic activity. He also remained deeply suspicious of the spiritual

---

[3]It was for this reason that the Jews were looked to as sources of capital; they were willing to lend money at interest because usury was not defined as immoral within the Jewish tradition.

effects of economic prosperity. At one point he commented, "Wherever prosperity flows uninterruptedly, its delight corrupts even the best of us," and at another time he suggested that "prosperity is like rust or mildew." In fact, Calvin's distrust of the influences of money is revealed in his considerable ambivalence about usury. One English clergyman who studied Calvin's theology commented that "Calvin deals with usury as the apothocarie doth with poyson" (cited by Tawney 1924/1954, p. 94). It seems clear that at the manifest level both Protestants and Catholics believed that "the spirit of capitalism is foreign to every kind of religion" (Samuelsson, 1957/1961, p. 19).

The reformers explicitly preached against certain practices that were later taken for granted as an integral part of the free enterprise system. In fact, he insisted that the massive accumulations of wealth that were acquired by later industrialists were not the result of the Protestant ethic or of thriftiness. Such capital was acquired by unscrupulous exploitation of people. Tawney agreed with Weber that Calvinistic theology and ethics subtly contributed to the advent of the free enterprise system, but he felt that early Protestant ethics were quite incompatible with the free-for-all capitalism of the 18th, 19th, and 20th centuries. Moreover, Tawney felt that capitalism would probably have made its entrance even without the contribution of Protestantism.

The most recent attempt to test the Weber thesis involved an effort to reduce Weber's broad-ranging theory to 31 specific hypotheses. (See the next "Doing Research on Religion" feature.) These were then tested by examining journals of 17th-century Puritan entrepreneurs, the writings and sermons of 17th-century Puritan clergy, and a review of previous analyses for data that might be relevant. Sociologist Jere Cohen (2002) concluded that some of Weber's assertions are supported by the evidence, but some specific hypotheses are not confirmed. In other words, the theory is not correct in all respects, but the entire thesis cannot be dismissed as invalid.

DOING RESEARCH ON RELIGION

The Protestant Ethic and the Spirit of Capitalism: Testable Hypotheses

Weber's theory paints a picture of the interaction of theology and economic activity. However, it paints with a rather broad brush. In order to test his theory, the ideas must be reduced to hypotheses or propositions that can be tested. Jere Cohen reduced the Protestant ethic to 31 hypotheses. A few of them are indicated here to show how abstract ideas are turned into specific statements that one can test with systematically gathered data:

The Work Ethic

- Ascetic Protestantism led its believers to work diligently at their occupations.
- Ascetic Protestantism made believers think of their work as a duty.

(Continued)

(Continued)

- Ascetic Protestantism's this-worldly approach focused believers' attentions on economic activities.
- Occidental culture's work ethic derived from Protestant teachings; Protestantism has helped to make work a cultural ideal.

Saving and Investment

- Ascetic Protestantism led its followers to save their money rather than spend.
- The money saved was often reinvested for capitalist growth.

The Spirit of Capitalism

- Protestant teachings, such as the duty to earn, helped to produce the spirit of capitalism.
- The spirit of capitalism appeared once the peak of religious piety had passed and the Puritan economic ethic had lost its religious component.

The Rationalization of Life

- An ascetic Protestant upbringing led to a rational, methodical life and to mastery of the world.
- Protestants' religious rationality carried over to their economic life in the form of rational business practices.

Wealth and Profit

- Protestantism saw wealth as God's blessing.

The Legitimation of Capitalism

- Protestantism legitimated profits as the fruit of an ascetic life.
- Protestantism legitimated inequalities of wealth based on different degrees of diligence.

Religious Anxiety

- Religious anxiety, born of the quest for certainty of salvation, was a driving force behind neo-Calvinist economic motivation.

The Quest for Salvation

- Neo-Calvinists worked hard to prove that they were saved; industry was a sign of election and idleness a sign of condemnation.

SOURCE: Cohen, Jere. (2002). *Protestantism and capitalism: The mechanisms of influence*. Hawthorne, NY: Aldine de Gruyter.

Cohen pointed out that religion may influence people by direct impact on their individual behavior or through shaping the larger cultural values over time. In the latter case, specific ideas from a theology may be selectively adopted and modified. Moreover, in this latter case it is all people in the culture who come to be influenced, not only those who belong to the group that spawned the ideas. Cohen found rather weak support for direct behavioral effect of the Protestant teachings on economic behavior. On the other hand, he believed a strong case can be made that Protestant ideas were selectively used and adapted in ways that legitimated certain economic behavior in the culture itself. So Protestantism seems to have had an impact mostly through cultural diffusion. For example, Protestantism seems to have strengthened the work ethic that had been a part of the Judeo-Christian tradition, but it was not the first or the only such force.

Moreover, even if a religious group does have a strong Protestant ethic, its members will not necessarily become highly affluent. Access to natural resources, willingness to use technology and other efficient sources of energy, and economic opportunities that are available to the group within the larger social structure all affect economic circumstances. For example, a very intense work ethic and an ascetic lifestyle has not been sufficient to bring extraordinary prosperity among the Old Order Amish—who reject modern technology such a tractors. No amount of asceticism and work ethic among African Americans has fully counteracted the structural discrimination they have experienced. A profound work ethic combined with an emphasis on asceticism may contribute to affluence, but they are certainly not sufficient.

So, Protestant theology was, at best, only one of many forces for capitalist expansion. It did not really embrace the pursuit of unbridled self-interest that the "spirit of capitalism" requires and is clearly not sufficient to bring about such a major economic shift. Jere Cohen (2002) pointed out that capitalism, instead, has "been legitimated

The Amish of Pennsylvania, Ohio, and Indiana have a very strong work ethic and live a simple life, but they do not become extremely affluent by contemporary American standards. They reject modern technology, such as tractors, and they define success in terms of faithfulness, community, and intragroup solidarity. The community support partially compensates for lack of technology, as we see in this 1-day barn-building project by a highly coordinated work crew. However, the strong Protestant ethic has not been sufficient to create great wealth. Photo by Brooks Kraft.

primarily by its ability to produce civic wealth. This justification has been chiefly secular in nature" (p. 142). Still, the ideas of several key clergymen helped to lessen the opposition to the principles of capitalism.

Cohen (2002) concluded the following:

There is something to the proposition that Protestantism influenced modern capitalism, but less than Weber claimed. Fewer mechanisms of influence operated than he innumerated; most that operated are weaker than he implied; and their impact was less revolutionary than he implied. The truth or falsity of the Weber thesis depends on how it is stated. At the most general level it is true that religious ideas can affect the economy; Protestantism aided modern capitalism . . . However, if the Weber thesis is stated in terms of the spirit of capitalism, one is on thin ice. . . . Weber's contribution to the Protestantism and capitalism question has been theoretical. He argued that religious ideas have an economic impact. Furthermore, he recognized that modern capitalism could not flourish without a supportive culture and institutional structure. (pp. 261–262)

Indeed, there is some evidence that in his later writings, Weber himself saw the economic ethic associated with the Protestant sects of Europe as playing an important but more limited role in the rise of modern capitalism than in his earliest writings on the topic.

Weber (1946) introduced the concept of *elective affinity*, the tendency for members of certain social and economic groups to be drawn to certain religious beliefs. He recognized that during the Reformation many people whose fortunes were rising may have been drawn to Calvinism because it justified (even sacralized) beliefs, behaviors, and outlooks that they already practiced. John Calvin's emphasis on individualism and his refusal to condemn usury may have attracted upwardly mobile people. Moreover, the lonely individual risk of salvation parallels the financial risks of venturesome entrepreneurs. This could be seen as a modification by Weber of his earlier suggestion that Calvinist beliefs caused upward mobility. Weber felt that both processes are at work; upward social mobility and Calvinistic beliefs were mutually reinforcing.

> *Critical Thinking:* Does Max Weber's argument—that religious ideas can cause social change—make sense to you? Why or why not? What do conflict theorists tend to say about the role of religious ideas in a society? Do you agree? Why?

After the Protestant Ethic

As we have seen, Weber sought to explain the role of religious values in the *rise* of modern industrial capitalism. He recognized, however, that once it was established, this economic system did not require the same religiously based ethic:

> The Puritan wanted to work in a calling; we are forced to do so. For when asceticism was carried out of monastic cells into everyday life, and began to dominate worldly morality, it did its part in building the tremendous cosmos of the modern economic order. This order is now bound to the technical and economic conditions of machine production which today determine the lives of all the individuals who are born into this mechanism, not only those directly concerned with economic acquisition, with irresistible force. Perhaps it will so determine them until the last ton of fossilized coal is burnt. (Weber, 1904–1905/1958a, p. 183)

What capitalism requires, then, is not asceticism but consumerism. It does not simply need people to save and invest their money; it needs people to spend their money buying all of the goods and services that are produced in a capitalism economy.[4]

Colin Campbell (1983) described the culture of consumer capitalism as one of extreme self-gratification:

> The revolution of rising expectations means that everyone not only expects to "better" himself but it is considered "immoral" not to strive to do so; this means an obligation to seek out and satisfy new "wants." . . . The obligation to satisfy want is linked to a money market economy embodying the principle of consumer sovereignty and mechanisms to guarantee the perpetual stimulation of new wants. (p. 281)

Extreme self-gratification has certainly not been one of the historical virtues of any of the

---

[4]If this seems contradictory to you, you are not alone. Some years ago, the great Harvard sociologist Daniel Bell (1976) wrote a book entitled *The Cultural Contradictions of Capitalism* in which he highlighted the desire not for self-denial but for self-gratification in late capitalist society.

world's religious traditions, and it has been anti-thetical to much of Christian teaching. Thus, in a reversal of the Weberian causal ordering—in which a religious ethic gave rise to an economic spirit—it seems that today modern consumer capitalism in the United States has given rise to a religious ethic that suits and supports it. A contemporary "Maxine" Weber studying American society might entitle her book *The Prosperity Ethic and the Spirit of Consumer Capitalism*.

In 2006, *Time* magazine ran a cover story called "Does God Want You to Be Rich?" (van Biema & Chu, 2006). The photo on the cover of the issue showed the front of a Rolls-Royce with a cross replacing the traditional Spirit of Ecstasy hood ornament. The story was about the rising number of American Christians who were adherents to what is known as Prosperity Theology (aka the Prosperity Gospel or the Health and Wealth Gospel). At the heart of this ethic is the idea that God provides material prosperity for those God favors. As the *Time* magazine story put it, "In a nutshell, it suggests a God who loves you does not want you to be broke" (van Biema & Chu, 2006).

Although it is has not been a historical affirmation of the Christian faith, the Gospel of Prosperity or Prosperity Theology has become a central message of a number of televangelists, including Joel Osteen of Houston, Texas. Prosperity theology embraces financial affluence and consumerism, with a message that God wants you to be rich and to have all the things you want. This provides sacred permission to engage in lavish consumerism.

Although historically Prosperity Theology grows out of the Pentecostal tradition, a national poll taken for the *Time* (van Biema & Chu, 2006) story found that the beliefs associated with it are more widespread:

- 17% of Christians surveyed said they considered themselves part of such a movement.
- 61% believed that God wants people to be prosperous.
- 31%—a far higher percentage than there are Pentecostals in America—agreed that if you give your money to God, God will bless you with more money.

Further analyses of these same poll data by sociologist Bradley Koch (2009) found that there was no relationship between income and adherence to the Prosperity Gospel, though African Americans and the less educated were particularly drawn to it.

At first blush, like their Calvinist forebears, adherents of the Prosperity Gospel seem to look to material success as a sign of their "chosen" status: a sign of their righteousness in the eyes of God. Unlike the progenitors of the Protestant Ethic, however, outward signs of material success are embraced by adherents to the Prosperity Gospel. They focus on the accumulation, not the ascetic self-denial. As Koch (2009) explained, adherents

> tend to interpret the New Testament as portraying Jesus as a relatively rich figure who used his wealth to feed the masses on several occasions and to finance what they argue to have been a fairly costly itinerate ministry. As such, Prosperity adherents argue that we should model our lives after Jesus' by living lavishly. (p. 1)

Thus, it is not uncommon to see preachers of the Prosperity Gospel driving Rolls-Royces and flying in private jets.

Interestingly, Pentecostals—the denominational group most associated with the Prosperity Gospel—are those on the *lowest* end of the economic spectrum. Also, those social groups that are most disadvantaged in American society—African Americans and Latinos—adhere to the Prosperity Gospel more than others. This gives credence to Marx's view that religious ideas serve to legitimate the status quo. As Marx would suggest, the belief that hard work will lead to God's favor and hence success serves to justify the current economic system. Individuals adhering to these beliefs, even if they are economically disadvantaged, will never rise up against the system. The belief becomes, in effect, an "opiate."

SOCIAL CLASS AND RELIGIOUS INVOLVEMENT: INTERACTIVE PROCESSES

Social Class and Religious Affiliation Today

As noted, Weber began his inquiry into the connection between religious beliefs and economic action by noting that Protestants tended to have higher class positions than Catholics at the time. Although the situation is more complex today, there remains a connection between denominational affiliation and social class. This is shown in Table 9.1, which looks at two key measures of social class—income and education—for the major denominations in American society.

Before examining this table, it is important to recall the discussion of denominations and congregations in Chapter 8. At the local level, congregations tend to be even more class-segregated than denominations as a whole. In other words, a large downtown Baptist church in a southern community may have a large percentage of upper- and middle-class people from that town. On the edge of town, a smaller Baptist church may be attended almost entirely by working- and lower-class persons. Likewise, the Congregational

Table 9.1   Socioeconomic Profiles of American Religious Groups: 2006

Religious Group	Educational Level (% with at Least a College Degree)	Religious Group	Annual Household Income (% Over $100,000)
Hindu	74	Jewish	46
Jewish	59	Hindu	43
Episcopal Church in USA	57	Episcopal Church in USA	35
Unitarian	51	Presbyterian USA	28
Buddhist	48	Orthodox	28
Presbyterian USA	47	Atheist	28
Orthodox	46	Unitarian	26
United Church of Christ	42	Buddhist	22
Atheist	42	United Methodist	22
United Methodist	35	Disciples of Christ	20
Disciples of Christ	35	Catholic	19

Religious Group	Educational Level (% with at Least a College Degree)	Religious Group	Annual Household Income (% Over $100,000)
Evangelical Lutheran in America	30	United Church of Christ	18
Nondenominational	29	Nondenominational	18
Latter-Day Saint (Mormon)	28	**U.S. Average**	**18**
U.S. Average	**27**	Evangelical Lutheran in America	17
Catholic	26	Latter-Day Saint (Mormon)	15
Muslim	24	Muslim	16
Southern Baptist	21	Southern Baptist	15
Seventh Day Adventist	21	Religious but Unaffiliated	12
Religious but Unaffiliated	17	Seventh Day Adventist	11
Historically Black Baptist	15	Jehovah's Witness	9
American Baptist	14	American Baptist	8
Assemblies of God	12	Assemblies of God	8
Other Pentecostal	11	Historically Black Baptist	8
Jehovah's Witness	9	Other Pentecostal	7

SOURCE: Pew Research Center Forum on Religion & Public Life (2008b). *U.S. religious landscape survey*. Retrieved from http://religions.pewforum.org

churches (United Church of Christ [UCC]) in big cities in the northeast tend to draw the more highly educated population, but in a small town in upstate New Hampshire, where only two or three congregations serve the community, the Congregationalists may draw heavily from the lower- and middle-class population. Probably no local congregation is totally class-exclusive, and some congregations are relatively well integrated in terms of class. Nevertheless, the tendency is for local congregations to be somewhat more segregated by social class than Table 9.1 indicates.

This caveat notwithstanding, there are clear differences between the denominations that call for some explanation. Relative to Weber's argument, we note that this hierarchy of denominations *cannot* be classified according to ascetic values and practices based in Protestant theology. Only one of the top 5 and 4 of the top 13 denominations in this hierarchy are Protestant. A group that clearly live out a this-worldly asceticism, Mormons, are below the national average for both income and college graduation rates.

How are we to explain, therefore, this religious stratification? Does religion continue to play a role, or are secular factors more important in producing this hierarchy? In explaining religious group stratification in America, we need to examine several broad categories of factors: historical effects, selection effects, and treatment effects. The first two sets can be considered predominantly secular factors and the last one the more truly religious factor.

Historical Effects

In the United States, the Episcopalians, Congregationalists, and Presbyterians are the denominations that have been established for the longest period of time. In fact, members of these denominations so dominated business and politics that they were known as "The Protestant Establishment" (Davidson & Pyle, 2006). Lutherans, who occupy higher status in certain European countries, represent somewhat more recent arrivals in this country. The highest-status positions were already occupied in North America when German and Scandinavian immigrants arrived, and the latter have had to work their way up the social ladder from the bottom. Persons from these ethnic backgrounds are also located in more rural areas of the United States. Hence, they tend to rank lower on such status indicators as educational level, income, and occupational prestige.

Roman Catholic immigrants from Europe in the late 19th century experienced the same pattern as Lutherans, and Catholic immigration from Latin American countries continues to shape the overall economic and educational achievement of Catholics. Indeed, both income and education levels for Roman Catholics vary substantially depending on ethnicity, with those who have been in this country longer generally ranking higher than more recent Catholic immigrants. For example, 31% of white Catholics are college graduates, while only 13% of Latino Catholics are, and 24% of white Catholics report household incomes above $100,000 per year, while only 9% of Latino Catholics report earning that much (Pew Research Center Forum on Religion & Public Life, 2008a).

There are notable exceptions to this pattern. Jews, for the most part, are more recent immigrants than the long-established Protestants, but they have obtained higher socioeconomic status (highest in terms of income and third highest in terms of education) than one might expect for fairly recent arrivals. Hindus as a group are even more recent immigrants than Jews and seem to follow the same pattern, particularly in their high levels of education but also in income. Their variation from the earlier immigrant pattern can be explained by the fact that they are more likely to be professionals or to have higher levels of education and capital *when they enter the country* than have most immigrants.

Muslims prove to be an interesting reversal of the Catholic pattern. Of the over 2 million Muslims living in America, 65% are foreign-born and 35% are native-born. Despite the fact that over half of foreign-born Muslims have immigrated since 1990, foreign-born Muslims on average have both higher levels of education and income than native-born Muslims (Pew Research Center Forum on Religion & Public Life, 2007b). This is due to the fact that a high percentage of native-born Muslims are African American converts to the tradition. The lower social class position of African Americans in general, and of those who are inclined to convert to Islam, explains the reversal in historic pattern for Christian immigrants from Europe.

Although there has been some mobility in the religious stratification system in America over the years, overall the socioeconomic inequality in the American religious system has been "quite persistent and stable" (Smith & Faris, 2005), whether we look over the past 20 years (Pyle, 2006) or the past 200 (Davidson & Pyle, 2006).

Selection Effects

Some scholars believed that the decline in the ascriptive basis of religion in American society—that is, the rise of religious voluntarism or the ability to pick and choose one's own religion—would make for a more fluid religious system and hence more mobility (Roof & McKinney, 1987). What they may not have accounted for, however, is that when people have the ability to choose with whom to associate, they often choose to associate with people like themselves.

Thus, a second reason for class differentials may be a simple matter of "like seeking like." Those of similar educational level may be drawn by common interests, common speech patterns, and other homogeneous characteristics. A highly educated person may not return to a congregation in which the religious leader demonstrates little scholarship and uses bad grammar. The visitor simply feels uncomfortable. Likewise, a person with little education may feel alienated and lost in a religious community where the clergy delivers a scholarly sermon based on tightly argued logic. Such a sermon may seem utterly irrelevant and uninspiring to the visitor. Moreover, when people move to a new community, they sometimes join local religious communities that are attended by their colleagues at work. To quote the title of a recent work by Timothy Nelson (2009), when it comes to religion and social class, we are "at ease with our own kind."

This view also suggests that as people move from one social status to another they tend to change their religious affiliation. Rather than suggesting that religious beliefs may cause social mobility (as Weber's thesis suggests), advocates of this explanation insist that changes in religious affiliation frequently *follow* changes in social status. This is so because people are more comfortable with persons of the same socioeconomic standing or because they are using religious affiliation as a way of reinforcing their upward social mobility.

This explanation—discussed in Chapter 6—is a controversial one. Some researchers have gathered data that do not support the idea that denominational switching follows social mobility (Nelsen & Snizek, 1976). Others insist that their evidence does support the theory that denominational switching follows social mobility (Newport, 1979). More empirical evidence is needed and perhaps a more sophisticated treatment of variables is necessary. It may well be that this explanation holds for some types of groups or for some types of people but not for others.

A second set of explanations that focus on selection effects deals with the relationship of the theology to the secular order. Some theological orientations endorse the present social order as ordained by God; others insist that religion has little or nothing to do with the social order and that religious ethics are concerned only with personal motives and intentions. In either of these cases, members of the privileged classes would not find their self-interests being threatened. On the other hand, some religious groups define affluence as the root of evil and glorify the life of poverty and self-denial. Still others attack the social order and advocate a restructuring of the socioeconomic system (much like the radical reformation of the Anabaptists). In these groups, members of the privileged classes might find themselves very uncomfortable, whereas the lower classes would find either solace or hope for the future. In this case, class differences between denominations would be due to "elective affinity"—people choosing a religious group that fits their own socioeconomic circumstances (Stark & Bainbridge, 1985; Weber, 1946).

Critical Thinking: Do you think the fact that affluent people tend to be attracted to certain kinds of religious communities is a simple matter of "elective affinity"? Is the correlation of social class and religious affiliation a matter of choosing religious communities compatible with one's financial circumstances, or does the religion tend to improve or to undermine one's socioeconomic standing?

Religions of the affluent tend to be very orderly and decorous in their ritual (sometimes called "high church"), and the theology is very logical or cognitively oriented. The formal garb and demeanor of the Anglican Archbishop of Canterbury, Rowan Williams, illustrates this pattern. The religious ritual of the disfranchised is more often spontaneous or lacking in a formal order, and the appeal is more emotional and expressive than logical–rational. Obviously the sort of pomp and circumstance of this photo is not a part of the worship liturgies of the poor.

Treatment Effects

Finally, treatment effects are those effects that have to do with the effect of belonging to a particular religious group on economic action and achievement. These can be considered the "true" causal elements of a religious tradition on the economic behaviors of adherents. They may help to explain the religious group stratification that was shown in Table 9.1.

Most of these studies highlight the *indirect* effect of religious tradition on educational and economic achievement. An indirect effect is one in which religious tradition affects some other behavior which in turn affects economic outcomes. Also, most of this body of research focuses its attention on explaining why certain Protestant groups fare so poorly in the religious stratification system.

The lowest positions in the education and income hierarchies are dominated by Protestant denominations that are sectarian in orientation and fundamentalist in belief. These groups are well known to demonstrate hostility toward secular education and an anti-intellectual orientation that reduces educational achievement (Beyerlein, 2004; Darnell & Sherkat, 1997; Sikkink, 1999). This has longer term effects since education itself is a key predictor of economic prosperity in American society (Sewell & Hauser, 1975). Sectarian Protestant affiliation also has a negative effect on verbal ability—at all levels of education—which can impede occupational success (Sherkat, 2010).

The exact opposite is the case for Jews, who are at or near the top of the education and income hierarchy in America. Some argue that Rabbinic Judaism promoted knowledge and learning, which translated into a positive embrace of schooling and literacy. Furthermore, "skills initially acquired for religious purposes (e.g., literacy in Hebrew) enhanced Jews' ability to acquire economically useful skills (such as literacy in local languages)" (Burstein, 2007, p. 215). Similarly, the Protestant elite (Episcopalians, Congregationalists, Presbyterians) have long been highly supportive of and tightly networked into the elite institutions of secondary and postsecondary education (Baltzell, 1964; Karabel, 2005).

Socioeconomic Status and Style of Religiosity

Turning the table around from considering the effect of religion on economic action, we can ask, what is the effect of economics on religious action?

There has been some research that has focused on ways in which one's social class seems to affect one's style of religiosity. Some of the early studies on religion and social class used attendance as the prime indicator of religious commitment. Because upper-class people tend to be more regular in attendance at worship, it appeared that the upper and middle classes were more "religious" than were the lower classes. When researchers began to use multidimensional measures of religiosity, a rather different pattern emerged. One analysis found that among Congregationalists, people who ranked high in

socioeconomic status also scored highest for religious knowledge (biblical teachings and the theological orientation of the denomination). On the other hand, lower-class members scored higher on devotionalism (personal religious experience, daily prayer, etc.) than did higher status members (Fukuyama, 1961).

Recent research continues to support these observations. For example, Joseph Baker (2008), using a nationally representative sample of Americans, showed the nearly linear inverse relationship between income and prayer frequency (see Figure 9.1). The connection between class and devotionalism remains strong. Studies of youth suggest that this connection may persist into the future. Research by Philip Schwadel (2008) using data from the National Study of Youth and Religion, shows that poor teenagers score high on devotionalism (they are especially likely to pray, read religious scriptures, and report high levels of personal faith) at the same time they are much less likely to participate regularly in organized religious activities.

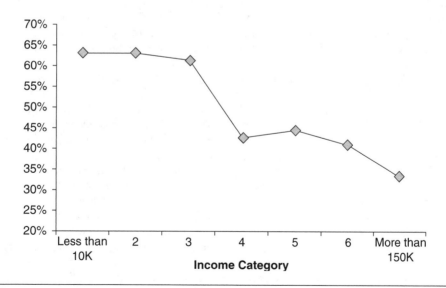

Figure 9.1   Percent Praying at Least Daily by Income

SOURCE: Baker, Joseph O. (2008). "An Investigation of the Sociological Patterns of Prayer Frequency and Content." *Sociology of Religion, 69*:169-85, used by permission of the Association for the Sociology of Religion.

Steven Cohen (1983) reported similar differentials among Jews. Highly educated Jews were more likely to have institutional affiliation and participation at the synagogue but were less likely to observe private family rituals (Jewish devotionalism) than were lower-class Jews. Family income was also positively correlated to institutional affiliation but negatively related to private family observances.

N. J. Demerath III (1965) added further insights on the relationship between social class and style of religious commitment. He began by distinguishing two types of religiosity: (1) "churchlike" and (2) "sectlike." Demerath found from his study of five Protestant denominations that *within* any given denomination, an individual's style of religiosity is highly correlated to his or her socioeconomic status. More important, Demerath suggested that members of different social classes who belong to the same congregation may be active in different ways and may have quite different needs met by the same local church. The higher a Lutheran's social status, the more likely he or she is to participate in churchlike ways. The lower the social status, the more likelihood there was of sectlike participation. The same pattern held for the other four denominations as well (Congregationalists, Disciples of Christ, Baptists, and Presbyterians).

Rodney Stark, using a different method of research, also found that the style of religious involvement was linked to social class. Stark (1972) noted that high-status church members participate to a greater degree in those activities that reinforce their respectability and confirm their worldly success (high levels of biblical or doctrinal knowledge and more voluntarism in the local congregation). Lower-class persons are religiously active in ways that will offer comfort and solace and will provide compensation for one's lack of worldly success. Several other empirical studies have also found both churchlike participation and sectlike participation within each denomination—and have found socioeconomic status to be the critical variable affecting a member's orientation (Dynes, 1955; Winter, 1977). Further, the bulk of empirical

research on social status and religiosity suggests that persons with high status are more likely than lower-class persons to be committed at an instrumental level—through investment of time and money in the formal structure (Estus & Overington, 1970). Lower-status persons are more likely than higher-status persons to be committed at the affective level—through close friendship networks. So differences in religiosity between the social classes are not ones of *degree* of involvement so much as ones of *kind* or *style* of involvement.

Religious perspective and moral standards can influence individuals in their economic choices in still other ways. Robert Wuthnow (1994a) also found that people whose moral values are shaped by theistic beliefs are less likely to compromise their conduct on the job than are those with other value orientations. People who are "theistic moralists" were less likely to bend the truth, cover for someone else who was not performing appropriately, or use office equipment for personal business. Further, people who were in religious fellowship groups were considerably more likely to have high standards of personal ethical behavior on the job. While Wuthnow concluded that religious orientation has less impact on economic behavior than religious officials would hope, it was clear from his study that religion is far from irrelevant to the way people behave and think about their work. Indeed, 30% of Americans feel they are called by God to their present line of work, and 46% of those who are frequent attendees expressed this sentiment. Religiously committed people therefore had more concerns about materialism and its consequences for society, for they saw other purposes that they thought transcended financial success. In another study, Wuthnow (1996) concluded that religious ideas provide an alternative way of thinking about economic behavior, a standard that contrasts with the simple utilitarian or "rational choice" approach of the larger secular society.

In conclusion, the sacralization of specific values or outlooks may affect one's economic behavior. While one's religious perspective may not be sufficient to change one's socioeconomic position, it may contribute to such a change. Religious ideas and communal networks can be an independent variable influencing socioeconomic behavior and circumstances.

> *Critical Thinking:* Does it make sense that style of religiosity would be influenced by socioeconomic status? Why or why not? Do the findings seem to apply to your own style of religiosity?

In the end, expressions of religious commitment tend to be significantly affected by socioeconomic status in a variety of ways. Not only do people of different social classes tend to affiliate with different denominations but members of the same denomination who are of different socioeconomic status tend to participate in the life of the religious community in different ways and for different reasons. Furthermore, the overall socioeconomic level of a local congregation affects the style of religiosity of that group, and members may be affected in their religiosity as they conform to the norms of their local congregation (Davidson, 1977). Socioeconomic status is certainly not the only factor affecting one's style of religiosity, but it has been an important one.

Throughout this text, we have pointed to religion as an open system, a system of meaning, belonging, and structure that impacts the larger society through outputs and is influenced via inputs from the larger society. This principle holds for socioeconomic processes as well as for other aspects of religion. Religion is both a dependent and independent variable as it relates to the stratification system of the society.

> *Critical Thinking:* Do you think religion is more an *independent* variable influencing socioeconomic standing of members of religious communities, or is religion mostly a *dependent* variable that is shaped by socioeconomic conditions? Why do you think as you do? What evidence is most compelling to you?

SUMMARY

Religion is correlated to socioeconomic status as both a dependent and independent variable. According to Weber's thesis, one's values and one's worldview may contribute to one's rise or decline in the socioeconomic system. If one is taught a sacred ethic that requires hard work and simple living, this may enhance eventual accumulation of wealth. However, other social conditions must be present for such social mobility to take place. The debate continues on the importance of a particular theological orientation having influenced the original formation of capitalism as a system. In the modern world, it is certainly very possible that certain religious values and perspectives may serve as contributing factors to upward mobility. Theological orientation is neither necessary nor sufficient, however, for economic prosperity.

Religion is also very much influenced by the socioeconomic circumstances of its adherents. People in the lower classes tend to have different styles of religiosity than do coreligionists of the upper classes. This variation shows up not only between denominations serving different socioeconomic groups but also between people of different socioeconomic status within the same denomination.

Religion both influences and is influenced by the socioeconomic system. At this point, the strong consensus among social scientists is that economics is the more powerful variable. In other words, a person's religiosity is significantly affected by his or her position in the stratification system. The effects of religion on the socioeconomic behavior of most people are considerably less pronounced. The Marxian perspective is that economic self-interests affect religion more than religion affects economic behavior. Much to the chagrin of many religious people, the empirical evidence tends to support this view. However, even the Marxian view is too simplistic, for religious principles do influence people's behavior on the job and in the marketplace.

10

RELIGION AND RACE

Here are some questions to ponder as you read this chapter:

- How is religion connected to race and ethnicity?
- Why is there so much racial segregation in congregations?
- What is the relationship between religious conviction and racial prejudice? Is religious commitment contrary to ethnic or racial bigotry?
- How might meaning systems, belonging systems, and institutional systems of a religious community contribute to racism in different ways?
- How might religion foster prejudice at the same time it combats it?
- Does religion cause bigotry or merely *justify* social inequality and discrimination after the fact?

Writing in *The Souls of Black Folks* at the beginning of the 20th century, the pioneering sociologist W. E. B. DuBois (1903) asserted, "The problem of the twentieth century is the problem of the color line" (p. 283). Over a century later, much has changed in American society. Racial minorities have increased their standing in almost every aspect of life, aided considerably by the civil rights struggles of the 1950s and 1960s (many of which were led by religious leaders as we will see in Chapter 13). For some, this racial progress culminated in 2008 when Barack Obama was elected president of the United States.

Despite this, racial inequality persists. For example, African Americans and Latinos continue to earn significantly less than whites even if they have the same level of education, as can be seen in Table 10.1. This economic inequality has a religious dimension. As we saw in the last chapter, denominations with a high proportion of African American members are lower on the socioeconomic hierarchy than historically white denominations.

Table 10.1   Income (in Dollars) by Race/Ethnicity and Educational Level

Education	White	Black	Hispanic
Not a high school graduate	22,289	17,439	21,303
High school graduate	32,223	27,179	27,604
College graduate	58,652	46,502	44,696
Master's degree	71,372	56,398	68,040
Professional degree	122,885	94,049	84,512

NOTE: Figures are for 2008.

SOURCE: U.S. Census Bureau. (2010). Mean earnings by highest degree earned. Table 227. Retrieved from www.census.gov/compendia/statab/cats/education/educational_attainment.html

Indeed, the very fact that we can talk about black and white denominations suggests the reality (or at least legacy) of religious racial segregation. Civil rights leader (and sociology major) Martin Luther King Jr. (2010) wrote,

It is appalling that the most segregated hour of Christian America is eleven o'clock on Sunday morning, the same hour when many are standing to sing, "In Christ there is no East or West." Equally appalling is the fact that the most segregated school of the week is the Sunday School. (p. 207)

King then went on to quote Liston Pope, who lamented, "The church is the most segregated major institution in American society" (King, 2010; Pope, 1957).

In this chapter, we ask and answer the question, is the same true today? If so, what accounts for that religious racial segregation? Further, what is being done to overcome it?

RACE, ETHNICITY, AND RELIGIOUS IDENTIFICATION

A key starting point here is to look at religious self-identification by members of different racial and ethnic groups.[1] We use data from the

---

[1]Race refers to a supposed biological difference that distinguishes a group, and although there is no scientific concept of race, it remains a social distinction because people treat certain physical differences as though they matter. An ethnic group is a group with a cultural difference—variations in language, symbols, norms, values, and beliefs.

Table 10.2   Religious Identification by Racial Group—Reported in Percentages

	U.S. Population	White	Black	Asian	Other/ Mixed	Latino
Evangelical Protestant	26	**30**	15	17	**34**	16
Catholic	24	22	5	17	14	**58**
Mainline Protestant	18	**23**	4	9	15	5
Historically Black Churches	7	<0.5	**59**	<0.5	2	3
Secular Unaffiliated	6	7	3	11	7	4
Religious Unaffiliated	6	5	8	5	9	8
Jewish	2	2	<0.5	<0.5	1	<0.5
Mormon	2	2	<0.5	1	2	1
Atheist	2	2	<0.5	3	1	1
Agnostic	2	3	1	4	3	1
Buddhist	1	1	<0.5	**9**	1	<0.5
Jehovah's Witness	1	<0.5	1	<0.5	1	1
Muslim	1	<0.5	1	**4**	1	<0.5
Orthodox	1	1	<0.5	<0.5	1	<0.5
Other Faiths	1	1	<0.5	1	5	<0.5
Hindu	<0.5	<0.5	<0.5	**14**	1	<0.5
Other World Religions	<0.5	<0.5	<0.5	2	<0.5	<0.5
Other Christian	<0.5	<0.5	<0.5	<0.5	1	<0.5
Total	100	100	100	100	100	100

NOTES: Due to rounding, figures may not add up to 100%.

We added the bold font for emphasis to key numbers.

SOURCE: Pew Research Center Forum on Religion & Public Life. (2008b). *U.S. religious landscape survey.* Retrieved from http://religions.pewforum.org

Pew Research Center Forum on Religion & Public Life's (2008b) U.S. Religious Landscape Survey to examine this issue. These data highlight the considerable differences in religious identification by race and how those differences are reflected in denominational racial composition. For example, Table 10.2 shows that more than three quarters (78%) of blacks are Protestant (including evangelical, mainline, and historically black churches), compared to just over half of whites (53%) and about a quarter of Asians (27%) and Latinos (23%). Twenty-four percent of all Americans are Catholic, but 58% of all Latinos are Catholic, compared with 22% of whites and only 5% of blacks.

Looked at from the perspective of religious traditions in Table 10.3, we see large concentrations of whites in Judaism, mainline Protestantism, Orthodoxy, Mormonism, and evangelical Protestantism. In addition to the historically black churches, African Americans are also overrepresented among Jehovah's Witnesses and Muslims, and Asians are dramatically concentrated in the Buddhist and Hindu traditions. Latinos are overrepresented among Catholics and Jehovah's Witnesses. Looked at either way, there continues to be a strong correlation between race/ethnicity and religion in American society.

> *Critical Thinking:* What additional pattern do you see in Tables 10.2 and 10.3 that might be relevant to understanding race, ethnicity, and religion?

These cross-sectional data do not reveal changes in religious affiliation by race or ethnicity over time, but using other research strategies, scholars have suggested that despite considerable stability in African American religious identification there has been some loosening of the connection between African Americans and specific denominational affiliations. For example, the American Religion Identification Survey (ARIS) found a decrease from 50% to 45% in African Americans identifying themselves as a Baptist, from 12% to 7% identifying with "Mainline Protestant," and from 9% to 6% seeing themselves as Catholic. These declines were offset by gains in the "Generic Christian" category from 9% to 15%, and in the "None" category from 6% to 11% (Kosmin & Keysar, 2009). This is not to say that African Americans are becoming less religious, because even among "Nones," African Americans are still highly religious. The Pew Research Center Forum on Religion & Public Life's (2008b) U.S. Religious Landscape Survey found that 71% of African American "Nones" report that religion is very important or somewhat important in their lives, compared to only 41% of all "Nones." Thus, most African American nones are what the Pew Research Center Forum on Religion & Public Life (2009) would call "religious unaffiliated."

Latinos in the United States are increasingly drawn away from Roman Catholicism and toward evangelical Protestantism. Fifty-one percent of Hispanic evangelicals are converts, and of that group, more than 80% are former Catholics (Pew Research Center Forum on Religion & Public Life, 2007a). This mirrors a worldwide trend, one that is especially prominent in the historically Catholic countries of South America (Martin, 1993; Stoll, 1990).

In Chapter 9, we noted that looking at socioeconomic differences at the denominational level obscures the higher levels of inequality at the local congregational level. The same is true for racial differences. The level of racial segregation we see at the level of denominational identification is even more pronounced in terms of congregational membership.

Table 10.3   Racial Composition of Religious Traditions—Reported in Percentages

	White	Black	Asian	Other/ Mixed	Latino	Total
U.S. Population	**71**	**11**	**3**	**3**	**12**	**100**
Jewish	95	1	0	2	3	100
Mainline Protestant	91	2	1	3	3	100
Orthodox	87	6	2	3	1	100
Mormon	86	3	1	3	7	100
Atheist	86	3	4	2	5	100
Agnostic	84	2	4	4	6	100
Evangelical Protestant	81	6	2	4	7	100
Other Faiths	80	2	1	13	5	100
Secular Unaffiliated	79	5	4	4	8	100
Other Christian	77	11	0	8	4	100
Catholic	65	2	2	2	29	100
Religious Unaffiliated	60	16	2	5	17	100
Buddhist	53	4	32	5	6	100
Jehovah's Witness	48	22	0	5	24	100
Muslim	37	24	20	15	4	100
Hindu	5	1	88	4	2	100
Historically Black Churches	2	92	0	1	4	100

SOURCES: Pew Research Center Forum on Religion & Public Life. (2008b). *U.S. religious landscape survey*. Retrieved from http://religions.pewforum.org/

Pew Research Center Forum on Religion & Public Life. (2007b, May 22). *Muslim Americans: Middle class and mostly mainstream*. Retrieved from www.pewresearch.org/pubs/483/muslim-americans

RACIAL SEGREGATION IN CONGREGATIONS

The leading scholar investigating congregational segregation is Michael Emerson. In a series of studies, books, and articles, Emerson has documented the persisting religious racial segregation in American society (Christerson, Edwards, & Emerson, 2005; DeYoung, Emerson, Yancey, & Kim, 2003; Emerson, 2006; Emerson & Smith, 2000). Emerson and his colleagues operationalized racially segregated congregations in two different ways: *binary* and *continuous*.

First, they operationalized racially segregated congregations in a binary (either/or) fashion. Those in which 80% or more of the members are of the same race are considered segregated (or homogenous), and those in which less than 80% are of the same race are considered multiracial (racially mixed) (Emerson, 2006, p. 36). Using data from a nationally representative sample of American congregations, they find that 93% are racially homogenous by this definition. Only 7% are multiracial. What about even more extreme cases of racial segregations? Emerson and Kim (2003) also found that 9 out of 10 American congregations have 90% of their members coming from one racial group, and roughly 4 out of 5 congregations have 95% of their membership from a single racial group.

Second, they examined racially segregated congregations in a continuous fashion by looking at the likelihood that any two randomly selected members of the congregation will be from different racial groups (Emerson, 2006, pp. 35–36). The method is called continuous because there is no single point at which we can say "this is a segregated congregation, and that is not." Instead, we get a continuous string of probabilities called a "heterogeneity index," which ranges from 0.0 (complete homogeneity) to 1.0 (complete heterogeneity). Put concretely, if a congregation has 50% of its members from one racial group and 50% from another, then the probability that any two

randomly selected members will be of different racial groups is 0.5. The more diverse a congregation is, the higher the probability will be, and vice versa. So, if there are four racial groups each constituting a quarter of the membership then the probability will be 0.75.

How do congregations fare in terms of this heterogeneity index? Not well. The average probability of randomly selecting two members of a congregation from different races is just 0.02—a 2% chance. This makes congregations far more segregated than even schools or neighborhoods. The heterogeneity index for schools in the U.S. is 0.40 and for neighborhoods is 0.20. Thus, Emerson, Mirola, and Monahan (2011) concluded, "religious congregations are *10 times less diverse* than the neighborhoods in which they reside and *20 times less diverse* than the nation's schools" (p. 161).

The bottom line of Emerson's studies is that no matter how you operationalize congregational racial segregation, Martin Luther King's and Liston Pope's observations about religion being the most racially segregated institution in American society (and religious education the most segregated schooling) is just as true today as it was 50-plus years ago.

EXPLAINING CONGREGATIONAL RACIAL SEGREGATION

As with racial segregation in society generally, explaining religious racial segregation is challenging in its complexity. We briefly highlight several explanations for congregational homogeneity and then turn to a more extended consideration of the relationship between religion and racial prejudice.

In the first place, individuals often choose to associate with people like themselves. As the saying goes, birds of a feather flock together. This is known as "homophily" (McPherson, Smith-Lovin, & Cook, 2001). As was noted in Chapter 9, this often takes place along social class lines, but it happens all the more so along racial lines. However, as sociological investigation has discovered, flocking together is not always a completely voluntary choice. Racial segregation in congregations, then, is a classic example of the mix of choice and constraint. As Karl Marx (1852/2008) said, "[People] make their own history, but they do not make it just as they please; they do not make it under circumstances chosen by themselves, but under circumstances directly encountered, given, and transmitted from the past" (p. 15). This fundamental sociological insight is highly applicable to this issue. The "black church" in America is a case in point, as we will see more fully in Chapter 12. It is an integral part of the black community in America, valued by its members and chosen by many of them. Still, the emergence of the black church is rooted in slavery and racial discrimination in the 19th century, not simply preference by blacks to be around people like themselves. Indeed, some have called the black church a "semi-involuntary institution" for this reason (Nelsen, Yokley, & Nelsen, 1971).

Immigrant groups have also often chosen to form and maintain segregated religious congregations when faced with language barriers and other forms of social exclusion. Historically, many Lutheran, Methodist, and Catholic congregations in the United States conducted worship services in German, Welsh, and Italian, and going to worship was a comfortable reminder of the home country. This pattern is also evident today in the vast number of Asian immigrant religious communities across the country, such as First Chinese Baptist Church (Walnut, California), Korean-American Calvary Baptist Church (Seattle, Washington), and Chinese American Family Bible Church (Virginia Beach, Virginia). Incredibly, in Columbus, Georgia, a mid-sized city of 250,000 people in southwestern Georgia, there are four separate Korean Protestant congregations, including two Presbyterian congregations: the Korean Full Gospel Church, Korean Presbyterian Church, Hallelujah Korean Presbyterian, and Korean Peace Baptist Church. Like the African American church, these Asian ethnic congregations also exemplify the principle of choice under constraint.

Whereas many Asians in America *become* Christian upon their immigration (Yang, 1999), other immigrants bring their religion with them. For some of these groups, it is impossible to separate their racial/ethnic identity from their religious identity. This is the case for Hindus from India. Consequently, peering into a Hindu temple in America today one is likely to find mostly individuals of Indian descent (Emerson & Kim, 2003, p. 219).

Uniracial congregations are not only created but they are also perpetuated by people's homophilous social networks. Since we already know that most people are recruited into religious groups through family and friendship networks (Chapter 6), and because family and friendship connections tend to be racially homophilous, recruitment into religious groups will also be racially homophilous.

These explanations for racial homogeneity apply to congregations from all religious traditions, so the processes that underlie them are not unique to any one faith tradition. On the other hand, racial homogeneity in congregations is not equally distributed across religious traditions. Emerson (2006) estimated that only 5% of Protestant congregations are multiracial, compared with 15% of Catholic and 28% of non-Christian congregations (p. 39). So while all religious traditions are racially segregated, they are not segregated to the same extent. Explanations of racial homogeneity in congregations, therefore, must include religious tradition as a factor.

Insofar as racial or ethnic segregation of religious groups is related to we/they categories and thinking, there are important issues raised about the role of religion in transcending or fostering prejudice. This brings us to a robust body of research on the relationship between religion and prejudice.

> ***Critical Thinking:*** What seems to you to be the most important factors in racial and ethnic segregation? What empirical data or studies support your position? What data or research provides the greatest challenge to your position? What might be some *consequences* of this segregation for the larger society?

RELIGION AND RACIAL PREJUDICE

In the 1950s, empirical studies showed that religiously affiliated people were more racially prejudiced than nonmembers. Despite the fact that Christianity claimed to enhance fellowship and love among people, the research indicated a correlation between Christianity and bigotry. A number of explanations were formulated to interpret this phenomenon. Some scholars attempted to identify factors in the belief system of Christianity that might contribute to prejudice. Others felt that the correlation was spurious, that the correlation of prejudice and religious affiliation resulted from some third factor (Eisinga, Felling, & Peters, 1991). The whole debate changed significantly, however, as more sophisticated and refined data were gathered.

In the 1960s, several survey studies revealed that although church members were more prejudiced than nonmembers, the most active members were less prejudiced than any other group. The earlier studies had lumped together all members without regard to level of commitment or amount of participation in the life of the religious community. Because there are larger numbers of marginal members in most religious groups than there are active members, the statistics were weighted heavily in

the direction of marginal member attitudes. The evidence indicates that infrequent attenders are more prejudiced than nonattenders but that frequent attenders are the lowest of all on scales of prejudice (Allport 1966; Gorsuch & Aleshire, 1974; Perkins 1983, 1985). In one study of respondents scoring very low on racism measures, the assertion that racism is incompatible with their religious beliefs was the most frequently cited reason for liberal attitudes toward others (Tamney & Johnson, 1985). In a recent cross-national study that compared prejudice, 11 European countries also found that adherence to the doctrinal beliefs of the Christian church was related to low levels of prejudice (Scheepers, Gijsberts, & Hello, 2002).

This finding of low levels of prejudice only among the highly committed led to several theories and hypotheses. One of the first asserted that some people are *intrinsically religious* and others are *extrinsically religious* (Allport, 1966). Intrinsically religious people join because their faith is meaningful to them in and of itself. Such people are committed at what Rosabeth Kanter would call the "moral level" (see Chapter 5). The meaning functions of religion are central for them. Individuals who are extrinsically religious tend to join religious congregations because of secular advantages. Those who join because of status factors, for example, would be considered extrinsically religious. To use Kanter's formulation of types of commitment, we might say that people who are committed primarily at the instrumental and/or affective levels would be extrinsically religious. The belonging, identity, and status functions are especially important to them. Gordon Allport insisted that persons who were intrinsically religious score low on measures of racial prejudice, while persons who are extrinsically religious score high. Most, but not all, subsequent studies have supported this claim (Batson, Schoenrade, & Ventis, 1993; Fulton, Gorsuch, & Manard, 1999; Jackson & Hunsberger, 1999; Kirkpatrick, 1993).

For some Americans, religious conviction *is* closely related to racism and even serves to sacralize white supremacy. In the rally depicted here, Ku Klux Klan (KKK) members display a burning cross (which they call a sacred "cross lighting"), historically, for blacks, a symbol of terrorism. It signaled that if they did not obey the norms of white superiority, blacks would be lynched on the nearest tree. The KKK identifies God with racial differentiation and white supremacy.

As research continues on this issue, it becomes obvious that the variables affecting the relationship between religion and prejudice are complex. For example, several scholars have found that the specific theological orientation of people was very important (Hadden, 1969; Quinley, 1974; Roof & McKinney, 1987; Kirkpatrick, 1993). In most studies, fundamentalists have been more likely than those of other theological persuasions to oppose civil rights for blacks, though one study finds a shift in that in recent years, with fundamentalism *not* correlated to racial prejudice. However, it did show fundamentalism as a positive predictor of hostility toward homosexuals (Laythe, Finkel, & Kirkpatrick, 2001). For many decades, theological liberals have been shown to be most sympathetic to granting equal rights to minorities, followed by the

neo-orthodox and then by the conservatives. One study even found that intrinsically religious members of one millenarian group—a group that anticipates the end of the world and the beginning of a new "millennium" soon— were *more* prejudiced than extrinsics (Griffin, Gorsuch, & Davis, 1987). These findings suggest that it is not enough to know whether a person is intrinsically religious; one must also understand something about the nature of the meaning system itself before correlations can be predicted. In an earlier study, Allport (1966) had found that the *most* prejudiced people were those who were "indiscriminately proreligious." That is, they agreed with all statements that were in any way supportive of religion, including those that contradicted one another. The desire to be generally proreligious, but without giving much thought to the specifics, was interpreted as a tendency to conform to whatever was acceptable in the larger society.

Clearly the relationship between religion and racial prejudice is complex. In the remainder of this chapter, we explore the possibility that religion may simultaneously contribute both to tolerance and to bigotry.

Critical Thinking: Why did the figures on the relationship between religion and racial prejudice change when more complex measures of religious commitment were introduced?

RACISM AS A WORLDVIEW

The belief that some categories of human beings are biologically or genetically less human than others is actually a modern phenomenon (Jordan, 1968; Kelsey, 1965; Mosse, 1985). Of course, people throughout history have believed that those who had different values, beliefs, or styles of life were stupid or inferior. Ethnocentrism (prejudice based on differences in cultures) is a universal phenomenon. One can also find occasions in history when "outsiders" have been excluded because they were not of the same lineage. The articulation of a systematic philosophical and "scientific" statement that divides humans into higher and lower orders of being, however, has been given credence only since the eighteenth century (Jordan, 1968; Mosse, 1985).[2] George Kelsey (1965) did an incisive analysis of the modern racist worldview and its fundamental difference from the basic worldview of Christianity.

The important characteristic of racism is that a person's inherent worth is judged on the basis of his or her genes. The philosophical foundations of racism are naturalistic. For the racists, a person's value is understood in terms of that which is below themselves—their animal nature and, specifically, their genes. The quality of a person's life is determined by ancestry and genetic structure. Persons who are racist feel superior to or of greater worth than someone who has a "less human" genetic

structure (different skin color, hair color, facial features, etc.). In other words, their sense of worth is centered on their genes (Kelsey, 1965).

Richard Niebuhr (1960), who uses a broad definition of relgion, points out that faith focuses on that which makes our life worth living. He insists that whatever provides one with a sense of worth and meaning is properly termed one's "god." Elsewhere, Niebuhr (1960) talked of faith as "trust in that which gives value to the self" (p. 16). In fact, the term *worship* refers to a celebration of the center of worth or the center of all other values. Literally, the word *worship* means a state or condition of worth. Kelsey cited Niebuhr's definition of faith and went on to explain how racism serves as a faith or worldview. Kelsey (1965) wrote the following:

> The racist relies on race as the source of his [or her] personal value. . . . Life has meaning and worth because it is part of the racial context. It fits into and merges with a valuable whole, the race. As the value-center, the race is the source

---

[2]The origins of racism are difficult to pinpoint with precision, for the emergence of the concept was gradual. The first systematic effort to classify all humanity into distinct groups based on physical characteristics was conducted by Francois Bernier in 1684. The first to place humans into a pseudoscientific ranked order (the chain of being) was Carolus Linnaeus in the 1730s. Many other efforts to develop a "scientific" system of racial classification ensued. Contributions to the idea that humans could be classified and valued according to physical characteristics were offered by such writers as David Hume in 1748, Johann Friedrich Blumenbach in 1775, Johann Kaspar Lavater in 1781, Christian Meiners in 1785, Peter Camper in 1792, Joseph Gall in 1796, and Charles White in 1799. Meiners was the first to suggest that civilization would decline because of interracial marriages and the degeneration of the white race. The full systematic articulation of modern racial thinking did not occur until the 1850s. In that decade, three major tracts were published that formulated the chief tenets of contemporary racism: (1) Robert Knox, *Races of Men* (1850); (2) Carl Gustav Carus, *Symbolism of the Human Form* (1853/1925); and (3) Comte Arthur deGobineau, *Essay on the Inequality of Human Races* (1853–1855). "Count" deGobineau is commonly cited as the "father of modern racism." Regardless of whether one identifies the beginnings of racism with the Linnean chain of being (1730s), with the more careful articulations of the late 1700s, or with the systematic pseudoscientific statement by deGobineau (1850s), racism is a relatively modern construct in the Western world. For an analysis of the history of modern racism, see Winthrop D. Jordan's (1968) *White Over Black* and George L. Mosse's (1985) *Toward the Final Solution*.

of value, and it is at the same time the object of value. (p. 27)

In fact, the logical means to improve humanity is, from this perspective, selective genetic breeding and maintenance of the purity of the superior race.

Kelsey went on to discuss the worldview of Christian theology. The source of personal worth for the Christian is not found in his or her biological nature but in one's relationship with God. In this sense, Christian theology has allowed for only one distinction between persons, that between the regenerate and the unregenerate. The means of saving or improving human life is not through biological controls but through divine grace. Humans have worth because of their relationship with that which transcends them, not because of something they inherit through their genes.[3] By contrast, racism assumes some segments of humanity to be defective in their essential being and thereby incapable of full regeneration (Kelsey, 1965). The assumed defect is not one of character or spirit but a defect of creation: Biologically, "they" are less human.

Of course, many persons in American society are racist and still consider themselves Christian. Kelsey suggested that such persons are, in fact, polytheists; they worship more than one god. The question is which center of worth predominates in any given situation? Such persons do not have a single worldview that gives unity, coherence, and meaning to life. Most of them are unaware of and unconcerned with theological contradictions in their outlook or the fact that they are actually polytheists.

The official position of all major denominations in the United States is that Christianity and racism are mutually exclusive. Racism is viewed as a form of idolatry (worship of a false god) that is utterly incompatible with Christian theology. The line of argument generally follows the same pattern Kelsey outlined. Yet, despite the fact that racism and Christianity involve assumptions that are logically contradictory and incompatible, the two ideologies have historically existed together and have even been intertwined. Let us investigate some of the ways that Christendom may have unconsciously contributed to racist thinking.

Critical Thinking: George Kelsey argued that racism is a religion and that many Americans are polytheists—having more than one "center of worth" that makes them feel valuable and that guides their evaluations of others. Does his argument have merit? What is your rationale for your judgment?

SOURCES OF RACIAL PREJUDICE IN CHRISTIANITY

Christianity may have unwittingly contributed to racism through its worldview (meaning factors), its reference group influences (belonging factors), or its organizational strategies (institutional factors). In the following pages, we explore how each of these types of factors may be conducive to the formation and/or the persistence of racism.

Meaning Factors

We will explore four meaning factors in Christianity that may have affected the development and persistence of racism, but first it is necessary to make a distinction between types of racism. In his psychohistory of white racism, Joel Kovel (1970) identifies two types of racist thinking. *Dominative racism* is the desire

---

[3]Of course a similar argument could be made about the theologies of other world religions—especially Islam and Judaism. So the point of the argument is the contradiction between racism and faith in God, not an exclusive Christian argument.

by some people to dominate or control members of another group. It is usually expressed in attempts to subjugate members of the outgroup. This is the sort of racism that historically has been predominant in the southern United States. White slave owners, for example, would live and work in close proximity to African Americans, even assigning black women to breast-feed and care for their children. White men also visited slave row for sexual purposes. Whites did not mind associating with blacks on a daily basis and having contact with them—as long as blacks knew their place! African Americans were not to get "uppity" or self-assertive.

This sort of racism is quite different from the racism of the North. Here the racism was of the aversive variety.[4] *Aversive racism* is expressed in the desire to avoid contact with African Americans rather than the desire to subjugate them. In fact, northerners have often been quite moralistic about the dominative racism of the South, while they were systematically restricting blacks to isolated neighborhoods and ghettos. Part of the reason that school desegregation has been more difficult in the North than in the South is that aversive racism has resulted in more isolated housing patterns. Therefore, desegregation has required a more significant cost. Meanwhile, with the decline of dominative racism in the United States in recent decades, other expressions of racism have been on the rise (Farley, 2010; Yancey, 1999).

Most measures of racism have not controlled for these two types of racism, but those studies that have made the distinction have found aversive racism even where people reject more blatant dominative racism (Gaertner & Dovidio, 1986; Yancey, 1999). Today, few Americans think it is acceptable to discriminate against a person on the job because of the color of her or his skin. Yet survey data suggest that at the turn of the millennium only 51% of white Americans

supported housing laws that would prohibit discrimination against African Americans in sale of a home, and only 50% are clearly opposed to laws that would *prohibit* interracial marriage (Davis, Smith, & Marsden, 2003).

The only study on religion and racial prejudice that controlled for aversive types of racism found lower levels of aversive prejudice among people who attend interracial congregations (Yancey, 1999). Unfortunately, as we have already seen, that is a very small percentage of the faith communities in North America.

Moral Perfection and Color Symbolism

The distinction between dominative and aversive racism is significant because Christian groups may at times contribute to aversive racism even while it fights against dominative racism. Gayraud Wilmore has suggested that certain strains of Protestant theology have placed heavy emphasis on the moral purity and perfection of the "saved." The desire for moral purity was especially strong among New England Puritans and later among the Perfectionists. An important aspect of this Puritanism was the desire to avoid contact with anything that was evil or could be polluting.

Wilmore (1972) related this to the cultural symbolism of European and American society. Perhaps this is most vividly seen in the color symbolism of the European languages, which is especially noticeable in English. For example, prior to the 16th century, the definition of *black*, according to the *Oxford English Dictionary*, included the following:

> Deeply stained with dirt; soiled, dirty; foul. . . . Having dark or deadly purposes, malignant; pertaining to or involving death, deadly; baneful, disastrous, sinister. . . . Foul, iniquitous, atrocious, horrible, wicked. . . . Indicating disgrace, censure, liability to punishment, etc.

---

[4]While a person may hold to both dominative and aversive racism, and they may be mutually reinforcing, the distinctions between the two is helpful for our present analysis.

In discussing this phenomenon, historian Winthrop Jordan (1968) went on to say, "Black was an emotionally partisan color, the hand-maid and symbol of baseness and evil, a sign of danger and repulsion. Embedded in the concept of blackness was its opposite—whiteness. . . . White and black connoted purity and filthiness, virginity and sin, virtue and baseness, beauty and ugliness, beneficence and evil, God and the devil" (p. 7). Even much of the art work of this period showed the Devil as dark-skinned and the saintly figures as white. This color symbolism cannot be traced particularly to Christian teachings, but Christian responses to persons with dark skin may have been influenced by this symbolic association. Use of language in shaping attitudes may be especially important in the case of children (Van Ausdale & Feagin, 1996).

Interestingly, there is empirical evidence supporting this claim of color symbolism. Athletic teams that wear black uniforms have more penalties called on them than teams with lighter colored uniforms (Frank & Gilovich, 1988). No one argues that this is a conscious process—in sports or in religion—but the association of blackness with something sinister (an association that does not occur in many other cultures) may have affected the way religious people perceived people with black skin.

Wilmore argued that areas of the United States that have been especially strongly influenced by Puritanism and Perfectionism are more likely than other areas to have strong aversive racism. Ownership of slaves was often condemned by these groups because it might compromise the moral righteousness of the owner—and moral purity was essential to these pietists. However, many of these same pietists did not want to have to associate with these dark-skinned people. Although the source of the feelings was probably only partially conscious to the individuals, they often felt that blacks were unclean—in body and in soul. They simply wished to avoid contact. Dominative racism was condemned, aversive racism was not.

This statue of Mary and baby Jesus is at the front of the sanctuary in Our Lady of the Sioux Church at the St. Joseph Indian School in Chamberlain, South Dakota. Jesus and Mary are clearly Native American, and the crucifix behind the alter at this same chapel has a Christ who looks very much like a Sioux on the cross. Europeans and Americans have done the same thing, making images of Jesus with white skin, classical Greek features, and even straight blond hair and blue eyes. Jesus was obviously Semitic and no doubt had rather dark skin and other Jewish features. Many people seem to be more comfortable with religious heroes who look like themselves. (Photo from St. Joseph Indian School.)

Jordan (1968) showed that, among some of the religious groups that opposed slavery, the opposition was based largely on concern for how slaveholding might corrupt the soul of the owner. In this case, the ultimate goal is the purity and righteousness of the dominant group member more than concern for minority group members who were suffering. For example, one of the primary reasons for the abolitionism of Quaker John Woolman in the mid-1700s was that slaveholding created a feeling of superiority and pride in the owners. At that time, pride was considered the most heinous of sins. Because slave ownership caused pride, it was a source of evil. By 1776, the Society of Friends was excommunicating any Quaker who owned slaves (Jordan, 1968).

The Quakers were more effective than most groups at setting up programs to educate African Americans and to provide them with resources for economic independence. Their stance against dominative racism was unparalleled among religious groups. Yet, very few blacks ever became members of the Society of Friends. No doubt this was partially because the quiet, contemplative style of worship practiced by the Friends was so unlike the emotional and enthusiastic style of African religious ritual. The emotional expressions of the Baptists and Methodists were more similar to the type of religious expression familiar to people of African ancestry. Beyond this, blacks may have felt unwelcome in many Quaker congregations; Quaker pietism may have created an unconscious aversive form of racism among some members of this group. This may be part of the reason for an almost total lack of African American converts (Jordan, 1968).

The Protestant emphasis on moral purity or perfection is perhaps the most important unwitting contributor to the formation of racist attitudes among American Christians. However, this doctrinal emphasis contributed primarily to just one type of racism and only in the context of certain other cultural attitudes and structural circumstances. Pietism in modern America may not have the same effect. There are other aspects of Christian thought, however, that may contribute to the *persistence* of racism in American society.

> **Critical Thinking:** Wilmore argued that when color symbolism of European languages combined with puritanical pietism, the result was *aversive* racism. What is most and least convincing about this argument? How would a conflict theorist critique this argument?

Freewill Individualism and Failure to Recognize Institutional Discrimination

Rodney Stark and Charles Glock, among others, have pointed to the importance of the Christian emphasis on free will and a radial individualism before God. The doctrines of sin and salvation are based on the assumption that humans are free and responsible beings. After all, if a person was entirely predetermined in his or her behavior, one could not hold him or her responsible or guilty for an act. Guilt implies freedom of choice, with the wrong choice having been made by an individual. The concept of individual freedom is also the foundation of such socioeconomic concepts as rugged individualism. In this latter case, each individual is viewed as getting his or her just desserts in society because one's circumstances are considered a result of personal choices, lifestyle, and willingness to work hard. Stark and Glock (1969) put it this way:

> Christian thought and thus Western civilization are permeated with the idea that [people] are individually in control of, and responsible for, their own destinies. If I am really the "captain of my soul" and "the master of my fate," then I have no one but myself to thank or blame for what happens to me. (p. 81)

This doctrine is significant in race relations because many conservative Christians put the blame for disadvantage on those who are disadvantaged. Although there is no evidence that the doctrine of freewill individualism contributes to the *formation* of prejudicial attitudes, it may disincline those who are affluent to help those who have been subjected to poverty and discrimination (Stark & Glock, 1969). The doctrine reinforces the attitude that those who are down and out are probably getting their just desserts. Further—the reasoning goes—if those who are impoverished are not receiving their due, they will better themselves without the help of anyone. Persons who hold such a view are not likely to want to change the institutional structures of society that systematically discriminate. In fact, they are not likely to recognize the existence of institutional discrimination at all. *Institutional discrimination* refers to policies that discriminate against members of a particular group. Frequently, this is not intentionally directed against members of a particular group, but because members of that group are disproportionately represented in a certain status, they are disproportionately affected.

One example of intentional institutional discrimination is the California Anti-Alien Land Act of 1913, which specified that foreign-born individuals could not own farmland unless they were American citizens or were eligible to become citizens. This in itself does not seem to discriminate unduly against any particular group. However, Congress had previously passed a law that specified that people of Japanese origin were not eligible for citizenship. Likewise, the literacy tests that required a person to be able to read and write before they were allowed to vote in certain southern states in the 1950s were disproportionately disfranchising to blacks. African Americans were not the only ones who could not vote, nor were *all* blacks prohibited, but there is no question that illiteracy was higher in the African American community and that African Americans were disproportionately affected.

Sometimes discrimination in one institution affects discrimination in another. This is called *indirect institutional discrimination*. For example, discrimination in education has often led to discrimination in the job market. Because African Americans, Native Americans, and Hispanics have poorer educational backgrounds (with poorly funded and staffed schools in their neighborhoods), they do not meet the job qualifications for the better-paying jobs in our society. The employer is not purposefully discriminating against members of these ethnic groups, but the net effect is that Native Americans and Hispanics are underrepresented in the professional positions in our society. Institutional or systemic discrimination is discriminatory in

effect, even if not in intent. This is a major reason for the substantial income gap between blacks and whites reported in Table 10.1 on page 237.

The important point here is that those who hold to a strong doctrine of rugged individualism usually deny the existence or importance of institutional discrimination. Stark and Glock found that those who hold to traditional Christian doctrines of total free will, individual responsibility, and moral retribution (punishment for lack of adherence to moral standards) are more likely to believe in rugged individualism and are less likely to work for the reduction of institutional discrimination. These religious traditionalists often maintain that the only factor necessary to reduce inequality between groups is to teach minority persons to be more responsible for their own lives.

Hence, a particular religious belief (freewill individualism) is correlated to other beliefs (such as rugged individualism) and the entire set of beliefs contributes to apathy about racial discrimination. It is not at all clear that religious beliefs cause complacency. In fact, it is likely that these beliefs simply justify complacency rather than cause it. They allow people to ignore structural inequality and to benefit from their own unacknowledged privileges without feeling guilty (Farley, 2010; Kinder & Sears, 1981; Lewis, 2003; Rothenberg, 2008). The religious beliefs do not contribute to the formation of racism, but they do contribute to its persistence. Many scholars who study race and ethnic group relations believe that the more virulent form of racism in American society is the color-blind claim that refuses to recognize the way in which racism operates at the macro levels of society as part of the structure and can therefore exist and affect lives entirely independently of individual prejudices (Ballantine & Roberts, 2011; Bonilla-Silva, 2003; Farley, 2010; Rothenberg, 2008). This is often called "symbolic racism," and it is the most common form of racism in the United States today (Farley, 2010).

Particularism and Antipathy to Outsiders

Another set of religious beliefs is also correlated with certain kinds of prejudice. This set revolves around the assumption that one's own religion is uniquely true and legitimate and that all others are false. Only members of one's own group are expected to be saved. Glock and Stark (1966) referred to this orientation as *particularism*.[5] Not all religious people are particularistic. However, some groups teach that persons who are members of any other denomination or any other faith are damned. Glock and Stark (1966) found that among Christians particularism is highly correlated with anti-Semitism. It is also highly correlated with antipathy toward atheists and agnostics. However, they did not find a correlation between particularism and white-black prejudice (Stark & Glock, 1969). This is no doubt due to the fact that both blacks and whites in the United States are predominantly Christian, though as we have seen, there is a substantial portion of the Muslim community in the United States who are African American. Moreover, a cross-national study of attitudes in nearly a dozen European countries found religious particularism to be the strongest predictor of prejudice against ethnic "others" (Scheepers et al., 2002).

Further, since the terrorist events of 9/11, there have been high levels of animosity toward any Arab Americans who are assumed to be Muslim (Farley, 2010). In many communities, the terrorist attacks of 9/11 resulted in two countervailing actions. Some people engaged in hate actions directed at Muslim people living in North America. The "we versus they" thinking was heightened and polarities between those identifying with Christianity and Islam were widened. Some churches offered adult education classes to warn people of the dangers of Islam and to stress that only the followers of Christ are chosen for heavenly rewards.

However, another interesting response was also widely seen following 9/11. Aziz Khaki, a prominent leader in the Islamic community in Vancouver, reported to Reginald Bibby that he was as appalled as anyone by the terrorist attacks. Khaki affirmed that no one who is truly Islamic could do such a thing. However, he also said that "from the beginning I was contacted by people from a variety of groups, including those who were Roman Catholic, United, and Anglican, who said they wanted to meet with me, that they stood in solidarity" (cited by Bibby, 2002, p. 245). In this case, religiously motivated people responded with compassion and sought to reach across the barriers that often separate one group from another. Education courses have been offered in some churches to stress that Christians, Jews, Muslims, and people of other faiths traditions are fundamentally alike as children of God.

So the tragedy of 9/11 can be understood in ways that enhance tolerance or bigotry. In its role of explaining the meaning of events, religious communities have an extraordinary role. Churches, mosques, and temples are capable of encouraging polarities between peoples, especially through particularism, or building bridges between people by downplaying differences.

Hence, particularism is a potential contributor to racial prejudice in certain circumstances, and it was definitely operative in the *formation* of racism in this country—especially before blacks were converted to Christianity (Jordan, 1968). However, particularism does not currently seem to be a significant factor in negative sentiment toward African Americans.[6]

---

[5]Particularism is a form of ethnocentrism—a concept that may be more familiar to sociology students.

[6]Because particularism does contribute to certain kinds of prejudice, a good deal of attention has been given to assessing its relation to hatred of Jews. There are several Christian doctrines that, if narrowly defined, may lead to a particularistic view. Gordon Allport (1966), for example, pointed to the doctrine of election as a belief system that may contribute to bigotry. The doctrine of election is the belief by some people that they are God's chosen people. The issue is whether this doctrine is interpreted as election for service and responsibility or election for salvation. Some Christians emphasize that only those who have experienced the holy in exactly the way they have are eligible for salvation. Others may view their religious experience as a unique call to serve humanity; they do not interpret the concept of election in particularistic terms. When the doctrine of election is interpreted in terms of exclusive salvation, it is highly correlated to particularism. Whether it is a cause, as Allport suggested, remains to be established by empirical research.

Despite the fact that official Christianity manifestly opposes racism and encourages a sense of the brotherhood and sisterhood of humankind, certain beliefs may have the effect of increasing certain kinds of prejudice. The meaning function of religion can bring bigotry or can build bridges between people. This brings us to the second process by which religion may contribute to prejudice: reference group loyalties and we–they categories of thought.

Belonging Factors

Religion may contribute to prejudice through its sense of community and the feeling of belonging. As the religious community becomes a major reference group, people want to conform to the norms of the community in order to feel accepted. Furthermore, as they begin to identify closely with the group, they develop a sense of "us" and "them." In fact, we discovered in Chapter 5 that the creation of strong group boundaries was one technique used to enhance commitment to a religious group (see the discussion of Kanter's theory of commitment, especially affective commitment). Now we explore two factors that may mean the belonging dimension of religion relates to intolerance toward ethnic minorities.

Informal Group Norms

The informal community of believers that provides individuals with a sense of belonging is a very important part of religion. However, the community develops unwritten norms and expectations—some of which may conflict with official religious policy. In an attempt to conform, members may adhere to the informal norms of the community rather than to the official policy of the formal religious organization. It is noteworthy that Gerhard Lenski (1963) found that "communal" members (who are influenced through the belonging function) were much more likely to be racially prejudiced than were "associational" members (who were formal members). Informal norms and values of

the community may be contrary to the official ones, but they may be vigorously enforced through informal sanctions (Lenski, 1963). A group of Lutheran laity at a Sunday afternoon picnic may tell ethnic jokes or may subtly reinforce negative images of blacks, regardless of the minister's sermon that morning to the contrary. The reference group norms are often more powerful in influencing behavior than are the idealized norms in the ideology. The friendship networks and unofficial norms may be more significant than official statements of religious bodies or sermons from the pulpit.

Group Boundaries and Identification With the In-Group

Another major theory of racial and ethnic prejudice is based on the tendency of people to accept those who are similar to them and to be suspicious of anyone who is defined as "different." This view of the cause of prejudice is sometimes called the we–they theory, perhaps best illustrated by an empirical study conducted by Eugene Hartley (1946). Hartley used a variation of the Bogardus Social Distance Survey, an instrument designed to measure prejudice toward various ethnic groups. A list of ethnic groups is provided and respondents are asked to rate the closest relationship that they would be willing to have with a member of that group. Seven categories are provided, ranging from "would marry a person who is a member of this group" and "would be willing to have a member of this group as a best friend" to "would allow only as visitors to my country" and "would exclude from my country entirely."

Hartley adapted the Bogardus instrument by adding three fictitious groups: (1) Danireans, (2) Pireneans, and (3) Wallonians. Using a random sample of college students at eight northeastern universities, he attempted to measure ethnic prejudice, including prejudice toward these fictitious groups. Hartley found a high level of prejudice toward these three nonexistent groups: more than half of the respondents expressed a desire to avoid contact with these people and some respondents wanted them expelled from the country. Moreover, nearly three fourths of those who were prejudiced against blacks and Jews were also prejudiced against Pireneans, Wallonians, and Danireans. He used these data as evidence that prejudice is not caused by *stereotypes* (stereotypes are rigid and exaggerated images of a particular group or category of people). After all, no one has a negative stereotype of a group that does not exist. The negative feelings of respondents toward Danireans, Pireneans, and Wallonians were based on the fact that these groups sounded unlike the respondents. In other words, their negative reaction was based simply on whether the people sounded similar to or different from the respondents; the name *Pireneans* sounded more like one of "them" than one of "us." Empirical studies by social psychologists have supported this we-they theory of the causes of prejudice (Farley, 2010).

One of the most important functions of religion, as we have seen at several points in this book, is to provide a sense of belonging, a sense of group identity, a sense of *we*. To add to this, a

number of Christian groups place a strong emphasis on particularism (discussed in the previous section). Such a belief tends to add to the in-group sense of superiority and to the distinction between "us" and "them." The belonging function of religion, then, is capable of contributing to antipathy.

The development of religiously based we–they prejudice is especially likely in situations where racial boundaries and religious boundaries are coextensive (see Figure 10.1). The importance of its influences is illustrated by several empirical studies that show high levels of prejudice against people with differing beliefs (e.g., atheists or Jews), often surpassing prejudice against those of another race (Byrne & McGraw, 1964; Byrne & Wong, 1962; Jackson & Hunsberger, 1999; Rokeach, 1968; Rokeach, Smith, & Evans, 1960; Smith, Williams, & Willis, 1967). Moreover, if "they" practice a different religion and also look different, the exclusionary tendencies are reinforced even further.

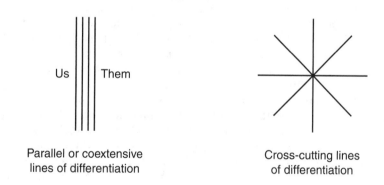

Parallel or coextensive
lines of differentiation

Cross-cutting lines
of differentiation

Imagine that each line represents a division in the society between groups. These divisions might be based on religion, ethnicity, political party, economic status, language spoken, skin or hair color, or other factors. Parallel lines of differentiation all divide people in each conflict along the same lines. Cross-cutting lines of we/they cut the differences so that people who were part of "them" in a previous antagonism become part of "we" in the present discord. This lessens the likelihood of deep and permanent hostilities within a social unit.

Figure 10.1   Lines of Differentiation Between "We" and "They"

The white Christians who encountered blacks from the 15th to the 18th centuries were meeting people who were different on several counts. In fact, in the 18th century, the words *white*, *Christian*, and *civilized* were used by many European writers as synonyms for "we," while the terms *black*, *heathen*, and *savage* were used interchangeably with "they" (Jordan, 1968). At that point, much of the denigration of blacks was because they were not Christians. We–they religious distinctions, then, may have contributed to the formation of racial prejudice, even if these factors are not a primary cause of racial prejudice today. Nonetheless, one can see the application of religio-racial we–they categories in contemporary KKK literature and in various types of anti-Semitic materials.

To summarize, the belonging functions of a particular religion may enhance antipathy toward others in at least two ways. First, the informal community may develop norms that encourage discrimination, despite the official position of the larger religious organization. Second, the development of high boundaries between groups can reinforce we–they categories of thought, and at least historically in North America, "we" among those in power included only whites who were also Christians. Currently, black/white distinctions in North America are not related to difference in religion. Still, persons of Arab background are often stigmatized and this polarization is especially strong when those Arabs are also Muslim. Religion can be related to prejudice against certain groups insofar as it creates and/or sanctifies we–they thinking.

Institutional Factors

Religion as an institutional structure can affect social behavior somewhat independently from the belief system or the reference group factors. This is perhaps best illustrated by a study of the clergy of Little Rock, Arkansas, in the midst of a major racial crisis.

Ernest Campbell and Thomas Pettigrew (1959) studied the role of Protestant ministers in the school desegregation controversy. They found that ministers of small working-class sects were supportive of segregation and that most of the clergy of mainline denominations were sympathetic to integration. However, very few of the ministers of mainline denominations spoke out or got actively involved in the conflict. The official positions of their denominations, their professional reference system (other mainline clergy), and their own personal convictions all supported integration. Still, they did not take a prophetic stance as one might expect.

When they did speak out, it was often in generalities, with references to "deeper issues" or with other techniques that would prevent anyone from taking offense (or from taking their statements too seriously). Social psychologists use the term *aligning actions* to refer to statements that modify the meaning of other statements or actions. When the clergy in Little Rock did discuss the racial crisis with their congregations, they would usually preface their remarks with a comment that softened the impact of the stance. An ethical view can be moderated and made more palatable when a minister begins by saying something such as this: "Everyone has to make up his or her own mind on moral questions. My opinion is simply the opinion of one person, and there is certainly room for other views." Such statements allowed ministers in Little Rock to speak briefly to the issue of segregation without raising too much opposition or offending people too deeply. Members of the congregation were, in effect, invited to ignore the comment or to view it as one opinion among many. This technique allowed ministers to feel that the issue had been addressed and that a prophetic stance had been maintained. Hence, they could avoid feeling guilty for having ignored the issue. They felt they had not compromised their principles.

One reason that the clergy avoided discussing segregation very forcefully was because their congregations overwhelmingly supported it. Because their congregations were important reference groups to the pastors, they were caught between reference groups (professional colleagues on the one hand and the congregation on the other). Such factors were no doubt part of the reluctance of clergy to take a strong stand, but the most important factors were institutional in character.

Although the official statements of most Christian denominations condemn segregation, the working propositions of the church bureaucracies are such that rewards come to those who do not "rock the boat." Concepts of success in the ministry are related to improvements in membership and in financial prosperity. The addition of a new educational wing or the need to build a new and much larger sanctuary are often viewed as signs that the pastor must be doing something right. Moreover, ministers often view harmonious, satisfied congregations as evidence of success. The minister whose church is racked with conflict and is declining in membership is certainly not a prime candidate for promotion to a larger congregation. Whether the minister is part of an episcopal system (where a bishop appoints clergy to a congregation) or of a congregational system (where a pulpit committee from the local congregation seeks out a pastor to hire), a reputation for controversy and uncompromising conscience is not an asset. The minister who wants to advance in his or her career keeps the congregation united, the funds flowing, and the membership stable and/or increasing.

This process provides an excellent illustration of Thomas O'Dea's "dilemma of mixed motivation." Clergy were torn between two motivations: (1) being faithful to the prophetic teachings of the denomination on the one hand and (2) enhancing their careers by doing that which was necessary to be "successful" on the other. Many felt that they would work on one of the goals first (building their careers) and would later take on the hard task of speaking their consciences forthrightly. Yet some

When the controversy about desegregation of this high school in Little Rock, Arkansas, arose, all major denominations had endorsed the position that racism is contrary to Christianity. Still, ministers in Little Rock were disinclined to speak out because they thought it would reduce membership in their congregations, damage the financial health of the church, and adversely affect their own careers. So despite official positions of the national denomination, pastors remained silent. The second statue on the right, above the front doors of Central High School, is named "Opportunity"—precisely the thing that was being denied blacks.

observers have been skeptical of such a strategy. One Little Rock minister who did speak out had this to report:

> I talk to the young ministers and I ask them why they aren't saying anything. They say no one will

listen to them, they aren't known and their churches are small. But wait, they say, until we get big churches and are widely known. We won't be silent then. Then I turn to the ministers in the big churches and I listen to them trying to explain why they have done so little. Their answer is a simple one; they say they have too much to lose. Only recently, one such man said to a group I was in, "I've spent seventeen years of my life building up that church, and I'm not going to see it torn down in a day." (Campbell & Pettigrew, 1959, pp. 120–121)

A complex set of factors plays on a minister and affects decision making. Some of these factors may create pressure that calls for contradictory behavior. Just understanding the belief system of a religion is by no means sufficient to understanding the behavior of religious persons. In the case of racial bigotry in Little Rock, the official position of most denominations called for prophetic forthrightness by the clergy, but at the same time they provided concrete rewards (promotions to bigger congregations) for taking the road of least resistance. In effect, the institutional procedures operated to reward those who were restrained in their comments or even complacent about the entire matter. The silence by the clergy, in turn, created an environment in which it seemed that everyone shared the same view. Conformity to norms of prejudice was easy, for few people were saying anything to challenge the norm. In such a case, religion is *not* a factor in *causing* prejudice, but in failing to oppose prejudice forthrightly, religious communities *contributed* to the *continued existence* of racism.

A study by Harold Quinley (1974) found similar patterns among Protestant clergy in California. He explored variables affecting the willingness of ministers to take prophetic stands on controversial social issues. The issues in California were civil rights legislation, support for a war or advocacy of peace, and organization of farm workers. Quinley found that support for prophetic ministry depended on at least three variables: (1) the relative liberalism or conservatism of the congregation, (2) the liberalism or conservatism of the denominational leaders, and (3) the organizational structure of the denomination (congregational, presbyterian, or episcopal). He did find, however, that regardless of

denominational structure, congregational members have more control over most minister's actions than do the denominational hierarchy. While the hierarchy may control appointments and removals from office in episcopal systems (with a bishop) and in presbyterian organizations (in which denominational leaders and the leaders of the local congregation share in decision making), the local congregation controls salary in virtually all Protestant denominations. So support from a powerful and like-minded denominational hierarchy could enhance the tendency to have a prophetic ministry, but in this study local congregations frequently had a moderating effect on the activism of clergy. Most parish ministers believed that their careers would suffer if they engaged controversial public issues. Ethical convictions regarding civil rights and other matters are often muted by institutional-maintenance concerns. These same forces were also found to be at work among clergy during conflicts in Rochester, New York (Martin, 1972), and in Boston (Thomas, 1985).

Of course, religious institutions are also very capable of using their influence to combat racism. For example, the Roman Catholic Church was instrumental in starting a program known as Project Equality. This involved churches using their buying power to reduce racial discrimination by requiring all their suppliers to adopt affirmative action policies. Some businesspeople became angry with their local congregation for such coercion and cut off their pledges, and some local religious communities discontinued the activity because of this response. Nonetheless, many national denominational offices remained members of Project Equality (as did some local congregations). This suggests only one of many ways in which religious groups have used their corporate influence to bring change, particularly at the level of the national denominational office. In fact, several studies have indicated that the more insulated the official hierarchy is from the local congregation, the more likely they are to emphasize egalitarian stances. Southern Catholics, for example, have historically been less likely to discriminate than members of congregationally based Protestant churches in the South. This has been largely because bishops, who are not directly responsible to the local congregation, have been willing to uphold the official position of the denomination (Beck, 1978; Wood, 1970, 1981).

Our point is that institutional factors may work to enhance racial prejudice and discrimination or to combat it. Simply knowing the official position of the denomination does not give the whole picture. Just as the religious belief system may have countervailing influences within it, so also may the religious institution provide motivations and influences that run counter to the official doctrine. It is simplistic to say that religion contributes to bigotry or that it contributes to tolerance. First, it is necessary to know what interpretation of Christianity one is talking about: pietist, fundamentalist, orthodox, liberal, and so on. Second, any given religious group may be contributing to tolerance in some respects while contributing to exclusivity and prejudice in other ways. The meaning, belonging, and institutional subsystems that comprise religion may work at cross-purposes.

Our focus on ways that religion sometimes contributes to bigotry is only part of the picture. Religious prejudice is frequently an expression of other conflicts within the society at large. On the one hand, bigotry within religious communities may be due to their being open social systems rather than the religion being the cause of bigotry (see Figure 10.2). In addition, some scholars argue that religious prejudice is frequently a consequence of other social forces (such as conflict over resources) rather than the cause of bigotry. This is a contribution of the conflict theory.

> *Critical Thinking:* In this chapter, the investigation has focused on racism directed by white Christians against African Americans. How might the causes of religious prejudice against Jews be different? Which theories might become more relevant for understanding this bigotry? Why? Which might be less relevant? Why?

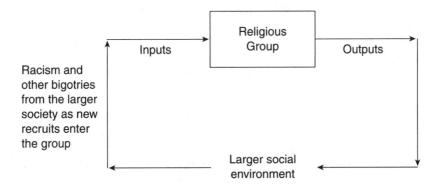

James Wood has suggested that prejudice within religious communities may be due to inputs from the larger society. Christian churches often try to recruit new members who are unchurched. Further, members of any religious community are bound to be profoundly influenced by attitudes in the larger society—their social attitudes shaped by mass media, work colleagues, and friendship networks outside the religious body rather than by the theology of the denomination. So, while much of the discussion thus far in this chapter has focused on ways that religion might influence the society, Wood's idea suggested that racism is an alien value that infiltrates the religious community.

Figure 10.2 Open Systems Model and Bigotry Within Religious Organizations

Social Conflict and Religious Expression: The Conflict Perspective

It often happens that religious prejudice is actually a reflection of larger social conflicts rather than their cause. In this case, religion may be acting purely as a justification for discriminatory behavior; the out-group is defined as spiritually inferior so that members of the in-group do not feel guilty about their blatantly unjust behavior. This perspective is an important contribution of the Marxian theory of social conflict. Marxian analysts look to nonreligious causes of prejudice, although they recognize the role of religion in maintaining social inequities.

Perhaps the most important theory of racial and ethnic discrimination is that which insists that prejudice and discrimination are caused by conflict over scarce resources. Some items in any society are scarce and nondivisible or nonsharable (or at least are viewed as such). The best jobs, the best housing, the best educational opportunities (admittance into a professional school), and social status are examples of items that are viewed as scarce and over which there may be conflict.

Three variables are central in discrimination and prejudice (Farley, 2010; Noel, 1968; Vander Zanden, 1983). First, when the two groups are visually distinctive either physically (skin color, facial features) or culturally (dress, language, beliefs), prejudice may occur, but it is not inevitable. Symbolic distinctions between groups may foster feelings of "we" and "they," but there is not necessarily hostility between the groups. Second, if those two groups are in conflict over some scarce resource, the likelihood of prejudice is very high. It is interesting to note that the contemporary high levels of anti-Arab sentiment in the United States began after the energy crisis came to awareness in the 1970s (with Arabs controlling oil resources). Likewise, anti-Asian sentiment is highest in industries where extremely qualified Japanese- or Chinese Americans are hired over whites in competitive high-tech jobs. Third, where one of the groups has more power to control access to scarce and valued resources, discrimination and prejudice are nearly universal. The more powerful group uses its power to control the resources. It then develops stereotypes of the out-group in order to justify discriminatory behavior and to make that behavior seem morally right. Religion is frequently a part of this moral justification.

An important point here is that prejudice and discrimination are enhanced if the groups in question differ not only in their economic interests but also in their physical appearance, their language, their culture, and their religion. In cases where two groups differ in all of these ways, it is common for the conflict to be justified and highlighted in religious terms. Native Americans were freely exploited and their lands taken from them because they were "heathens" who supposedly needed to be "civilized" (which usually meant "Christianized"). Likewise, many southern states had laws that forbade conversion of slaves during the early period of the slave trade. The slave owners were afraid that, if blacks held the same religious beliefs as whites, owners would be forced to free their chattel. Once this idea passed, justification for discrimination came to be based more intensely on color than on religious affiliation.

The central point is this: When lines of differentiation between people in racial characteristics, cultural backgrounds, language, religious orientations, and economic self-interests are coextensive and mutually exclusive, antipathy is likely to occur (Farley, 2010). Although religion is one cause, it is not necessarily the primary cause. However, religion may be used as the primary justification for hostility. After all, if "they" are immoral heathens and infidels, treating them with something less than respect seems quite reasonable and moral. Thus, what seems on the surface to be religious prejudice may, in fact, be caused by conflicts between groups that seek control over the same scarce resources. The source of the conflict is primarily economic.

The conflict in Northern Ireland represents a situation in which economic, religious, and ethnic boundaries are coextensive or parallel. The Protestants are the landowners and are of Scottish descent. The Catholics tend to be poor laborers who are fiercely loyal to the concept of an independent and united Ireland. The conflict is referred to as one between Protestants and Catholics, but religion is not the sole or even the primary cause of the conflict. The medieval crusades of Christians against the Moors were also justified on religious grounds, but the conflict was rooted in economic and ethnic interests as well.

The contrary situation is where lines of demarcation are crosscutting. In social circumstances, where members of an ethnic group are not necessarily of the same social class or the same religious group, the likelihood of religiously based prejudice is reduced. Likewise, where members of a particular religious group do not share the same ethnic background or economic interests, the likelihood of prejudice is somewhat diminished, and religious justifications for discriminatory behavior are also reduced. Hence, religion often acts to reflect larger social conflicts. The student of religion must be wary of overgeneralizing about religion as a *cause* of social conflict. In any given situation, religion may well be a cause of conflict, but each case must be evaluated in terms of the specific social setting and the specific worldview of the religious group.

> *Critical Thinking:* What, in your opinion, are the four most important factors in the existence of racism among Christians? Explain how these factors work and provide your rationale for why you think these are the most important variables. Do you think any of these variables are necessary or sufficient, or are they all just contributing factors? Why?

SUMMARY

Michael Emerson, the leading sociologist of religion and race today, along with William Mirola and Susanne Monahan (2011) summarized their findings this way: "Forget the color line. When it comes to religion in the United States, we have a color wall" (p. 161). There is a strong correlation between racial identity and religious preference, and that translates at the local level to extremely high levels of racial segregation in congregations. There are many explanations for this racial homogeneity, but given the history of race in American society, we cannot overlook the importance of racial prejudice in driving religious racial segregation.

Christianity may contribute in subtle ways to the development and persistence of racism and other ethnic prejudice, even if its official posture mitigates against such prejudice. Although there are differences between members of different denominations and between theological liberals and conservatives, for the most part, the most religiously committed people tend to rank lower on scales of racism than less active members or than those disaffiliated with religion. Further, the logic of Christian theology is such that racism is really incompatible with Christianity; racism is, in fact, a theology with its own internal logic and value structure. Despite all of this, Christianity has sometimes contributed, even unwittingly, to the formation and continued existence of racism.

First, aversive discrimination may have been unconsciously fostered through a meaning system that linked concern with purity to color symbolism (with blackness associated with evil). Other beliefs emphasizing individual autonomy and free will may also have contributed to the persistence of racism because of a failure of Christians to recognize the realities of systemic or institutional racism.

Second, prejudice may be passed on through the we–they boundaries and through informal norms of the community. The belonging function of religion can be a source of exclusivity and hostility.

Finally, formal institutional structures may contribute to the *persistence* of prejudice by rewarding behaviors other than prophetic ministry. Ministerial or organizational definitions of success may reward (with promotions and professional advancements) those whose style is noncontroversial and unobtrusive.

In short, the three subsystems of religion may sometimes work at cross-purposes, contributing simultaneously to tolerance of others and to bigotry.

Not only is Christianity capable of contributing to racial prejudice but religious prejudice is also often a reflection of larger conflict in the society. This is especially true if color lines, socioeconomic stratification, political parties, and religious boundaries are coextensive and mutually exclusive. In any case, our discussion points to the interactive relationship between religious values and the social conditions at large. We have also found that religions are themselves complex entities that can have many countervailing forces within them.

11

RELIGION, GENDER, AND SEXUALITY

Here are some questions to ponder as you read this chapter:

- Does religion—and Christianity in particular—contribute to negative attitudes and discriminatory treatment of women?
- How can the meaning systems of religion contribute to gender equity and gender bias?
- Why do structural components and meaning systems of a religion sometimes have contradictory effects relative to gender equity?
- Why do some women actively choose religious communities that have high gender barriers and limited roles for females?
- How do religious systems relate to homosexuality, homophobia, and heterosexism?
- How are same-sex marriage and ordination of gays and lesbians playing out in religious communities?

When examining the relationship among religion, gender, and sexuality, one confronts an interesting paradox. Religion is historically connected to sexism[1] and gender inequality, and yet in most cases women exhibit higher levels of religiosity than men. Why would this be the case, and why would a group whose identity is stigmatized in most major religious traditions—homosexuals—remain religiously active, often times in religious communities whose teachings stigmatize them?

Women's groups that are committed to changing gender roles and attitudes have often charged that religion perpetuates traditional role expectations of men and women. Historically, many groups have refused to ordain women—barring them from the major leadership positions in the religious group. A few denominations continue to maintain this position today. Studies have suggested that this is not only true of the established religious bodies; women have often been singled out and accused of sexual weakness, sinfulness, and impurity in the new religious movements (NRMs) as well (Jacobs, 1987). Some pagan religious traditions, such as Wicca, have a much more positive attitude toward the feminine, with female gods and leadership positions for women. However, as John Hawley (1986) concluded, "Theological appreciation of the feminine does not necessarily lead to a positive evaluation of real women" (p. 235). Moreover, religious ideas about gender and religious standards of sexual ethics have often influenced attitudes toward same-sex relationships. In the following pages, we explore the role of religion in sacralizing gendered behaviors and roles and in shaping attitudes toward homosexuals.

To keep the scope of this chapter manageable, we again focus primarily on the relationship between Christianity and gender prejudice. Although the coverage is not comprehensive of all religious traditions, students should be developing skills in analysis and should gain a sense of how sociologists approach the issue for any religious tradition. We begin by exploring the empirical evidence about whether organized religion really does stigmatize women and legitimate male dominance.

GENDER AND RELIGIOUS INVOLVEMENT

We begin by documenting and explaining the higher levels of religious involvement among women. Table 11.1 shows the gender difference in identification by religious tradition. The most heavily female religious traditions are the historically black churches and Jehovah's Witnesses. The members of the largest religious traditions in the United States also are more than half women: Catholic (54%), Mainline Protestant (54%), and Evangelical Protestant (53%). Indeed, over three quarters of Americans belong to religious traditions in which the majority of members are women.

Men, by contrast, dominate among the nonreligious: atheists (70%), agnostics (64%), and secular unaffiliated (60%). Men are also the majority of members in the major non-Christian traditions in the United States: Jewish, Buddhist, Muslim, and Hindu.

---

[1]Sexism refers to any behavior that discriminates against a person because of sex and any ideology that maintains that one sex is intrinsically and immutably superior to the other. We will be dealing in this chapter primarily with ideologies or belief systems that define women as inherently inferior. Sex bias may be manifested in either of these related concepts: rigid *gender role expectations* or *concepts of gender*. Gender roles are tasks that are assigned to males or females. Concepts of gender are ideas about the personality characteristics and capacities of the "normal" female (femininity) or of the "normal" male (masculinity). Gender roles are usually justified on the basis of gender concepts.

Table 11.1 Gender Composition of Religious Groups in the United States (by percentage)

	Women	*Men*	*Share of Total U.S. Pop.*
Historically Black Churches	60	40	6.9
Jehovah's Witnesses	60	40	0.7
Mormon	56	44	1.7
Catholic	54	46	23.9
Mainline Protestant	54	46	18.1
Orthodox	54	46	0.6
Evangelical Protestant	53	47	26.3
Total	52	48	**100.0**
Jewish	48	52	1.7
Religious Unaffiliated	48	52	5.8
Buddhist	47	53	0.7
Muslim	46	54	0.6
Secular Unaffiliated	40	60	6.3
Hindu	39	61	0.4
Agnostic	36	64	2.4
Atheist	30	70	1.6

SOURCE: Pew Research Center Forum on Religion & Public Life. (2008b). *U.S. religious landscape survey.* Retrieved from http://religions.pewforum.org

Beyond religious affiliation, women also exhibit higher levels of religiosity than men across a number of measures (see Table 11.2). In terms of religious participation, women attend religious services and pray more frequently than men. Women also report higher levels of belief in a personal God and that religion is more important in their lives than do men. All of these differences are considerable, and they hold true across almost all religious affiliations (and non-affiliation)—even in those religious traditions (Jewish, Buddhist, Muslim, and Hindu) in which a majority of affiliates are men.[2]

---

[2]The only differences are 86% of Mormon men claim a belief in a personal god compared to 85% of women, and Muslim and Buddhist men attend religious services more frequently than women (Pew Research Center Forum on Religion & Public Life, 2008b).

Table 11.2   Gender Differences in Religiosity (by percentage)

	Women	Men	Gender Gap
Are affiliated with a religion	86	79	7
Absolutely certain belief in a God or universal spirit	77	65	12
Pray at least daily	66	49	17
Say religion is very important in their lives	63	49	14
Attend worship services at least weekly	44	34	10

SOURCE: Pew Research Center Forum on Religion & Public Life. (2008b). *U.S. religious landscape survey*. Retrieved from http://religions.pewforum.org

There is no definitive explanation for this gender difference in religion. Some argue that differential gender roles explain the gender gap in religion (Levitt, 1995). Women are responsible for child rearing, so the argument goes, and this encourages greater religious affiliation and involvement. Although it is true that parenthood has a positive effect on religiosity, the effect of parenthood on religiosity is stronger for men than for women; parenthood alone does not fully explain the gap. Others argue that not being in the labor force gives women more free time for attention to faith issues. This is also partially true. Full-time employment does have a negative effect on religiosity, but the gender gap persists even when controlling for women's employment status. That is, women who work full-time are more religious than men who work full-time.

In the next two sections, we consider two other ways of understanding the gender gap in religion: a more cultural approach that focuses on the "feminization" of religion and an approach grounded in a rational choice perspective which focuses on the relationship between religion and risk tolerance.

Feminization of Religion

Although women were viewed for centuries as spiritually more vulnerable to sin and corruption than men and as a source of evil influence, an interesting shift occurred in America in the 19th century. As opposed to the sternness and harshness of 18th-century American Christianity, the sentimentality of the 19th century was clearly more "feminine" in tone. Barbara Welter (1976), who referred to this as the "femininization of American religion," wrote, "When . . . a more intuitive, heartfelt approach was urged, it was tantamount to asking for a more feminine style" (p. 94). Welter pointed out that the most popular hymns written at that time stressed passive and accepting roles. Such hymns as "Just As I Am, Without One Plea," "To Suffer for Jesus Is My Greatest Joy," and "I Need Thee Every Hour" illustrate a pattern in hymnody of exalting dependency, submissiveness, and a willingness to suffer without complaint as Christian virtues. These were also viewed as feminine characteristics and as feminine virtues in 19th-century American culture. In contrast, the ideal characteristics of the male were embodied in the aggressive, independent, self-sufficient industrialist. This transformation of American religion was perhaps most noticeably seen in the imagery of Jesus. In the 18th century, he had been viewed as the stern taskmaster and as the exalted ruler of God's kingdom. In the 19th century, the major characteristics attributed to Jesus were loving self-sacrifice, tenderheartedness, and willingness to forgive those who injured him.[3]

Given the mixed signals that men were receiving about masculine and Christian virtues, it is not surprising that male church attendance dropped off and religion came to be viewed as a woman's concern. As depicted in almost any popular novel or sermon of that day, women were more spiritual, more noble, and more generous than men. As Welter (1976) put it, "Womanhood was believed to be, in principle, a higher, nobler state than manhood, since it was less directly related to the body and was more involved with the spirit; women had less to transcend in their progress" (p. 95). Religion also came to be viewed as less rational (Ruether, 1975). Rationality was associated with science, technology, and industry—all of which were male-dominated spheres. Women come to be viewed as more religious than men and as the transmitters of morality but still not as highly rational. The clergy, who were still almost entirely male, were sometimes viewed as naive, unknowing, and incapable of understanding business practices. As religion was feminized, the leaders of the church were also attributed with feminine virtues and vices (Ruether, 1975).

Although women came to be seen as more spiritual, this happened at a time when religion was having decreased influence on the affairs of the world. Commerce and politics were increasingly secular and governed by principles of secular rationality. At the very time when women were being identified with religiosity, religion was being demoted to a less influential position in society. In fact, the feminization of religion may have occurred precisely because religion was being dislodged from direct access to political power; it was becoming identified as a concern of the home, the hearth, and the individual. This loss of social power was equated to taking on a more feminine role in society. To state it in its most negative form, to be religious was to be unknowing, lacking in power, and guided by naive sentimentality rather than by realism and reason. The stereotypes of women as irrational and emotional had remained constant, but the role of religion in society—and hence its image—had changed significantly.

Research in the 1990s indicates that persons—male or female—who have more "feminine" value orientations (valuing feelings and expressiveness) tend to rate higher in various measures of religiousness than persons of either sex who have more "masculine" (instrumental) orientations. In fact, gender orientation in one's values is a better predictor of religiousness than is the sex of the person (Thompson, 1991). At least in the contemporary United States, religion continues to have feminine overtones, despite the fact that it has often been less than sympathetic to women.

Critical Thinking: Do the faith traditions with which you are familiar have more feminine than masculine values—stressing on cooperation, support, nurturance, and relationships as central, rather than competition and individualistic achievement? If so, does this explain why fewer men affiliate and attend services?

Religion and Risk Tolerance

Sociologists Alan Miller and John Hoffman (1995) developed a rational choice explanation for the higher levels of religiosity of women relative to men that focuses on risk aversion. Recall that the basic premise of the rational choice perspective is that in making decisions people try to minimize their costs and maximize their benefits. In terms of deciding whether to engage in religious practices, one of the potential

---

[3]Research in the latter half of the 20th century found that girls are more likely to view God as loving, comforting, and forgiving, while boys tend to view God as a supreme power, forceful planner, and controller (Batson, Schoenrade, & Ventis, 1993; Cox, 1967; Wright & Cox, 1967).

benefits of being religious is the possibility of a better next life, and one of the potential costs of irreligion is the possibility of eternal damnation. From this perspective, *irreligion is risky behavior*. Because women have a lower tolerance for risk than men, they are more religious.

Much of the research in this area since the original statement of the theory has focused on trying to understand the origins of risk tolerance and aversion. Not surprisingly, the major competing explanations are *nature* and *nurture*, biology and socialization. On the side of nature, Miller and Stark (2002) tested socialization explanations of gender differences in risk preference and concluded that "physiological differences related to risk preference appear to offer the only viable explanation of gender differences in religiousness" (p. 1401). This position argues that men have a biological propensity for risk taking owing to higher levels of testosterone.

Others have focused on nurture, examining how socialization processes could lead to a gender difference in religiosity. One promising line of explanation is based on what is known as "power–control theory" (PCT). PCT was originally developed to explain gender differences in propensity to engage in risky behavior—such as committing crime. PCT maintains that variations in the social control of sons and daughters within a household are linked to risk preference and behavior outside the household. In more patriarchal households, daughters are more subject to social control than sons, and so sons develop more preference for risky behavior than daughters. In more egalitarian households, there is less of a gap in the risk preference between sons and daughters, just as there is less gender-based difference in the freedom or control of children.

Jessica Collett and Omar Lizardo (2009) applied PCT to the gender difference in religiosity and found, in line with PCT, that gender differences in religiosity are greater for individuals raised in "traditional" (patriarchal) households than those raised in more gender-egalitarian households. They also found that women raised in more egalitarian households tend to be more irreligious than women raised in more patriarchal

households. These findings suggest that there could in fact be a connection between the socialization of women into risk tolerance or aversion that affects their subsequent levels of religiosity.

Although the gender difference in religiousness is one of the most consistent findings in the sociology of religion, we still have no solid explanation for it. The explanations previously given have been suggested but not definitively confirmed. Socialization explanations for the gender difference have been questioned. The connection between risk preference and gender differences has also been challenged (Freese, 2004). In the end, this remains a finding in search of an explanation.

> *Critical Thinking:* Do religious affiliation and behaviors by women seem to you more about female risk aversion (fear of consequences), or does socialization seem to be the better explanation for gender differences? Why? Which explanation best takes into account the fact that *male* involvement is higher than that of females among Jews, Buddhists, Muslims, and Hindus?

THE RELIGIOUS MEANING SYSTEM AND SEXISM

As we noted at the outset, one of the interesting paradoxes we encounter when examining religion and gender is women's higher levels of religiosity *despite* the historical connection between religion and sexism.

One might expect George Kelsey's thesis, that racism and Christianity are incompatible, might apply to sexism and Christianity as well. After all, Christian theology has stressed that one's worth is not founded in one's chromosomes but in one's relationship to God. Basing one's sense of self-worth on gender rather than on skin color would seem to be no less a form of idolatry. Indeed, logic would seem to require such a position. However, the history of sex bias is much longer than that of racial bias. Sexism is not a modern phenomenon; it is deeply rooted in the religious and philosophical traditions of the Western world. Sexual differences have been viewed as creations of God, and sexual inequality has frequently been viewed as God-ordained. Hence, the meaning system has often reinforced gender roles (tasks designated as "women's work" or "men's work") and concepts of gender (concepts of "masculinity" or "femininity"). Because inequality based on a person's sex is so much a part of the history of the Western world, it is appropriate that we briefly explore sex-role attitudes in Western philosophical and religious thought. Obviously we can only scratch the surface in the space we have here, but it is instructive to note some of the views held at various points in history by influential religious and philosophical thinkers and by religious bodies.

A Historical Overview of Gender Attitudes in Western Christianity

Because of the increase of women theologians and biblical scholars, more attention has been given in the last few decades to gender role and gender assumptions in the biblical tradition. Familiarity with the original Hebrew, Greek, and Aramaic has allowed feminist scholars to find patterns that were previously overlooked. For example, the Hebrew language has two different words that have commonly been translated as *man*. One of these words, *'adham*, is a generic term that refers to all of humanity. The other word, *'ish*, refers specifically to males. Because both terms have been translated into English as *man*, some of the subtle implications are lost in translation of biblical verses. The passage in the first chapter of Genesis that says that God created man is written with the term *'adham*. Hence, it should read "God created humanity in his own image, male and female he created them." Here the scriptures are less sexist than

some people assume when they are limited to English versions (Bird, 1974; Trible, 1979).

Elsewhere it is clear that the scripture was written for males, as in the books of law (Bird, 1974). The legal code is divided into apoditic law (moral commandments) and casuistic law (case law). The apoditic law is written primarily to men in that the literary voice and the examples are relevant to males. The Tenth Commandment provides a good example: "You shall not covet your neighbor's house; you shall not covet your neighbor's wife, or his manservant, or his maidservant, or his ox, or his ass, or anything that is your neighbor's." Not only does the passage specify *wife* rather than the more general term *spouse*, but the wife is included in a list of property that was owned exclusively by men. Furthermore, the Hebraic second person pronoun, which we translate as *you*, had masculine and feminine forms (similar to the way our third person pronoun has masculine and feminine forms: he and she). In the Ten Commandments, as in most of the apoditic law, the masculine form of the second person pronoun was used. Note as another example the audience that is being addressed in Exodus 22:22–24: "You shall not afflict any widow or orphan. If you do . . . then your wives shall become widows and your children fatherless." The intended audience, in much of the Bible, was clearly male.

Similar assumptions are part of the casuistic law, in which punishments for committing proscribed acts are spelled out. Most casuistic law begins with the formula "If a man does X, then. . . ." However, the term *man* here does not translate as the generic term *person*, for it is the masculine term *'ish* that is used. In those instances in which laws were articulated for women, they often served to remind them of their inferior position. Women were defined as unclean during their menstrual period and were unfit to enter the temple for 7 days after the birth of a son. By comparison, women were unclean for 14 days after the birth of a daughter. Such contamination was not normally associated with the natural bodily processes of men.

Of course, attitudes toward women are not consistent in the Hebrew Bible (called the Old Testament by many Christians), for it was written over a period of many centuries and contains many types of literature. The book of Proverbs, for example, depicts women as sources of great wisdom. There are also ancient stories about women, such as Ruth and various women in the book of Judges, that depict women as strong and courageous. Still, the overall effect of the Hebrew Bible is that women are subordinate. Even much of the symbolic action is limited to males; the primary symbolic act that represents the covenant with God is circumcision. Hence, only men could be ritually inducted into the covenanted community of God's people. Clearly, there were elements of intrinsic sexism in the early biblical period. Such attitudes are taken particularly seriously today by those Christians who accept the totality of the Bible as literally true.

Christian thought has, of course, been more than just a continuation of ancient Judaism; historically, it represented a synthesis of Hebrew and Greek worldviews. Many of the early theologians drew heavily from Greek philosophers. The Greek tradition was actually more explicitly sexist than was Hebrew culture. Aristotle's biological and political sciences, for example, depicted free Greek males as the embodiment of rationality. Such rationality was to be the ruling force in the good society, and the "spirit people" (males) were obligated to subjugate the "body people"—slaves, barbarians, and women (Ruether, 1975). Furthermore, Aristotle taught that every male seed should normally produce its own image in another male. Females were the result of an accident or aberration in the womb in which the lower material substance of the female womb subverted and warped the higher characteristics of the male. Women, clearly, were viewed as defective human beings (Ruether, 1975).

Even before Aristotle, the common belief was that reason and affectivity were mutually exclusive and were associated with good and evil, respectively. Plato had taught that the sexual act lowered people to the frenzied passions characteristic of beasts. The world of nature and of natural impulses was viewed as depraved and corrupted, whereas reason was viewed as the path to true goodness and spirituality. Moreover, reason was identified with males; passion and natural creation (in the form of childbirth) were associated with females.

This sort of dualistic worldview is characteristic of nearly all philosophical systems in classical Greece. Religious historian Rosemary Ruether (1975) maintained that such *hierarchical dualism*[4] is universally associated with sexist thinking (Shields, 1986). Women are invariably associated with the "lower" processes and with worldliness. Women are viewed as the cause of passion and are believed to be preoccupied with it; furthermore, they are identified with worldly creation because of their biological function in childbirth. Ruether insisted that dualistic thinking—in which the empirical world is defined as evil and the spiritual world is viewed as good—may have an inherent sexist bias. Ruether maintained that much of the sexist bias in Christian history comes from the Greek, not the Hebrew, legacy. She wrote that, unlike Christianity and Greek philosophy, "Hebrew religion, especially in its preexilic period, is not a religion of alienation that views nature as inferior or evil" (Ruether, 1975, p. 187).

The New Testament Gospels depict a much more positive view of women (Parvey, 1974; Ruether, 1975). Jesus himself violated many of the gender role taboos of his day. He allowed women to join his traveling group (Luke 8:1–3) and encouraged them to sit at his feet and learn (traveling and studying with a rabbi were viewed as very improper for women in that day). Jewish law also defined any woman with a flow of blood as unclean and polluting, and it forbade any Jewish male from speaking alone to a woman who was not his wife. Jesus deliberately disregarded both taboos (Mark 5:25–34; Matthew 9:20–22; Luke 8:43–48; John 4:27). He frequently contrasted the faithlessness of the religious leaders with the profound faith of poor widows and

---

[4]Hierarchical dualism refers to any belief system that divides all of life into two distinct realms, one of which is higher than the other (this world/other world, darkness/light, carnality/spirituality, emotion/ reason, etc.).

outcast women. In that day, unattached women were considered suspect and were to be avoided. His comments would certainly have been insulting and sacrilegious to many people (Luke 4:25–29).

Even his own ministry was shaped by values and positions that were widely identified with women, for Jesus taught that the role of the faithful was not one of glory and fame but of service. He capped his ministry by washing the feet of his disciples, a task normally assigned to women or servants, and at the time of his greatest disappointment outside the gates of Jerusalem, he described himself as feeling like a "mother hen"—an interesting analogy because of its feminine connotation. Finally, after Jesus had been killed, it was only the women followers who remained faithful. Other examples could be used, but few scholars question that the Gospels are among the least sexist books in the Bible.

If the Gospels were remarkable for their lack of sexism, the New Testament epistles are more ambivalent. Some people have a view of Saint Paul as one of the world's worst misogynists. He ordered women to obey their husbands, he told them not to speak in church, and he held to the old Hebrew belief that women serve men and only men could serve God.[5] Much has been written on these passages; they are frequently used today in conservative congregations to reinforce traditional roles. However, there are numerous references to women preachers who were sent by Paul or who accompanied him. In actual practice, he did not prohibit women from leadership roles.

Paul was a complex personality who often lacked consistency among his theology, his social teachings, and his behavior. It was Paul who asserted that in Christ, "There is neither male nor female." The society in which he lived was much more extreme in its sexism, and by comparison he appears liberal (Parvey, 1974). For example, some Christians of that day were much influenced by Gnostic philosophy and attempted to synthesize it with Christian doctrine. Some of these Gnostics taught that women were not worthy of becoming Christians, at least not unless they first became males.[6] The Gnostic Gospel of Thomas (which was never accepted as part of the biblical canon) stated boldly, "For every woman who makes herself male shall enter the kingdom of heaven" (cited in Bullough, 1973, p. 113). It was this sort of influence from Greek philosophy that Paul combated in stating that in Christ there is no distinction between men and women. Early Christianity was arguably the least negative in its attitude toward women of any religion in the Roman Empire (Stark, 1995).

Despite Paul's efforts, this ancient Greek view that women are defective humans found its way into the scriptures. Vern Bullough (1973) pointed to this when he concluded, "the most misogynistic statements in the scriptures appear not in the Epistles of Paul but in the Apocalypse (the Book of Revelation)" (p. 103). In that book, John the Elder describes the procession of the redeemed as a company of virgin *men* "who have not been defiled with women." Women were viewed as lesser beings who were not capable of being saved and might even interfere with men being saved.[7] The influence of Gnosticism can be clearly seen in this final book of the New

---

[5]Some scholars have argued that Paul's statement about women not speaking in church was his summary of current practice, which Paul set out to refute (see Iannaccone, 1982).

[6]Some works have attempted to depict Gnosticism as a profoundly feminist religious tradition (e.g., Pagels, 1979). Gnosticism was not a unified movement or philosophy. Some Gnostics appear to be very sympathetic to women, but others are clearly hostile, viewing sexuality and women as intrinsically evil. One must be wary of sweeping statements about gender images within Gnosticism as though it were a single, coherent philosophical tradition (Williams, 1986).

[7]It is noteworthy that early Christianity was not dualistic (O'Dea, 1966). In fact, the book of Revelation was held to be heresy by many church leaders for the first few centuries of Christendom. It was excluded from the Canon (those writings officially accepted as sacred scripture) until a vote by church leaders at the Provincial Council in AD 392 made it an official part of the Christian scriptures (Borg, 2001; Lightfoot, 2003; Weaver, 1975).

Testament. In the context of this extreme antipathy to women in the larger culture, Paul appears quite liberal on the gender issue.

While he rejected Gnostic misogynism, Paul was the one who emphasized so heavily the sin of Adam and Eve, and he believed that the responsibility for original sin lay with Eve. The implication is that women are an easier mark for the forces of evil. Interestingly, the idea that Eve introduced original sin is discussed only in the first few chapters of Genesis; it is never mentioned again in the Hebrew Bible. Yet, centuries later this was to become a major theme in the teachings of many Christian evangelists and theologians. Tertullian, one of the early Christian leaders, picked up on this theme and continually reminded women that each one of them was an Eve, "a devil's gateway." He held women responsible for being "the first deserter of the divine law" and wrote to them, "You are she who persuaded him who the devil was not valiant enough to attack. You destroyed so easily God's image, man. On account of your desert—that is, death—even the Son of God had to die" (cited by Bullough, 1973, p. 114). Hence, women were even held responsible for the crucifixion. At a much later time, Martin Luther also emphasized that a woman was responsible for the Fall. Although he generally opposed ridicule of women in public, he did on one occasion follow his comments on the story of humanity's Fall with the directed observation, "We have you women to thank for that!" (cited in Bullough, 1973, p. 198). So it was that the Adam and Eve story became a much more important justification of misogynism in Christian history than it ever was in the ancient Hebrew tradition.

Although the story of the Fall and the guilt of Eve became a justification for sexist attitudes, such scholars as Ruether have pointed to dualist theology as a more basic cause. When a worldview separates the world into two distinct realms, with the worldly realm governed by the passions and being inherently evil and with the heavenly realm governed by rationality and being inherently good, women and men are commonly identified as beings of one or the other of these realms. Such dualistic theology also frequently associates the sexual drive as worldly, ruled by passion, and evil. As a cause of sexual arousal in men, women have often been identified with the evil, worldly, and passionate side of the polarity. This occurs because men usually have had the power to apply the labels of good and evil and to make their labels stick.

Saint Augustine, a theologian who has influenced Christian thought for centuries, held to the Platonic view that passion was evil and that the sexual act lowered humans to the frenzied and unthinking level of beasts. Hence, he taught that when the sexual act was performed it should be done without emotion or feeling. The man was to plant his seed in the woman with the same dispassion as a farmer sowing seeds in the furrow of a field. So appalled by passion was Saint Augustine that he held the male erection to be the essence of sin. He waxed at some length and in horrified disgust about the "hideous" and uncontrollable (irrational) nature of the male erection. Moreover, if the erection was the essence of sin, it was clear who was responsible for causing it: women (Ruether, 1974). In fact, this sort of projection of sexual lust on women and a general view of women as temptresses was so common among Christian theologians that Bullough (1973) concluded the following:

> Sometimes it almost seems as if the church fathers felt that woman's only purpose was to tempt man from following the true path to righteousness. . . . Many of the church fathers seemed to find it difficult to follow their ascetic ideals and obviously felt the task would have been somewhat simpler if women did not exist. (p. 98)

Because sexual activity was viewed as corrupting (if not outright evil), a life vow of chastity was considered a more holy mode of life. Yet Ruether has studied the rationale for virgin lifestyles in the early Christian community and has found a consistent pattern: Virginity caused women to *rise above* their (innately evil) natures, but it caused men to *fulfill* their (innately good) natures (Ruether, 1974). Further, celibacy was *one* Christian lifestyle for men. For women, it was the *only* path to holiness. In fact, Augustine and Jerome suggested that for women the choice between childbearing and celibacy was a choice between shame and glory (Bullough, 1973). Eventually, the feeling that sexual activity was depraved and evil led to a new doctrine about the birth of Jesus. Because Jesus was to have been born of an uncontaminated womb, Mary herself came to be viewed as a source of purity. In the medieval period, the doctrine of Immaculate Conception was articulated. According to this doctrine, Mary was herself born without original sin; hence, her womb was a sinless environment. The obsession with sex as evil had become extreme.

One can also see the increase in misogynism at the time of the Protestant Reformation as it is manifested in Christian art. For many centuries, the Prince of this world (Satan) was sculpted as a handsome, attractive man as viewed from the front. As one walked around the statue, one would see that the back was a hollow shell, eaten by worms, frogs, and snakes. The imagery was powerful in its condemnation of this-worldly values. In the 14th century, however, the image of this-worldliness became "Frau Welt," a beautiful and alluring young woman from the front; the back side of the figure was decayed and infested with snakes, frogs, rats, and other vermin. The image of evilness and worldliness had become female (McLaughlin, 1974). Furthermore, the physical attraction of a beautiful woman was identified with baseness and corruption. This trend in Christian art was expressed in both Protestant and Catholic circles.

Sexism continued in the Protestant Reformation. The Protestant reformers abolished the requirement of celibacy among the clergy and reemphasized the childbearing role as a holy vocation, and one might expect that this would result in a lessening of sexism. Nonetheless, sexism can be seen in the writings of Martin Luther, John Calvin, John Knox, and other prominent reformers. In fact, *The First Blast of the Trumpet against the Monstrous Regiment of Women* was published by Knox in 1558 and stands to this day as one of the most misogynistic statements in Christendom. Moreover, the removal of Mary as a primary religious figure of adulation left Protestantism without a major saint or model who was female (Douglass, 1974; Ruether, 1975).[8]

> *Critical Thinking:* There are strong antifemale images and rules in both Christianity and other faith traditions, but there are also affirmations of the dignity of all children of God. How important is the *meaning system* in creating and preserving gender inequality? Why?

RELIGIOUS BELONGING SYSTEMS AND GENDERED WE VERSUS THEY THINKING

When members of conflicting ethnic or racial groups also belong to different religious groups, the likelihood of religiously based prejudice increases. However, in the industrialized world, men and women are not segregated as happens with race. Men and women are members of the same families and are members of the same religious organizations. Hence, it might seem that we–they distinctions would not occur. Nonetheless, the distinction of "we" and "they" between male and female is often given sanction in religious groups. Among Orthodox and Hasidic Jews, men and women sit separately during worship. The same pattern has prevailed among certain Protestant groups in the United States, such as the Shakers and the Old Order Amish, and among the Eastern Orthodox in places like Greece. Such religiously sanctioned segregation of the sexes would seem to reinforce we–they distinctions that are part of the larger

---

[8]Of course, not every religious tradition is equally sexist in its view of gender. Indeed, some traditions are actively pro-woman, notably the "goddess feminist movement" associated with the writer Starhawk, Wicca, and some types of neo-paganism more generally (Berger, 1999; Starhawk, 1989).

culture: Males and females are viewed as different species. Belonging has to do with a particular sex group.

Of course, sexual segregation was not an entirely negative factor; for example, many of the women in Roman Catholic convents were able to work in jobs that otherwise would have been denied to women. Because they were free of childbearing and child rearing responsibilities, they could devote their lives entirely to careers (teaching, nursing, etc.). Moreover, many nuns advanced to positions of organizational leadership and responsibility within the convent that would have been denied them in the larger secular society. Hence, the effects of such segregation were mixed: some women gained freedoms and opportunities otherwise unavailable, but the system may also have heightened sexism among the men by highlighting we–they distinctions.

It is also important to point out that there is considerable disagreement about whether segregation, emphasis on "difference," and gender-specific roles constitute sexism or gender inequality. Conservative Christian, Jewish, and Muslim women have argued that while there is role segregation, they do not experience it as inequality. They make a distinction between inequality and equity (balanced treatment where the needs of each are met), the latter of which they believe is most important (Purvis, 1995; Wallace, 1992; Williams & Vashi, 2007). Others believe that when "we" versus "they" thinking is fostered, and when one group has more power, the problem of sexism is inherent in the situation. This difference between these two views is partially a matter of definitions and it is unlikely that empirical evidence will soon resolve the matter. The point made so far is only that "we–they" thinking is enhanced by some religious practices, and the reader can decide whether that is problematic.

Beyond the issue of gender segregation, however, informal norms within the religious community may cause people to adhere to prejudicial attitudes in order to feel accepted—regardless of the position of the clerical hierarchy or of official denominational theology. To feel included—a sense that one belongs—one may feel compelled to laugh at jokes about someone out of the proper gender role. The penalty for ignoring the informal, unwritten standards is social exclusion; the reward is a sense of belonging. The official positions of the denomination—which may downplay gender difference and gender roles—frequently lack such immediate and concrete reinforcements. Hence, the informal culture of the group and the desire to belong may be more important in shaping attitudes than the group's formal or official doctrines.

The belonging function of religion, then, may foster gender attitudes in several ways. Insofar as the religious outlook stresses the fact that males and females are utterly different, the sense of "we" versus "they" may be heightened and the sense of belonging may be tied to sex group rather than to faith community. This may lead to suspicion and prejudice. Beyond this, the informal community that provides the sense of belonging may have informal norms that foster sexism. This leads to the next major way in which religion may enhance sex bias—the formal organizational structure itself.

RELIGION AND INSTITUTIONAL SEXISM

Perhaps the most vexing problems of religious prejudice toward women lie in institutional patterns—the structural dimension of religion. After a study of Christianity and Islam, Bullough (1973) concluded, "Regardless of what a religion teaches about the status of women, or what its attitudes toward sex might be, if women are excluded from the institutions and positions which influence society, a general misogynism seems to result" (p. 134). We have already seen that denigration of women became part of the theological system (meaning system) of western Christianity. This would be less likely to have occurred if women had been in positions to formulate and shape the official theology of the faith community. For this reason, denial of ordination becomes a significant issue. Furthermore, as long as a significant distinction is made between clergy and laity and as long as the clergy are looked up to as leaders, as those most in tune with God and as the legitimate messengers of God, the denial of ordination to women affirms their inferior position among the "people of God."

Ordination of Women

Many women scholars have seen ordination of women as a key issue: "By this exclusion the church is saying that the sexual differentiation is—for one sex—a crippling defect which no personal qualities of intelligence, character, or leadership can overcome" (Daly, 1970, p. 134; see also Ruether, 1975). When the Church of Jesus Christ of Latter-day Saints (Mormons) refused to ordain African Americans, it was widely recognized as a statement about the inferiority of black people.[9] Yet, some of the same denominations that were critical of the Mormons have opposed ordination of women. In this latter case, the argument that denial of ordination was a statement of inferiority was vigorously denied. Theologically trained women who have sought ordination have not found the denial convincing. Furthermore, the lack of women in leadership positions can subtly influence attitudes, especially of small children. The absence of women in important positions often communicates to children—much more vividly than any words to the contrary—the social inferiority of females.

The reasons for exclusion of women from ordination are noteworthy. Some scholars trace this exclusion back to Saint Paul who, in keeping with the contemporary attitude toward women, wrote that women were not to speak in church. However, we have already pointed out that he did not vigorously follow that policy himself. At least one historian has traced the original restriction on ordination to Constantine. When

This clergywoman was ordained into the Presbyterian Church USA (PCUSA) in 2008. Ordination of women is an important issue because when women are in leadership positions and involved in formulation of the official theology, there is much less likelihood of misogynistic policies and ideologies. Within the Episcopal tradition, ordination of women into the priesthood was so controversial that it led to the only schism in the Episcopal Church and the founding of the conservative Anglican Church in America.

Christianity became the official religion of the Roman Empire, it came under some of the cultic attitudes then prevalent. Specifically, religious leaders were to maintain ritual purity. Because women were viewed as unclean at particular times of the month because of natural biological processes of menstruation, they were unfit for ministry (Cox, 2009; Ruether, 1975). Because males dominated the priesthood and formulated the theology, women were continually defined as unfit. In fact, in the 13th century, Thomas Aquinas adopted Aristotle's view that women are defective males—biologically, morally, and intellectually. Hence, he reasoned that only men could fully represent Christ (the perfect human being) in the ministry. Such a rationale would not likely have occurred if women had been part of the hierarchy all along.

---

[9]The Mormon Church began to ordain African Americans in 1978.

Interestingly, although the Catholic Church does not ordain women as priests, many of the most active members of Catholic parishes are women and women "religious" ("nuns") have long played a key role in Catholic life (Fialka, 2003). Even more dramatically, in some small parishes today, where the church cannot supply a priest, women sometimes act as functional equivalents of priests at the local level. Indeed, Ruth Wallace (1992) found that in some Catholic churches, these women are called "pastor." Given the declining number and advancing age of Catholic priests in America, if the church continues to deny ordination to women, there may be more and more women acting as de facto leaders of their Catholic congregations (Schoenherr, 2002).

Most major denominations now ordain women, and the official pronouncements of most churches reject sexism and endorse equality for women at all levels. There is also a willingness by rather large majorities of members of congregations to have a female minister (Lehman, 1980, 1981). Despite all this, few theologically trained women are receiving appointments to large congregations. They continue to be placed in positions as assistant ministers, as directors of religious education, or as the sole minister in small, struggling congregations that have very low salary scales. If the official position of these church hierarchies is that women are equally competent and should be placed in positions of leadership, why is this not happening? The answer seems to lie in the inherent biases and goals of complex organizations.

The Organizational Survival Impulse and Women Clergy

Edward Lehman Jr. (1981, 1985, 1987b) approached the issue of women in the ministry from the resource mobilization perspective. Once an organization is in existence, it tends to take on characteristics and needs of its own, the most important being viability. If the organization is to survive, it must mobilize and control critical resources: the financial support of members; the skills, time, and energy of members; members' compliance with role requirements; and so on. However, in such voluntary organizations as churches and temples people must be convinced rather than coerced into compliance. Members who are not convinced can simply withdraw their support from the organization. (See the discussion on the dilemma of power in Chapter 6.) Part of the problem for voluntary organizations (including religious entities) is that members experience few negative consequences for *not* participating. Political and economic organizations control many basic resources, so one can be severely incapacitated by refusing to play by the rules. It is the lack of coercive control that makes commitment mechanisms so important for voluntary organizations (see the discussion of commitment in Chapter 5). Regardless, the leaders and committed members of an organization will hold the survival of that organization to be the highest priority. This sort of built-in value system tends to play against women in the ministry.

Although Lehman's research on American Baptist churches indicated that a majority of church members had no objection to having a woman minister, three fourths of the members believed that "most *other* members" were opposed to having a woman pastor. Furthermore, the more active a member was in the life of the church, the more likely he or she was to be negative about the idea of a woman minister. The primary concern was that having a woman as minister would cause controversy and conflict, and such conflict might result in members leaving the church or withholding financial support. Because this sort of action could threaten the continuation of the entire organization, the controversy is studiously avoided. In virtually all churches, the pulpit committee or search committee is elected from representative areas of the church life. In almost all cases, the members of the committee are chosen from among the most active members in the church—people who have made an investment in the church and care about its health and survival. Hence, pulpit committees normally have a built-in bias to avoid anything that might threaten its vitality or viability—including conflict.

Lehman found that churches that were growing were the least likely to consider a clergywoman. He also found that the larger the congregation, the more resistance there was to having a woman as a senior minister (Lehman, 1985). Where things were going well, members saw little reason to risk everything on a potentially divisive action. So the churches that were "healthiest" from an organizational standpoint were least likely to accept a woman pastor. The churches already in serious trouble, with declining enrollment and dwindling financial resources, had little to lose in accepting a clergywoman. In fact, their members often reasoned that it was better to have a first-rate clergywoman than a second-rate clergyman. Moreover, because women have harder times finding positions, they will more often accept lower-paying appointments than will equally qualified men.

Although the official position of the denomination may encourage local congregations to accept women in positions as senior pastors, organizational concerns regarding viability often play against that. The local congregation is simply more interested in promoting and maintaining its cohesiveness and stability than in responding to denominational resolutions. When denominational statements are perceived as being in conflict with local congregational needs, there is seldom much question about which will prevail in the end (Lehman, 1981). Still, the fear of declining membership and reduction in giving was not justified. Lehman (1985) found that the installation of a clergywoman to a faith community actually brought substantial increases in giving, attendance, and membership.

Virtually all Christian denominations have some sort of executive minister who is responsible for a geographic region. These area ministers or district superintendents are supposed to act as "pastors to the pastors" and implement official church policy. Furthermore, these organizational functionaries are usually the ones to whom pulpit committees go when they seek a new minister and to whom ministers appeal when they desire a move. The specific organizational pattern and the official titles vary from one denomination to another—as does the amount of authority of these officials over the congregations in their district. Although the role and job descriptions vary, one expectation remains rather constant: A major function of officials in this position is to maintain harmony in the congregations. Furthermore, an area minister may appear to be doing a poor job if a large number of congregations in that area withdraw from the denomination or radically reduce their pledges to the national organization. To be recognized as successful and to be a candidate for career advancement, the regional official must cultivate intrachurch harmony. The institutional reward system tends to favor caution. Although most of the executives of the American Baptist churches Lehman interviewed felt that clergywomen were just as competent as clergymen, they also shared the perception that laypeople would balk at the placement of a woman pastor. Their tendency was to avoid rocking the boat. So although these executives were charged with responsibility for implementing official denominational policy, they often did not press potentially controversial issues. Lehman (1981) wrote, "An important theme in these patterns is the executives' role of maintaining organizational viability, especially in the local churches" (p. 114).

In a follow-up study, Paul Sullins (2000) has argued that this conflict-aversive behavior continues and is more severe in religious communities than in other types of organizations. The reason for this is that faith communities are more like families than they are like typical bureaucratic organizations—with sensitivity to feelings of others and concerns about not alienating anyone being a core value of the organizational culture. This makes churches, temples, and mosques more change resistant than most organizations.

The problem of organizational maintenance and conflict aversion has continued to be a critical factor in resistance to the appointment of clergywomen. In subsequent studies of lay members of four denominations in England (Lehman,

1987a), of the Presbyterian Church in the United States (PCUSA) (Lehman, 1985), and of the Episcopal Church in the United States (Sullins, 2000), concern for the harmony and vitality of the organization and anticipation of destructive conflict over the appointment of a female minister was the primary barrier to appointment of a clergywoman. However, Lehman did find that over time in PCUSA denominational placement personnel became sympathetic to the placement of women and were less likely to be fearful of local backlash. Further, as congregations have more contact with clergywomen, resistance does decline significantly (Lehman, 1985; Purvis 1995). Attitudes and behaviors are changing, but they are changing slowly; meanwhile deployment of women into major leadership positions is moving at an even slower pace (Sullins, 2000).

Currently, seminaries are educating increasing numbers of women. Because the number of male applicants to seminaries has declined in the past several decades and because the average level of academic ability of male candidates has also declined, a tremendous increase in women theology students has allowed theological schools to continue without lowering their standards. In effect, this same organizational need—survival—has driven many seminaries to admit women to preministerial programs. Once granted a degree, however, women have difficulty finding a suitable appointment, and seminaries do not have sufficient influence on local congregations when it comes to helping their graduates find positions.

In the 21st century, ordination of women and the frequency of women clergy lag far behind the involvement of women in other professions, such as law and medicine. While the ratios of men to women in law and medicine and ministry had been very comparable up until 1970, there are now significant gaps. Roughly 34% of lawyers (U.S. Bureau of Labor Statistics, 2009) and 32.3% of physicians and surgeons are women (U.S. Bureau of Labor Statistics, 2006); by contrast women are only a bit more than 10% of the clergy (Chang, 1997). Women are the primary minister of about 8% of congregations, while roughly 5% of American Christians attend a congregation led by a woman (Faith and Leadership, 2010). While the numbers vary considerably by denomination, women clergy in virtually every denomination still lag behind the numbers of women in most other professional fields. In a recent analysis, Mark Chaves (1997) has offered an insightful analysis, one that is consistent with our open systems model.

Open Systems, Comparison Communities, and Religious Gender Roles

Chaves discussed ways in which outside forces may shape or have inputs into decision making in religious organizations. Chaves (1997) started with the issue of legitimacy of organizations in a society:

> [W]hen an organizational practice or structure becomes commonly understood as a defining feature of a "legitimate" organization of a certain type, organizational elites feel pressure to institute that practice or structure. If there is a cultural norm that says "In order for an organization to be a good organization, it must have characteristic X," organizations feel pressure to institute characteristic X. (pp. 32–33)

The more an organization is assimilated into the mainstream of the society, the more pressure it experiences to conform to the norms of other legitimate organizations and corporations. When the larger society did not stress gender equality and when patriarchy was the norm, it was often the more avant-garde or more sectarian groups that violated social expectations by recognizing women as ministers; the mainstream faith communities conformed to the patriarchal model. Of course, in the 19th century, ordination of women even among sectarian groups was not linked symbolically as a statement of gender equality as it is now. Sectarian groups simply did not care about credibility in the eyes of the secular society. The mainstream religious groups did. As greater gender equality has become the norm, there is greater pressure on organizations that care about legitimacy in the dominant secular society to accept women as equals. This would cause us to predict a greater likelihood of ordination of women among Unitarians, Congregationalists (the United Church of Christ), United Methodists, Lutherans, Presbyterians, Episcopalians, and Reconstructionist and Reform Jews than among more sectarian or isolationist groups. This is indeed what the evidence shows. As ordination has come to symbolize a secular principle (gender equity), the sectarian groups have become more resistant to it and mainline groups have become more open.

One reason that ordination and numbers of active female clergy lag behind other professions is the role of the state. Discrimination against women can be prohibited by the state in the fields of medicine and law. By contrast, Chaves (1997) wrote, "Religious organizations, as a population, are slower than other types of organizations to institute formal gender equality because of their greater autonomy from the state" (p. 42). In Norway, Denmark, and Sweden, where the state church is run by the government, sex-based restrictions on ordination were banned by the parliaments. In the United States, where there is a separation of church and state, ordination of women has occurred mostly during one of the two great waves of the women's movement: the first beginning with the Women's Rights Convention in Seneca Falls, New York, in 1848 (peaking in the 1880s) and the second beginning in the 1960s (hitting full stride in the 1970s). Indeed, more denominations began to ordain women in the 1970s than during any other decade in the previous 40 years. The cultural climate of the larger society involved "inputs" into the cultural and structural system of religious communities, even if the government was not mandating it. Indeed, even in denominations that oppose ordaining both genders, the terms of the debate have changed so that women's inferiority is no longer an acceptable argument. The

women's movement has changed the grounds for the debate even in those associations that resist female clergy (Chaves, 1997).

Furthermore, most denominations think of themselves as similar to certain others; they tend to form alliances and even formal organizations of denominations. Denominations or local churches that are very evangelical or fundamentalist do not generally use the United Church of Christ or the PCUSA as models of how to do things. Indeed, the more liberal denominations are often vilified and used as anti-models by very conservative groups. Chaves suggested that in very broad terms we can identify three sorts of Christian religious organizations that look to one another as comparison communities: (1) mainstream liberal denominations (United Church of Christ, PCUSA, United Methodist, Evangelical Lutheran, American Baptist, and the majority of groups that belong to the National Council of Churches), (2) scriptural inerrancy churches (Southern Baptist, Church of God, Missouri Synod of the Lutheran Church, and virtually every other group that belongs to the National Association of Evangelicals or the Council of Bible Believing Churches), and (3) sacramentalist churches (predominantly Roman Catholic and Eastern Orthodox, but to some extent Anglican, Episcopalian, and some Lutheran groups). When they adopt new policies, denominations often do research on other churches; however, their search is often limited to churches "like" themselves. Thus, groups seek credibility in large measure in relationship to other groups in the same general family. Even if their histories are quite different in origin and organizational style, religious communities are aware of their kinship with other groups that share much in the way of theological outlook.

An issue, then, is the symbolic meaning of ordination of women within a denomination. For scriptural inerrantist churches, literal reading of certain selected passages of the scripture have defined what is or is not acceptable, and ordination of females is interpreted to be beyond the pale. While many of those passages could be interpreted in several ways, faith communities tend to be influenced by the interpretations of kindred groups. Indeed, a number of Christian denominations that believe strongly in the inerrancy of scripture—but that are not highly integrated into the fundamentalist denominational subculture—do ordain women and have done so for more than a century (Evangelical Covenant Church, Church of the Nazarene, Salvation Army, Church of God—Anderson, and The Pillar of Fire). They simply focus on different scriptures that stress gender equality, believing that the churches prohibiting ordination of women are out of line with scriptural teachings. Most inerrantist churches, however, have defined ordination as unbiblical, with the central issue being that to be a Christian is—in their eyes—to be antimodernist (Chaves, 1997). Those fundamentalist religious bodies that do ordain women typically began doing so in the 19th century—before the matter was defined as an issue of gender equality (a secular and therefore modernist theme).

Sacramentalist churches have historically insisted that the person distributing the elements in the mass or Eucharist (communion) is symbolically in the role of Christ and therefore must be "like" Christ. Now there is no compelling reason that gender must be the defining feature of being "like" Christ, since each person has many qualities that are quite unrelated to her or his sex. However, the Catholic Church has defined sex as highly salient, and other Christian groups with sacramentalist leanings have tended to follow the lead of the Vatican. As one bishop said during the National Conference of Catholic Bishops in 1992, "A woman priest is as impossible as for me to have a baby" (quoted by Chaves, 1997, p. 88). Indeed, Chaves believed that sacramentalism has historically been characterized by a measure of antimodernism, and since gender equality is a key element of modernism, rejection of that symbol of modernism becomes an affirmation of orthodoxy. The fact of the matter is that churches that deviate from the expected norm within a family of denominations may be threatened with a kind of "excommunication" from that family of churches. The

pressures on denominations from other denominations are sometimes quite overt, as when the Roman Catholic Church has made it explicit to other sacramentalist religious organizations that ordination of women would destroy any prospects of ecumenical union in the future (Chaves, 1997).

A key element of Chaves's analysis is that we cannot understand ordination of women as solely an internal decision within a denomination. It is influenced by "inputs" by the larger society (norms about the conduct of business within legitimate organizations) and by pressures from comparable religious groups. Most ordinations of women occurred during one of the two major waves of women's rights activism, demonstrating the influence of secular social movements within religious social systems (Chaves & Cavendish, 1997). As open systems theory would predict, any organization exists in dynamic interaction with its larger environment.

The major issue of this section is how the ordination of women and their deployment in the workforce may perpetuate assumptions of female inferiority in our culture, and there is little evidence of substantial improvement since the late 1970s (Chang, 1997; Faith and Leadership, 2010; Nesbitt, 1997; Sullins, 2000). The lack of women in positions of congregational or denominational leadership may affect the way people think about males and females, a fact that is especially likely to influence the perceptions and attitudes of children since they learn attitudes toward people largely by seeing how people are treated (Farley, 2010). Furthermore, the lack of women in the hierarchy has allowed sexist policies and beliefs to be expressed by religious officials and to go unchallenged by anyone with similar credentials.

The current trend (at least as it is expressed in the official resolutions of most major denominations) is in the direction of ordaining and appointing clergywomen. Some denominations continue to deny ordination to women, the most notable example being the Roman Catholic Church and the Southern Baptist Convention. The first Protestant denomination in North America to ordain women was the Congregational Church (now called the United Church of Christ) in 1853 (Chaves, 1997). The first African American group to ordain women was the African Methodist Episcopal (AME) Zion Church in 1898 (Lincoln & Mamiya, 1990). The last three U.S. Protestant denominations to ordain women were the Protestant Episcopal Church in 1976, the Reformed Church in America in 1979, and the Christian Reformed Church in 1995. As we have seen, however, even in those denominations that have been more aggressive in endorsing ordination of women, clergywomen have a hard time gaining positions in larger churches, unless they settle for an assistantship or a directorship of religious education. The problem is not one of outright prejudice as much as one of organizational goals and assumptions taking precedence over other considerations. In short, the church has been guilty of institutional sexism. Clergywomen have encountered an "invisible ceiling" that limits their opportunities. The ceiling is invisible because it is an *indirect* result of policies and attitudes within the church.

One fact is highlighted by this exploration of institutional factors: Religion is both a worldview (set of beliefs, attitudes, and outlooks) and an institution in society. Understanding religious behavior can never be limited to one or the other. As an organization in society, religious bodies are often influenced by purely organizational considerations. In many ways, the organizational considerations take precedence. Hence, a strictly philosophical analysis of belief systems can never provide anything more than a partial analysis of religious behavior. It is precisely the interplay between social forces and belief systems that fascinates the sociologist of religion.

The central theme of this unit thus far has been that Christianity has often contributed to sex bias through its meaning system, its belonging system, and its institutional reward system. Many denominations are currently trying to reverse these trends, but the countervailing forces of gender inequality are deeply embedded. Still, the empirical question remains regarding the connection between the outlooks and policies of religious organizations and the values and perceptions of individual members of those organizations. Further, some women intentionally join groups that appear to be oppressive to them. These create intriguing questions for sociologists.

Critical Thinking: How can religious organizational structures combat gender inequality while simultaneously contributing to it?

RELIGION, INDIVIDUAL GENDER ROLE ATTITUDES, AND THE SOCIAL CONSTRUCTION OF DIFFERENCE

The Catholic Church has remained steadfastly opposed to the ordination of women, and yet the majority of individuals who identify as Catholic support women's ordination (Goodstein & Sussman, 2010). This suggests that sexism in the teachings and the institutional practices of churches, temples, and mosques may not simply translate to sexist attitudes in individual members.

We have much less empirical research available on the connection between religion and gender attitudes of individual members than we have on religion and racial attitudes. Most of the studies available contrast the religiously affiliated with the unaffiliated, or they compare gender role attitudes between denominations. Nonetheless, researchers have turned up some interesting findings. Generally those who are nonaffiliated tend to score lower on measures of sexism than those who are members of faith communities (Bayer, 1975; Dempewolff, 1974; Henley & Pincus, 1978; Lipman-Blumen, 1972; Martin, Osmond, Hesselbart, & Wood, 1980; Mason & Bumpass, 1975; McMurry, 1978; Meier, 1972; Tedlin, 1978). Moreover, unlike the studies of religion and racial prejudice (in which the most active members of faith communities were least prejudiced), the most active church members in the United States have been found to be most likely to hold negative attitudes toward women and oppose equity (Lehman, 1985; McMurry 1978; Roof & McKinney, 1987). Interestingly, the correlation does not seem to hold in Canada. The religiously devout appear no more likely than other Canadians to hold traditionalist views of gender roles, although our data are very limited for that country (Bibby, 1987a).

One study in the United States compared the effects of 13 different variables on gender role traditionalism and found religious affiliation to be the most important single factor in predicting gender attitudes (Martin et al., 1980). Even cross-national studies of citizens in eight European countries found that religious affiliates were consistently more opposed to female employment outside of the home (Hayes, 1995). Other studies indicate that individuals who are literalistic and authoritarian in their religious orientation are more likely than other religious persons to insist that women "stay in their place" or to show less respect for women (Hesselbart, 1976; Laythe, Finkel, & Kirkpatrick, 2001; Thornton & Freedman, 1979), though some recent studies indicate that conservative evangelical Christians are increasingly open to the two-career family (Bartkowski, 2001; Bloch, 2001; Lockhart, 2001; Williams, 2001).

There are significant variations between denominations, with active members of liberal denominations (United Church of Christ, Episcopal, and PCUSA) more supportive of gender equality than inactive members and far more egalitarian than conservative Protestants (Southern Baptists, Evangelicals, Fundamentalists, Pentecostals, Nazarenes, and Adventists). Indeed, in every American religious group *except* the liberal denominations,[10] level of activity in religious life has been positively correlated to gender role traditionalism (Roof & McKinney, 1987).

Another study used the distinction between intrinsic and extrinsic religious orientations as

they apply to gender bias. Gordon Allport (1966) had found that intrinsic religiosity (valuing religious experience for its own sake and not because of secondary rewards) was related to lower rates of racial antipathy. However, in the case of sexism, it was intrinsic religiosity that correlated highly with sex bias—attitudes that privileged males (Kahoe, 1974). Finally, one's present religiosity and affiliation have been found to be more important predictors of gender role attitudes than one's childhood religious orientation (Lipman-Blumen, 1972; Welch, 1975).

Despite all of this, some women intentionally join faith communities that have high boundaries between men and women and seem to offer limited options for females. This in itself is intriguing.

Critical Thinking: How is religiously based differentiation similar in the cases of racial and sex-based inequality? How is it different?

Negotiating Gender Barriers in Religious Communities

Of course, social reality is more complex than closed-ended surveys of religion and gender role attitudes can capture. Several scholars have used qualitative research methods such as ethnography and open-ended interviewing to understand in greater depth the process by which women negotiate gender barriers in organized religion. Women are not simply pawns of patriarchal religious leaders but actively engage conservative religion in a way that leads them to find personal fulfillment and empowerment.

Lynn Davidman (1991) sought to understand why young, secular women—women who have come of age since the feminist revolution of the 1960s and 1970s—would choose to convert to

---

[10]Liberal denominations in this analysis included liberal and moderate white denominations (e.g., PCUSA; Episcopalians; United Methodists; American Baptists; Christian Church), mainstream black Protestants, Roman Catholics, and Jews.

Orthodox Judaism, a tradition with very restrictive gender roles for women. To do so, she immersed herself in the lives of those she wanted to understand by joining a Hasidic community in Minnesota and an Orthodox synagogue in New York City. In addition to this participant observation, she also formally and informally interviewed over 100 women who were turning to Orthodox Judaism. The title of Davidman's book based on this study, *Tradition in a Rootless World: Women Turn to Orthodox Judaism*, captures her main argument: Orthodox Judaism is attractive to women who feel a lack of fulfillment in the contemporary world and seek to become grounded in an ancient tradition. Part of that grounding is in the clarity of the gender roles Orthodox Judaism provides, but that combines with the rootedness of those roles in long-established religious traditions and family life. This contrasts with the gender role confusion in contemporary secular society and the emptiness of working life in the modern corporate world.

Following Davidman's work, several scholars interviewed conservative Protestant women about their understanding of the subservient role of women in their religious traditions' theology and practices. In *Godly Women: Fundamentalism and Female Power*, Brenda Brasher (1998) reported on her ethnographic study of women in two conservative Protestant churches in Southern California. Like Davidman's converts to Orthodox Judaism, these fundamentalist women (also largely converts) were seeking refuge from a secular world in which the competing demands of work and home were often overwhelming. Fundamentalist teaching offered some relief to these women by providing clear dictates to focus on the domestic sphere. The women responded to their exclusion from formal positions of power by erecting a "sacred wall of gender" behind which they created single-sex ministries that ultimately became sites of empowerment for them. They were able to develop supportive bonds and exercise leadership in the women-only settings. That sacred wall of gender is writ even larger in *God's Daughters: Evangelical Women and the Power of Submission*, R. Marie

Griffith's (2000) 2-year long ethnographic study of Women's Aglow Fellowship International—the world's largest interdenominational evangelical women's organization. In observing this sex-segregated organization, Griffith found that when it is at its best, the women experienced considerable psychological healing, satisfaction, and even transformation from the group processes of prayer and testimony.

Christel Manning's (1999) research covers some of the same ground as Davidman, Brasher, and Griffith but has the advantage of putting women in conservative religious traditions in a comparative frame. Manning spent 2 years observing conservative Catholic, Evangelical Protestant, and Orthodox Jewish women in three Los Angeles-area congregations. In each case, the women in these patriarchal religious traditions do not see themselves as victims, but as active agents in defining gender roles for themselves. Catholic women are excluded from the priesthood, but they look to the many women saints in the Catholic tradition as role models for their own leadership. Evangelical Protestantism strongly professes the submission of women to men, but the Protestant traditional also emphasizes the individual relationship of each believer to Jesus Christ which can be used to justify women's empowerment. Likewise, Orthodox Judaism requires sex-segregated worship in the synagogue, but it gives women a central role in the domestic rituals that are so important to the Jewish tradition. In the end, although they would never accept the label, the women Manning studies are in certain ways feminists. Indeed, one of Manning's points is that in the contemporary United States, even patriarchal religious traditions must engage feminist ideas, since these ideas have become mainstream.

The negotiation of gender and religion takes place in every religious tradition, as is shown in the next "Illustrating Sociological Concepts" feature. The bottom line is that we cannot simply assume a mechanical relationship between patriarchal religious teachings and the acceptance or interpretation of those teachings by individuals, especially in the American context where voluntarism is strongly emphasized.

ILLUSTRATING SOCIOLOGICAL CONCEPTS

Is Veiling Oppressive to Muslim Women?

As the number of Muslims in American society has increased, so has the visibility of the traditional veil (known as a *hijab* in Arabic). In a society that values clothing as form of personal expression and individuality—indeed, even as a right—the veil is profoundly countercultural. Because only women are required to veil in Islam, it is seen by some observers to be an oppressive means of controlling Muslim women.

Recent research in the sociology of religion, however, has called into question that outsiders' view of the hijab. Scholars are taking seriously the experiences of Muslim women who choose to veil, and their findings are quite surprising.

Women in fundamentalist Muslim societies are separated from men (except for fathers and brothers) in work and in worship, and they generally remain covered. When they come to North America, the women are often accustomed to the practices and continue with them in the new country. *Purdah*, meaning curtain, refers to practices of seclusion and separate worlds for women and men. Screens in households and veils in public enforce female modesty and prevent men from seeing women who are not part of their families (Ward & Edelstein, 2006). Today, some Muslim women argue that the head scarf they wear is for modesty, cosmetic purposes, or to protect them from stares of men. They see it as a statement of identity as a Godly person and as a protection against intrusion by men. It is often a way to negotiate between the ethnic identity of their parents and the open, free, and sexually promiscuous society around them. It becomes a symbol of transition and dual identity, with some feeling that wearing a hijab is part of their free expression, at the same time it expresses their moral standing. Interestingly, Williams and Vashi also found that the style of wrap around the head, the material and color of the scarf, and the way it drapes over shoulders becomes a "fashion statement." So while some people claim that the veil is a symbol of oppression and subservience, Orthodox Muslim women often do not experience it the same way (Bartkowski & Read, 2003; Knickmeyer, 2008; Read & Bartkowski, 2000; Williams & Vashi, 2007).

Critical Thinking: So far in this chapter, the focus has been on ways that religion might be a causal or contributing factor in gender inequality. However, religion may sometimes simply be the arena in which larger social conflicts are expressed and justified. In this case, religion is primarily a *reflection* of conflict rather than a *cause*. A number of scholars feel that this perspective is more helpful in understanding the witch-burning craze of Europe, which you may explore in the website for this book at **www.pineforge.com/rsp5e.** Which causal sequence—religion as cause of bigotry or religious attitudes as effect of other social conflicts—seems more plausible to you? Why?

RELIGION, GENDER, AND HOMOSEXUALITY

Heterosexism, Homophobia, and Religious Systems

A number of commentators have noted that homophobia (intense fear and hatred of homosexuality and homosexuals) is related to sexism, for homophobia is highly correlated with and perhaps a cause of traditional notions of gender and of sex roles (Lehne, 1995; Morin & Garfinkle, 1978; Pharr, 1997; Price & Dalecki, 1998). *Homophobia,* a concept from the discipline of psychology, suggests that fear or hatred of homosexuals is a personality trait or even an abnormality (Weinberg, 1973). *Heterosexism,* a more sociological concept, is the notion that the society in many ways reinforces heterosexuality and marginalizes anyone who does not conform to this norm. Heterosexism focuses on social processes that define homosexuality as deviant and sacralize heterosexuality (Oswald, 2000, 2001). Most established religious groups have had a strong assumption about the normality of heterosexuality. Marriage is sanctified by ceremony in all major religious groups, but it has only been recently that some denominations have even openly debated the possibility of officially authorizing weddings for gays and lesbians. Regardless of whether we focus on psychological or sociological approaches, homosexuality has become a hot issue in faith communities, threatening to split some denominations.

Table 11.3 highlights three facts of interest here. First, homosexuality is more accepted by the American population as it has ever been (cf. Loftus, 2001), with 50% of Americans saying that society should accept homosexuality. However, that acceptance has not yet extended to allowing gay and lesbian couples the legal right to marry, something only a minority (39%) of Americans favor. Third, both of these views vary greatly by religious tradition, ranging from 12% of Jehovah's Witnesses to 82% of Buddhists agreeing that homosexuality should be accepted, and from 17% of evangelical Protestants to 45% of Catholics favoring same-sex marriage.

In the Christian tradition, there are six passages in scripture that condemn homosexual relationships: Genesis 19:1–28; Leviticus 18:22, 20:13; Romans 1:26–27; 1 Corinthians 6:9; 1 Timothy 1:10. However, it is interesting that only one of those condemns lesbian relationships. The others focus on male/male couplings. In any case, there is a correlation between religious involvement and opposition to homosexuality, with opposition to homosexuality especially high among religious fundamentalists. In recent years, there have been strident conflicts in the Presbyterian (USA) and United Methodist denominations, but in the most conservative denominations the questions raised by homosexuality (such as ordination of clergy, marriage of homosexuals by an ordained minister, and formal planks that either condemn homosexuality as a sin or suggest a more open approach that embraces any relationship characterized by love and support) are not even debatable. The rightness of absolute heterosexuality is taken for granted.

Theological liberals tend to see this issue as one of prejudice against persons for a characteristic that is immutable; they assume that homosexuality is an inborn trait. Conservatives argue that homosexuality is a choice that has moral implications. They tend to see homosexuality as a behavior that is acquired through socialization; therefore, acceptance of homosexuality will likely increase the numbers of people who engage in this lifestyle. The positions themselves are interesting in that they reverse the assumptions liberals and conservatives tend to make about gender and social roles. Conservatives are more likely to posit that gender is innate and inborn, whereas liberals argue that gender and the social roles associated with each gender are learned.

Table 11.3 Attitudes Toward Homosexuality

Religious Groups in the United States	Homosexuality should be accepted by society (percent agreeing)	Allowing gay and lesbian couples to marry legally (percent favoring)
Buddhists	82	nd (no data)
Jews	79	nd
Catholics	58	45
Mainline Protestants	56	39
U.S. average	50	39
Hindus	48	nd
Black Protestants	39	25
Muslims	27	nd
Evangelical Protestants	26	17
Mormons	24	nd
Jehovah's Witnesses	12	nd

NOTE: The Pew Research Center did not publish data for attitudes on gay marriage for every religious group.

SOURCE for acceptance of homosexuality: Pew Research Center Forum on Religion & Public Life. (2008b). *U.S. religious landscape survey*. Retrieved from http://religions.pewforum.org

SOURCE for gay marriage: Pew Research Center for the People & the Press. (2009). Majority continues to support civil unions. Retrieved from http://people-press.org/reports/

In any case, religious conservatives tend to see liberal notions about homosexuality and gender roles as a threat to the viability of society and as clear debasement of the moral social order. By contrast, liberals often feel that sexuality within a committed relationship is not a moral issue at all. Indeed, it is lack of tolerance of other lifestyles that liberals deem morally reprehensible.

Some mainstream Christian denominations have created provisions for congregations to proclaim they are officially hospitable and supportive of lesbian, gay, and bisexual persons. These designated congregations go by various names: More Light (Presbyterian), Open and Affirming (United Church of Christ and Christian Church), Reconciled in Christ (Lutheran), and Reconciling Ministries (Methodist). Official numbers are hard to come by for some of these traditions but, for example, the More Light Presbyterian (MLP) website (www.mlp.org) lists 124 member churches in the United States that affirm the MLP mission statement:

Following the risen Christ, and seeking to make the Church a true community of hospitality, the mission of More Light Presbyterians is the full participation of gay, lesbian, bisexual, and transgender people of faith in the life, ministry, and witness of the Presbyterian Church (USA).

This represents just over one percent of the 10,751 PCUSA congregations in America. Similarly, only 295 United Methodist congregations (out of more than 30,000) are part of the Reconciling Ministries Network within that tradition (www.rmnetwork.org).

In these churches, homosexuals are recruited and are told that they can be out of the closet and still be accepted as members. Efforts are made to make them feel welcome as part of the family of God. However, some gays and lesbians have felt that this does not go far enough to be really supportive of homosexuals. Thus, a new denomination has developed: the Universal Fellowship of Metropolitan Community Churches (UFMCC). The UFMCC, which has almost 300 local churches in 22 countries, is a "gay/lesbian positive" church that affirms homosexuality as a legitimate lifestyle for Christians (www.ufmcc.com). See the next "Illustrating Sociological Concepts" feature for some history of this denomination. One study conducted on a UFMCC congregation in New York found that homosexuals who are affiliated with this church have exceptionally positive personal adjustment; they are more likely to have an integrated self concept as both a gay person and as a religious person (Rodriguez & Ouellette, 2000). Religious persons who are homosexual and members of traditional congregations tend to experience less sense of integration in their self-identity.

RELIGIOUS DIVISIONS OVER HOMOSEXUALITY

Two of the most divisive issues in contemporary religion have to do with homosexuality: the ordination of homosexuals as ministers and the performance of marriages or "holy unions" of

ILLUSTRATING SOCIOLOGICAL CONCEPTS

Origins and Development of the Metropolitan Community Church

One of the hallmarks of American religion, according to R. Stephen Warner (1993), is that it is a locus of individual and group empowerment. Religious involvement, Warner writes, "has historically been one way that groups have improved their lot" (1993:1068). Warner goes on to observe that "the gay liberation movement is itself a practitioner of the art of church-based mobilization," that the Metropolitan Community Church is "the organizational center of the attempt to legitimate gay culture in the United States," and (quoting Altman 1982:127) that founder Troy Perry is "perhaps the most charismatic leader yet produced by the American gay movement" (1993:1068–69).

(Continued)

(Continued)

The Metropolitan Community Church began in 1968 with a gathering of 12 people in Rev. Troy Perry's living room. Perry had pastored Pentecostal churches in Florida, Illinois, and California. Then in the early 1960s, Rev. Perry was defrocked as a minister by a Pentecostal denomination. The issue was his homosexuality. He spent several years struggling to reconcile his commitment to Christianity and his sexual inclination.

Some years later he had a bout with depression—a deep despair in which he questioned the very meaning of life. In that context, his religious faith was unable to help him. After recovering from a nearly successful suicide attempt, he began to experience a renewed sense of his own spirituality. When a friend was arrested as part of a familiar pattern of police harassment in gay bars, his friend commented that no one, not even God, cared for gay and lesbian people. Perry was inspired by this incident to mobilize a church that served the homosexual community. He writes that "God wanted me to start a new church that would reach into the gay community, but that would include anyone and everyone who believed in the true spirit of God's love, peace, and forgiveness" (Perry, 2010). This was to be not a gay church but a Christian church.

Given the individualism and independence of the lesbigay community, his friends were skeptical that anyone would come. Further, the longstanding hostility of the Christian church to homosexuals created a huge barrier—for many people felt they could not be a homosexual and a child of God. Still, many members of that community had grown up in religious homes, and the liturgy and hymns of Christianity were a deep part of their memories and their identities.

There were 12 people who met in his living room in Huntington Park, California, for the first worship service in October of 1968. The experience was deeply moving to those who came, and the gatherings grew. In less than three months the attendees for services had tripled—too many for his residence. The worshippers had come from many traditions, and the services needed to appeal to their childhood traditions, but the diversity meant that they could not be limited to one particular faith community. Perry developed a synthesis of prayers, hymns, and liturgies drawn from Episcopal, Presbyterian, and Lutheran denomination resources. Still, Perry believes it was not the mechanics of worship that drew people, but the authenticity for the feeling of worship and the deep sense of belonging that was compelling in the lives of the new members. Perry commented, "People came out of the shadows, out of the closets, out of the half-world. They were drawn to the Metropolitan Community Church. For what? Some were curious. Some were incredulous. We were new. We were a novelty. We were an item in the gay world. . . . We excluded no one. We welcomed everyone. We still do." Indeed, roughly 20% of the membership is heterosexual.

The new movement was so novel that it got a good deal of press within the lesbigay community, and in forty years the group has expanded to forty-three thousand adherents and 300 congregations. It is a vivid example of a denomination being created in response to experiences of rejection and stigma—similar to other ethnic religious communities.

SOURCE: Metropolitan Community Church. (2010). *History of MCC*. Retrieved from http://ufmcc.com/overview/history-of-mcc/

homosexuals by clergy (Koch & Curry, 2000; Wood & Bloch, 1995).

A Gallup poll taken in the late 1990s already showed a shift toward acceptance of homosexual clergy by the majority of Americans: 53% approving in 1996 as opposed to 36% in 1977 (Gallup & Lindsay, 1999, p. 87). Nonetheless, in most religious traditions, the issue of ordaining LGBT (lesbian, gay, bisexual, and transgender) individuals as clergy is a nonstarter. This is true of Roman Catholics, Eastern Orthodox Catholics, Evangelical Protestants, Orthodox Jews, Muslims, Mormons, and Jehovah's Witnesses. Even in a more moderate tradition like United Methodism, the official position of the denomination is to deny ordination to "self-avowed practicing homosexuals."

In those denominations in which LGBT ordination has been considered, controversy has raged. The case of the PCUSA is instructive here. The ordination issue among Presbyterians focuses on what is called the "fidelity and chastity amendment" to the Book of Order passed in 1996. This amendment addresses sexually appropriate behavior for ministers, insisting that clergy must either be faithful within heterosexual marriages or remain celibate in singlehood (www .pcusa.org). Four times since 1997—including a 373 to 323 vote in June 2010—the PCUSA General Assembly (the denomination's highest legislative body) has voted to repeal the ban on noncelibate homosexual clergy. In each of the first three instances, a majority of individual presbyteries[11] did not ratify the General Assembly vote, and so the old standard remained in place. Whether the 2010 vote will be ratified remains to be seen.

The Evangelical Lutheran Church in America (ELCA) has become the largest Protestant body in the United States to ordain noncelibate gay clergy (Goodstein, 2010). Its path to this position was many years in the making. The denomination appointed a task force in 2001 to study the issue and debated the issue over the following 8 years before voting in 2009 to allow noncelibate but monogamous gays to be ordained. In July 2010, the first seven gay Lutheran ministers were welcomed onto the denomination's clergy roster and approved for service, though the decision to hire individual clergy is left to the congregation. The decision has not come without some cost to the denomination. One hundred and eighty-five ELCA congregations have voted to leave the denomination (out of 10,396 total congregations), and a coalition of theologically conservative Lutheran churches may form a new denomination (the North American Lutheran Church) in response (recall discussions of schisms in Chapter 8).

In the Episcopal Church, an even bigger step has been taken when in 2003 the Right Reverend Gene Robinson—an openly gay Episcopal priest—was consecrated as bishop in New Hampshire. This created great controversy. Some Episcopal congregations and even some dioceses have withdrawn from the denomination and the worldwide Anglican Communion (the global network of Anglican ecclesiastical groups with which the Episcopal Church has a longstanding connection) have also rebuffed the liberal U.S. church ("Illinois Diocese Votes to Split From Church," 2008). African Anglican bishops were especially upset by the decision and many have severed ties (Harrison, 2007). The Episcopal Church has not faltered in its support for Bishop Robinson, however, and in 2009 yet another diocese elected the Reverend Mary Glasspool to be their next bishop, believing that her openly lesbian identity was irrelevant

---

[11]Presbyteries are the formal associations of perhaps 200 Presbyterian congregations in a geographic area. They are the next step above local congregations in the hierarchy of the Presbyterian denomination.

("Los Angeles Diocese Elects Second Gay Episcopal Bishop," 2009). That appointment was approved by a majority of dioceses in the United States, and she was consecrated as the bishop for the Los Angeles diocese in May 2010. Churches and dioceses that have been displeased by the decision and have left the Episcopal Church have usually affiliated with the conservative Anglican Church of North America.

Another controversial issue in denominations is same-sex marriage. Although same-sex marriage is legal in many countries (including the Netherlands, Spain, Canada, Norway, and Argentina), in the United States, jurisdiction on this issue resides with individual states. Therefore, the issue has been debated and decided on a state by state basis, beginning with the decision of the Supreme Court of Hawaii in 1993 that laws denying same-sex couples the right to marry violated their constitutional rights to equal protection under the law (a decision overturned by the voters of Hawaii in 1998). Estimates are that 99.3% of all counties in the United States have same-sex couples, and about 3.1 million people live in same-sex relationships. One out of every nine unmarried cohabiting couples are gay or lesbian, and one in three lesbian couples and one in five gay male couples are raising children (Benokraitis, 2010; Human Rights Campaign, 2003). As of the summer of 2010, Connecticut, Massachusetts, Iowa, New Hampshire, Vermont, and the District of Columbia are issuing marriage licenses to same-sex couples. The increasing legal recognition of same-sex marriage has forced religious groups to consider whether to allow their clergy to preside at same-sex marriage ceremonies.

Few denominations in the United States allow their clergy to perform same-sex marriages. Even the PCUSA, which recently voted to ordain practicing homosexuals as clergy voted at the same General Assembly by a 51% to 49% margin to retain the definition of marriage as between a man and a woman, rather than accepting new language approved by a policy committee that would define marriage as between "two people." The issue has been tabled for 2 years. Among the denominations that do sanction their clergy performing same-sex marriage ceremonies are the Metropolitan Community Church, Unitarian Universalist Association, and United Church of Christ.

Gene Robinson (left) was the first openly gay priest to be elected and then consecrated as bishop in the Episcopal Church. Some people have been enraged by the decision, as we can see in the photo on the right.

Of course, just because a denomination does not allow its clergy to perform these ceremonies does not mean that its clergy do not do so. The controversy in the United Methodist Church has centered on ministers who have performed holy unions (a commitment ceremony for gays or lesbians that is similar to a marriage ceremony). A number of ministers have been performing these ceremonies, and the Methodist Church's official position is that a clergyperson can be defrocked from the ministry for doing so. When the General Conference of the church (which convenes every fourth year) voted to affirm this ruling in 2001, a demonstration ensued, and four bishops were arrested for their participation in the protest. Some ministers continue to perform such ceremonies as a protest against a policy they think is immoral. Further, one of the authors (Roberts) has been told by several Methodist pastors that a significant number of clergy continue to perform holy unions privately, taking a "don't ask; don't tell" position with their bishops. One retired minister (who declined to be identified) estimated that even in his very conservative Midwestern state that there are roughly 50 holy unions performed by Methodist pastors each year. He had this to say:

Back in the 1950s, ministers could lose their ordination and ministerial privileges if they married someone in a church wedding who had been divorced. Performing a marriage of someone who had been divorced was a serious offense in the Methodist church—and in most other Protestant churches, too. If a divorced person wanted to marry a second time, a civil wedding was supposedly the only option. Many of us in the pastorate thought that the policy was inhumane. These were people whom we pastored; they had made a mistake in a previous selection of a spouse and wanted to start again. Many clergy secretly went against formal policy of the church, but we did not tell our bishops. By the 1960s, if the church purged all of us who had done this, the Methodist church would have lost more than half of its ministers. In another

20 years, the same thing will be true with holy unions. If you have two people in your congregation who really care for each other and want to ritually affirm their lifelong commitment, how can we not honor that commitment and caring? I think it is a matter of time before this will become widely accepted.

Whether this retired pastor is correct or not remains to be seen. One study has documented a softening of opposition to homosexual rights among religiously committed people (Petersen & Donnenwerth, 1998), while others find that religious affiliation and religiosity continue to be powerful predictors of public opinion about same-sex marriage (Olson, Cadge, & Harrison, 2006; Whitehead, 2010).

A helpful perspective on this issue is the open systems analysis. Open systems theory says that religious organizations influence society (in this case, creating more traditionalism and therefore more resistance to homosexuality) and that the larger society influences the religious group. Chaves (1997) argued that attitudes toward gender issues are shaped by other similar denominations; each denomination looks around to comparable others to see what is defined as acceptable in those denominations. From this point of view, very conservative organizations are less likely to change because the other conservative denominations serve as "comparison communities"; however, if either the PCUSA or the United Methodist changes its policies, this may well have a liberalizing effect on the counterpart since they tend to see themselves as similar. Indeed, since the United Church of Christ (Congregationalists) already allows same-sex marriage, and the Presbyterians tend to see the Congregationalists as a comparison community, the movement toward more open policy by Presbyterians is not surprising.

Moreover, attitudes in a local community or a region of the country would be expected to influence attitudes within a congregation—irrespective of the positions of the official denomination

(Koch & Curry, 2000). People living in a local community are in regular contact with neighbors, friends, and coworkers, and the attitudes of those community people are likely to shape outlooks of those within a religious community. In addition, many new recruits to a congregation involve people whose views were shaped—prior to their joining the faith community—by local customs. One study that sought to test this open systems proposition found that most resolutions to the Presbyterian synod seeking to restrict sexual behavior to heterosexuality come predominately from congregations located in religiously conservative regions (Koch & Curry, 2000). Likewise, we could predict that regions of the country with more liberal attitudes are more likely to have "Open and Affirming" or "More Light" congregations.

Homosexuality is often defined as a moral issue. Gender, too, is usually laden with normative expectations. Religious bodies as institutions have historically been conservative or traditional forces in society, even if the meaning system of the faith is not. Two catalysts are creating pressure for change: (1) the emphasis on the equality of *all* people before God in Christian theology and (2) an impetus in the larger society that normalizes and legitimates women in leadership roles and affirms alternative types of sexual relationships. Much research is needed to understand the dynamic changes happening in these areas.

While this chapter has focused on gender and sexuality issues within the Christian tradition, the same sorts of issues face many religious traditions. For example, Mary Jo Neitz (2000) and Helen Berger (1999) had found that some American spiritual communities within Wicca were highly focused on heterosexual tension and attraction, emphasizing the polarity and complementary of the genders. So even Wicca has sometimes assumed the normality of heterosexuality and provided a sacralization of such relationships. However, one of the major annual celebrations of Wicca has undergone significant change in the past decade as homosexuality is becoming more visible and more accepted within

the movement (Neitz, 2000). Some Wiccan covens are exclusively female, and lesbian relationships have come to be more normalized. All religions experience influence due to inputs from trends and issues in the larger society.

> **Critical Thinking:** How is hostility toward homosexuals similar and how is it different from negative attitudes toward women? Do you think homophobia contributes to sexism? Is it fair to say that religion sometimes fosters heterosexism and homophobia? Why or why not?

SUMMARY

Although the ways Christianity fostered racism were for the most part subtle and indirect, Christianity has encouraged sexism much more directly. Except for the most liberal churches and temples in the United States, traditional images of women are correlated to high levels of congregational involvement and religiosity. Christianity has contributed to inequality and stereotypes of women in a variety of ways. First, women have been conceived as subordinate citizens in the meaning systems of most religions and philosophies of the Western world, and Christianity is no exception. Second, we–they thinking is sacralized in some religious communities and informal norms of the community that communicate women's inferiority may actually trump official statements of the denomination or the theology. Finally, formal institutional concerns, such as concern for maintenance of the health and vitality of the local congregation, can predispose lay leaders against having a clergywoman. When women are not in leadership roles, it can contribute to a perception of inferiority of women, especially in the eyes of children. More important, if women are not ordained and in leadership positions, there is a greater likelihood that the meaning systems will develop with a strong bias against women. The fact of the matter is that ordination itself has become a symbol of modernity, for equal treatment of women is increasingly expected—defined as rational behavior—in secular organizations; this adds pressure for ordination of women on those religious groups that want legitimacy in the eyes of secular institutions. By contrast, antimodernist groups reject the idea precisely because it symbolizes modernism.

One cannot make generalizations about Christianity as a whole on any of these characteristics. Some Protestant Christian groups, for example, have been remarkable for their lack of prejudice. The Shakers believed in a female Christ figure (Mother Ann Lee), and the Christian Scientists were founded by a female charismatic leader (Mary Baker Eddy). One also cannot conclude that sexual segregation of men and women into monasteries and convents in the Roman Catholic tradition was always bad for women. It did allow for leadership roles otherwise denied to women.

The foregoing discussion should serve to illustrate the complex way in which any given religious body can have countervailing influences. This is precisely why the sociologist who studies religion is not satisfied with an investigation only of the beliefs of a religious group. Religion can influence human behavior in a variety of ways.

Just as gender equality has become a hot issue, so has homosexuality become a tinderbox. The questions of ordination of homosexuals and performance of sacred ceremonies to celebrate homosexual unions divide some denominations and local congregations. Religions have historically sanctioned only heterosexual conduct and commitments. As the society comes to see this as discriminatory, conflict reins in many groups. Some denominations now ordain gay and lesbian clergy and some support same-sex marriage, or holy unions. Clearly, religious values and behaviors are intertwined with social conditions in the larger society. Moreover, religious organizations affect secular ones and secular organizations influence religious ones: the causal connection runs both directions.

12

RELIGION, INEQUALITY, AND SOCIAL ACTIVISM

Here are some questions to ponder as you read this chapter:

- How is one's position of privilege or disprivilege related to one's theology?
- Is the fact that affluent and disfranchised people have different theological explanations of their circumstances a matter of *elective affinity*, or is theology used as a tool to prevent social activism?
- Does religion mobilize minorities to protest against an unjust system or does it act as an opiate to pacify them?
- What are some unique features of African American Christianity?
- How has religion served as a resource to minorities in coping with oppression?

In this section, we have been exploring the relationship between religion and social inequality, but an important issue regarding religion and social status remains: What is the relationship between religion and social activism aimed at redressing social inequality?

Karl Marx (1844/1977) set a tone for much sociological work on this question when he famously asserted that religion "is the sign of the oppressed creature, the heart of a heartless world, and the soul of soulless conditions. It is the opium of the people"[1] (p. 131). In his view, religion served to pacify individuals so that they did not come to see the true basis of their oppression, which for Marx was economic. Religion, therefore, was viewed as inherently conservative because it serves to legitimize an unequal social order.

At the same time, history has repeatedly shown that religion can play an important role in motivating people to address social inequities. As Christian Smith (1996a) has put it, religion can be "disruptive" of the status quo. We see this in the case of abolitionist (antislavery) and civil rights movements in the United States, the anti-apartheid movement in South Africa, the Solidarity Movement in Poland, Liberation Theology in Latin America, and others.

As with most issues, then, the ultimate answer is not that religion is an opiate or a stimulant, but that under certain social circumstances religion can act in either capacity. Thus, the general question previously posted suggests several more specific questions: How does theology serve as a social ideology appropriate to one's economic circumstances? How is the religion of the affluent different from that of the poor? What are the consequences of the religions of the affluent and of the disfranchised for social stability or transformation? How does the religion of the disfranchised affect the larger society? Does it always act to justify the status quo, or does religion sometimes inspire oppressed people to militant resistance and advocacy of change? Why do affluent people sometimes become proponents of change that would reduce inequality—and support policies for diversity? We address these questions in this chapter.

THEOLOGY AS SOCIAL IDEOLOGY

Early in this study, we discovered that one of the functions of religion is to address issues of meaning. When people experience suffering or encounter injustices, they want to know why—or why me? Why is it that the good often seem to encounter suffering and hardship, while the evil seem to flourish like the green bay tree? On the other hand, when a family member or a good friend meets tragedy or death, people sometimes question, "Why couldn't it have been me?" The arbitrariness of suffering causes people to want an explanation; the world should make sense—it should have some ultimate meaning. If these events do not make sense, then life somehow seems a cruel joke. Any belief system that attempts to explain the reasons for evil, suffering, and injustice by placing them in a divine master scheme is referred to as a *theodicy*.

In trying to make sense out of the world and out of human experience, religious ideologies frequently provide explanations for the inequalities that exist in the social system. Sometimes religious beliefs endorse the current social system as established under divine will. For example, the present social arrangements may be viewed as God's divine plan, or the structures of this world may be viewed as a testing ground established by God to determine the truly faithful. Alternatively, the structures of this world may be viewed as the province of an evil force (e.g., Satan). Because lower-class persons are

---

[1]Marx (1844/1977) poetically concluded, "The abolition of religion as the illusory happiness of the people is the demand for their true happiness. The call to abandon illusions about their condition is the call to abandon a condition which requires illusions. Thus, the critique of religion is the critique in embryo of the vale of tears of which religion is the halo" (p. 131).

more likely to experience frustrations with the existing social system and to feel it is unjust, they are likely to have a rather different theodicy—one that views current social arrangements as something other than God's will. Max Weber (1922/1963) maintained that the lower their social class, the more likely people are to adhere to an otherworldly faith system.

Alignment of Theodicy and Economic Position

The theodicies of the lower classes are essentially "theodicies of despair" or "theodicies of escape," whereas those of the upper classes tend to be theodicies of "good fortune." People who are socially oppressed and who are experiencing a great deal of suffering need some explanation of a deeper justice or a deeper meaning that will ultimately prevail. In many lower-class religious groups, financial affluence is defined as the root of avarice and as a sign of evil. Human experience is divided into worldly and spiritual realms, and attachment to the latter requires a rejection of worldly success. This rejection makes economic deprivation much easier to bear, for deprivation is espoused as a noble choice; it is made a virtue. In fact, a frequent text for preachers in lower-class Christian sects is the saying by Jesus, "It is easier for a camel to go through the eye of a needle than for a rich man to enter the kingdom of heaven." The saying is likely to be understood literally and as applicable to the present day. (In affluent churches, this saying is often treated as a comment directed specifically at the rich young ruler Jesus was addressing or strictly at the rich people of Jesus' day and age. Its application to today's world is minimized.)

Other devices can also be used to make deprivation and suffering meaningful. When the ancient Hebrews were exiled from their homeland and were made slaves of the Babylonians, their plight seemed hopeless, and many of them felt their God had forsaken them. These people awaited a Messiah who would rescue them and take them back to their homeland. (The expectation of a Messiah was a distinctly Jewish theodicy.) The Messiah, they believed, would be a military general who would show forth the power of God and liberate them from captivity. However, the Messiah did not rescue them from defeat and exile to Babylonia. Eventually, a prophetic genius came along who reformulated the theodicy. He insisted that the ultimate victory would come not through military power but by the ability to endure suffering. The suffering of the Jews was part of God's master plan to provide salvation to all of humanity. The Messiah was recast as a "suffering servant." Suffering was not meaningless; indeed, it was a high calling. Release from the suffering would come in Yahweh's time—only after the divine purposes had been fulfilled. The idea of being a chosen people was preserved by being redefined. Rather than being chosen for privilege, the Jews were chosen for service. They were handpicked to be God's tools to transform the world. Jews have been persecuted and oppressed for 2,500 years, yet their theodicy of suffering has sustained them and held them together. Second Isaiah (Isaiah 40–55) was the first to articulate a theodicy for their oppression.

This provides only one of many possible examples of religious groups that recast their theology and worldview to fit their social circumstances. Groups that are composed of members of the privileged class also develop theodicies to justify their good fortune. As Weber (1922/1963) put it, "the fortunate is seldom satisfied with the fact of being fortunate. Beyond this one needs to know that one has the right to good fortune, ... that one 'deserves' it" (p. 107). Elizabeth Nottingham (1971) elaborated the following:

Almost equally important for a society [as a theodicy of disprivilege] is a morally acceptable explanation of its successes. Since a successful society often enjoys its worldly accomplishments at the expense of less fortunate peoples, its members are frequently driven to find a moral formula that will not only provide positive meaning for their own good fortune but also will help diminish any guilt they feel about the less happy situation of other groups. (p. 126)

One reason for differential denominational affiliation by social class may be that the theodicy of one denomination may have a better fit with the needs and concerns of those in certain circumstances. Weber (1946) introduced the concept of elective affinity, the tendency for members of certain social and economic groups to be drawn to certain religious beliefs.

As we found in Chapter 3, Karl Marx believed that economic self-interests are the driving forces of social behavior. He was convinced that the economic self-interests of the rich—those who own the industries and corporations—caused them to develop a religious view that justified their wealth and alleviated any sense of guilt. Because those with wealth and power can do much to control the belief systems of the society, Marx viewed religion as a system that sacralized the current forms of inequality and even oppression. For conflict theorists, theodicies of privilege are not so innocent as to be called an "elective affinity." They are, instead, insidious tools of the "haves" that "mystify" the true causes of inequality and serve to keep an unfair social system in place.

Whether the theodicy of privilege simply attracts the affluent because of compatibility with their own circumstances or is a consciously developed instrument of the wealthy to help them maintain their privileged position, the theological virtues and vices of the different socioeconomic groups do tend to be different.

Liston Pope (1942) found that for lower-class Protestant mill workers the world is a battlefield where God and Satan struggle for each individual soul. The sacrificed "blood of Jesus" and Bible reading by the faithful are the critical elements that allow God to be victorious in any given situation. Moreover, the lower-class concept of the deity is one of a comforter, protector, and savior. The role of God is to take care of His people—who are powerless to do much in this world of woes (Niebuhr, 1929/1957).

The chief decisions that control the lives of unskilled laborers occur at a level that they cannot control directly. They are often bystanders as the owners of corporations close plants, move factories to new locations, and make other decisions that profoundly shape the lives of wage laborers. Likewise, Pope found that mill workers viewed their role in the supernatural realm as one of observer, cheerleader, and marginal supporter. The battle is viewed as one between superpowers (God and the Devil), and it is largely the action of a third party (the sacrifice of Jesus) that will determine the outcome. This passive observer posture is typical of lower-class religiosity, especially for those groups stressing a theodicy of escape. Note the impact of this on people's sense of efficacy to bring change. A content analysis of hymns used in lower-class congregations illustrates the emphasis on dependency, alienation from this world, and blood sacrifice. Notable also is the negative concept of human character and of one's self evaluation ("such a worm as I"). (See the next "Illustrating Sociological Concepts" feature for examples.)

ILLUSTRATING SOCIOLOGICAL CONCEPTS

Hymns in Churches of the Less Affluent

In lower-class and working-class churches, the hymns frequently depict the world as a place of suffering and hardship, and the inherent worth of the individual is viewed rather dimly. Sin is viewed

(Continued)

(Continued)

as a state of being rather than as a specific action. A major focus is comfort in this world combined with hope for the next. Finally, the decisive action that determines one's changes is not accomplished by the individual but by some external force or action (e.g., the sacrificial blood of Jesus).

"Nothing But the Blood"

What can wash away my sin? Nothing but the blood of Jesus;
What can make me whole again? Nothing but the blood of Jesus.
Oh! precious is the flow, That makes me white as snow;
No other fount I know, Nothing but the blood of Jesus.

Nothing can for sin atone, Nothing but the blood of Jesus;
Naught of good that I have done, Nothing but the blood of Jesus.
Oh! precious is the flow That makes me white as snow;
No other fount I know, Nothing but the blood of Jesus.

This is all my hope and peace, Nothing but he blood of Jesus;
This is all my righteousness, Nothing but the blood of Jesus.
Oh! precious is the flow That makes me white as snow;
No other fount I know, Nothing but the blood of Jesus.
(Baptist hymn)

"Remember Me"

Alas! and did my Savior bleed? And did my Sov'reign die?
Help me, dear Savior, Thee to own, And ever faithful be;
Would He devote that sacred head, For such a worm as I?
And when Thou sittest on Thy throne, Dear Lord, remember me.

Was it for crimes that I have done, He hung upon the tree?
Help me, dear Savior, Thee to own, And ever faithful be;
Amazing pity! grace unknown! And love beyond degree!
And when Thou sittest on Thy throne, Dear Lord, remember me.
(Nazarene hymn)

"No One Understands Like Jesus"

No one understands like Jesus, He's friend beyond compare;
Meet Him at the throne of mercy, He is waiting for you there.

No one understands like Jesus, When the days are dark and grim;

No one is so dear as Jesus—Cast your ev'ry care on Him.

No one understands like Jesus, Ev'ry woe He sees and feels;

Tenderly He whispers comfort, And the broken heart He heals.

No one understands like Jesus, When the days are dark and grim;

No one is so dear as Jesus—Cast your ev'ry care on Him.

(Country hymn used in Independent Baptist and Methodist Sects)

Moreover, for the disfranchised it is not simply the individual who needs saving but the whole social system that is corrupt. The current social arrangements are often viewed as unjust, inequitable, and corrupted. In fact, sin is not just thought of as wrong actions but as a pervasive depraved condition that infects and becomes an inherent quality of both the individual's soul and the fabric of society. Whether this means the people of God should be activists for societal change depends, in part, on whether God's kingdom is this-worldly or is in another realm of existence—as part of the afterlife.

In the upper classes, different ideas about sin and the social order generally prevail. Economic prosperity is defined as a blessing of God or even as a sign of divine favor. Note that members of the upper classes are accustomed to controlling their own destinies. Their outlook stresses individual accomplishment, a positive assessment of their ability to control things in this world, and a high valuation of individual initiative. In fact, Niebuhr suggested that the American middle and upper classes characterize the deity as "energetic activity" or "dynamic will," and God expects the same sort of productive activity from humans. This set of values is vividly illustrated by the hymns that are typically sung in upper- and upper-middle-class churches (see samples in the next "Illustrating Sociological Concepts" feature).

ILLUSTRATING SOCIOLOGICAL CONCEPTS

Popular Hymns in Affluent Denominations

In upper- and middle-class churches, the hymns frequently express a positive value of this-worldly activity, an affirmation of individual self-worth, a high valuation of individual initiative and accomplishment, and a sense that persons are in charge of their own destinies. In one of these hymns, Jesus is depicted as an *example* to humankind rather than as a bloodied sacrificial Lamb. Furthermore, the saints of God are depicted in one hymn as common folks rather than as a highly committed elect.

"O Brother Man, Fold to Thy Heart"

O brother man, fold to thy heart thy brother;

Where pity dwells, the peace of God is there;

(Continued)

(Continued)

> To worship rightly is to love each other,
> Each smile a hymn, each kindly deed a prayer.
>
> Follow with reverent steps the great example
> Of him whose holy work was doing good:
> So shall the wide earth seem our Father's temple,
> Each loving life a psalm of gratitude.
> *(Congregational hymn)*

"I Sing a Song of the Saints of God"

> I sing a song of the saints of God
> Patient and brave and true,
> Who foiled and fought and lived and died
> For the Lord they loved and knew
> And one was a doctor, and one was a queen,
> And one was a shepherdess on the green:
> They were all of them saints of God,
> And I mean, God helping, to be one too.
>
> They loved their Lord so dear, so dear,
> And his love made them strong;
> And they followed the right, for Jesus' sake,
> The whole of their good lives long.
> And one was a soldier, and one was a priest,
> And one was slain by a fierce wild beast:
> And there's not any reason, no, not the least,
> Why I shouldn't be one too.
>
> They lived not only in ages past,
> There are hundreds of thousands still:
> The world is bright with the joyous saints
> Who love to do Jesus' will.
> You can meet them in school or in lanes, or at sea,
> In church, or in trains, or in shops, or at tea;

For the saints of God are just folk like me,

And I mean to be one too.

(Episcopal hymn)

"Rise Up, O Men of God"

Rise up, O men of God!

Have done with lesser things;

Give heart and soul and mind and strength

To serve the King of kings.

Rise up, O men of God!

His Kingdom tarries long;

Bring in the day of brotherhood

And end the night of wrong.

Rise up, O men of God!

The Church for you doth wait,

Her strength unequal to her task:

Rise up, and make her great!

Lift high the cross of Christ!

Tread where His feet have trod.

As brothers of the Son of Man,

Rise up, O men of God.

(Presbyterian hymn)

Equally important is the way in which conventional secular norms and values are embraced among upper-class Christians. Pope and Niebuhr each pointed out that while lower-class churches focus on "sin" as a state of being, the more affluent Christians tend to limit their concept to the plural sins—specific traits or behaviors, such as failure to pay ones debts. Among the affluent, sin is not so much the innate state of the soul or of the social system but the personal choices of an individual. This difference in definition of sin can affect how one thinks a society needs to change to make it better. If the problem is poor decisions by avaricious or ignorant people, social problems become individualized and change of the social system is irrelevant at best.

In middle- and upper-class religious communities, individuals are encouraged to cultivate a sense of self-worth and self-esteem; members do not often sing on Sunday morning about grace that "saved a *wretch* like me" or about how the blood of Jesus was shed "for a *worm* such as I." Such negative self-images are characteristic of lower-class hymnology where the conception of sin is more pervasive. Thus, a theologian who has strong appeal to the more affluent and well educated can write about sin:

> I believe a careful examination of Biblical sources will indicate that humanity's most debilitating proclivity is *not* pride. It is *not* the attempt to be more than human. Rather it is sloth, the unwillingness to be everything humanity was intended to be. (Cox ,1964, p. xi, 2009).

As one reviews the hymns in the previous two features, one can see this difference in emphasis. These distinctions are not a categorical or exclusive difference between upper- and lower-class religious groups. Clearly, there is wide variation of religious expression and theology within denominations and between persons of the same social class. Here we describe some broad patterns to understand that socioeconomic status does affect religiosity and how one is likely to try to change society, if at all.

Theological and Class Realignments in a Postindustrial Age

The specific issues have changed over time, but differences continue between the faith communities of the privileged and those of the lower and working classes. In the modern economies of the United States and Canada, the working class has had some successes, and while they are not highly prosperous, they are able to feed, clothe, and shelter their families. Many even have some luxuries, such as a boat or a camper. The society looks less menacing, and they have interests that they do not want threatened. On the other hand, as we will see, diversity and openness to change has come to serve the interests of at least some relatively affluent people, including many in the professional classes. Interestingly, the positions of religious groups parallel some of these shifts.

In the postindustrial United States, the religious bodies of the less affluent are those who are normally identified as "conservative." They include Southern Baptists, Churches of Christ, Nazarenes, Pentecostal and Holiness sects, Assemblies of God, Churches of God, Adventists, Jehovah's Witnesses, and various evangelical and fundamentalist groups. Conservative white Christians tend to define the moral issues of our day as secular humanism and moral relativism, which includes the ban on prayer in the schools; the teaching of evolution rather than creationism in the schools; changing roles of women; increases in the divorce rate; and the new morality (which includes nonmarital sexual behavior, legalization of abortion, tolerance of pornography, and tolerance of homosexuality). The U.S. government—which is viewed as too secular and too supportive of diversity and pluralism—is seen as too permissive to have any say in moral issues.

These conservative churches stress absolutism and obedience to rules. However, notice the congruence of that view with low-prestige jobs. People in lower- and working-class jobs usually find that obeying rules of the workplace and adhering to the instructions of the employer or supervisor are qualities that are central to success on the job. Lack of conformity can have serious consequences, and parents socialize their children to develop those attitudes and behaviors that they believe will serve their children well in life (Bowles & Gintis, 1976; Kohn, 1989; McLeod, 1995). Further, if one works in an occupation where brute strength is essential and often gets the job done, one may look for clear, decisive, and forceful solutions to religious problems as well (Batson, Schoenrade, & Ventis, 1993). Clearly if one is to become a conservative activist, the core value issue is not social inequality.

On the other hand, the most affluent Christian churches in the United States are also those that are the most theologically and ethically liberal: Unitarian, Episcopalian, United Church of Christ (UCC), and Presbyterian. (Methodist, Lutheran, Disciples of Christ, American Baptist, and reformed churches serve as a middle-class buffer and are often referred to collectively as "moderate Protestants.") The moral issues of our day for these churches—at least as identified by their clergy and their official denominational boards and agencies—tend to be war and peace issues; protection of the environment; social justice (defined largely in terms of elimination of discrimination against women and minorities and support for more diversity); and the lack of tolerance of those who are "different" (including those from other cultures and often those with different sexual orientations). Pluralism and tolerance seem to be central moral themes of the churches of privilege (Roof & McKinney, 1987).

It is important to note that many affluent professional people are paid to be divergent thinkers—to be creative and to solve problems. They often manage organizations and are frequently rule makers. They will not do well in their fields if they merely obey rules. They must respect divergent or nonconformist thinking because creativity and innovation are often necessary to solve new problems. Acknowledgment of ambiguity and utilization of careful analysis of problems are essential (Batson et al., 1993; Bowles & Gintis, 1976; Kohn, 1989). It should not be surprising that in a postindustrial world, the conception of good and evil among the affluent embraces tolerance of differences and condemns rigidity, absolutism, and conventionalism.

Like many professional positions, being a scientist requires tolerance for ambiguity, creativity, and divergent thinking in order to solve complex problems and to be successful in their careers. This acceptance of uncertainty and respect for diverse views often carries over into what they expect in their spiritual lives and their religious communities.

These people also are self-conscious about socializing their children to develop those characteristics that they believe will help their progeny in the world of work—including openness to diversity of all types. Experience tells these professionals that critical thinking, creativity, and even a streak of independence are valuable characteristics. This becomes a part of what they expect in their religious life as well. Their social values are congruent with their financial self-interests, and both are served by more liberal theologies.

Table 12.1 illustrates the divergence in social and political attitudes between religious traditions in the United States. The table includes two measures of tolerance (liberal attitudes on abortion rights and homosexuality) and two measures of political liberalism (self-described ideology and view of government's role in enforcing moral views). In every case, more affluent traditions (Jews, Buddhists, Hindus, and mainline Protestants) are more tolerant and liberal than less affluent traditions (Black Protestants, Evangelical Protestants, Mormons, and Jehovah's Witnesses)—in some cases by wide margins.

Theology and Financial Self-Interests in Conflict

Of course, this discussion would be incomplete if we pointed merely to economic self-interest as the sole determinant of religious ideologies. An important and interesting phenomenon in today's world is the fact that many denominations whose members are affluent have directly and rather aggressively challenged the structures of inequality and privilege—as their scriptures and religious traditions urge them to do. For example, Congregationalists, Presbyterians, Unitarians, Episcopalians, and Catholics all have commissions or task forces on racism, poverty, and social inequities. Some task forces do not just attack the problem at an individual level; they challenge the very structure of society and point to systemic causes of poverty and racism. They issue statements calling for change—sometimes for radical change—in the basic social and economic structures of society.

Table 12.1 Social Attitudes of Members of Various Religious Groups (in percentages)

U.S. Religious Traditions	Abortion should be legal in all or most cases	Homosexuality should be accepted by society	Government is too involved in morality	Political views are liberal or very liberal
U.S. Average	51	50	52	20
Buddhists	81	82	67	50
Jews	84	79	71	38
Hindus	69	48	45	35
Mainline Protestants	62	56	58	18
Catholics	48	58	49	18
Muslims	48	27	29	24
Black Protestants	47	39	42	21
Evangelical Protestants	33	26	41	11
Mormons	27	24	39	10
Jehovah's Witnesses	16	12	36	17

SOURCE: Pew Research Center Forum on Religion & Public Life. (2008b). *U.S. religious landscape survey*. Retrieved from http://religions.pewforum.org

Many of the liberal positions taken by the National Council of Churches and by various denominational boards, of course, have not been supported by a majority of the lay constituencies of the local congregations (Hadden, 1969; Jenkins, 1977).

There are many—but by no means all— relatively affluent members who are highly supportive of denomination programs that challenge the status quo (Wuthnow & Evans, 2002). This concern for the less privileged among affluent congregations appears to be stimulated by religious teachings. The prophets of the Bible consistently called for social justice as the primary indicator of true religious expression. James Wood (1981) has found that congregations will often act collectively in ways that members would reject individually. This is largely because the clergy are able to call on beliefs that are commonly held but that do not coincide with the self-interests of the members. In the context of a group that they view as important, these members will support policies and actions that might not benefit their own portfolios.

An ethical or prophetic theme is also central to the Reform branch of Judaism. Although Jews are generally among the more affluent members of American society, they have often championed the cause of blacks and other minorities. For example, Jews were instrumental in founding the National Association for the Advancement of Colored People (NAACP) and the National Urban League, two organizations that have had crucial roles in fighting for equal rights for African Americans.

Migrant farm workers (predominantly poor Mexican Americans) have received support from affluent Christian and Jewish congregations, which send financial aid to the United Farm Workers and encourage their members to participate in boycotts of grapes, lettuce, and other products. The boycotts end only after wages are increased, conditions have improved, and a union contract is signed. However, this usually means increases in the cost of the produce. In other words, members of affluent congregations knowingly boycott goods with the ultimate result that they must pay higher prices. Such behavior is not in the narrow self-interests of these affluent individuals, yet they persist in this behavior because they are convinced that it is the moral thing to do. If faith is truly the organizing principle of one's life, it can inspire actions beyond self-interest.

The theodicy that people adopt may be related to their own socioeconomic standing, but the theodicy is also related to their inclination to social activism on issues of social inequality. In the following section, we will see how a theodicy may mobilize people to action or encourage passivity. Since racial and ethnic minorities experience special issues of disfranchisement and destitution, we focus here on the role of religion in how they cope with inequality.

Critical Thinking: Theology often seems to parallel the economic circumstances and even the self interests of people. What factors do you think best explain the correlation? Equally intriguing is when people are motivated by faith and moral convictions to do something that is directly contrary to their self interests? What factors might explain that pattern?

RELIGION AND MINORITY STATUS

Is Religion an Opiate for the Oppressed?

A good deal of sociological research has focused on whether religion among the disfranchised acts as an opiate of the masses or inspires the dispossessed to militancy. As we have seen, Karl Marx maintained the former: that religion gave the poor a feeling of solace and a hope of compensation in the afterlife so that they would not rebel in the present. They were, in essence, drugged. Those who hold to this position insist that religion serves as a tool of control for the dominant economic class and ethnic group. Critics of religion have maintained that when European missionaries went to regions that were later colonized, they had the Bible while the natives had the land. When the missionaries left, the natives had the Bible, and the Europeans had the land (Marx, 1967). Although examples of this view that religion is a tool of exploitation abound, examples neither provide proof nor establish causality.

Other researchers point to contrary data, such as the facts that most civil rights leaders have been members of the clergy and that high percentages of the members of civil rights groups attend worship services every week. Many historical analyses have also found that black religion was frequently a motivating force in slave revolts, since religion asserted the intrinsic human worth of slaves and raised their hopes for more dignity in life (Lincoln & Mamiya, 1990; Wilmore, 1972). Scholars from several disciplines have used a variety of research methods to clarify the relationship between religion and activism regarding social injustices. Our treatment in this section focuses on the function of religion in minority groups—specifically ethnic minorities.[2]

[2]The sociologist uses the term *minority* to refer to groups that have less power to control their destiny than do others. It does not mean the group is necessarily smaller in numbers. (Blacks constitute three fourths of the population of the Republic of South Africa, but they are still referred to as a minority group.) In the United States, women, homosexuals, the disabled, and the elderly are often referred to by sociologists as minority groups. In suggesting that we will be focusing on ethnic minorities, we wish to make clear that we will not be treating these other groups in this section.

One landmark empirical study of the effects of religion on blacks in the United States was conducted by Gary Marx (1967). Using a sample survey method of research, Marx discovered that blacks who were members of higher-status (and predominantly white) denominations were more likely to be militant in their civil rights positions than were those from lower-class churches or sects. A militant black was defined as one who actively and consistently opposed discrimination and segregation. Blacks who were militant for each denomination were as follows:

Episcopalian	43%
Congregationalist (UCC)	42%
Presbyterian	36%
Roman Catholic	36%
Methodist	28%
Baptist	25%
Sects and cults	15%

For most scholars, the fact that blacks in lower-class churches were more passive was not really unexpected, but Marx's finding that blacks in *entirely black congregations* were more passive than blacks in predominantly white congregations was something of a surprise.

Gary Marx also found that infrequent religious attendance was positively correlated with militancy. Eighteen percent of those who attended worship more than once a week were militant, while 32% who attended less than once a year were activists (Marx, 1967). Of those who said religion was "extremely important" 22% were militant, while 62% of those who answered "not at all important" were militant (Marx, 1967). When he combined all his factors into an overall index of religiosity, he found a negative relationship between religiosity and militancy. Furthermore, the finding held even when he kept certain key variables constant: age, sex, denomination, and region of the country in which the respondent was raised. Marx concluded that there apparently is an "incompatibility between piety and protest."

Seymour Lipset, by doing a cross-cultural political analysis, came to a similar conclusion but offers an added dimension to Gary Marx's investigation. He maintained that rigid religious dogmatism is based on the same underlying personality characteristics, attitudes, and dispositions as political radicalism. However, he also pointed out that religious fanaticism and political fanaticism tend to serve as functional alternatives; one usually finds only one or the other at any given time (Lipset, 1960). Lipset's study adds a new insight, however, because he found that religious groups are often spawning ground for political militancy, but in the process of development, these militant groups frequently become more secular.

Eric Lincoln and Lawrence Mamiya (1990) maintained that regardless of the belief system, African American religious structures have enhanced resistance to subjugation. These scholars insist that faith communities can empower people simply by establishing stable institutions, fostering networks, and forging a sense of common identity in suppressed peoples. Regardless of the intent of the religious group, those interested in change are able to use the networks and sense of common interest to mobilize a social movement. Since religious bodies were the first black organizations locally and the first national institutions, they contributed to social activism and sense of potency. Religious institutions spawned various types of protest organizations by providing networks, leadership training in organizational management, and an overall sense of organizational competence (Morris, 1986). This sort of contribution by religion to the civil rights movement is not assessed by survey research like that of Gary Marx.

Follow-up studies have indicated a varied effect of black religiosity, demonstrating that black religion sometimes does stimulate social change (Nelsen, Madron, & Yokley, 1975). Two other studies found that, regardless of denominational affiliation, blacks who are churchlike in their religiosity are more likely to be inspired to militancy than are blacks who are sectlike (Hunt & Hunt, 1977; Nelsen & Nelsen, 1975). When this variable is held constant, the inverse relationship between militancy and religious attendance disappears. Still, some black "sects" are highly militant (Baer, 1984). A more in-depth understanding of the worldview or the theodicy of a subjugated group is necessary before one can predict passivity or militancy.

> *Critical Thinking:* In what ways can religion be an "opiate" to oppressed people? What kinds of evidence support this view of religion? In what ways can religion mobilize suppressed people to greater self-respect and to work for social change?

Theodicies and Levels of Activism

The worldview of many sectarian groups is otherworldly. The reality or at least the importance of this world is denied except for its function as a testing ground. Only the faithful will be saved and will reach heaven in the afterlife; only the true believers will have "pie in the sky in the sweet by and by" (as it is sometimes referred to affectionately by believers and derisively by skeptics). This afterlife experience is expected to commence for each individual immediately after death. This sort of otherworldly religion is frequently associated with passivity in this-worldly affairs. Members of the dominant social groups are usually more than happy to have their subordinates believe that vindication will come only after death. Slave owners in this country often had that sort of doctrine preached to their slaves and coupled it with warnings that the saved would be those who lived out their status in this world without causing any trouble.

Another sort of worldview is eschatological. In the eschatological worldview, the ultimate victory over suffering and death will commence at some future time in history. Eschatology may be expressed in either of two forms: (1) progressivism or (2) millenarism. Furthermore, millenarism sometimes has a subtype known as *apocalypticism.*

In the progressive view, the day of perfection will be reached when God and humanity have worked together to attain it. This involves a gradualistic concept of social evolution. The view is based on the idea that God is the creator and rules over the earth. God's master plan is for the evolution of the world into an ever more humane, just, and godly kingdom (the kingdom of God on Earth), but it is believed that God will not establish this without human effort and participation. According to this evolutionary eschatology, trust in God is often equated with trust in the goodness of God's creation. Believers are to look for signs of God at work in this world and are to become actively involved in the material world. Progressive eschatology was characteristic of much of the Social Gospel movement in the United States at the turn of the 20th century. In a somewhat modified form, this sort of outlook—with its positive view of this world—continues to be a force in many mainline denominations and is the predominant view of Reformed Jews.[3] If God expects human participation in creating the "kingdom," it becomes an obligation for believers as well as a way of creating a heaven on earth for all. Therefore, a progressive view of history is often tied to liberal activism by some people who are affluent. The progressive view of history is depicted in Figure 12.1.

---

[3]Actually, this-worldly eschatology is the predominant theodicy in the Hebrew and Christian Bibles. In the entire Jewish Bible (the Christian Old Testament), the idea of life after death is mentioned only four times. Salvation was expected to be this-worldly and was anticipated within history. God was thought to be in charge of creation and of history, both of which were viewed as good. The idea of the soul, as something separate from the body that would live on after the material body died, was an idea introduced by the Greeks and is found in later rabbinical writings and in the Christian New Testament.

The Kingdom of God on earth

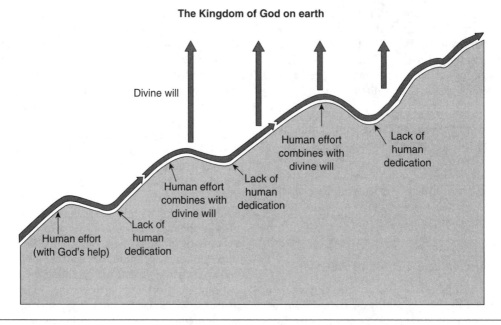

Divine will

Human effort
combines with
divine will

Lack of
human
dedication

Human effort
combines with
divine will

Lack of
human
dedication

Lack of
human
dedication

Human effort
(with God's help)

Figure 12.1   The Progressive View of History

A progressive view of history—with a perception that things are getting better and more fair—does not fit the experiences of the dispossessed and the victims of racial antipathy. Other views of history resonate better with their experiences. Another form of eschatology, millenarism, is more compatible with lives of social outcastes, assuming as it does that the transformation of the world will be sudden rather than gradualistic and will be inaugurated primarily by supernatural powers. Norman Cohn, a historian who has done extensive comparative studies of millenarism, points to five defining characteristics of these movements. The millenarian vision is as follows:

1. Collective, in the sense that it is to be enjoyed by the faithful as a group;

2. Terrestrial, in the sense that it is to be realized on this earth and not in some otherworldly heaven;

3. Imminent, in the sense that it is to come both soon and suddenly;

4. Total, in the sense that it is utterly to transform life on earth, so that the new dispensation will be no mere improvement on the present but perfection itself;

5. Accomplished by agencies that are consciously regarded as supernatural (Cohn, 1964, p. 168).

The word *millennium* means a thousand years and refers to the new era to come. Although the term originates from the New Testament prediction (in the book of Revelation) that Jesus will return and rule for 1,000 years, the word is also used to refer to non-Christian groups with this sort of worldview. Hence, millenarians are those people who await a future event by which the kingdom of God or the new era will begin. Many millenarian Christians have their own life-after-death scenario. This involves a bodily resurrection of the dead at the time the new kingdom begins. This is a rather different concept from the belief that eternal life for an individual begins immediately after death and that the spiritual world is coexistent with the material one. Nonetheless, many people hold some combination of both beliefs and are not much troubled by the need for coherence and consistency in their theodicy.

The concept of salvation for millenarians, then, is time-oriented, terrestrial (although life on earth in the new era will be quite different from present life on earth), and collective (rather than individualistic). Although the transformation will ultimately be accomplished by supernatural forces, humans do have an active and important role in preparing the way. Life in this world is viewed as a time of suffering and of being tested, but the new era will mean the advent of a new social order where justice prevails.[4]

For disadvantaged groups, the millenarian view offers great hope for the future (see the next "Illustrating Sociological Concepts" feature). Groups that hold this view are frequently highly emotional in their religious expression and become fanatical in their efforts to inaugurate the new kingdom. Hence, it is not uncommon for millenarian groups to precipitate active revolt against the established authorities. Yonina Talmon (1965) wrote the following:

> Comparative analysis seems to indicate that, generally speaking, the more extremely millenarian a movement is the more activist it is. . . . There seems to be a correlation between the time conception of each movement and its position in the passivity-activity continuum. Movements which view the millennium as imminent and have a total and vivid conception of redemption are, on the whole, much more activist than movements which expect it to happen at some remote date. . . . It would seem that truly great expectations and a sense of immediacy enhance the orientation to active rebellion while postponement of the critical date and lesser expectation breed passivity and quietism. (p. 527)

ILLUSTRATING SOCIOLOGICAL CONCEPTS

Millenarian Anticipation Expressed in Hymns

These two hymns express vividly the revolutionary and earthly expectations of the millennial vision. The first stresses the collapse of the present social order and the destruction of kingdoms. The second expresses a mood of waiting until the supernatural does its work. Compare the messages to those in the hymns in earlier boxed features.

"Jesus Comes"

Watch ye saints with eyelids waking, Lo, the pow'rs of heav'n are shaking;
Keep your Lamps all trimmed and burning, Ready for your Lord's returning.

(Continued)

---

[4]Each of these theodicies (salvation in an afterlife, progressive eschatology, and millennial eschatology) are within the mainline tradition of Christian theology and are given different emphasis by various denominations. For example, one study revealed that 94% of the Southern Baptists felt that Jesus would definitely return to Earth some day. By contrast, only 13% of the Congregationalists fully believed in that prediction (Stark & Glock, 1968). This is one reason that it is suspect to operationalize orthodoxy only in terms of one of these views and then assume that those who are more "orthodox" are also more religious.

(Continued)

Lo! He comes, lo! Jesus comes; Lo! He comes, He comes all glorious!

Jesus comes to reign victorious, Lo! He comes, yes, Jesus comes.

Kingdoms at their base are crumbling, Hark, His chariot wheels are rumbling;

Tell, O, tell of grace abounding, Whilst the seventh trump is sounding.

Lo! He comes, lo! Jesus comes; Lo! He comes, He comes all glorious!

Jesus comes to reign victorious, Lo! He comes, yes, Jesus comes.

Nations wane, tho' proud and stately, Christ His Kingdom hasteneth greatly;

Earth her latest pangs is summing, shout, ye saints, your Lord is coming!

Lo! He comes, lo! Jesus comes; Lo! He comes, He comes all glorious!

Jesus comes to reign victorious, Lo! He comes, yes, Jesus comes.

"Our Lord's Return to Earth Again"

I am watching for the coming of the glad millennial day

When our blessed Lord shall come and catch His waiting Bride away

Oh! my heart is filled with rapture as I labor, watch and pray

For our Lord is coming back to earth again.

Oh! our Lord is coming back to earth again,

Yes, our Lord is coming back to earth again,

Satan will be bound a thousand years,

We'll have no tempter then,

After Jesus shall come back to earth again.

Then the sin and sorrow, pain and death of this dark world shall cease

In a glorious reign with Jesus of a thousand years of peace;

All the earth is groaning, crying for that day of sweet release,

For our Lord is coming back to earth again.

Oh! our Lord is coming back to earth again,

Yes, our Lord is coming back to earth again,

Satan will be bound a thousand years,

We'll have no tempter then,

After Jesus shall come back to earth again.

Talmon pointed out that millenarism has enjoyed popularity at all levels of society at one time or another—including relatively affluent middle- and upper-middle-class people. Nonetheless, it has normally appealed to deprived people—oppressed peasants, the poorest of the poor in cities and towns, and populations of colonial countries. The millennial outlook usually develops as a reaction to especially severe hardships and suffering. Talmon (1965) wrote, "Many of the outbursts of millenarism took place against a background of disaster—plagues, devastating fires, recurrent long droughts that were the dire lot of the peasants, slumps that caused widespread unemployment and poverty and calamitous wars" (p. 530). In most cases, millenarism is a phenomenon of ethnic groups that have endured sustained subjugation.

For example, when simple tribal societies encounter complex ones and the people in the simpler society are attracted by advanced technologies and tools, there are generated enormously inflated expectations without an adequate development of institutional means for their satisfaction. This was the case in the Cargo cult of Melanesia where exposure to an American military base caused poor indigenous peoples to feel frustrated with their own lack of possessions. Their response was to develop a mystical cult around a flag pole. They believed that if they cracked the mystical marching code of the soldiers, an airplane loaded with cargo for the natives would arrive. The discrepancy between desire and reality is often bridged by millenarian hope. The millennial hope, in turn, sometimes leads to action.

Unlike the progressives who see the present order as good and getting better, millennial movements usually seek total transformation of this world—which they view as unjust or even inherently evil. Many groups that begin with a rational approach become progressively more strategic and more rational (more secular?) if they begin to meet with success. This is illustrated by the tendency, which

Lipset described, for religious movements to spawn radical political movements and the political movements, in turn, to lose much of their religiosity (Lipset, 1960). Of course, for sociologists who insist that a worldview may be religious without being supernaturalistic, these secular political movements are no less religious than their predecessors. They simply have this-worldly and rationalistic systems of faith.

To complicate matters further, there are actually two types of millenarism. Within Christian circles, *postmillenialism* holds that Christ will come to reign over the earth, but only *after* humans have prepared the way. This is sometimes thought to involve a 1,000-year period of justice and peace prior to Christ's arrival. Obviously, this stimulates this-worldly activism and is what has been described so far.

The other type of millenarism is common among the most destitute—those who feel utterly vulnerable in relationship to another group of people. This form emphasizes much more strongly the idea that the world is evil and controlled by Satan. Contrary to the evolutionary eschatology, this view describes human history as being on a hopelessly downward spiral. Ironically, this depressing circumstance is viewed as a sign of hope because it indicates that the end is near. God will intervene in history and bring forth the new era. Nothing humans can do will significantly alter the course of history. The believer can only be ready for the day of judgment, preparing his or her own soul and perhaps engaging in mystical ritual action, such as dancing around a fire (the Ghost Dance), marching around a flag pole (the Cargo cult), or spreading the word until everyone is informed of the noble story (Seventh-day Adventists and Jehovah's Witnesses). This extreme form of millenarism is called apocalypticism, or *premillenialism* (see Figure 12.2). This latter phrase means that Christ will come before the 1,000-year period of justice, peace, and divine rule.

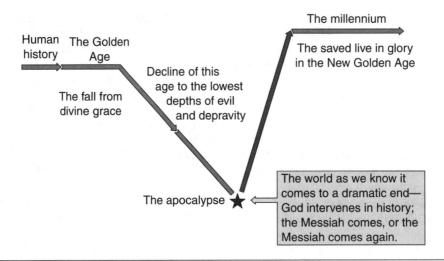

Figure 12.2   The Apocalyptic (Premillenial) Worldview

Within Christianity, apocalypticism is normally based on a literal interpretation of the book of Revelation. Its utter rejection of the present age and present world disallows any attempt to bring change. For this reason, it usually leads to passivity in terms of the social structure. Groups or individuals are more likely to adopt this posture and this worldview if they are powerless and utterly despairing. If some hope exists of social change through human action, the group or individual is more likely to develop a rational strategy.

Christianity, then, has within it several theodicies, which in numerous ways are quite different. Some congregations and some denominations stress one of them exclusively. Most congregations have some people who do not believe in a Second Coming or in an imminent end to the world; rather, they believe in a spiritual world that is coexistent with this world and that is attained by individuals if they have a right relationship with God. Others do not believe in life after death, at least not immediately after death; rather, they expect the millennium to occur sometime in the future. Many Christians believe a little bit in each of these outlooks but have no coherent explanation of how the views fit together.[5] The important point is that, depending on which worldview is stressed among an oppressed minority group, one may expect very different levels of activism or militancy. Those whose hope lies in a coexistent spiritual world that is attained by individual means are frequently passive. Those who hold an apocalyptic view are also usually passive. However, millenarians are frequently activist and militant. For them, religion is not an opiate; it is the inspiration that gives them hope, provides them with vision, and shores up their courage. The

---

[5]Lack of logical coherence or consistency is not uncommon; many people hold more than one worldview, even though those worldviews may be contradictory in many respects. Melford Spiro (1978) pointed out that some Chinese say devotions at both Taoist and Buddhist temples; some Japanese worship the gods of both Shinto and Buddhism; certain Singhalese pay homage to both the Hindu *deva* and the Buddhist Gautama; and many Burmese Buddhists believe firmly in the Thirty-Seven Nats (the folk religion of Burma). In many cases, the outlook on life of two theodicies is utterly different, yet local people claim allegiance to both worldviews.

progressive view, with its suggestion that the present social system is already a good one and is constantly improving, has little appeal to the oppressed. This view is frequently held by social activists in more affluent religious groups.

Judaism also has within it several theodicies. Reform Jews await a "messianic age," which they believe will be established by combined human and divine effort. Orthodox Jews hold to a belief in a coming Messiah who will bring the kingdom of God into being, will reunite the Jews, and will rebuild the temple. This millenarian view has been especially emphasized in times of Jewish history when oppression of Jews was most severe. A belief in resurrection from the dead was introduced to the Hebrews through the Persians nearly 2,000 years ago and is now part of the messianic expectation as well. Another theodicy—the belief in life after death in heaven or purgatory[6] was also introduced by the Greeks. Orthodox Jews affirm this doctrine as well, even though it represents quite a different theodicy from the more this-worldly messianic expectation. Conservative Jews vary a great deal from one congregation to another in which of these theodicies they emphasize.

Reform Judaism (at least in its official form) does not accept the beliefs in resurrection or in heaven and purgatory. Its members are messianic, but they anticipate a messianic age—a time of peace and justice—which will be inaugurated due to the work of God and of many people. They do not expect a single individual to arrive who will solve the world's problems. The confidence of Reform Jews that human action can be effective in bringing significant change is itself a product of a people who are not destitute and powerless. The social circumstances of Reform Jews has allowed for and encouraged this modification of the traditional theodicy. Likewise, the theodicy has justified and encouraged social activism. The Reform branch of Judaism is often noted for being especially prophetic and activist in its attention to social and political change.

The Jews have been subjugated and persecuted throughout much of their history, and the theodicies that they have developed have usually been ones of disprivilege. Sometimes the theodicies have motivated Jews to militancy and activism (such as the times of the escape from Egypt and the Maccabean revolt), and sometimes they have called for passivity—waiting for the Messiah. There can be little doubt, however, that these theodicies have served to bond the Jews together and have helped to sustain them through incredible hardships.

What we have found is that oppressed people frequently adhere to a worldview that is either otherworldly or is millenarian; the answer to life's frustrations is sought in a transformed future or in a different realm of existence. Given the subjugation of women throughout the history of the Western world, the role of women in millenaristic movements is especially interesting. Cohn (1964) pointed out that millennial movements are common when there is a substantial group of wealthy, leisure-class women who are without social function or prestige. He pointed out that a number of millennial reformers during the Reformation were able to survive because they were sheltered and supported by women of the nobility. These women were experiencing extreme status inconsistency; they had high social status in terms of wealth and family political prestige, yet as individuals they had no respected function and could demand little personal prestige or respect. Weber (1922/1963) also noted that women showed a great receptivity to all religious prophecy except that which is exclusively military in

---

[6]Jews believe that no soul is so evil that it deserves a permanent condemnation. Hence, they have no concept of hell. The worst that can happen to an utterly evil person is that his or her soul will cease to exist and the person will not be remembered among the living. A person who has lived a life that is less than holy may spend up to 11 months in purgatory but is eventually united with God. Jewish theology spends much less time on speculations about the nature of the afterlife than does Christian theology.

orientation. He emphasized that prophets challenge the status quo and are usually rather egalitarian in their relationships with women. Jesus and Buddha are both cited as examples of charismatic figures who ignored many traditional gender role norms. The point here is that women—like many oppressed groups—are disproportionately attracted to world-transforming charismatic religious leaders and movements. This should not surprise us since women have actually experienced minority status—a position of less power—across the world.

One conclusion can safely be drawn from our foregoing discussion: The experience of social and political disprivilege can have a significant impact on one's worldview and on one's style of religious expression. Likewise, the worldview and sacred ethos of a group may have a significant effect on how its members respond to the experience of social and political subordination. To understand better the workings of religion in the experience of a specific subjugated group, let us turn to a brief overview of religion in the African American experience.

Critical Thinking: If you were a dictator who had total power—which you did not want challenged—what kind of theological orientation and what sort of theodicy should you encourage among the people? What would be the ideal theological system and the most powerful sacred justification to preserve the status quo? What kind of theology would be most threatening to your power and privilege?

RELIGION IN THE AFRICAN AMERICAN COMMUNITY

The vast majority of African Americans who are religiously affiliated are Christian (85%), and most of those are members of historically black churches (69%). Although the identification of African Americans with the big seven historically black denominations has decreased over the years, two thirds of the members of historically black churches still identify with the Baptist tradition. Churches in the Methodist tradition, as well as Pentecostal churches like the Church of God in Christ, also constitute a large share of the black church.[7] Nonetheless, one must be cautious about making generalizations about African American religion in the United States, for there is tremendous diversity of religious expression among African Americans. There is a strong local autonomy emphasis in the independent black denominations, and this means that local churches within the same denomination may vary considerably in theology and in style of ritual. Of course, there is also wide variation in the white denominations, but these denominations tend to be somewhat more centralized. The local bias of black churches tends to facilitate local variations and to exaggerate diversity within a denomination.

In Cincinnati in the 1930s, only 10.6% of the population was black, but black churches accounted for 32% of all churches in that city. The same pattern was true in Detroit and Philadelphia (Wilmore, 1972). More recently, in his study of the predominantly African American Four Corners neighborhood of Boston in the 1990s, Omar McRoberts (2003) found 29 churches in a half square mile area. Studying the 98% black Eastside neighborhood of Charleston, South Carolina, in

---

[7]Data are from the Pew Research Center Forum on Religion & Public Life (2008b). The seven most prominent historically black denominations include three Methodist ones (the African Methodist Episcopal (AME), the AME Zion, and the Christian Methodist Episcopal) and four Baptist associations (National Baptist Convention, National Baptist Convention, U.S.A., National Baptist Convention of America, and the Progressive National Baptist Convention). The proportion of African Americans who are members of these seven denominations has declined some over the years, with more African Americans claiming nondenominational church membership and no religious preference in recent years (mirroring the American public). For more information and a history of these black churches, see Lincoln and Mamiya (1990).

the same time period, Timothy Nelson (2005) identified 15 active black congregations in an area 1 mile long and a half mile wide. These large numbers illustrate the high degree of divisiveness and separation that often exists in the black religious community.

Perhaps the most important variation in black religiosity is that due to socioeconomic standing. The lower-class religious groups are characterized by emotionalism and fundamentalism, and the minister is unlikely to have had any formal theological training. On the other hand, the religious expression of the black professional class is quite similar to the religiosity of the white middle class. It is characterized by orderly and rational worship conducted by a well-educated, theologically trained minister. Beyond the variations of Christianity in the African American community, there also exist a large number of black religious cults. Regardless of variations, we can still safely say that African American religiosity tends to be more emotional in character than white religiosity (Chaves, 2004; Nelson, 2005).

We are able to discuss the existence of black religion as a distinct phenomenon in large part because the vast majority of African Americans do belong to all-black churches. This is not due to theological differences but to the castelike nature of American society relative to African Americans. During the time of slavery, black Christians were required to occupy the balcony while whites were seated on the main floor. This allowed slave owners to keep track of their slaves and ensure that religious meetings were not used to incite rebellion. Gradually some "trusted" black preachers were allowed to meet separately with slaves to have religious services. This was virtually the only official leadership role slaves were allowed, and it should not be surprising that the African American pastor enjoyed tremendous prestige and occupied the primary leadership position in the black community for more than two centuries (Lincoln & Mamiya, 1990).

However, it was free blacks who actually founded the first independent all-black denominations.[8] In Philadelphia, two African American clergymen, Richard Allen and Absalom Jones, started the AME Church in 1794, and a group of free blacks in New York started the AME Zion Church in 1820 (Wilmore, 1972). The segregation in the white churches was unbearable for free blacks. To Allen and Jones, it seemed to be a direct contradiction of the Christian faith. Furthermore, predominantly white denominations were very little concerned about the needs of African Americans. The ability of all-black churches to minister effectively to the social and religious needs of their people resulted in a tremendous increase in the membership of these denominations and eventually in the spawning of others. The result is that most African American Christians have continued to this day to worship in segregation from European American Christians. As we noted in Chapter 10, "the most segregated hour of Christian America is eleven o'clock on Sunday morning." It is precisely this separateness that makes it possible to speak about black religion and the black church as a distinct entity.

The Unique Character of the Black Christian Experience

There are two different schools of thought regarding the origins and central character of African American religion. One group of scholars believes that the conversion of blacks was a final step in obliterating any remnant of African culture among the slaves. The experience of being torn from their homelands and their families and being involuntarily relocated on a new continent was a stunning experience for the first generation of African Americans. Furthermore, slaveholders frequently had policies that blacks from the same African culture or who spoke the same African language were not to be placed on

---

[8]The first black Baptist churches predate the Methodist ones—in 1758 the African Baptist or "Bluestone" Church was founded in Mechlenberg, Virginia, but the Baptists did not organize a denominational network or structure until long after the AME and AME Zion churches were established.

the same plantation. Hence, no common language, religion, or culture could enhance communication and solidarity among slaves. In fact, members of enemy societies were sometimes thrown together in the most unhappy of circumstances.

At first, the white masters refused to allow preachers access to their slaves, for most Christian denominations maintained that a Christian could not own another Christian. Hence, if a slave converted, the master would either have to give up his or her own church membership or would have to free the converted slave. For this reason, slave owners prohibited proselytization. Because many Christian missionaries wanted to preach to the Africans, they gradually compromised their position: Conversion did not automatically require manumission. Eventually, slave owners found that religion could be a powerful tool for controlling slaves, for they could use the aura of sacredness to reinforce desirable behavior patterns: submissiveness, industriousness, and obedience. By providing blacks with a worldview that is profoundly otherworldly, slave owners hoped to replace the last vestiges of African hopes for freedom with a sacred system that the whites could control.

E. Franklin Frazier (1957, 1963) and Arthur Fauset (1944) first developed the thesis that the Christianization of blacks was the final step in the deculturation of Africans. However, they also pointed out that these African converts used the imagery of Christianity to forge a religious expression appropriate to their own needs. In this view, Christianity became functional for the slaves in that it established a common base for unity and solidarity among otherwise disparate peoples. This view of African American religiosity stresses the fact that slave religion was a synthesis of white religion and black experience, and it developed its own unique character and history. The emotionalism that is characteristic of African American churches is attributed to the fact that it was mostly Baptists and Methodists who evangelized the slaves. Because the revival-meeting style of those denominations was highly enthusiastic and emotional, it is not surprising that black Christianity was also revivalistic. Furthermore, poorly educated and economically impoverished groups are frequently more emotional in religious expression than are more educated and affluent coreligionists. This would also add to the highly emotional tenor of African American religion (Baer, 1984).

Other scholars, such as Melville Herskovits (1958), Gayraud Wilmore (1972), Joseph Washington (1972), Peter Williams (1980), and Eric Lincoln and Lawrence Mamiya (1990), have insisted that slave religion was influenced by certain patterns of religiosity that are common to many African religious ceremonies. According to Herskovits, the tendency to turn to religion rather than to political action to alleviate frustrations is typical of African cultures. Moreover, the rhythm and motion that characterize the singing, preaching, and congregational responses in black churches is also common in Africa. Hence, the style of religiosity of the African American church is viewed as a survival of previous cultural patterns in much the same way that the sentence structure of "ebonics" (the African–English dialect) is viewed by linguists as a survival of African languages. The specific religious belief systems are granted to be a product of the new world and of contact with white missionaries, but the mode of expression is viewed as uniquely African.

The debate continues between those holding each of these views. We cannot expect to resolve the issue here, but we do want to highlight the significance of the alternative views. For those who deny that African culture had any real influence on slave religiosity, the characteristic features of the African American church are thought to originate in socioeconomic subordination. If this view is correct, the emotionalism of black religiosity may fade, even in all-black churches, as members of those congregations improve their socioeconomic standing.

On the other hand, African American religiosity will continue to have a different flavor from European American religiosity if the variation is rooted in ethnic differences that are preserved in segregated religious organizations. There is empirical evidence to support both views, but we do not believe that one can fully understand the religiosity of the first slaves without some understanding of their previous experiences of religion. Through the transmission of religion from one generation to the next, the remnants of African religiosity have been passed down to the contemporary black church. Only history will be able to answer the question about whether assimilation to white American culture and changes in socioeconomic status will result in a more subdued and rationalized religious style among blacks. Our expectation is that the unique, expressive style of the black church will continue.[9]

Regardless of the causes of black religious patterns, there is a consensus that black religion in the United States does have a unique character (Cone, 1972; Johnstone, 2001; Lincoln & Mamiya, 1990; Washington, 1964, 1972; Wilmore, 1972; Winter, 1977). In fact, Washington referred to black Christianity in the United States as a form of folk religion. First of all, the Christianity that was preached to blacks was a truncated and manipulated version of Christian theology; it was designed to help pacify and compensate them for their inferior position in life (see the next "Illustrating Sociological Concepts"

feature). Furthermore, religion became the means by which slaves could express their frustrations and their hopes—both of which emanated from their subordinate standing in society. Other scholars point out that all world religions adapt to local needs and that while African American Christianity has its unique character, it is no different than other expressions of Christianity in adapting a local flavor. Like Christianity in Poland, Brazil, or Anglo America, African American Christianity has both folk elements and strong universalistic themes (Lincoln & Mamiya, 1990).

> *Critical Thinking:* Which argument is more convincing to you: African American religious style is rooted in religious patterns and styles of Africa or it is shaped by socioeconomic circumstances? Why is this position more persuasive to you?

ILLUSTRATING SOCIOLOGICAL CONCEPTS

Excerpts From a Catechism for American Slaves

These questions from a catechism designed for slaves illustrate the way Christianity was twisted to serve the interests of the powerful group. Several denominations established special catechisms for slaves which were, at best, truncated interpretations of the faith.

Question: What did God make you for?

Answer: To make a crop.

Question: What is the meaning of "Thou shalt not commit adultery?"

Answer: To serve our heavenly Father, and our earthly Master, obey our overseer, and not steal anything.

SOURCE: Wilmore, Gayraud S. (1972). *Black religion and black radicalism*. Garden City, NY: Doubleday.

---

[9]It is interesting to note that middle-class black theologians, such as James Cone and Gayraud Wilmore, continue to support the emotionalism of the black church. Although they are not personally impoverished, they do not conform to the subdued style of worship of white churches. Most black preachers—despite their economic standing—take pride in the distinctiveness of black preaching and the emotional expressiveness in black worship services.

Religion of the Oppressed and Coded Messages: The Black Spiritual

One of the characteristic expressions of this African American religiosity was the black spiritual. Surprisingly, few African American spirituals were Christ-centered; in fact, many do not even mention God (Wilmore, 1972). The message of most spirituals was an expression of hardship and a hope of freedom. Those spirituals that do focus on Jesus stress his suffering, his experience of being scorned, and his role as liberator. Some spirituals were based on biblical stories (such as "Joshua Fit the Battle of Jericho"), but many were commentaries on contemporary events. "Oh, Lord, What a Morning, when the Stars Begin to Fall" emerged right after Nat Turner's insurrection (1831) when slaves were under the tightest scrutiny. The slaves hoped for that day when the Apocalypse would come, the revolution would be successful, and the sky would fall on slave owners. Washington insisted that slaves used the vocabulary of white ministers and of the Bible and that whites believed the slaves were being socialized in the values that owners wanted. However, content analysis of these hymns, combined with reports from former slaves on the role of the spirituals, has suggested a different interpretation. Washington (1964) wrote the following:

> The popular view that [the] spirituals are of Christian origin is based upon the preponderance of otherworldly themes, Biblical words, and the instruction and messages of the missionaries. These were the tools the [slaves] had at hand, but this view assumes the credulity of the slave. It overlooks the awareness of [blacks] that religion was methodically used to hold them in check, and their capacity to use it for other purposes than worship. Thus, the distinction between spirituals being forged from materials presented by Christians and forged from the Christian faith itself is essential . . . . (p. 218)

Scholars have found that many spirituals were, in fact, "code songs" that communicated one thing to blacks, while white masters sat by—content that their slaves were getting a heavy dose of otherworldly religion. For example, the spiritual "Let Us Praise God Together, on Our Knees," which is included in the hymnbooks of many mainline white denominations and many sects, was actually a call to a secret meeting of slaves at dawn. The chorus of that spiritual is as follows: "When I fall on my knees *with my face to the rising sun,* Oh, Lord, have mercy on me." Likewise, when a slave working in the fields began singing "Steal away, away home; I ain't got long to stay here," he or she was indicating *this-worldly* intentions to other slaves. Participation of the other slaves in the chorus was a way to wish the person well and a promise to try to cover for the slave's absence as much as possible. Some of the spirituals were rather thinly veiled codes, such as the one that went like this:

> I am bound for the promised land;
>
> I am bound for the promised land;
>
> Oh who will come and go with me?
>
> I am bound for the promised land!

"Canaan, Sweet Canaan" did not point only to an otherworldly realm, but it referred to Ohio, Indiana, Illinois, and even Canada. Similarly, references to the Jordan River usually meant the Ohio River. "Swing Low, Sweet Chariot" provides an example. When the Underground Railroad was ready to take another group of escapees north, African Americans could let others at a worship gathering know about it without giving themselves away to white attendants who came to ensure that nothing subversive happened at these religious gatherings. Someone would begin to sing, with great emotion:

> I looked over Jordan [the Ohio River] and what did I see;
>
> Coming for to carry me home;
>
> A band of angels [Harriet Tubman or another conductor of the Underground Railroad] coming after me;

Coming for to carry me home [freedom in the North]

Swing low [deep into the South] Sweet Chariot [the Underground Railroad]

Coming for to carry me home.

The slaves at those worship services understood the symbolism and double meanings very well. (Cone, 1972). In some cases, it is hard to know whether a particular spiritual was otherworldly in its meaning or a code song. Some spirituals, like "When the Saints Go Marching In" had a definite otherworldly character. Other songs had a here and now double meaning, but they were not necessarily calls to action. A biblical theme was being rehearsed, but contemporary characters were clearly identified with historical figures in the story. The spiritual "Go Down, Moses" emerged at the time when Bishop Francis Asbury of the Methodist Church was instrumental in formulating antislavery planks in the Methodist code of discipline. Asbury had himself referred to South Carolina as "Egypt" when he had been there to preach (Washington, 1964) and that theme was expressed in the chorus of this popular hymn:

When Israel was in Egypt's land, let my people go;

Oppressed so hard they could not stand, let my people go.

Go down, Moses, way down in Egypt land;

Tell old Pharaoh, "Let my people go!"

* * * * * * *

Oh let us all from bondage flee, let my people go;

And let us all in Christ be free, let my people go.

Go down, Moses, way down in Egypt land;

Tell old Pharaoh, let my people go.

It doesn't take much imagination to figure out who represented "Moses," "the Israelites," and the "Egyptians" in the eyes of those slaves.

The use of religious language for coded communication is certainly not a new phenomenon, for there are other reports of oppressed people communicating in a similar manner. Most biblical scholars believe that the Apocalypse (the book of Revelation) was a coded message from John the Elder (a prisoner on the Island of Patmos) to his people in the churches of Asia Minor. At that time, one was required to worship the emperor of Rome. Because Christians refused to do so, they were persecuted. The book of Revelation is an encoded book that is very difficult to translate because many of the symbolic meanings of that day have been lost. An understanding of the double meanings requires fluency in Aramaic and Greek. The book was not destroyed by the Romans, for they viewed it as a harmless fantasy about another world. They never recognized the political references that abound in the book and that served as a resounding criticism of Rome.

All of this simply serves as a warning against facile generalizations regarding the otherworldliness of the religion of the oppressed. Of course, some blacks did understand Christianity in otherworldly terms, and it served to compensate them and to discourage any rebellion in this world. They believed they would get their just desserts in the next world. The important point here is that slave religion was not a simple adoption of white Christianity; it was a reworked Christianity that had its own character, style, and outlook. Much of the black church today has been influenced by this heritage. Even where the message is otherworldly, political issues have never been entirely foreign to black churches. Most black Christians today feel that it is utterly appropriate to use the faith community for political purposes: 92% of African American clergy surveyed nationally support the idea of churches expressing their opinions on social and political questions, 91% of black laity believe it is appropriate for clergy to participate in civil rights protests, and nearly a third of the black churches house civil rights organizations and/or voter registration programs. The African American church has historically been much more involved in political affairs than its white counterpart (Lincoln & Mamiya, 1990; Wilmore, 1972). It has been a source of activism.

> ***Critical Thinking:*** Slave owners used religion to manipulate and control slaves, while slaves used religion as a coded system to help them with escape plans. Is it at all troubling to you that religion was used as a tool to prevent or to stimulate change? Why or why not?

The Leadership Role of the African American Minister

One reason the African American church has been involved in political matters has been the fact that the black preacher was the main spokesperson for the African American community. During the period of slavery, the role of plantation preacher was often the only leadership role afforded southern blacks. Hence, it became a position of considerable prestige within the African American community. Following emancipation, this position continued to be the most important leadership role; the black preacher became the spokesperson for the community and the liaison between the dominant white class and the subordinate black one. Because whites often owned the buildings where blacks worshipped and could impose various sanctions on the black community, the black preacher had to be sensitive to the interests of both whites and blacks. This liaison role was certainly not just religious in character; it was often explicitly political.

The preacher's role as the central spokesperson for the community has largely survived to this day. The African American church has served since emancipation as the heart of that ethnic community. It has sponsored social and cultural affairs, established insurance programs for members who did not qualify for insurance under white-controlled corporations, started schools and colleges to educate young people, sponsored political debates held in church sanctuaries, initiated economic recovery and growth programs for African Americans, and generally served as a community center. In fact, Gunnar

Myrdal (1944) called the African American church a "community center par excellence" (p. 938). The preacher was the person who gave impetus to most of these programs and thus came to be highly esteemed in the community.

Because the preacher held a position that afforded leadership opportunities and offered status in the community, his was a highly coveted position. Hence, there have often been young would-be preachers waiting in the wings to have their chance to preach and start their own congregations. Perhaps this is another reason for the large number of small black churches, each with its own semiautonomous preacher. The ministry was attractive to energetic African Americans because other professions were essentially closed to them. However, there has been a significant decline over the past 50 years in the ratio of blacks in the ministry (per 10,000 people in the population). Increasing numbers of African Americans now enter law, medicine, politics, and other professions. The black minister no longer holds a monopoly on leadership and status as was once the case. Nonetheless, the preacher still holds a more substantial position within the African American community than does his or her white counterpart. Many African American political figures have begun their careers as ministers.

Partially because of the acceptance of clergy being involved in political affairs, ministers in the black community were able to gain ready acceptance as civil rights leaders. Martin Luther King Jr. provides a particularly good example. King's civil rights speeches were constructed and delivered in the style of African American preaching. His nonviolent resistance strategy of the 1950s and 1960s required resistance to injustice but forbade participants from using violence. He insisted that blacks would change social structures by appealing to the conscience of the nation and by economic boycotts. If a white police officer struck a black protestor, African Americans were to resist the temptation to strike back. They were instructed to turn the other cheek and to love their enemies. They were told to hate injustice but not the person who perpetuated it. They were taught that love would be the weapon by which opponents would be transformed (King, 2010). The mixture of religious teaching and political action did not seem at all inconsistent to King's followers, for they were used to African American preachers also being political figures.

It is hard to imagine how the many rallies that King led could have remained nonviolent without the influence of religious teachings. Moreover, when people's homes and churches were being bombed and rallies were dispersed by police officers swinging billy clubs, religion served to shore up the courage and conviction of the people, for they were assured that God was on the side of justice and ultimate victory was certain. Some African American critics of nonviolence felt that King's religion was another form of opiate and that blacks should fight back. Nonetheless, there is little doubt that the civil rights movement of the 1950s and 1960s gained much of its impetus from a black Baptist preacher—Dr. Martin Luther King Jr. Clearly, the role of the minister has been a central one in the activism of the African American community.

The role of the pastor was not the only leadership role that the black church has afforded. In a society where African Americans have often held menial and low-status jobs, the roles of elder, deacon, Sunday school superintendent, or choir leader have provided leadership opportunities and respected positions in the community. Often these positions have been given exalted and lengthy titles. Such roles in exclusively black denominations enhance the self-esteem of individuals who have rather humble standing in the larger society (Baer, 1984). This self-confidence and these leadership skills, again, are often used in the service of activism for change.

New Religious Movements in the African American Community

African American religion in the United States has also included many black sects and new religious movements (NRMs). There are a great variety of them, each emerging out of the common black experience of subjugation but each offering its own unique characteristics

(Fauset, 1944; Lincoln, 1994; Washington, 1964, 1972; Wilmore, 1972). Two of the best-known African American NRMs were Father Divine's Peace Mission and the Nation of Islam (Black Muslims). The Father Divine movement started in 1932 when a man by the name of George Baker opened a mission in Harlem. He took the name of Major J. Devine and quickly gained notoriety by distributing alms among poor blacks. He taught that God is everywhere, everything, and everyone, and eventually his followers came to believe that Devine was God incarnate. Baker then came to be known as Father Divine. Under his direction, the mission developed into a communal living settlement and maintained extremely high standards of morality. Although the group was highly ascetic in character, the emphasis was not otherworldly. It was directed toward changing the socioeconomic system. The strategy of change focused primarily on benevolence for the poor rather than radical political

Dr. Martin Luther King and his wife, Coretta Scott King, lead a civil rights march in Alabama. Black clergy have considerable social status and are often looked to for leadership on political and social justice issues within the larger community. This is another way in which a religious group—if it is an open system—can have "output" on the larger society.

changes. Because of its moderate stand, it did not incur the opposition of powerful conventional organizations. The movement spread across the country and was one of the larger African American NRMs in the United States.

The Lost/Found Nation of Islam was more radical in its outlook and its activism, and it has had considerable influence in many of America's prisons—where much of the proselytization took place. The movement started in the summer of 1930 when an Arab peddler, known as Wali D. Fard, came to the ghetto of Detroit. He sold silks and other materials and preached that black people in Africa and the Middle East were Islamic, not Christian. Christianity was depicted as the religion of the white people with a white God and a white savior. Being a Christian was equated with worshiping white people and was described as the white society's way of duping blacks into subordinate roles.

Fard insisted that whites are incapable of telling the truth, and he sought to tell the real story of black civilization in Asia and Africa. He told fantastic stories about black culture on other continents. All African Americans were depicted as Muslims in their origins and were referred to as the "Lost Tribe of Shebazz." Many African Americans were delighted with this stranger's stories of sophisticated and advanced black culture and his insistence on black superiority. It provided a basis for a sense of dignity and pride, which was often denied to poor ghetto blacks. Furthermore, the preacher's claim of Arab heritage lent credibility to his claims of firsthand knowledge of Africa and Arabia.

Fard developed a substantial following—estimated at 8,000 adherents. One of his devotees was a dynamic African American whom Fard renamed Elijah Muhammad. (Black Muslims refused to accept the names they received from slavery and were given a new name when they joined the Islamic temple.) When Fard mysteriously disappeared in 1934, Elijah Muhammad was named Minister of Islam. One of Muhammad's main disciples, in time, was Malcolm X.

The Black Muslims originally insisted that only black people could join. This political–religious group experienced its heyday in the 1960s. Malcolm X became the primary spokesperson and gained national attention as a militant civil rights leader, being among the first to stress "black pride" and "black is beautiful" as central themes. Propelled by his forceful and articulate speeches, the Black Muslim movement grew rapidly. Attention was also focused on the movement when such notable sports personalities as Muhammad Ali and Kareem Abdul-Jabbar converted to the Nation of Islam.

The Black Muslims have been a religious group that is as thoroughly political as it is religious. For them, the distinction between politics and religion is meaningless. They explicitly reject any otherworldly views, so the socioeconomic–political structures of this world are of central importance to their view of "salvation." In fact, the ultimate goal of the original "Nation of Islam" was an autonomous and separate black nation in the United States. (Whites, they believed, belong in Europe.) The movement advocated a program of social, economic, and political segregation of blacks from whites. Because the "original man" was declared by Allah to be black, whiteness meant a lack of purity and truth. In short, whiteness was a sign of evil.

The larger Islamic world did not recognize the legitimacy of the original Black Muslim theology. In terms of official Islamic orthodoxy, the teachings of the Lost Nation of Islam were heretical. In short, the Lost Nation of Islam was a folk religion, which had grown out of the experience of black America (Washington, 1972). However, to call such a movement a folk religion is not to denigrate its importance as a religious movement. Scholar Eric Lincoln has described the Black Muslims as one of the most important religious developments in 20th-century America. Because of the ascetic teachings, the emphasis on industry and hard work, and the extreme sacrifice and devotion to the cause, the Lost/Found Nation mobilized a significant amount of financial and personal resources on its behalf. Moreover, because of high levels of internal discipline, Muslim ministers have also been able to deliver a significant block of votes to politicians. By acting as a unified front, they were able to make their presence felt in the larger society (Lincoln, 1994). They became a force for change.

Malcolm X formally broke with Muhammad in 1964 after a trip to Mecca. On that trip, he learned that the teachings of the Black Muslims were quite different from Orthodox Islam. He established the Muslim Mosque, Inc., that same year but was assassinated in 1965. Before he died, he repudiated the views of whites as intrinsically evil and introduced a movement toward a more orthodox Islamic faith. When Elijah Muhammad died in 1975, his son Warith (Wallace) Deen Muhammad became the new leader of the Nation of Islam. He modified doctrine to bring it more in line with Sunni Islam and changed the name of the group twice. It is now known as the American Society of Muslims. This mission was recognized in the Islamic world as a branch of orthodox Sunni Islam and its North American membership has been estimated to be as high as 100,000. A splinter group continues to call itself the Nation of Islam and is led by Minister Louis Farrakhan, having a following of somewhere between 20,000 and 70,000. Farrahkan continues the black separatist teachings of Wali Fard and Elijah Muhammad (Lincoln, 1994; Lincoln & Mamiya, 1990).

In earlier chapters, we discussed the fact that sectarian movements are more common among socially and economically disfranchised groups. Because blacks in this country are disproportionately represented in this category, it is not surprising that African American religion is characterized by a large number of sectarian movements. Movements like the Black Muslims and the Father Divine Mission actually comprise a relatively small percentage of the religiously affiliated African Americans in this country. Although 24% of Muslims in America today are black, only 1% of blacks are Muslim (Pew Research Center Forum on Religion & Public Life, 2007b). Nonetheless, they express an alternative mode of coping with the experience of being a subjugated minority.

African American Religion: Present Themes and Future Trends

James Cone (1969, 1970) and Gayraud Wilmore (1972) are among many black theological scholars who stress the eschatological theme in Christianity and de-emphasize the otherworldly one. Cone and Wilmore are among a core of black theologians who have been articulating a "liberation theology," an emphasis on social and economic liberation at some future time in history. Otherworldliness is viewed by Cone, especially, as an opiate to black people. The true worldview of Christianity, according to these theologians, is eschatological. In fact, a Christianity that is not supportive of black power is viewed as not being Christianity at all. Liberation theologians insist that because Jesus advocated for the poor and the oppressed, so also must all true Christians. This has also led black theologians to seek common voice with peoples who are oppressed in other parts of the world. This sort of unified and liberation-oriented Christianity would certainly incline believers toward activism rather than passivity. It is noteworthy that Cone finds a rejection of otherworldliness and an endorsement of eschatology (millenarism) to be a first step in making the black church an effective tool of social and political change.

Even a brief look at African American faith communities is enough to demonstrate that the religion of any group is affected by the socioeconomic status of its members. Furthermore, the outlook the religion fosters may motivate people to seek change, or it may enhance acquiescence. If African American theologians who teach at major seminaries—scholars like Cone and Wilmore—are any indication of what the future will be, black ministers are likely to continue to be more politically involved than their European-American counterparts. Moreover, if these theologians do set a course for the black church (and that is a big "if"), then the black church is likely to be increasingly an inspiration to social and economic militancy rather than an opiate. However, one empirical study indicates that while many African American pastors are interested in instilling ethnic pride and identity, the majority are not actively asserting liberation theology, as such, in their preaching (Lincoln & Mamiya, 1990). Indeed, as we saw in Chapter 9, the "prosperity gospel"—a theology that seems to legitimate the American status quo—is particularly strong in African American congregations, notably the megachurches of Creflo Dollar and T. D. Jakes.

> *Critical Thinking:* Do you think that the black intelligentsia will ultimately have much influence on local black churches? Why or why not?

SUMMARY

In providing meaning in life, theology must address the real issues that face people. Because the problems of meaning in people's lives are different for the affluent and the disfranchised, it is not surprising that theodicies are different. The theodicy of privilege tends to justify one's good fortune, whereas theodicies of despair tend to provide certain psychological compensations for and a sense of victory over one's adversities. Even the basic values and sense of morality tend to be different. Those in occupations where obeying rules is fundamental to success are more comfortable with religious systems that are precise and absolute about moral expectations. Those who are in fields where creativity and divergent thinking are the keys to occupational success are more comfortable with relativity, ambiguity, and open-endedness in systems of meaning.

Religions of the disfranchised can either cause passivity and inaction in the face of social injustice, or they can inspire people to work for change. Theologies that are either otherworldly or apocalyptic (premillennial) tend to cause passivity, whereas postmillennial theologies tend to inspire activism.

African American religion provides an interesting example. Slave Christianity often inspired change and was used to transmit coded messages for escape, despite the efforts of slave owners to control slaves with religion. Further, the black church has often provided roles and leadership opportunities that enhanced self-esteem. African American pastors often became the central leaders in civil rights struggles and were the spokespersons for the African American community in the political arena. This opportunity for leadership by the black pastor made the role so attractive that a number of aspiring charismatic blacks started their own sects and NRMs as channels to gain prestige and power within their own community.

Clearly, religion is correlated to socioeconomic status, and theology often serves as an ideology that is appropriate to one's economic circumstances.

PART VI

SOCIAL CHANGE AND RELIGIOUS ADAPTATION

One of the realities of the modern world is social change. Some changes occur so rapidly that we can observe them over a few years. Others are long term and have been so gradual that individuals hardly see any change within their own lifetimes. Religion can be a source of stability and certainty in times of change, but some people also claim that if religion does not change with the times, it will become alien to the lives and experiences of the people. In these last three chapters, we explore ways in which religion has responded to shifts and modifications in the larger society.

In Chapter 13, we discuss the theory of secularization that has guided sociological research on religious change since the discipline's inception, as well as criticisms and revisions of that theory. In Chapter 14, we examine the various ways in which religion manifests itself "outside the (God) box." That is, we explore evidence of some alternative forms of religion that seem to be emerging in our rapidly changing society and that may be replacing traditional expressions of religion. In Chapter 15, we consider the religious implications of globalization—the trend toward the world becoming "one place."

13

SECULARIZATION

Religion in Decline or in Transformation?

Here are some questions to ponder as you read this chapter:

- What is *secularization*?
- Is secularization inevitable? Is it even happening?
- How is secularization linked to other social processes, such as pluralism?
- What is *civil religion*, and how does it function in a society? What are some different forms or styles it can take?
- What are some significant stages in the evolution of civil religion in the United States?
- Are the consequences of secularization destructive or beneficial to a democratic society?

In earlier chapters, we discussed the role of myth and explored the importance of worldview in religion and in the society at large.

Until the 19th century, religion—in virtually every society—fundamentally shaped the worldview of the entire culture and profoundly influenced most other social institutions. In our contemporary world, other forces come into play, especially the advent of science as an institution, the greater sophistication of technology, and the higher levels of formal education in the general population. The scientific method and rational–legal analysis become the arbitrators of truth. Further, other institutions in the society appear to become less dependent on religious legitimation. The concept of secularization was developed originally by Max Weber (1958a) and is closely tied to ideas of rationalization and bureaucratization that were part of his *The Protestant Ethic and the Spirit of Capitalism* (see Chapter 9).

Industrialized societies tend to generate their own worldviews, independent of religious myths and symbols, and these secular worldviews may even begin to shape the traditional religious systems of belief. This may spell conflict not only between religious groups and the larger society but also within religious groups themselves. Rationalization or *modernization* is a long-term process in the classical theories.

TRADITIONAL MODELS OF SECULARIZATION

Traditional ideas of secularization held that secularization is of considerable import, but various models emphasized different processes. One point of agreement seems to be that traditional religious symbols are no longer a unifying force for industrialized society and religious ways of knowing are no longer given priority in all circumstances. The modern empirical scientific worldview has largely replaced the miraculous religious worldview, especially in institutions that are not explicitly religious. Truth and knowledge are gained through double-blind, experimental studies rather than through nonrational visions or religious intuition. People look to science and technology to solve problems. As a means of compensating, some clergy and philosophers have developed systematic theologies that are logical, coherent, and rational. Mythology has given way to empiricism, both in the secular world and within some religious groups. Another point of agreement is that this change of worldview has been accompanied by structural changes in society, with institutions (family, government, economics, religion, education, science, health care, etc.) becoming separate and differentiated bodies, and with various institutions gaining greater autonomy from religious authority.

We might, then, define secularization as transformation of a society involving (1) greater *institutional differentiation* and organizational autonomy from religious authority and (2) a more rational, utilitarian, and empiricist outlook on life and on decision making by individuals and groups. When the same sort of transformation occurs *within* a religion—either in the form of a theology that emphasizes logical reasoning and acceptance of scientific method and discoveries or as an organizational change using rational, utilitarian business practices in running the institution—it is sometimes called modernization.

That a process of change has taken place in the worldview and social organization of the Western world is unquestioned.[1] How to understand and evaluate it is more problematic. A number of writers in the middle of the 20th century felt that this meant the decline of religion or even danger to the entire society. Others maintained that secularization is a healthy process that will strengthen the influence of religion. They acknowledge religious change but do not see this change as decline. To provide some sense of the traditional secularization debate, we will begin by exploring the issues as outlined by

---

[1]Secularization has different effects in different cultures and even on different religions in the same society. For a discussion of political and other sociocultural variables, see Martin (1969, 1978). For a discussion of secularization in Judaism, with its emphasis on orthopraxy (conformity to ritual and ethical behaviors) rather than orthodoxy (conformity to beliefs), see Sharot (1991).

several key theorists who represent very different points of view.

Secularization as Religious Decline: Loss of Sacredness and Decline of Social Consensus

Peter Berger (1967, 1979) was very much alarmed by the process of secularization early in his career. In the 1960s, he depicted this process as having many adverse effects on individuals and on the society as a whole. Berger (1967) defined secularization as "the process by which sectors of society and culture are removed from the domination of religious institutions and symbols" (p. 107).

In Berger's view, the function of religion in providing unifying symbols and a unifying worldview is extremely important. He insists that a society's *constructed world* (his term for worldview) is very fragile. It must be protected by being clothed in an aura of sacredness. The critically thinking empiricist allows that nothing is sacred; that is, nothing is beyond question and analysis. The world construction of the scientist is based on causality and logic. Because individual thinking is valued, the scientifically oriented society allows and even encourages a plurality of worldviews. In fact, Berger (1967) wrote that "the phenomenon called pluralism is a social-structural correlate of the secularization of consciousness" (p. 127). However, pluralism is not necessarily a good thing in his view. A multiplicity of worldviews may make all worldviews seem relative. None of them can be taken as absolute and above doubt, and this raises anxiety about life's meaning and purpose (Berger, 1979). This lack of meaningfulness in life can be extremely disorienting to individuals, for critical thinking about one's worldview "was bought at the price of severe anomie and existential anxiety" (Berger, 1967, p. 125).

The dysfunctions of allowing pluralism of worldviews and of reducing the realm of the sacred is not limited to individuals. Pluralism also poses threats to the stability and workability

of the entire society. Berger is a functional or consensus theorist; he believes commonly held beliefs, values, and symbols are the glue that holds a society together. Religion has always played this role, and Berger wrote with alarm that the absolute and uncompromising imperative of religion has been relativized. Because of this, individuals become aware of the fact that there is a plurality of possible religious views—each potentially legitimate—from which they must choose. The fact that one consciously selects a religious orientation (rather than being compelled by the conviction that there is only one possible view) automatically makes the choice relative and less than certain. Berger did not view this situation as one in which the individual is *free* to choose—an option now available to individuals. Rather, each person *must* choose; that is, one is coerced into doing so. Berger (1979) called this the "heretical imperative," because the Greek root of heresy (*herein*) means "to choose." The net effect, he believed, is a diminishing of the power of religion in the lives of people. To use Clifford Geertz's (1968) phrase, it is the difference "between holding a belief and being held by one" (p. 17).

Berger (1967) concluded that religion must either accommodate, "play the game of religious free enterprise," and "modify its product in accordance with consumer demand," or it must entrench itself and maintain its worldview behind whatever socioreligious structures it can construct (p. 153). A religious group that takes the first course tends to become secularized from within and to lose its sense of transcendence or sacredness. It focuses on "marketing" the faith to a clientele that is no longer constrained to "buy." In the process, the faith may be severely compromised and changed. By contrast, if a religion fails to accommodate, the religion may be charged with being an "irrelevant" minority faith that does not respond to the needs of the society.

Berger described the effects of secularization as primarily destructive. The unifying power of religious symbols and the integrating function of a common religious worldview—a "sacred canopy"—are very important for social stability.

The lack of a single worldview, couched in sacred symbols and an aura of absoluteness, is a fundamental concern for those approaching secularization from this point of view.

Finally, for Berger this secularization process was viewed as so powerful and sweeping that it was pretty much irresistible and inevitable in the Western world. Like it or not, secularization was the wave of the future.

This was his position in the 1960s, and it represented well the traditional view of secularization. In 1999, Berger reversed his position regarding the trend being religious decline, as we will see later. Still, his articulation of secularization was one of the more thorough statements of secularization theory for several decades.

Secularization as Religious Evolution and Development: Increased Complexity of Thought and Greater Religious Autonomy

Talcott Parsons (1964) and Robert Bellah (1970c) have treated secularization as part of the increasing complexity and diversity of modern industrial society. Religion, both as institution and as individual belief system, has become increasingly differentiated from the rest of society. According to Parsons and Bellah, it has simultaneously become more of a private matter. However, they do not see this as a negative change. The possibility of consciously choosing one's religious outlook, rather than being inculcated with a theology, may make religion more important to the individual. Parsons insisted that Christianity might still have a great effect on Western society; after all, the institutions of the Western world were developed by people who were under the influence of the Christian ethos and the Christian worldview. Moreover, private religiosity will continue to affect public behavior; ultimate values cannot help but affect an individual's "system of action." In this view, much of the influence of religion will be unconscious. Greeley (1972) put it this way:

Modern man, working in a large corporate structure that is infused somewhat by Christian ideas, with himself directed somewhat by the Christian ethic, may be behaving religiously a good deal of the time, though the fact that he is influenced by religion may not be immediately clear to him because the influence is both indirect and implicit. (p. 134)

Bellah would agree with Berger that religious institutions exert less direct influence on secular institutions than in the past, but he explained this change as a process of religious evolution rather than decline. Bellah (1970c) used a very inclusive definition of religion—"a set of symbolic forms and acts which relate [people] to the ultimate conditions of [their] existence" (p. 21). His focus on evolution involves an increase in the complexity of symbols over time, which is correlated with an increase in the complexity of social organization. Through this evolution, religion is able to do more than reaffirm the present social structure; it can challenge the current norms and values of the secular society and offer an alternative culture. Hence, the latter stages of evolution represent an increased autonomy of religion relative to its social environment. Bellah did not suggest that the evolutionary process is irreversible or unidirectional. He simply suggested a general trend of increasing differentiation of religion from the rest of the society. On the other hand, the trend is rooted in large-scale structural changes in the society, so one might reasonably predict that it will continue in democratic societies with high levels of educational attainment.

Bellah specified five stages of religious evolution: (1) primitive (e.g., Australian aborigines), (2) archaic (e.g., Native American), (3) historic (e.g., Ancient Judaism, Confucianism, Buddhism, Islam, Early Palestinian Christianity), (4) early modern (e.g., Protestant Christianity), and (5) modern (religious individualism). He argued that beginning with the single cosmos of the undifferentiated primitive religious worldview in which life is a "one possibility thing," evolution in the religious sphere is toward the increasing differentiation and complexity of symbol systems.

In the modern stage of religious evolution, the hierarchic dualistic religious symbol system that emerged in the historic epoch has collapsed and the symbol system that results is "infinitely multiplex."

In the midst of this transformation, new forms of religiosity are emerging. These new forms are less dualistic (with the material world which is evil opposing spiritual existence which is good) and more this-worldly but still involve a symbol system that "relates people to the ultimate condition of their existence." The new expressions of religion involve spirituality but not necessarily other-worldliness. Bellah (1970c) wrote, "The analysis of modern [humanity] as secular, materialistic, dehumanized, and in the deepest sense areligious seems to me fundamentally misguided, for such a judgment is based on standards that cannot adequately gauge the modern temper" (p. 40). The attempts by several key social scientists to discover "invisible religions" is in keeping with this emphasis on new forms of privatized religion (Luckmann, 1967; Wuthnow, 1973, 1976b; Yinger, 1969, 1977) and will be explored in the next chapter.

Religious evolution is seen by Bellah as an increase in the complexity of symbol systems and in the complexity of the organization of society. The later stages of the evolutionary process allow individuals to make choices about which worldview they will accept. Furthermore, individuals have more autonomy in being able to think for themselves and to create their own system of meaning. In this post-traditional situation, the individual confronts life as an "infinite possibility thing" (p. 40), as Bellah (1970c) put it. Each person is "capable, within limits, of continual self-transformation and capable, again within limits, of remaking the world, including the very symbolic forms with which he deals with it . . . the forms that state the conditions of his own existence" (Bellah, 1970c, p. 42).

This allows for greater freedom and in this sense may be viewed as an *advancement* for humankind. Several other researchers have pointed out that the idea of a "religious marketplace" can mean inexhaustible variety rather than watering down and marginalizing of religion (Finke & Stark, 1992; Stark & Bainbridge, 1985). As Stephen Warner (1993) put it, when it comes to religious commitment, achievement may be far superior to ascription.

Parsons and Bellah each treated secularization as the process by which religion has become a private matter, and they believed that such secularization is taking place. Bellah would later give a prime example of religious privatization in *Habits of the Heart: Individualism and Commitment in American Life*, a best-selling book he wrote with four coauthors. Bellah, Madsen, Sullivan, Swidler, and Tipton (1985) discussed a young nurse they interviewed, Sheila Larson (actually, a pseudonym), who told them the following:

> I believe in God. I'm not a religious fanatic. I can't remember the last time I went to church. My faith has carried me a long way. It's Sheilaism. Just my own little voice. (p. 221)

Noting that Sheilaism raises the possibility of over 200 million religions in America—an "infinite possibility thing" indeed—the authors conclude that "'Sheilaism' somehow seems a perfectly natural expression of current American religious life." (p. 221).

However, if secularization is referred to as a process by which religion decreases in importance and by which it has less influence on one's worldview and on social behavior, then Bellah and Parsons would deny that secularization is occurring. Differentiation of worldviews and pluralization of the culture are evaluated in positive terms. They represent progression of society rather than regression and dissolution.

Many theological liberals have de-emphasized the absoluteness of biblical authority (at least as literally understood) but have emphasized that all life is to be understood in theological terms. Liberal theologies tend to reformulate the tradition so that it will be in harmony with the most recent scientific findings and other humanistic values in contemporary society. Scientific evidence may even be considered a legitimate starting point for "doing theology."

Furthermore, a norm of logical consistency and coherence becomes a measuring stick for truth. Such theologies are rational meaning systems; they sometimes involve recasting biblical perspectives so that they are consistent with the modern scientific worldview. For example, one strand of Methodist theology—developed by theologians Borden Parker Browne, Albert Knudson, Edgar Brightman, and Harold DeWolf—maintains that because God is the creator of all of life and all that exists in this world, we must remain open to truth wherever it leads. Defending old dogma when new scientific methods prove the dogma wrong is irreligious, in this view. They insist that scientific methods were merely discovering what God had created (Brightman, 1940; Burrow, 1999, 2006). This sort of interpretation, which invites rational and empirical outlooks into the heart of the faith, is referred to as theological modernism. Process theology also emerged in the 20th century, at a time in which social change seems to be the only certainty. One of the fundamental principles of process theology is that change is indisputable, inescapable, and eternal. Not even God is viewed as immutable and absolute but rather as changing and evolving.

Such theologies tend to be intentional in assimilating the religion to the values and perspectives of the larger society. Bellah and Parsons are more in tune with this more liberal strain of theology, and they are not inclined to think of it as less religious or as a sign of the decline of religion. Secularization and spirituality are *not* seen as opposites.

THE "NEW" PARADIGM

"Secularization" has been part and parcel of the sociological lexicon from the discipline's inception. Indeed, the assumed validity of the secularization perspective was so taken for granted for so many years that it became "part of the conventional sociological wisdom" (Lechner, 1991b, p. 1103). This conventional wisdom has been challenged in a series of insightful, if highly controversial, analyses by Rodney Stark, William Bainbridge, Roger Finke, and Stephen Warner, among others (Hadden, 1987; Iannaccone, 1995; Warner, 1993). Indeed, the idea that secularization is *not* inevitable now seems to have become the conventional position—a "new paradigm," as Stephen Warner (1993) has called it—at least in the United States. Even Peter Berger (1999) claimed that "the assumption that we live in a secularized world is false. The world today, with some exceptions . . . is as furiously religious as it ever was, and in some places more so than ever" (p. 2). Indeed, Rodney Stark (2000b) now claims that secularization theory is dead—may it rest in peace.

Although the "new paradigm" advocated by Warner (1993) cannot be reduced to a rational choice perspective, it is the case that the most severe critics of the "old" secularization paradigm have been rational choice theorists (Yamane, 1997; Young, 1997). Rational choice theorists begin with the observation that "humans seek what they perceive to be rewards and try to avoid what they perceive to be costs" (Stark & Bainbridge, 1985, p. 5). They add that some of the rewards are scarce and unequally distributed within society and some desired rewards (such as life after death) may not be available from the society at all. When humans do not have access to desired rewards, they seek compensators. "A compensator is a belief that a reward will be obtained in the distant future or in some other context which cannot be immediately verified" (Stark & Bainbridge, 1985, p. 6).

The affluent in society are able to satisfy most of their material wants and needs with empirically available methods. Those who are at the lower end of the socioeconomic hierarchy may desire compensation for their *inability* to meet their material needs, including everything from food and shelter to health care. Compensators who address needs that could be satisfied through empirical means but are denied because of scarcity or social competition are referred to as *specific compensators*. Supernaturalism that is geared exclusively to specific compensators is a form of "supernatural technology," a method of manipulating nonempirical forces in the service of this-worldly needs. (In Chapter 1, this orientation to the supernatural was called "magic" rather than religion.)

Stark and Bainbridge believe that religion, properly understood, deals only with *general compensators*. Yet they also point out that most religious organizations in the past and a great many in the present offer a mixture of general and specific compensators. Religious groups are being discredited, they believe, because empirical scientific methods are disproving many specific compensator claims (for example,

see the "Illustrating Sociological Concepts" feature). This creates a problem of plausibility for many religious groups. Doubts about the ability of supernatural methods to satisfy everyday empirical needs have caused skepticism about religious claims to satisfy general or nonempirical needs (e.g., hope for life after death). Stark and Bainbridge insist that science will ultimately drive out magic because modern science and technology prove more effective in addressing this-worldly needs. However, science can never replace religion, per se, because empirical methods are incapable of addressing nonempirical (spiritual) needs (Stark & Bainbridge, 1985).

ILLUSTRATING SOCIOLOGICAL CONCEPTS

Scientific Discovery and the Plausibility of Religious Claims

Magical explanations about how to gain a desired reward (or avoid a damaging cost) will tend to be discredited by scientific test and to be discarded in favor of scientifically verified explanations. This tendency has serious consequences for religions that include a significant magical component. Consider the case of the lightning rod. . . . For centuries, the Christian church held that lightning was the palpable manifestation of divine wrath and that safety against lightning could be gained only by conforming to divine will. Because the bell towers of churches and cathedrals tended to be the only tall structures, they were the most common targets of lightning. Following damage or destruction of a bell tower by lightning, campaigns were launched to stamp out local wickedness and to raise funds to repair the tower.

Benjamin Franklin's invention of the lightning rod caused a crisis for the church. The rod demonstrably worked. The laity began to demand its installation on church towers—backing their demands with a threat to withhold funds to restore the tower should lightning strike it. The church had to admit either that Ben Franklin had the power to thwart divine retribution or that lightning was merely a natural phenomenon. Of course, they chose the latter, but, in so doing, they surrendered a well-known and dramatic magical claim about the nature of the supernatural. Such admissions call into question other claims made by a religion, including even those that are eternally immune from empirical disconfirmation.

SOURCE: Stark, Rodney, & Bainbridge, William Sims (1985). *The future of religion: Secularization, renewal, and cult formation.* Berkeley: University of California Press.

> **Critical Thinking:** Is it true that religion and science address very different purposes and that one cannot ever fully replace the other? *Why* do you think as you do?

These "rational choice" theorists also maintain that the effectiveness of the scientific view of the world has created a problem for many traditional religious organizations. People in modern societies have come to depend increasingly on scientific empirical methods as sources of validation rather than on supernatural explanations (as the case of Ben Franklin's lightning rod indicates). Many religious leaders have coped with the scientific revolution by lowering tension with the secular (this-worldly) culture: They adopt a more secular perspective. They develop a theology that is consistent with scientific empiricism, often by shying away from supernaturalism altogether. In the 17th century, European scientists like Galileo were forced by the Christian Church to recant their findings, but "in this century it has been the religious intellectuals, not the scientists, who have done the recanting" (Stark & Bainbridge, 1985, p. 434). These rational choice theorists believe that when religious organizations abandon supernaturalism, they are only able to offer very weak general compensators (Finke & Stark, 1992; Stark & Bainbridge, 1985). The result of secularity even within religious systems is not a society of empirical secularists who eschew any sense of supernaturalism, but a

> **Critical Thinking:** Stark and Bainbridge say that a few centuries ago scientific innovators had to recant before ecclesiastical tribunals but that today it seems that theologians recant and adapt the findings of the empirical sciences. Do you agree? If this is so, what caused the shift in power?

restoration of more otherworldly expressions of religion. These come in the form of renewals of traditional faiths (sects) or in the formation of entirely new religious movements (NRMs).

A key contribution of this critical analysis is that two misconceptions have been put to rest: (1) There clearly is no wonderful mythical past when everyone was religious, and (2) religion is not in an inexorable retreat from those good old days. As we will see later in this chapter, the religious membership statistics in the United States suggest that affiliation with religious organizations has *not* been declining over the past 200 years. There has also been little decline in England. In Britain in 1800, 12% of the population belonged to a religious organization; by 1990, 17% belonged. During the entire year of 1738, only 5% of the population took communion in the Oxford diocese in England. This supposedly was the part of the grand past when people were more devout than today (Stark, 2000b). The next "Illustrating Sociological Concepts" feature provides other evidence cited by Stark that the days of profound piety were not so reverent after all.

The movement away from supernaturalism and toward a purely secular outlook is not an inevitable process that spells a substantial and long-term decline of religion, as some forms of secularization theory had implied. Indeed, for Stark, it is not clear that supernaturalism has declined over the past few centuries at all. The inability of secularism to provide general compensators (a sense of meaning in life, the hope for an afterlife, and so forth) means that secularization is self-limiting. Secularization does not mean a decline of religion, the rational choice theorists tell us, but it may entail a reformation of traditional religions and the spawning of NRMs. In the region of the continent where traditional forms of religion are weakest (the Pacific Coast), secular humanism does not abound; cults do. They develop precisely to fill the vacuum of general (supernatural) compensators. Stark (2000b) concluded that secularization is no longer a useful concept. Vigorous religious group identification, affirmation of beliefs in the

ILLUSTRATING SOCIOLOGICAL CONCEPTS

Irreligiosity and Theological Ignorance in the "Good Old Days"

Rodney Stark believes that not only were most people not highly religious in centuries past but the clergy themselves were often ignorant of basics of Christian theology, ritual, and scripture. In this passage, Stark illustrated the low level of functioning within Christian churches. This kind of evidence, he said, indicates that the "good old days" of religious piety were not so good after all.

That religious participation was lacking [in fifteenth and sixteenth century Europe] is not very surprising when we realize that going to church . . . required the average person to stand in an unheated building to hear a service which was conducted in incomprehensible Latin by priests who may indeed not have been speaking Latin at all, but many of whom were mumbling nonsense syllables. The Venerable Bede advised the future bishop Egbert that . . . few English priests and monks knew any Latin . . . William Tyndale noted in 1530 that hardly any of the priests and curates of England knew the Lord's Prayer or could translate it into English. This was confirmed when in 1551 the Bishop of Glouster systematically tested his diocesan clergy. Of 311 pastors, 171 could not repeat the Ten Commandments and 27 did not know the author of the Lord's Prayer. Indeed, the next year Bishop Hooper found "scores of parish clergy who could not tell who was the author of the Lord's Prayer, or where it could be found." . . . [It is clear from the evidence] that church officials thought most serving clergy knew considerably less than a modern 10-year-old attending parochial school [does today].

It must be noted, too, that when people back then did go to church they often did so unwillingly and behaved very inappropriately while there . . . Presentations before ecclesiastical courts and scores of clerical memoirs report how "members of the population jostled for pews, nudged their neighbors, hawked and spat, knitted, made coarse remarks, told jokes, fell asleep, and even let off guns." Church records tell of a man in Cambridgeshire who was charged with misbehaving in church in 1598 after his most loathsome farting, striking, and scoffing speeches had resulted in . . . the great rejoicing in the bad [behavior]." A man who issued loathsome farts in church today surely would not draw cheers from part of the congregation (material in quotes from Keith Thomas 1971: 159–162).

SOURCE: Stark, Rodney. (2000b). Secularization, R.I.P. In W. H. Swatos Jr. & D. V. A. Olson, *The secularization debate* (pp. 41–66). Lanham, MD: Rowman & Littlefield.

supernatural by large numbers of people, and high weekly attendance rates are among his strongest evidence.

At the macro level, rational choice theorists advocate a "religious economies" approach. According to this model, a religious economy consists of all the religious activity going on in any society. Religious economies are like commercial economies in that they consist of a market of current and potential customers, a set of firms seeking to serve that market, and religious "product lines" offered by various firms (Stark & Bainbridge, 1985, 1996). Also like commercial economies, religious economies thrive when they are allowed to operate without government interference. Finke (1990) summarized the logic of the model: Deregulation leads to pluralism, pluralism to competition, competition to specialization of product (catering to a market niche) and aggressive recruitment, specialization and recruitment to higher demand, and higher demand to greater participation. Thus, as a "natural" consequence of the invisible hand operating unencumbered by state regulation, "over time the diversity of the religious market will reflect the very diversity of the population itself" (Finke, 1990, p. 622).

This basic model has been applied in several different studies, some of which take the unit of analysis for comparison to be the nation-state (Chaves & Cann, 1992; Iannaccone, 1992; Stark,

The oldest Jewish synagogue in the United States is in Newport, Rhode Island, where separation of church and state and tolerance of other religious traditions was a founding principle. After George Washington was elected president of the new nation, he received a letter from that early Jewish congregation in Newport asking about his policies of pluralism. In response in 1790, Touro Synagogue received a handwritten letter signed by President Washington (prominently displayed in the synagogue) embracing an open and "liberal" policy to all American citizens, regardless of origins or religious affiliation. In this letter, George Washington affirmed a policy of religious pluralism from the very beginning of the country's existence as a nation.

1992) and some of which use subnational units of analysis such as cities (Finke & Stark, 1988) or dioceses (Stark & McCann, 1993). In the breakthrough article for the economics of religion, Finke and Stark (1988) tested perhaps the central hypothesis derived from the religious economies model, namely that "religious pluralism" contributes to higher levels of religious participation (what they call "religious mobilization"). Using quantitative data from the 1906 Census of Religious Bodies, Finke and Stark studied the impact of "adherence" (their indicator of mobilization) on "pluralism" and found a positive relationship. While Finke and Stark (1988) and others provided empirical support for the religious economies perspective, the data on which they are built do not give any evidence for a trend over time. Thus, the empirical jewel in the religious economies crown is the award winning book, *The Churching of America*. Among other things, in this book Finke and Stark (1992) argued that between 1776 and 1990, religious "adherence" in the United States grew from 17% to 60% and that this linear, upward slope is exactly the opposite of what is predicted by secularization theory.

Thus, contrary to British scholar Steve Bruce (1999) and to Peter Berger's thesis of the 1960s, rational choice theorists argue that pluralism in the society actually makes the religious market competitive, and therefore invigorates religious participation. They do not believe it undermines plausibility or commitment. Recent research, however, has questioned the positive connection between pluralism and participation in the religious economies model. This controversy is discussed in the "Controversies in the Field" feature.

Critical Thinking: Some scholars maintain that ascription (birth into a religion where no options exist) makes for stronger religious commitment; others argue that achievement (choosing one's faith in a competitive marketplace) is best for religion. With which position do you agree? Why?

CONTROVERSIES IN THE FIELD
Secularization and Pluralism

One of the hotly contested points in the secularization debate is whether pluralism undermines religious commitment by making the faith position seem relative and less than certain—as Berger argued—or whether pluralism creates competition between religious groups—as rational choice theorists assert. The latter argue that religious pluralism creates more options for people so that they can choose from an array of religious products. Further, pluralism generates more vitality and energy among "religious entrepreneurs" as each tries to recruit members. The competition makes the entrepreneurs hungry and aggressive, thereby finding new niches in the market. Pluralism prevents religious leaders from becoming complacent. The latter happens where competition is missing. Pluralism, therefore, creates religious vigor according to rational choice theorists.

In 1988, Roger Finke and Rodney Stark published a major article in support of the idea that pluralism and participation are positively related. They examined data from the 1906 U.S. Census of Religious Bodies on the 150 largest cities in America to test the following hypothesis: "The more pluralism, the greater the religious mobilization of the population—the more people there who will be committed to a faith" (Finke & Stark, 1988, p. 43). Their independent variable, pluralism, was measured using a religious diversity index that accounts for the number and size of different denominations. Using advanced statistical analyses (called multiple regression models), Finke and Stark found a strong, positive relationship between religious diversity and religious participation. From this they criticized the secularization thesis.

Kevin Breault (1989a) responded to Finke and Stark's work using more recent data—the 1980 Glenmary data on Churches and Church Membership—in which he found the exact opposite: "a highly significant, consistently negative relationship between religious pluralism and religious participation" (p. 1049). In a comment on his article, Finke and Stark (1989) rejected Breault's findings. They noted that they asked a colleague—fellow rational choice theorist Laurence Iannaccone—to replicate Breault's statistical models using the same Glenmary data and that Iannaccone found a highly significant *positive* relationship between pluralism and participation (correlation of 0.21). For his part, Breault (1989b) replied with a defense of his methodology and conclusions. He, too, recalculated the pluralism index and religious adherence rates and again found a relationship of almost exactly the same magnitude as Iannaccone's, only *negative* (correlation of -0.22). The exchange ended at an impasse.

Almost a decade later, something very interesting happened. Another sociologist interested in the debate over pluralism and participation, Daniel Olson (1998), tried to replicate the findings and found his results exactly in line with Breault's: a negative relationship (correlation of -0.22) between pluralism and participation. How could Finke/Stark/Iannaccone and Breault/Olson come to exactly the opposite conclusions using the same data and methods? Olson explained that when he inspected the statistical analysis software program files that Iannaccone used—which was provided to Olson by Iannaccone in the spirit of scientific objectivity and empiricism that we discuss in Chapter 2—he discovered a simple mathematical error in the programming. Without going into all of the details (interested readers can see Olson, 1998, p. 760), Iannaccone's analysis in support of

(Continued)

(Continued)

Finke and Stark mistakenly yielded the exact opposite of the actual finding. The relationship between pluralism and participation in the 1980 Glenmary study was in fact negative. In their response to Olson's finding, Finke and Stark (1998) marshaled other studies in support of their perspective and suggested that Olson's single finding was not typical.

Does the preponderance of the data support a positive or negative relationship between pluralism and participation? By extension, does the evidence support the secularization or new paradigm perspective on the effect of pluralism on religion more generally? Mark Chaves and Philip Gorski (2001) have done a secondary analysis of 193 empirical tests of the relationship. After careful critique of the methods of research in each study, they concluded that the large majority of studies indicate that pluralism in itself does *not* increase religious vigor or commitment in most social settings. On the other hand, competition due to a plurality of religious groups does seem to increase religious commitment in *some* situations. More research is needed to understand the circumstances that create a growth situation and those that do not. Still, this review of an extensive body of empirical literature does indicate that no dependable *general law* can be supported that identifies pluralism as a consistent cause of religious vitality.

NEOSECULARIZATION THEORY

Although Rodney Stark has boldly declared secularization theory dead, many other scholars hold to a new version of secularization theory that they claim is well supported by the evidence. These scholars argue that decline of faith, of attendance at worship services, or of membership in religious communities are inadequate measures of a complex phenomenon that occurs at several levels (Beyer, 2000; Chaves, 1993, 1994; Chaves & Gorski, 2001; Dobbelaere, 1981, 2000; Yamane, 1997).

Secularization at Various Levels of Social Analysis

Secularization, according to sociologist Mark Chaves (1993, 1994), is a process that can occur at the individual (micro) level, the organizational (meso) level, and the societal (macro) level. The extent of secularization varies at each level and in various social contexts. Indeed, in the United States secularization may be occurring at the societal level but not at the individual level (Chaves & Gorski,

2001). It is important to note that secularization is conceived as the declining scope of religious authority; that is, the range of issues to which religious authority applies is narrowing as secularization occurs.

Thus, when individuals do not believe that their faith is relevant to everyday life—to their conduct on the job, in political choices, in regard to sexual behavior, in the science classroom, or in the attitudes toward racial relations—we might say that secularization at the individual level has occurred. Within faith-based agencies (Baptist hospitals, Presbyterian colleges, Jewish social service foundations), if the decisions about how to deliver services or who will be hired or fired are based on systematic policies designed for organizational efficiency, then we might say that meso level secularization is present within a religious organization (Chaves, 1993). If policies in the society at large are made with little discussion of theological implications—decisions being based on human rights arguments rather than on what is sinful—then we might say that the society has become secular at the societal level (Chaves, 1993; Dobbelaere, 1981, 2000; Yamane, 1997).

Macro Level Secularity

Although secularization theory views religion on three levels of analysis, the most important level of analysis is the macro level (Tschannen, 1991). The neosecularization paradigm emphasizes the centrality of institutional differentiation at the societal level. Institution differentiation refers to the process by which "specialized institutions develop or arise to handle specific features or functions previously embodied in, or carried out by, one institution" (Wallis & Bruce, 1991, p. 4). As a consequence, in a highly differentiated society, the norms, values, and practices of the religious sphere have only an indirect influence on other spheres such as business, politics, leisure, and education (Wilson, 1976). It is for this reason that we can point to differentiation as leading to a decline in the scope of religious authority: Specifically religious institutions have only a limited (or no) control over other institutional spheres.

Debates about prayer in schools, court decisions about "one nation under God" in the Pledge of Allegiance, and involvement of religious leaders in political issues suggest that societal-level secularization is a point of controversy in the United States. It is interesting, however, that the nature of various policy debates seems to be increasingly secular—based on rational consideration of social consequences and civic rights and responsibilities rather than on matters of sin or scriptural authority. As a Moral Majority lobbyist told researcher Allen Hertzke (1988), "We can't afford to say, 'God settled it, that's it'" (p. 196). Televangelist Jerry Falwell also argued that in discussing abortion, the issue must be debated in terms of rights rather than theological imperatives: "We are reframing the debate. It is no longer a religious issue, but a civil rights issue" (quoted by Hertzke, 1988, p. 196). This framing of issues in terms of rational–legal authority rather than charismatic or traditional–religious authority suggests considerable secularization at the societal level. The discourse following the terrorist attacks on 9/11, was also interesting, for media accounts often depicted governments that were not secular as being the most dangerous. Being "secular" was equated in the media with being democratic, rational, civil, and governed by law.

Meso Level Secularity

Secularization at the meso level occurs when secular transformations take place within religious organizations. For this reason, Chaves (1993) also called this "internal secularization." At this level of analysis, we see religious authority playing a diminished role in controlling the resources of religious organizations (including their core values and practical priorities), particularly the agency structures of religious organizations (e.g., religious boards, associations, lobbying arms, fundraising units, schools, hospitals).

The secularization of higher education provides a good example of meso level secularization. Most private universities in the United States began as religiously sponsored institutions. Prominent examples include such Ivy League schools as Harvard (Calvinist), Yale (Congregationalist), Princeton (Presbyterian), Brown (Baptist), and Dartmouth (Congregationalist), as well as other prestigious universities like Chicago (Baptist), Duke (Methodist), Vanderbilt (Methodist), and Wake Forest (Baptist). All of these institutions are now secularized in terms of their organizational structure and curriculum (Marsden, 1994).

In the United States, some religiously related colleges, especially within fundamentalist or evangelical traditions, may make personnel decisions and curriculum decisions based on theological issues rather than on standard bureaucratic procedures and guidelines designed to determine functional competence. In so doing, they continue to assert religious authority in organizational deliberations. This would indicate resistance to meso level secularization. However, in a society that has experienced considerable institutional differentiation of religion from education, the resistance to internal secularization can come with a cost: diminished academic reputation (Burtchaell, 1998). Of nationally prominent universities in the United States, few maintain a strong religious identity. Exceptions include Baylor (Baptist), Notre Dame (Catholic), Georgetown (Jesuit/Catholic), Brigham Young (Mormon), and Pepperdine (Churches of Christ), but none of the 20 private institutions with the highest measures of academic reputation in the *U.S. News and World Report* ranking had a strong religious identity (Mixon, Lyon, & Beatty, 2004).

Similarly, many other hospitals and social service agencies founded by religious bodies operate on the basis of management and administrative procedures used for any modern organization that has similar goals. Some religious agencies take pride in the fact that they are secularized in their operations—that they use conventional business practices and nonreligious organizational theories to operate.

Earlier in this chapter, we discussed the fact that liberal religious communities and theologians tend to adapt the faith system to accommodate science and other components of the modern world. In Chapter 8, we examined megachurches that focus on "marketing" the faith. An explicit marketing of religion was very much a part of 19th- and early 20th-century evangelism (Frankl, 1987). Further, contemporary churches and televangelists who tend to employ marketing strategies for growth also tend to be rather conservative in their theologies, but they have adapted the principles of scientific management and the business world to the administration of their congregations and sometimes have adjusted the music and theology based on what would "sell." This, in its own way, is a different form of secularization of religious organizations at the meso level—in this case done by conservative groups (Frankl, 1987; Miller, 1997; Sargeant, 2000).

Micro Level Secularity

At the individual level, what needs to be assessed in terms of secularization is the orientation people have to religious authority structures. A secularized society is one in which people will feel free to believe and act in ways which disregard, differ from, or even go against the prescribed views of religious authority structures. People's views and behaviors will be characterized by autonomy and choice. Echoing Sheila Larson of *Habits of the Heart* fame,

supermodel Cindy Crawford has given a very succinct statement of modern religious autonomy, reminiscent of Sheilaism: "I'm religious but in my own personal way. I always say that I have a Cindy Crawford religion—it's my own" (Spitz, 1992, pp. 110–116).

Although we have an abundance of survey data on individuals' religious beliefs and practices, surveys rarely ask citizens whether and how religion affects their everyday decision making. One bit of longitudinal data that is useful comes from the "Middletown" studies. Stark and colleagues often cited evidence from Middletown as disconfirming secularization theory. For example, Finke and Stark (1988) pointed out that "in 1931 there was one 'house of worship' for every 763 residences of Muncie, Indiana (sociology's famous Middletown). By 1970, there was one church or temple for every 473 residents—a pattern of growth that applies across the nation" (p. 47). Stark and his colleagues failed to consider other findings from the Middletown studies, however, such as Caplow, Bahr, and Chadwick's (1983) data that suggested individual-level secularization may be occurring alongside the growth pattern in ratio of religious organizations per capita. The Middletown data in Table 13.1 reveals a *truncation* of the applicability of religious beliefs beyond the narrowly religious sphere. This indicates a decline in the scope of religious authority at the individual or micro level.

OVERVIEW

The primary assertion of the new secularization theorists is that some form of secularization is occurring at societal, organizational, and individual levels. However, there are several important caveats. First, neosecularization theorists do not necessarily think that religion is declining. They believe it is changing. Second, most such theorists concede that religion continues to be a powerful force at the individual level and that it even has some influence in the larger culture (the

Table 13.1 Religious Beliefs of "Middletown" High School Seniors, 1924 & 1977

Belief	% Agreeing		
	1924	*1977*	*% Change*
Christianity is the one true religion, and all people should be converted.	94	41	−53
The Bible is a sufficient guide to all problems of modern life.	74	53	−21
It is wrong to attend movies on Sunday.	33	8	−25
Evolution is more accurate than [the book of] Genesis.	28	50	+22

Adapted from Caplow, T., Bahr, H., & Chadwick, B. (1983). *All faithful people: Change and continuity in Middletown's religion*. Minneapolis: University of Minnesota Press.

macro level).[2] Third, they maintain there is a trend toward secularization in most Western societies but that the trend is neither inevitable nor uniform in all societies and at all levels. This drift toward secularization may be counteracted by other social processes. Fourth, for most neosecularization theorists, secularization is *primarily* about reduction of influence of religion at the *macro* or societal level, not the individual level (Dobbelaere, 2000; Lambert, 2000; Martin, 1978; Sommerville, 2002; Wilson, 1982; Yamane, 1997). Dobbelaere (2000) wrote, "religiosity of individuals is not a valid indicator of societal secularization" (p. 36). Secularization, he said, is primarily about separation of the state, business, education, health care, science, and other realms of society from ecclesiastical and biblical authority.

On the other side of the argument, Rodney Stark, the most strident critic of secularization, has insisted that secularization is mostly about individual religious commitment. He grants that societal level secularization—understood simply as the liberation of other institutions from the dominance of religious institutions—has indeed occurred (Stark, 2000b). Of course, secularization at the *most* macro level—global—is just beginning to be analyzed (explored in the "Global Perspectives" feature).

One point raised by neosecularization theory is especially troubling. The claim by Chaves, Yamane, and Dobbelaere is that secularization may occur at the macro and meso levels while not really occurring at the micro level. Sommerville (2002) described this as "a secularization of things [structures and processes], but not of people" (p. 371). This may be true, but one would expect considerable strain and even overt conflict in the society in such a situation. One wonders how long such tension may continue. The United States seems to have just such strain, expressed in the conflict between conservatives and liberals on public issues about prayer in schools, reference to God in the pledge, prayer at the outset of high school athletic events, and the presence of the Ten

---

[2]Some European sociologists, like Britain's Steve Bruce (1995, 1999, 2002), are convinced the secularization process is an overwhelming force in much of Europe (certainly in England) and that individual religious participation shows severe reduction over time—much higher than in the United States.

GLOBAL PERSPECTIVES
Secularization and Global Perspective

The relatively recent emphasis on viewing social issues from a worldwide perspective raises some interesting questions for those interested in secularization. Canadian Sociologist Peter Beyer points out that the largest unit of social cooperation is not the nation, but the globe as an interactive human enterprise. Thus, we might ask whether religion has an influence on policy decisions and policy deliberation at the global level. Clearly, if secularization were interpreted as religion having no role and no influence, it would be obvious that secularization has not occurred globally. The global spread of Pentecostalism as a Christian movement in the last decade of the 20th century has been phenomenal. Moreover, certain Islamic extremists have defined their faith as relevant to international relations, as the events of 9/11 made clear. Religion is hardly irrelevant across the globe.

Having said this, we must also conclude that at a global level, no particular theological authority has power to define reality or to determine policies. The use of language to describe developed or disadvantaged societies is itself interesting. If a country has weak political or economic systems, poor educational attainment, poor health care institutions, and unsophisticated technologies, most people and global organizations would describe this country as "disadvantaged" or "underdeveloped." However, would the same happen if the *religious* organizations were weak or deteriorating? Peter Beyer (2000) does not think so. Further, when it comes to the deliberations of global agencies, international courts of justice, or the world trade organization, the guiding principles are based on utilitarian or functional rationality; for Weber, this was at the core of modernization. Thus, while the extent of secularization seems to be variable at national, organizational, and individual levels, Beyer (2000) concluded that "Religion . . . is *optional* at the level of global society" (p. 90). This is another way of saying that at the most macro level of human interaction, religious authority structures are minimal, and secularization is well established.

Perhaps this is another reason why conservatives of nearly every religious stripe are leery of global processes and global organizations, like the United Nations; our global organizations are governed by rational–legal or rational–functional authority.

Commandments in courthouses and city buildings. How will this stress be resolved? The trend toward individualization of religion (for example, individuals saying they are "spiritual but not religious") may be one type of solution, since it involves separation of individual "spirituality" from "religious authority." In the view of neosecularization theorists, this is secularization at the individual level.

Critical Thinking: Would it, indeed, cause tensions or strains in the society if the society were secularized at the macro level but not at the micro level? Why or why not? If it would be expected to create tensions, how severe are those likely to be for the society? Why?

Public policy and the operation of institutions like the government are secularized—based on principles of rational deliberation rather than doctrine and scientific evidence rather than scripture or dogma. Controversies arise over court challenges when religion does enter the public sphere. This sign was put up in Kentucky regarding whether the Ten Commandments could be posted in city hall or the courthouse. The courts ruled against it. This illustrates micro level dissatisfaction with secularization occurring at the meso level.

While secularization may be complex in that it operates at various levels (at least three), Tschannen (1991) suggested that secularization is also complex in having more than one dimension. One dimension, the one stressed by most neosecularization advocates, is institutional differentiation and increased autonomy of various institutions from religious domination. A second dimension is development of a rational, utilitarian, and empirical/scientific approach to decision making, so that the world becomes "disenchanted."

This rational–legal approach to determining what is true and authoritative is often correlated with a decrease in otherworldliness or supernaturalism. Table 13.2 illustrates the complexity of secularization once both levels of analysis and dimensions of the phenomenon are taken into account.

Some scholars think changes in thinking patterns or "consciousness" were primary and caused institutional changes; others stress structural changes causing change of outlooks. In any case, the dimensions of secularization seem to be closely linked at any one level of social analysis. By contrast, secularization at various levels may occur somewhat independently.

We might conclude with four observations about secularization: (1) Secularization at the macro level is clearly at work in all Western democratic societies; (2) secularization at the individual level is variable and is far from universal; (3) secularization is not inevitable or unstoppable; and (4) secularization is a useful concept for describing a social process, but since it no longer *predicts* future change, its utility as a social *theory* is severely limited. Secularization may be a useful descriptive concept, even if the idea of its eventual complete subversion of religion is no longer tenable (Bibby, 2002; Sommerville, 2002).

Decline in religious practice and belief, then, is not necessarily an element of secularization. Indeed, at least one scholar thinks that *failure* of religion to modernize—failure to adapt to the secular trends—is a sure formula for religious decline (Tamney, 1992). A few scholars even believe that the conservative trend in mainline denominations may be disastrous for them, since they are alienating many young, urban professionals who are offended by the swing away from rational secularity (Hout & Fischer, 2002).

In the end, religion is changing. Change in itself, however, is not necessarily declining. Rather, adaptation is characteristic of all religion that survives over time. (Whether the adaptation is "good" or "bad" is a normative judgment rather than an analytical one.) Some sociologists have argued that one significant way religion has

Table 13.2 The Complexity of Secularization

	Institutional Differentiation	*Rational/Utilitarian Decision Making*
Macro level	Institutions in the society, including government, science, education, and the economy, are functionally independent and autonomous of religious organizations.	Decision making about social policies are based on cost–benefit analyses, using logic and empirical data rather than scripture or proclamations of religious authorities.
Meso level	Organizations look to other social associations and units that have similar functions and goals for accepted practices of how to operate the organization, not to religious organizations and authorities.	Decision making about an organization's policies and goals are based on utilitarian analyses and consequences, rather than scripture, theological arguments, or proclamations of religious authorities.
Micro level	Individuals emphasize being "spiritual rather than religious," formulate their own meaning system or theology, and may believe that spirituality has little to do with other aspects of their lives.	Decision making about life decisions is based on individual self-interest without concern for the teachings of the religious group or the clergy. Calculations of cost and benefit to the individual are foremost in individual decisions.

Neosecularization advocates believe that while not all societies in the world are secularized, most Western democratic societies show a great deal of secularization at the macro level. At the micro level, there is less secularization, especially in the United States, though it is evident in the United States at the micro level as well. Some places in Europe have rather high levels of secularization at the micro level.

changed as society has become secularized at the institutional level is that it has become a generalized aspect of society itself, rather than a specific part of religious institutions and organizations. This is known as civil religion.

CIVIL RELIGION

Civil religion refers to the cultural beliefs, practices, and symbols that relate a nation to the ultimate conditions of its existence. The idea of civil religion can be traced to the French philosopher Jean-Jacques Rousseau's *On the Social Contract* (1762/1954). Writing in the wake of the Protestant–Catholic religious wars, Rousseau maintained the need for "social sentiments" outside of organized religion "without which a man cannot be a good citizen or faithful subject." The broader question motivating Rousseau

concerned political legitimation without religious establishment.

Although he did not use the term, Emile Durkheim's work in *The Elementary Forms of the Religious Life* (1912/1995) was clearly influenced by his countryman's concern for shared symbols and the obligations they articulate. Recognizing that "the former gods are growing old or dying," Durkheim (1912/1995) sought a more modern basis for the renewal of the collective sentiments societies need if they are to stay together. He found that basis in the "hours of creative effervescence during which new ideals will once again spring forth and new formulas emerge to guide humanity for a time" (p. 429). Civil religious ideals arise from national civil religious rituals.

Robert Bellah's (1970b) 1967 essay, "Civil Religion in America," brought the concept back into contemporary sociology. Like Rousseau and Durkheim, Bellah saw legitimation as a problem faced by every nation, and civil religion as one solution—under the right social conditions. Following from his perspective on secularization in the course of religious evolution, Bellah argued that in premodern societies the solution consisted either in a fusion of the religious and political realms (in the archaic period) or a differentiation but not separation (in the historic and early modern periods). Civil religion proper comes into existence only in the modern period when church and state are separated as well as structurally differentiated. In other words, a civil religion that is differentiated from both religion and the government is only possible in a modern, secularized society (Bellah, 1970a, 1975).

Civil Religion in the United States

Bellah (1970b) argued that civil religion exists alongside but distinct from formal organized religion. It is actually a religious "dimension" of society, characteristic of the American republic since its founding. Civil religion is "an understanding of the American experience in the light of ultimate and universal reality" and can be found in presidential inaugural addresses from Washington to Obama, sacred texts (the Declaration of Independence) and places (Gettysburg), and community rituals (Memorial Day parades) (Bellah, 1970b, p. 186). It is especially evident in times of trial for the nation like the Revolution and the Civil War.

The mythology of American civil religion began early in the nation's history. In the speeches of some of the first presidents and in sermons of some of the colonial preachers, America was treated as the "Promised Land." In fact, in his second inaugural address, Thomas Jefferson explicitly compared the founding of the new nation to the founding of Israel, and Europe was defined as the contemporary Egypt—from which God's people had fled. This began a long process of myth development that has grown through the past 200 years. The belief in the American Dream, the American Way of Life, and the fundamental goodness of America is expressed with reference to a supernatural blessing: "America,

America, God shed His grace on thee, and crown thy good with brotherhood from sea to shining sea." Hymns such as these evoke a profound sense of reverence, for they express a deeply ingrained mythology. "God Bless America" serves much the same function.

Civil religion is also expressed through the rituals that take place on such "high holy days" as Memorial Day, the Fourth of July, and presidential inauguration days. The ceremonials on these days express central American values and inspire a feeling of unity and a sense of transcendence (greater purpose). The meaning of the nation is believed to transcend individual lives and is important in understanding the significance of contemporary events. The effort to stop Hitler and "make the world safe for democracy" provides one example. In the 1960s and 1970s, the space program and moon landings provided a sense of national accomplishment and collective identity. Currently the mission of the United States as a collective people seems to be focused on ridding the world of terrorists. Following the events of 9/11, freedom has become a uniting value that bonds people of many backgrounds and calls them to sacrifice and commitment for a shared purpose.

National ceremonies tend to emphasize the transcendent purpose of the nation. Most such celebrations occur around national shrines, which are themselves capable of eliciting a feeling of awe, for they symbolize both the ideals of the nation and the sacrifices made on the nation's behalf. Examples of national shrines are the Washington and Lincoln Memorials in Washington, D.C., the Capitol, the Tomb of the Unknown Soldier, war cemeteries, and the birthplaces or burial sites of American presidents. In an excellent analysis of the symbolism of a Memorial Day celebration in an American town, Lloyd Warner (1953) discussed the unifying quality of these rituals and symbols:

> The cemetery and its graves become the objects of sacred rituals which permit opposing organizations, often in conflict, to subordinate their ordinary opposition and to cooperate in expressing jointly the larger unity of the total community through

Civil religion blends reverence for the nation with more traditional symbols of faith. Pictured is the chapel at Punchbowl, the Pacific cemetery for U.S. military personnel, located in Hawaii. Note that two U.S. flags are inside the altar area and are more prominent than two of the three religious symbols that also adorn the chancel: the Christian Cross, the Jewish Star of David, and the Buddhist wheel of Dharma.

the use of common rites for their collective dead. (pp. 24–25)

Because some sociologists view these ceremonials as the central expression of civil religion, content analyses of speeches and newspaper articles on Memorial Day and the Fourth of July and analyses of central themes in presidential inaugural addresses have become an important means of studying this phenomenon (Toolin, 1983). There are also other national holidays, but they tend to be of somewhat lesser significance as expressions of civil religion: Thanksgiving (which George Washington first made a national holiday in 1789 so that citizens might thank God for the blessings of this land and this nation), Presidents' Day (birthday celebrations for admired Presidents), and Labor Day (a time to celebrate the accomplishments of American labor and the upward social mobility provided for Americans through the labor movement).

The paramount sacred object in this religion is the American flag. The importance of this symbol can be seen not only in the prescribed handling of the flag (see the "Illustrating Sociological Concepts" feature in Chapter 4, pp. 72–73)) but also in the intensity of the outrage when the stars and stripes are "desecrated"—treated inappropriately.

Daily recitation of the pledge to the flag by schoolchildren became a major theme of George Bush Sr. in the 1988 presidential election. Particular national heroes, or saints, also serve as focal points for veneration and myth development. Washington and Lincoln are the most important and most widely recognized "saints," and are paid homage on Presidents' Day. For some people, Presidents Jefferson, Wilson, Franklin Roosevelt, or Kennedy are key figures. Sometimes folk heroes (like Betsy Ross, Daniel Boone, or Charles Lindberg), business tycoons (who symbolize the rags-to-riches mythology), and military heroes (who symbolize courage and a willingness to sacrifice for the nation) are given honored status and held up to children as exemplary of the American Way. In many quarters, Martin Luther King Jr. has become a saint of the civil religion because of his efforts to apply the motto "freedom and justice for all" to all Americans. These national "saints" serve as inspirations and as behavioral models, much as Saint Francis, Mother Seton, or Saint Teresa of Avila do in the Christian tradition.

---

[3]Shintoism is a Japanese religion (the state religion prior to 1945) that involves worship of ancestors and ancient heroes, a glorification of national accomplishments, and a deification of the emperor of Japan.

Many observers of this American civil religion have been appalled by it, labeling it as idolatry of the nation or as "American Shinto" (Marty, 1958).[3] Such scholars believe that, although many Americans attend weekly worship and have memberships in traditional religious groups, the central meaning system is Americanism. For these people, the flag is a more important symbol of their religion than is the Star of David or the Cross, and the Fourth of July is a more celebrative and meaningful holiday than Easter or Passover. This sort of analysis is by no means limited to social scientists, but it does suggest that civil religion sometimes replaces more traditional theological beliefs. Theologian H. Richard Niebuhr (1960) believed that nationalism was a greater threat to Christianity than atheism.

Although he has a more benign view of civil religion overall, Bellah (1970b) partially agreed with Niebuhr when he insisted that "without an awareness that our nation stands under higher judgment, the tradition of the civil religion would be dangerous indeed" (p. 185). It would be dangerous because it would serve only to sanctify the status quo and the current social structures, regardless of whether they are just. He denied that the conservative function is the only role of civil religion in the United States. Its structural position relative to both church and state allows civil religion to act not only as a source of legitimation but also of prophetic judgment. That is, American civil religion proclaims judgment on the United States when the country fails to live up to its creed. The ideals of the nation provide a foundation for criticism and improvement.

For example, Martin Luther King Jr. delivered a critical speech about his hopes for the nation, his "I Have a Dream" speech delivered at the March on Washington in 1963. It proved deeply moving to the American people because his dream was "deeply rooted in the American Dream":

I have a dream. . . . deeply rooted in the American dream . . . that one day this nation will rise up and live out the true meaning of its creed: "We hold these truths to be self-evident: that *all* men are created equal."

I have a dream that my four . . . children will one day live in a nation where they will not be judged by the color of their skin. . . . when all of God's children will be able to sing with new meaning "My country 'tis of thee, sweet land of liberty, of thee I sing. Land where my fathers died . . . from every mountainside, let freedom ring." (quoted in Lincoln, 1968, pp. 65–66)

That speech became a major formulation of American civil religion. Although King was a Baptist preacher, he appealed not to values that are uniquely Christian but to ones that would be compelling to Americans of any religious stripe.

Over 40 years later, then-presidential candidate Barack Obama addressed the issue of race in his 2008 "more perfect union" speech. His language, many observers immediately recognized, was civil religious:

Two hundred and twenty one years ago, in a hall that still stands across the street, a group of men gathered and, with these simple words, launched America's improbable experiment in democracy. . . . The document they produced was eventually signed but ultimately unfinished. It was stained by this nation's original sin of slavery. . . . Of course, the answer to the slavery question was already embedded within our Constitution . . . a Constitution that promised its people liberty, and justice, and a union that could be and should be perfected over time. . . . Yet words on a parchment would not be enough. . . . What would be needed were Americans in successive generations who were willing to do their part . . . to narrow that gap between the promise of our ideals and the reality of their time.[4]

According to Bellah, at its best civil religion guards against governments doing whatever they want to do and then sanctifying their actions; civil religion provides a standard of judgment for national policy. Unfortunately, civil religion is not always at its best. By 1975, Bellah declared in *The Broken Covenant:*

---

[4]This excerpt from Obama's speech is from Gorski, P. (2010, January 8). *A neo-Weberian theory of American civil religion.* Retrieved from http://blogs.ssrc.org/tif/2010/01/08/a-neo-weberian-theory-of-american-civil-religion/

American Civil Religion in Time of Trial that American civil religion was "an empty and broken shell" (p. 142) because it had failed to inspire citizens and lost its critical edge.

In colonial days, said Bellah, a legal and economic system was devised that would protect the rights of the individual to pursue his or her own self-interests. However, a moral ethos counterbalanced narrow selfishness by proclaiming the obligation to act in the public interest. This moral climate was reinforced by two sources: the biblical concept of covenant with God and the secular republican political philosophy that called forth commitment to the common good. He documented the self-sacrifice of many of the founders of the country, even while they defended the legal right of individuals to act in their own self-interest. Without this moral tone and sense of obligation, unfettered self-interest would destroy both the cohesion and the spirit of goodwill within the country. An endorsement of individualism, said Bellah, must be countered with some sense of obligation to the larger community. Bellah and his coauthors believed that both the republican spirit and the sense of covenant have so deteriorated over the past two centuries that the nation now faces a national crisis. A revival of civil religion is needed. Yet the nature of that civil religion—the way we define the meaning of the nation and the core values that unite us—will be of critical importance in the development of the American character in the next century.

It is always difficult to tell for sure whether a major event will be a defining moment for a nation when one is very close to it. Still, the events of 9/11, when planes crashed into the Twin Towers in New York, the Pentagon in Washington, D.C., and a field in Pennsylvania, the mood and focus of the entire country shifted dramatically. The sense of national identification and coherence was aroused, new heroes who sacrificed their lives for others became dominant, and stories/myths abounded about the courage and values of firefighters and other public servants. Common citizens, who forced the crashing of a plane they were on so that it

would not crash into the White House or a highly populated area, became celebrated figures. Heroes and idols were no longer those who sought their own fortune, as was common in the 1980s, but those who sought a common good beyond their own self interest. It appears that a new sense of national purpose may arise from the rubble of 9/11, but it is too early to analyze the character of this apparent new phase of American civil religion.

Regardless of one's personal evaluation of civil religion, the student of religion must keep in mind the key legitimating function it serves for societies characterized by differentiation of political and religious institutions. In these pluralistic, secular societies—societies in which no one religion can serve to sacralize the existing social arrangements and provide a common core of values—civil religion emerges to fill that void.

> *Critical Thinking:* Robert Bellah said that the founders of the United States stressed individual freedom but counterbalanced that individualism with biblical and republican emphasis on community and the "common good." Bellah claimed that the counterbalance has faded, and Americans are now engrossed in a radical individualism that lacks the prophetic tradition of civil religion (calling people to a higher purpose and a greater generosity). Do you agree with Bellah? What is your rationale? What evidence supports your view?

SUMMARY

Secularization has been viewed by sociologists as one of the most powerful forces in the modern world. Yet there are marked differences in what are considered the core characteristics of secularization and in the causes and effects of the process. Berger has defined it as loss of sacredness; he has depicted secularization as causing a decline in religion. Parsons and Bellah viewed secularization as evolution, as an increased sophistication in religious symbolism and religious structures. They understand secularization as involving change in religious worldviews but not necessarily as decline in religion.

Stark and Bainbridge agree with Berger that secularization is antithetical to religion, but they do not think that religion is therefore necessarily declining. They are convinced that secularization is "self-limiting" and is not an inevitable and overpowering force. They believe that magic, or supernatural solutions for this-worldly problems, will fade as secular scientific empiricism replaces it. Religion, properly understood, addresses general needs that cannot be solved by this-worldly methods.

Neosecularization theorists insist that it is a complex problem that has several dimensions: (1) institutional differentiation and increased autonomy of various aspects of life from religious authority and (2) a rational, utilitarian, and empirical/scientific approach to decision making and to truth, so that the world becomes "disenchanted." This latter dimension typically involves a decrease in otherworldliness or supernaturalism.

Whether secularization entails a decline in religion may depend in part on whether one uses a substantive or functional definition of religion, since not all scholars view supernaturalism as central to religion. More importantly, secularization can occur at several levels of social analysis: micro (individual), meso (organizational), and macro (societal, or even global). Neosecularization theorists tend to argue that secularization is occurring in most Western societies at the macro and meso levels (societies' institutions are gaining independence from religious authorities and are making decisions on the basis of rational–legal criteria) but is more variable at the micro level. However, they no longer claim that it is an irresistible and inevitable force; secularization is seen more as a variable.

As Bellah argued, under conditions of societal-level secularization, a meaning system that explores the ultimate significance of the nation can emerge: civil religion. Civil religion helps modern societies solve the problem of political legitimation that did not exist in societies not characterized by institutional differentiation of religion from political life. In that sense, it can be seen as a cultural response to secularization.

14

RELIGION OUTSIDE THE (GOD) BOX

Here are some questions to ponder as you read this chapter:

- What effect have technologies like the printing press and the radio had on faith and faith communities?
- Who is using television for religious purposes, and what impact might this have on religion in the future?
- How might the Internet affect religion as a communal/shared experience, and how might it influence individual faith systems?
- Is sport an avenue to enhance faith (and does a religious faith contribute to athletics)—or are they competing systems, each needing the time, energy of participants, and socializing participants in quite different values?
- Are there other forms of religion that operate subtly in the society—forms of religiosity that are implicit or taken for granted?

- Is religion becoming privatized and individualized? If so, what might be some consequences of this?
- What is the place of movements like astrology, scientology, and similar quasi-religious organizations in society?

Students are often told to think "outside the box"—to think broadly, critically, unconventionally. Sociologists of religion are well served to do the same. Not all religion or religiousness is to be found inside the "God Boxes" we think of as churches, synagogues, temples, and mosques. Thinking "outside the God Box" allows us to understand some important social changes that have taken place in modern society and the religious adaptations that have resulted.

Two of the most important aspects of contemporary society are the media and sport. These arenas are often not taken seriously by students of religion, however, because they are "mass" or "popular" cultural phenomena. Religion is seen as serious and important while popular culture is silly and frivolous, yet the two commonly intersect. One cannot understand some very important developments in religion without understanding its relationship to the media and sport.

Excessive association of "religion" with church, temple, or mosque (institutionalized forms of religion) is the starting point of Thomas Luckmann's (1967) analysis of what he called "invisible religion." The identification of religion with the God Box has narrowed the field of sociology of religion, which is especially problematic because, in his view, organized religion is becoming more marginal in modern societies. Luckmann argued that emphasis should be placed on the individual's struggle for a meaningful existence in society, a struggle which is fundamentally "religious." If we focus only on institutionalized forms, we will miss the key religious activity going on in modern society. Much of religion has become privatized, as individuals work out religious solutions for themselves, on their own terms.

Last, a number of social scientists maintain that religion is undergoing significant transformation as new forms of religion are emerging. Some of these new forms are nontheistic, and some even lack a supernatural dimension. For this reason, many sociologists prefer to call these processes "quasi-religious phenomena" or "functional alternatives to religion." Regardless of what one calls them, these value perspectives provide many people with a sense of purpose in life and with a center of worth (which is the etymological basis for the word *worship*). When any ideology or value system becomes a meaning system—one that defines the meaning of life, death, suffering, and injustice—it usually takes on a sacred cast in the eyes of the adherents. By looking only inside the God Boxes, we overlook these new phenomena.

RELIGION AND THE "OLD" MEDIA

Although a study of radio/television and religion had already been published by the mid-1950s (Parker, Barry, & Smythe, 1955), sociological study of media and religion really took off with the rise of modern "televangelism" in the 1970s and 1980s. Those who lived through those decades could not escape the sight (if not the influence) of "TV preachers" like Billy Graham, Oral Roberts, Rex Humbard, Robert Schuller, Pat Robertson, Jimmy Swaggart, and Jim and Tammy Faye Bakker. Even as those pioneers have passed from the earth or at least public consciousness, broader concerns about the relationship between religion and media have been addressed by sociologists.

As Stewart Hoover (2009) observed, "media are fundamentally technological in origin, and technological change plays an important role in their development and evolution" (p. 689). Our survey of the relationship between religion and the media, therefore, will look at the evolution of media technologies from print and publishing to radio and television to "new media" like the Internet. Of course, technological developments cannot be understood independent of their economic, political, and social environments, so we pay some attention to those factors as well.

Print Publishing

The development of the mechanical (movable type) printing press by Johannes Gutenberg in 1440 was arguably the most important technological innovation of the second millennium. It had a monumental effect on all of the major institutions of society: education, medicine, politics, and, of course, religion. Indeed, the so-called Gutenberg Bible was the first major book printed using the press. It played a central role in ushering in the print revolution and also had a significant effect on the practice of religion itself (Man, 2002).

Prior to the wide distribution of religious texts to people who were not ordained ministers, the hierarchies of Christendom controlled what was disseminated as Truth. The common (and typically illiterate) member of the local church did not have any basis for challenging the Pope or other ecclesiastical leaders. Those leaders were the authority. However, Martin Luther used the printed word in many powerful ways. He and other reformers claimed that the Bible alone was the ultimate source of Truth and religious authority. The church leaders were to be believed only insofar as they were faithful to the scriptures. Luther himself used the printed word to spread his version of Christian Truth, and he did so with a vengeance. He not only wrote more than other dissenters but he outpublished the entire legion of Vatican defenders (Brasher, 2004). He published in the common languages of the people rather than in Latin, and the Protestant Reformation was launched. It is doubtful that

this could have happened without the printing press. The printing press had similar revolutionary effects within Judaism (Brasher, 2004).

Religious publishing today remains a vibrant cultural and economic phenomenon. According to the Association of American Publishers (2010), net sales of books in their "religious category" grew 2.4% from $557 million in 2002 to $658 million in 2009. Although this rate of growth was modest due to the challenging economic conditions of the time, it more than doubled the overall rate of growth in the book publishing industry over the same time period (1.1%). These sales are supported by organizations such as *Christian Retailing* magazine (which serves the $4.6 billion industry in "Christian products"), the Evangelical Christian Publishers Association, Religious Book Trade Exhibit, and the Association for Christian Retail.

As in Gutenberg's day, Bibles lead the way. Indeed, nearly 28% of individuals surveyed by *Publishers Weekly* reported purchasing a Bible in the previous year (Elinsky, 2005, p. 24). Recently, religious fiction and self-help blockbusters have also contributed to this market segment. The "Left Behind" series of 16 novels, written by Tim LaHaye and Jerry Jenkins, have sold over 60 million copies. This is remarkable for a series of books that focus on the "end times" from the premillennial dispensationalist perspective of some sectarian groups' of the book of Revelation. Even more remarkable is that five of the books were *New York Times'* best sellers, indicating their mass appeal. Another book that has dominated the best seller list for years is megachurch pastor Rick Warren's *The Purpose-Driven Life.* Published in 2002, it has sold over 25 million copies and is the best-selling hardback book in American history. It has also been translated into 30 languages (Pew Forum on Religion & Public Life, 2005). Highlighting the articulation between different media platforms, megachurch pastor and televangelist Joel Osteen's book, *Your Best Life Now,* sold 700,000 copies in the first month after its publication in 2004, and his follow-up, *Become a Better You*, was also a *New York Times'* best seller.

Radio

This history of the development of radio technology is complex, but for our purposes we can note that in the 1920s commercial radio broadcasting began. Despite the fact that in 1930 the average cost of a radio receiver was $78 (equivalent to $1,019 in 2010 dollars), 40% of all U.S. households owned one. As the price dropped, that proportion more than doubled to 83% by 1940, and by 1950 nearly every household in the United States (96%) had a radio (Craig, 2004).

Because broadcasting licenses issued by the Federal Communications Commission (FCC) required broadcasters to carry some programming that was directed to the public interest, from the start radio programming was often religious in nature, because religious programming was deemed to be for the public good. In the late 1920s, S. Parkes Cadman had a Sunday afternoon radio program on the NBC radio network. In the hour before Cadman, one could listen to the National Youth Conference of Dr. Daniel Poling, and Dr. Harry Fosdick's National Vespers service followed in the hour after. The Columbia Broadcasting System (CBS) also provided considerable religious programming. Together, they broadcast religious content from Protestants, Catholics ("Catholic Hour"), and Jews ("Jewish Art Program Sundays"). By 1931, *Time* magazine had taken note of the phenomenon, writing this:

Enterprising evangelists and regularly employed clergymen snapped at radio's religious opportunities quickly after Westinghouse began broadcasting ten years ago. Three years ago first N.B.C., then Columbia systematized radio religion and offered time to Protestants, Catholics, and Jews. The three creeds took advantage of their opportunity . . . Apart from such chain broadcasting are individual stations operated by churches, societies and evangelists. They number about two score. ("Religion: Air Worship," 1931)

By 1946, *Time* characterized "radio religion" as "a national institution," one that was "preached to an estimated congregation of ten million" ("Religion: Radio Religion," 1946).

Excluded from this national institution, however, were evangelical Protestants. Radio stations gave their free public interest time only to mainstream Protestant, Catholic, and Jewish groups. Unable to take advantage of the free airtime, entrepreneurial Protestants like Paul Rader, Aimee Semple McPherson, and Charles Fuller bought airtime for their programs and thereby helped to transform the religious orientation of the American mass media in an evangelical direction (Hangen, 2002). In less affluent parts of the world, like Latin America, radio remains the predominant vehicle for the dissemination of religious ideas.

Televangelism

In 1950, 96% of American households owned radio receivers. In that same year, only 9% had television sets. The decade of the 1950s was the key period of transition from one technology to the other. As televisions became widely available in the 1950s, what was done by radio previously began to migrate to television. (Note that just as the advent of radio did not eliminate religious publishing, neither does the advent of television eliminate religious radio.)

By 1960, an important change in the regulation of the airwaves took place when the FCC ruled that television stations could count paid programming toward their "public interest" contribution. This effectively opened the door for those who carried on the entrepreneurial spirit Rader, McPherson, and Fuller had brought to radio. The first three mass-media evangelists to have a major television presence were Billy Graham, Oral Roberts, and Rex Humbard. They built not only on the radio evangelists who preceded them but also on the contributions of revivalists Charles Grandison Finney (1792–1875), Dwight L. Moody (1837–1899), and Billy Sunday (1862–1935).

Critical Thinking: The marketing perspective that drives much of the televangelism industry has its foundations in late 19th- and early 20th-century urban evangelism. At the website for this book (**www.pineforge.com/rsp5e**), an essay on "Evangelical Foundations of a Marketing Perspective on Conversion and Recruitment" examines this foundation. This essay may help you address the question of why evangelicals have been more inclined to a marketing perspective and more entrepreneurial as ministers of the gospel. How does this historical perspective help you understand the contemporary patterns of television broadcasting?

Billy Graham (born in 1918) never established a weekly program, and his televised programming remained in the format of a worship service or a revival. Still, he was the first evangelistic fundamentalist to gain national recognition. Much of his early notoriety was because of an editorial directive from publisher William Randolph Hearst, who sent a simple memo to the editors of his nationwide chain of newspapers: "Puff Graham." Graham made the most of the subsequent publicity and became a role model for younger evangelists. More importantly, his career linked the evangelical tradition to the new media of television and radio. Eventually, he came to be a spiritual adviser to a number of U.S. presidents. Graham was the first evangelical to gain national credibility on television, and he was especially influential in challenging evangelicals to adopt the newest mass media technology to spread the faith.

Unlike Graham, Oral Roberts (1918–2009) was an innovative showman, introducing a variety of formats, using a wide range of entertainment forms, and showing the value of flexibility of material to meet changing markets. Most importantly, however, Roberts discovered that if a televangelist wants to raise money, the key is to build buildings, even if they seem outlandish. He discovered that people will contribute far more money to build a university (Oral Roberts University was founded in 1963) or a hospital (ORU School of Medicine operated from 1981–1989) than they will to buy airtime. In the 1980s, Roberts made an infamous plea to the viewers of his television ministry that if he did not raise $8 million by March 31, 1987, to help fund his medical center, God would "call him home." His followers donated $9.1 million (Ostling, 1987). The excess money brought in for building projects can then be used to support other projects such as broadcasting costs. With the oligarchical structure of the organization (with family members serving on the oversight board), Roberts and some other televangelists found they could make such transfers without challenge.

Rex Humbard (1919–2007) was the third of the original televangelists with nationwide programming. Humbard contributed two innovations. First, he created an intensely personal style, using a family format (telling stories to his gathered children and to the audience who became part of his family), sharing personal troubles, and airing feelings publicly. Humbard often cried on television and would bare his soul for his national audience. Whatever else this process did, it was highly profitable as people responded with their hearts to a person they had come to know intimately—more intimately than members of their own congregation. Later on, other televangelists also found that open displays of emotion build intense audience loyalty.

Second, Humbard built a cathedral especially designed and equipped for broadcasting. Graham and Roberts initially simply brought cameras into their existing services. By contrast, Humbard invested in the most sophisticated modern technology; he also established television effects as the highest priority in the architecture and decor of the cathedral.

Other televangelists have developed their own distinctive styles, theological orientations, and formats, yet each has built on these early foundations. Without these foundations, televangelism empires would probably not exist. Further, this legacy has encouraged religious leaders to be entrepreneurs. Interestingly, most of them would find that word—entrepreneur—to be a compliment. This is in sharp contrast to many mainline clergy who find this an insulting metaphor for the caregiving of clergy.

Television is medium in which the viewer has much control—via the remote channel changer—so there is pressure on televangelism programs to entertain and to avoid too many complexities, which might turn the viewer off. This creates issues of marketing strategies and requires an entrepreneurial spirit, both of which have been more compatible with how evangelicals tend to see their ministries.

Critical Thinking: Has televangelism been an asset to the promotion of traditional orthodox Christianity, or does it threaten to undermine the faith through introduction of other motives and the aspirations of "entrepreneurs"?

The Televangelism Audience

The importance of televangelism as a resource to shape the larger culture depends in part on the nature of the audience: How large is the audience, and who is watching? Clearly, televangelists have the potential for an extensive audience, with some programs translated into other languages and broadcast into a dozen or more countries. The high estimate for the electronic church in the U.S. alone is 130 million viewers; the low is 10 million. Televangelists tend to exaggerate the size of their audiences, while their detractors tend to underestimate. Even the "hard data" are difficult to interpret. Nielsen and Arbitron ratings indicate how many televisions are turned on and to which channels. On the other hand, they do not necessarily tell us whether anyone is in the room or is paying attention and whether those television sets were turned to that channel for 5 minutes or for the duration of the program. Pollsters ask people whether they watched religious television in the past month, but surveys leave it up to the viewer to decide what is religious programming (they may be including a movie with a religious theme), and they allow respondents to recall their behavior from the past month (which may or may not be accurate) (Gerbner et al., 1984; Hadden & Shupe, 1988; Hoover, 1987).

While we really do not know exactly what the viewership is for televangelism, several efforts have been made to control for the many variables and the different sources of information. In 1986, more than 15 million households a week tuned into religious broadcasts; by 1992, only 9.5 million were viewing each week. Still, the number of local religious TV stations grew from 25 in 1980 to 339 in 1990. By 1996, one report indicated that 16% of all television stations were religious in character. However, because of separation of religion and state, these broadcasters are not overseen by the government, and many are not even part of the National Religious Broadcasters, an industry body that oversees most religious stations. Thus, obtaining complete figures is difficult, at best. Still, the viewership appears to be considerably higher and religious broadcasting more warmly accepted in the United States than elsewhere in the Global North, though growth is apparent in the Global South. (The affluent or "developed" countries of the world are almost all North of the 25th degree latitude and are therefore called Global North countries; Global South refers to less economically and technologically prosperous countries—the areas that during the Cold War era with the Soviet Union were called "Third World.")

Those viewers who send money to support the programming tend to increase their levels of commitment. By making an investment they develop a sense of loyalty to the organization. The top televangelists bring in tens or even hundreds of millions of dollars each year to finance their programs and their projects (medical centers, universities, etc.). The electronic church has a committed core who provide a base of support for some truly enormous empires (Hadden & Shupe, 1988; Peck, 1993).

Another issue concerning the televangelism audience has to do with who regularly views and supports the electronic church. Viewers tend to already be evangelical in orientation. One rationale often given by televangelists for using television is to fulfill "the great commission" of Christ to spread the gospel. The idea is that television allows preachers to reach people who might otherwise never have heard of Christ. Yet empirical evidence suggests that televangelism has little persuasive impact in terms of converting degenerates; most viewers are already sympathetic to the basic philosophy (Gaddy, 1984). This is not to say that religious broadcasting is insignificant; it serves as a reinforcer and a plausibility structure for those already committed.

Does Televangelism Undermine Commitment to Local Congregations?

Televangelism has come under heavy criticism for several reasons and from a number of quarters. Former evangelist Charles Templeton, who was himself involved in the production of religious television programming, assessed contemporary televangelism in this way:

> Television Christianity is an undemanding faith; a media apostasy that tells listeners that to become a Christian all they have to do is "believe." . . . The offerings, [extracted] mostly from the poor, the elderly, and lonely women, amount to millions of dollars annually. Few of these dollars are used to give succour to the needy, to put food in empty bellies or to help the helpless and dispossessed. . . . There are, among the host of televangelists, exceptions to those I have described, but they are a minority. . . . On balance I think the contemporary television evangelist is deleterious to society. (cited in Bibby, 1987a, p. 36)

No less a figure than Billy Graham (1983) has spoken out against abuses, the most severe of which he believes is the constant harangue for money. However, the most common complaint of critics has been that televised Christianity erodes support for local churches.

The most consistent and most strident attacks on televangelism have come from mainline denominations and from the National Council of Churches. The primary conflict has been over resources: money and members. The first concern here is that significant numbers of people may stay home and watch televised worship services rather than attend and contribute in other ways to their local congregations. Televangelism is feared by mainline denominations as an alternative to activity in the local faith community and therefore as a threat to the viability of individual churches.

Yet research shows that the gratifications one receives from watching a televised worship service are different from those gained by attending worship services. Among those who are regular viewers of television, experiencing solemnity and atmosphere, praying to God, experiencing God's presence, feeling forgiven for sins, demonstrating solidarity with Christian values, and getting a sense of distance from the worries of everyday life are all facilitated more by attending services in a local house of worship than by watching religious television (Petterson, 1986). Indeed, viewers of religious television attend community worship services more frequently than nonviewers, and only one viewer in seven reports watching services on television rather than supporting the neighborhood congregation (Wuthnow, 1987). Bibby found the same pattern in Canada, with religious television serving for most viewers as a supplement to attendance at worship rather than as a substitute (1987a). Only among the very elderly is religious television a substitute.

The second issue, of course, is whether financial contributions to television evangelists are replacing support to local faith communities. Most of the research indicates that money given to religious television programs is not in lieu of contributions to nearby congregations. They are donations made over and above the local church pledge and do not normally involve a decline in hometown contributions. The only category of viewers that seems to send money to televangelists that would otherwise be given to local churches are the very elderly or disabled persons who find it difficult to get out of the house to attend services (The Gallup Organization, 1987; Petterson, 1986).

It seems clear that the charge of mainline denominations—that televangelism is siphoning off support for the local religious communities in terms of both "bucks and bodies"—is unfounded. Although there may be some dent in the coffers of local congregations because of televangelism, the impact appears to be quite small. From a market standpoint, the two do not seem to be in a win/lose competition.

Arguably, the heyday of American televangelism was the 1980s, and thus much of the scholarly research on televangelism is about that period. By the 1990s, some serious scandals had undermined some of the cultural influence and social import of televangelism. Jimmy Swaggart, whose television ministry was transmitted to thousands of stations every week, admitted in 1988 to transgressions with a prostitute and tearfully apologized on television in his well-known "I Have Sinned" speech (available on YouTube). Also in 1988, Jim Bakker, host of *The PTL Club* with his wife Tammy Faye, was convicted on 24 counts of fraud and conspiracy for embezzling donations to their ministries. In the course of this investigation it also came out that he had an affair with a supposed employee of PTL and paid hush money to cover it up.

The 2000s marked the passing of the founding generation of televangelists with the deaths of Rex Humbard and Oral Roberts and the gradual withdrawal from ministry of nonagenarian Billy Graham. Also, the ministry built by televangelist Robert Schuller—whose *Hour of Power* has broadcast from the Crystal Cathedral in Garden Grove, California, for three decades—is being torn apart by debt and family division as he tries to transition out of the role of head minister. According to Quentin Schultze, who specializes in the study of Christian media, "I don't see a scenario for maintaining a TV-based megachurch anymore. The days of doing that in the models of Schuller and Jimmy Swaggart and Oral Roberts are over" (Associated Press, 2009). The most famous and influential evangelical Protestant minister today, Rick Warren, has purposely chosen not to broadcast services regularly from his Saddleback Church, favoring attendance at live services instead.

Despite this, televangelism lives on, especially in the Global South (Leslie, 2003; see also the next "Global Perspectives" feature). It also lives on in the United States, if in an altered form. The new model of televangelism may be seen in the work of Joel Osteen, as well as African American ministers like T. D. Jakes and Creflo Dollar (Walton, 2009). As was noted in Chapter 8, each of these ministers leads one of America's largest megachurches. Although they do engage in televangelism, their ministries seem as much oriented toward the local congregation as their worldwide television audience.

GLOBAL PERSPECTIVES

Televangelism Beyond the Borders of the United States

The globalization of religious broadcasting was itself a product of a global event: World War II. Broadcast stations and equipment were constructed and distributed around the world to conduct the war. Following the war, many of these stations were abandoned and appropriated by Christians with evangelical intents (Hadden, 1990). There are now more than two dozen international radio broadcasting stations with evangelical Christian sponsorship, with the largest three producing

(Continued)

(Continued)

20,000 hours of programming weekly in 25 languages. This makes Christian stations the largest block of international radio broadcasters in the world (Hadden, 1990). Televangelist Pat Robertson's Christian Broadcasting Network (CBN) is the largest of these, with offices in Africa, Asia, China, Europe, Indonesia, Latin America, the Middle East, Siam, and Singapore. An international edition of CBN's flagship program, *The 700 Club*, can be seen in more than 180 countries (Leslie, 2003).

Support for televangelism is much stronger in the United States than in most other parts of the Global North. In 1958, 1 out of 10 Canadians were regular patrons of religious broadcasts, but within 30 years, the count had dropped to 1 in 25 (Bibby, 1987a). Even a report in 1998 on The Christian Channel in Canada indicated an audience of only 300,000 viewers nationwide (Gross & English, 1998). This pattern seems to be consistent with the lack of growth of conservative evangelical Christianity generally in Canada.

Televangelism is also not widespread in Britain or elsewhere in Europe. Kay (2009) noted that the United Kingdom's highly regulated broadcasting industry (dominated by the state-run British Broadcasting Corporation or BBC) severely curbed the rise of televangelism until recent years. While there are religious broadcasting efforts in the Netherlands and Germany, most broadcasting is Sunday morning services of mainline churches. Viewers are primarily people who already attend church, and those numbers are very small in virtually every country in Europe (Davie, 2000).

In Latin America, by contrast, where Pentecostalism is growing rapidly, locally owned and produced religious broadcasting began to take off in the late 1980s. Prior to that, American Pentecostal televangelists like Jimmy Swaggart and Jim Bakker were dominant. Given the comparative economic underdevelopment in much of Latin America, it is radio rather than television, which continues to dominate in religious broadcasting, since even the poor can own radios. For example, the Catholic Church operates 181 radio stations in Brazil and 7 in Guatemala, and much of the programming on those stations is produced by those involved in the Catholic Charismatic Renewal—basically Catholic Pentecostalism (Chesnut, 2010).

The Indian government's openness to satellite television makes it a fertile ground for televangelism. Looking just at networks that broadcast 24 hours a day, there are four Christian, one Islamic, and one Hindu (James, 2010). A recent study of Christian Broadcasting Network India's programming finds a blending of American and Indian influences. The authors cleverly call this "Masala McGospel." (In Chapter 15, we will examine the concept of "glocalization" to describe the blending of the global and local in the process of globalization.) If masala here represents the Indian culture and McGospel the homogenized, globalized ("McDonaldized") version of Christianity exported from America to India, the McGospel is trumping the masala.

Christian televangelism in some Islamic countries is highly controversial and perhaps even dangerous for the sponsors. Muslims want the Christian televangelism to end, viewing it as a foreign intrusion into their cultures. In Nigeria, there have been concerted efforts to stop these broadcasts ("Muslims Aim to End Televangelism," 1998). Some converts to Christianity have received death threats, and Christians at Nigerian television stations have been threatened, with assertions that they are fomenting religious crises in families and in the society at large. The hard sell evangelical approach has come to be viewed as a kind of religious colonialism (Hadden, 1990).

In response to the constant barrage of messages from outside, a number of developing nations (Muslim and others) have signed a declaration objecting to broadcasts into their countries. They believe that this kind of conduct destroys their culture, undermines family relationships and social stability within their countries, and acts as an assault on their national sovereignty (Hadden, 1990, 1991).

RELIGION AND THE "NEW MEDIA"

If you are reading this textbook, chances are that you were "born digital"—that is, you are part of the first generation in human history to have been born into and come of age in a postdigital revolution social environment. Some have suggested that this will create a generation gap like never seen before between "digital natives" and their older kin (Palfrey & Gasser, 2008). This raises the possibility of a religious gap based on orientation to "new media."

According to Wikipedia,

New media is a broad term that emerged in the later part of the 20th century to encompass the amalgamation of traditional media such as film, images, music, spoken and written word, with the interactive power of computer and communications technology, computer-enabled consumer devices and most importantly the Internet. New media holds out a possibility of on-demand access to content any time, anywhere, on any digital device, as well as interactive user feedback, creative participation and community formation around the media content. ("New media," 2011)

Although many faculty will tell their students of the dangers of using Wikipedia as an authoritative source in research, we use Wikipedia here because it is, in fact, an example of "new media" itself.

As the Wikipedia definition makes clear, new media are not reducible to the Internet, but the Internet is the most important aspect of new media.[1] Internet usage has been skyrocketing. According to the Pew Internet & American Life Project, in 2000, 55% of U.S. adults said they used the Internet the previous day. By 2009, that had grown to 73%. Internet use by teens and young adults is even higher, with 93% of 12- to 17-year-olds and 18- to 29-year-olds accessing the Internet daily.[2]

Of course, our interest here is in how the technological changes represented by the new media affect the mass mediation of religion and in particular the new opportunities these media create for the practice of religion. As was the case with radio broadcasting, people have been using the Internet, and especially the World Wide Web, for religious purposes from the beginning. Just 6 years after Tim Berners-Lee *proposed the idea* of the web, *Time* magazine ran a story entitled "Finding God on the Web." The story reported the following:

Like schools, like businesses, like governments, like nearly everyone, it seems, religious groups

---

[1]Although they are very closely connected, the Internet and the World Wide Web are not the same thing—despite the fact that many people use the two terms interchangeably. The Internet is the network created by the linking of millions of computers worldwide. The term is an abbreviation of "inter-networking," which was meant to describe networks of networks. These computers can communicate with one another by passing information over this network. The World Wide Web is a particular way of sharing information over the Internet based on the "hypertext transfer protocol" (HTTP) developed by computer scientist Tim Berners-Lee. HTTP specifies how information is shared between computers. Incorporated into web "browsers," HTTP combines with web programming languages like hypertext markup language (HTML) to create a visually appealing, easy to use experience of connecting to other computers on the Internet, accessing information stored on them, and sending information to them. Therefore, "surfing the web" is just one way in which people use the Internet. People also use the Internet to connect to others using electronic mail (e-mail) and instant messaging (IM). The rise of interactive applications on the web has led some to speak of the second generation of the WWW, or "Web 2.0." Unlike the first generation of the WWW, in which most users simply accessed (consumed) content created by others, Web 2.0 allows users to create (produce) content themselves, often in collaboration with others. Well-known examples are social networking sites (Facebook, Twitter, and MySpace), blogs, wikis, and video-sharing sites like YouTube. The situation is rendered still more dynamic by the fact that people access the Internet not only through computers but increasingly through personal digital media devices (smartphones, iPads).

[2]Calculated by the authors using data provided on the Pew Internet & American Life website: http://www .pewinternet.org/Static-Pages/Trend-Data/Usage-Over-Time.aspx

are rushing online, setting up church home pages, broadcasting dogma and establishing theological newsgroups, bulletin boards and chat rooms. Almost overnight, the electronic community of the Internet has come to resemble a high-speed spiritual bazaar, where thousands of the faithful—and equal numbers of the faithless—meet and debate and swap ideas about things many of us had long since stopped discussing in public, like our faith and religious beliefs. (Ramo, 1996)

Books titled *The Soul in Cyber-Space* (Groothius, 1997) and *Cybergrace: The Search for God in the Digital World* (Cobb, 1998) soon followed. If today's traditionally aged college students are "digital natives," a significant part of their native culture is religious. One religious organization has now tried to claim that that adoption of technology is an inherent part of their Lutheran heritage, as the screen shot from lutheran.org makes clear.

Gary Bunt (2009), a sociologist who specializes in the study of religion on the Internet, recognizes that the Internet "holds transformative potential for religions," but he also cautioned that "there are dangers in overemphasizing the transformative powers of the Internet in the area of religion" (p. 705). In examining religion and the Internet from the perspective of both producers and consumers of content, we find that (so far, at least) much religion online is quite conventional.

Religion Online: Producing and Accessing Information

Much of the total online activity around religion is what scholars call "religion online." Religion online is "information about religion that is accessed via computer-mediated networks" (Cowan, 2007, following Helland, 2000). From both the supply and demand sides, this is the bulk of the religious/spiritual activity that takes place online. Religious organizations put information online, and Internet users seek out that information. In this sense, the Internet stands in a continuous relationship with other media that came before it—a medium for the transmission of information to consumers.

The National Congregations Study (NCS; discussed in Chapter 8) provides solid data on the growth in new media use by congregations in

If Luther were alive today ... he'd be on the Web

Martin Luther's impact as one of the greatest religious reformers was due in part to his use of emerging technology. It wasn't quite Cyberspace. In the 15th and 16th centuries it was movable type, and through it he gave common people easy access to his translations of the Gospels, his writings and sermons. The rest, as they say, is history.

In the spirit of Martin Luther, this page is provided to allow easy access to some Lutheran organizations on the Web.

Evangelical Lutheran Church in America
God's work. Our hands.

THE
LUTHERAN
CHURCH
Missouri Synod

WELS
Wisconsin
Evangelical
Lutheran
Synod

America. This was the area of greatest change between Wave 1 and Wave 2 of the NCS. From 1998 to 2006 and 2007, the proportion of congregations with a website grew from 18% to 44%. As Chaves and Anderson (2008) emphasized, this large change in less than a decade suggests that some 10,000 congregations per year are adding websites. Because larger congregations are more likely to have websites, 74% of regular participants are in congregations with a web presence. Even greater growth was found in the use of electronic mail to communicate with members. In 1998, 21% of congregations used e-mail in this way, and by 2006 and 2007, the percentage had grown to 59. Looked at from the individual perspective, the growth is even more dramatic: from 31% of attenders in congregations using e-mail in 1998 to 79% in 2006 and 2007.

Because of their greater resources, their evangelistic orientations, and their embrace of what is current in the culture for marketing purposes, we would expect megachurches to be even more connected to the Internet than other congregations. This was true even in early 2000 when Scott Thumma (2001) found that nearly every megachurch surveyed had an e-mail address, and 99% had a website. This is rather remarkable when considering that as recently as 2010 less than half of all congregations had websites.

The growing online presence of congregations raises the question of how congregations use the Internet. An early study of 1,300 congregations by the Pew Internet & American Life Project found that congregations were largely using the Internet for informational purposes (Larsen, 2000). Congregations most commonly viewed their websites as places to encourage visitors to attend (83%); to post mission statements, sermons, or other texts about their faith (77%); to link to denominational and faith-related sites (76%) or to scriptural and devotional material (60%); and to post schedules, meeting minutes, or internal communications (56%). The more interactive capacities of the web, by contrast, were the least common features of congregational websites:

Prayer requests	18%
Sign-up feature for classes/programs	8%
Online fund-raising	5%
Webcast worship services	4%
Discussion space for study or prayer groups	3%

The Pew survey also allowed congregational respondents to indicate what they would like to add to their websites in the future. The responses were quite conventional, emphasizing the informational (photos, youth group material) over the interactive. Thus, the Pew survey did not exactly put congregations at the center of Web 2.0 developments. We await a follow-up study to see if and how things have changed since this original research.

Of course, congregational websites are just one way that religion goes online—and maybe not the most important. Stewart Hoover (2009) wrote of the media generally, "The social and institutional autonomy of the media sphere can thus be said to constitute a location for the making of religion that is in many ways independent from the religious doctrine, institution, and history" (p. 694). Just as the advent of satellite broadcasting and cable democratized television, so too does computer technology and the Internet democratize the ability to produce and consume information.[3] The new media lowers barriers to entry, and consequently it is easier for all

---

[3]As in the political sphere, the ideals of democracy often run up against the realities of socioeconomic inequality. Thus, in thinking about the democratizing potential of the new media, we must remain cognizant of the "digital divide" and those who may not be able to fully participate in the digital revolution for economic or other social reasons. See the Pew Internet & American Life's "digital divide" page (http://www.pewinternet.org/topics/Digital-Divide.aspx).

groups—including religious ones—to have a public media presence. In this view, the World Wide Web is the ultimate "spiritual marketplace." In her 2004 book, *Gimme That Online Religion*, Brasher (2004) reported that there were more than a million religion websites, and they covered an extraordinarily wide range of religious beliefs and practices. The democratizing potential of the web may be seen most clearly in the robust web presence of new religious movements (NRMs) and nonmajority religions (Dawson & Cowan, 2004).

Religion online is a mode of information dissemination called "one-to-many." There is one source and many recipients. The other side of the religion online equation, therefore, is how people access the information that is provided. One way to approach this issue is to look at how people use search engines to access religious information on the web. Using very new information science methods to examine over 5 million web searches from three search engines, Bernard Jansen and his colleagues found that most search terms employed in religious searches are associated with established, mainstream religions (Jansen, Tapia, & Spink, 2010). Web search data do not support the idea of significant religious "seeking" online.

Another approach is to ask people how they use the Internet. A 2003 survey by the Pew Internet & American Life Project found that 64% of online Americans use the Internet for faith-related reasons (Hoover, Clark, & Rainie, 2004), including the following:

- 38% have sent and received e-mail with spiritual content.
- 35% have sent or received online greeting cards related to religious holidays.
- 32% have gone online to read news accounts of religious events and affairs.
- 21% have sought information about how to celebrate religious holidays.
- 17% have looked for information about where they could attend religious services.
- 7% have made or responded to online prayer requests.
- 7% have made donations to religious organizations or charities.

Although the number of individuals who have used the Internet for faith-related reasons is impressive—some 82 million Americans—it is important not to overestimate this level of activity. As Cowan (2007, p. 362) has commented regarding this particular survey, although 64% have used the intent for faith-related reasons, less than 5% do so for those purposes on any given day.

In the end, the researchers conclude that most of the individuals are using the Internet to develop and express their personal religious and/or spiritual beliefs, but they are not true religious seekers. Instead, the cultivation of their personal faith online helps cement their preexisting beliefs and institutional attachments. The authors find this

> interesting because many analysts have assumed that the Internet would make it more likely for people to leave churches in favor of more flexible online options for religious or spiritual activity. Faith-related activity online is a *supplement* to, rather than a *substitute* for offline religious life. (Hoover et al., 2004, p. ii)

It seems to function much the same way televangelism does.

This is perfectly in sync with the way most religious groups are currently using the web: as online supplements to their "brick-and-mortar" operations. Although Gary Bunt (2009) noted that the "spiritual supermarket now has an online checkout" (p. 706), most religious groups are not yet ready to give up the shop and move their entire operations online, nor must they do so in order to meet the market demands.

Critical Thinking: Although more than 50% of residents in North America have Internet connections, only 5% of the world population has access to the Internet and the World Wide Web. Moreover, in some poor Global South countries, less than 1% of women have ever even heard of the Internet. What does this suggest about religion online?

Online Religion: Practicing Faith on the Web

In contrast to "religion online" is what scholars call "online religion." Cowan (2007) defined this as "the various ways in which religious faith is practiced over [computer-mediated] networks" (p. 361). The Pew Internet & American Life data on the religious uses of the Internet suggest some of the ways in which individuals engage more interactively while online. For example, some people send greeting cards related to religious holidays (35%); some make or respond to prayer requests (7%). Although these more interactive modes of online religion are not as common as the informational mode of religion online, the examples are interesting and suggestive of a possible future as the younger "digital generation" comes of age.

One of the most common forms of online religion is prayer sites. Many of these simply allow people to submit prayer requests, but one prayer site that has a strong interactive element is "Light a Candle." Entering this site (http://www.gratefulness.org/candles/enter.cfm?l=ENG), the visitor is told, "In many different traditions lighting candles is a sacred action. It expresses more than words can express. It has to do with gratefulness. From time immemorial, people have lit candles in sacred places. Why should cyberspace not be sacred?" Clicking Begin, the visitor is asked to "take a deep breath and quiet your thoughts" and then continue on. The visitor is then taken to a page with 30 candles on it, some lit, others not. The virtual flames on the candle flicker slightly and the lit candles actually get smaller over time as they "burn." The visitor is able to light any additional candles by hovering the cursor over the wick and clicking on it. Before the candle is actually lit, however, the visitor is told on an intermediary screen, "Before continuing, reflect for a moment on your reason for lighting a candle." On the subsequent screens, the visitor has the opportunity to add a dedication for the candle, enter initials to be able to find the candle again later, give a country of origin and state (for United States), and enter a name.

On the final page, the unlit candle shows alongside the dedication. Clicking on the wick "lights" the candle. The visitor is then encouraged to "Stay here quietly, as long as you wish, before continuing." In addition to lighting her own candle, the visitor can also click on any of the "burning" candles to see the source and dedication. The day we visited, 13,541 candles from 134 countries were lit, and it is possible to scroll through all 452 pages to look at them. Not all of the candles had a written dedication, but many did, including the following:

"Thinking of you and praying that you can find peace during this horrible time. My heart aches for you."—Anna from IL, United States

"I pray that I can be a good father and that my divorce will not harm my children or my ex-wife."—BT from CA, United States

"Ancient Gods and Goddesses, I invoke thee. Waters from the sky, Let it be." "I command thee now, to thee all. Listen to my desire, Rain fall!"—Mother Nature from Jeffersonville, IN, United States

Although the candles on this site exist only virtually, the sentiments expressed in all of the candles we examined seem very real, and the process by which one proceeded to light the candle was thoughtful and deliberate—if the user brought that spirit with them to the action.

Another example of a website for prayer actually has a real world dimension to it. The site virtualjerusalem.com has a 24-hour live "wallcam," a streaming video webcam fixed on the Western Wall (Kotel) in the Old City of Jerusalem. The Kotel is the last remnant of the Second Temple and therefore seen by many as the most important site in the world for Jews. Millions visit the Kotel every year, but those who cannot can still partake of the tradition of writing prayers to be placed inside the cracks of the walls by filling in a form on the Virtual

Jerusalem website, providing name, e-mail address, and a prayer of up to 200 characters. As the website explains, "Whatever your message, we will place your prayer among the countless others: a testament to the connection between God, Israel, and the Jewish People that refuses to be extinguished. Your prayers will be collected every week and taken down to the Wall by Virtual Jerusalem staffers."

According to Christopher Helland (2007), the virtual world known as Second Life (www.second life.com) has an increasing religious presence within it. Virtual worlds are three dimensional online spaces that people access in real time. In the case of Second Life, people create "avatars" (three-dimensional characters that represent themselves) who live "second" lives inside the virtual world. They have houses and jobs, they shop and party, they have relationships, and sometimes they attend religious services. Because the content of Second Life is user-generated, these services take place in meeting places designed by other users who can purchase and build on land within the virtual world. Religious groups from all traditions have begun to explore the possibilities of having a presence in virtual reality. There are Buddhist and Hindu temples, LGBT (lesbian, gay, bisexual, and transgender) friendly churches, Unitarian Universalist services, Wiccan covens, mosques, and synagogues (see Radde-Antweiler, 2008).

Of course, some Second Life religions are bigger than others. With an average attendance of almost 27,000 a week, LifeChurch of Edmond, Oklahoma, is the second largest megachurch in the United States, according to the Hartford Institute for Religion Research's continually updated database. Also known as LifeChurch.tv, it describes itself as a "multisite" church with campuses in six states and one that meets exclusively via the Internet (called "Church Online"). In addition, LifeChurch.tv has spent thousands of dollars developing a

presence in Second Life. The same simulcast video that unites the church's physical campuses is streamed into the Second Life sanctuary on Experience Island and projected onto large screens at the front of the auditorium so that those in the virtual world can worship in sync with those in the real world. One story on this development reported 3,912 Second Life avatars visited the church in a single day (Simon, 2007). Not long after LifeChurch was established, the Anglican Cathedral of Second Life opened it virtual doors. It offers seven services a week, including a Sunday Bible study, which people from 20 nations attend. In addition to its main cathedral, it also has a community center, peace garden, and labyrinth. The website Islam Online created a replica of the holy site of Mecca (Saudi Arabia) in Second Life so that individuals could learn about (and perhaps even perform) the Hajj, a pilgrimage that a Muslim is expected to undertake, according to the five pillars (duties) of Islam, at least once in a lifetime.

These online religion developments are still relatively new, so not much research has been published examining how people experience this form of virtual faith and practice. Some questions do necessarily arise, though. First, are these practices likely to generate authentic religious experiences? Second, are these religious practices and communities sustainable over time in the absence of face-to-face interactions? Third, how will the economic costs and technological limits of the Internet affect the durability of these sorts of interactive websites?[4] These are among the most important questions about online religion that sociologists will be attempting to answer in the future.

Although this discussion has treated religion online and online religion as mutually exclusive categories, Cowan (2007) has argued that they are actually ends on a continuum (see Figure 14.1). Most websites fall somewhere in the middle: They

---

[4]For example, in his scholarly analysis of virtual pilgrimages on the Internet, MacWilliams (2002) discussed Joseph Rice's Croagh Patrick (Ireland) Holy Mountain virtual pilgrimage. That website no longer exists.

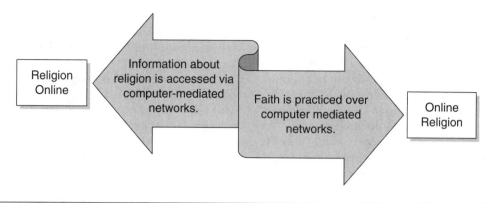

Figure 14.1   Continuum of Types of Internet Religion

SOURCE: Cowan, Douglas. (2007). Religion on the Internet. In J. Beckford & N. J. Demerath III (Eds.), *Sage handbook of the sociology of religion.* Thousand Oaks, CA: Sage.

provide some information and allow for some interactivity. This is especially true as more and more religious groups use Web 2.0 social media like Facebook and Twitter to connect with the people. The exact ratio of information to activity varies from site to site, though we think it is safe to say that at this point in time, the balance tips heavily in the direction of religion online over online religion. Nonetheless, it is important to bear this continuum, and not fall into false binary thinking, in mind in considering any incarnation of Internet religion.

> *Critical Thinking:* How "real" do you think online religion can be for people? What are some of the drawbacks? How might this new phenomenon affect religious communities that exist in your own community?

RELIGION AND SPORT

Another "outside the box" arena of social life in which religion figures prominently is sport

(Yamane, Mellies, & Blake, 2010). Any casual observer surveying the world of sport cannot help but notice scenes like these: With a gold cross dangling from his ear, Barry Bonds hits another home run; when he touches home plate, he points to the sky. U.S. soccer star Landon Donovan makes the sign of the cross prior to taking penalty kicks, as do countless baseball players before stepping into the batter's box, American football kickers prior to attempting field goals, and basketball players before shooting free throws. After leading the St. Louis Rams to a Super Bowl victory, quarterback Kurt Warner was asked by an interviewer on national television, "Kurt, first things first. Tell me about the final touchdown pass." Warner responded, "Well, first things first, I've got to thank my Lord and Savior up above. Thank you, Jesus!" After getting injured in the first game of the season, Wake Forest University quarterback Riley Skinner thanked God for the opportunity to play college football.

Sociologists, however, are not casual observers of the social world. The vast amount of religious activity we see in and around sport today can obscure from view the long-term trend toward the secularization of sport. A comment by sports commentator Frank Deford made the point well: "Sport owns Sunday now, and religion is content to lease a few minutes before the big games" (Price, 2005, p. 198). In fact, secularism is a key characteristic that distinguishes modern sport from ancient athletics (Guttmann, 1978). Secularism is particularly evident in the changing purpose of modern sport. From the beginning of human history, people aimed to please the gods through ceremonies, dance, and athletic activity. Originally, athletic competitions were fundamentally religious enterprises, meant to show special talents to the gods, express thanks to them, or implore them to take certain beneficent actions such as assuring the earth's fertility (Guttmann, 1978).

Examples abound. The Mayans and Aztecs erected stone ball courts next to their places of worship and often used stories of athletic competition to explain nature, believing, for example, that the sun and moon resulted from a game between the gods and a set of twin brothers. As a result of losing, the twin brothers lost their heads as a sacrifice. This tradition continued, with one player from every game being sacrificed. The secular Olympic Games we know today were created as an exercise of devotion to the Greek god Zeus. Athletes had to swear on the highest deity that they had been training for at least 10 months and would abide by the rules of competition. Violations of this oath led to fines that were used to construct statues of Zeus. The original Olympics lasted 5 days, of which only half of the time was used for competition; the entire first day was devoted to religious ceremonies. Native American tribes used sport to explain nature and to please the gods. Southwest Apaches used unwed males in relay races in honor of the masculine sun and the feminine moon (Baker, 2007).

In the Western world, athletics began to be approached more secularly by the Romans, and their principal purpose became not religious expression but entertainment. Roman athletes focused on fighting, as in gladiator contests, and Roman sport often pitted the members of the lower classes against each other as entertainment

for the elites. Victory usually allowed the competitor to survive, so winning was pretty important (Guttmann, 1978)! Modern sports, though less extreme, have adopted this Roman value as part of their secularization. According to Kliever (2001), the secularization of sport parallels the general pattern of societal-level secularization (discussed in Chapter 13)—religion "has lost effective control over vast areas of cultural life that were once conducted under its watchful eye" (p. 43). As with other institutional spheres— science and the arts, politics and economics, health care and social welfare—sport "operates under its own rules and pursues its own ends" (Kliever, 2001, p. 43).

This societal-level secularization changes the overall relationship between sport and religion as social institutions; it diminishes the religious significance of sport. At the same time, the institutional differentiation of religion from other social institutions facilitates the development of a free market for religion. This free market is a fertile soil in which organizations can freely compete for attention and adherents. Evangelical Christians in the United States in particular have been very adept at using sports organizations to advance religious ends—from the YMCA of yesterday to the sports ministries of today.

Muscular Christianity and the YMCA

The idea of "Muscular Christianity," a Christian commitment to health and "manliness," can be found in sections of the New Testament, but the term was not coined until the mid-1800s. Indeed, many Puritans were suspicious of sport for moral reasons. Like dancing, playing games was considered sinful—as an idle waste of time that could be better spent working or worshipping. Eventually, even these Protestant sects came to embrace sport, in part through the ideology of Muscular Christianity. Religious leaders who supported connecting religion and sport promoted the idea of the body as a temple. This notion provided a framework in which both a sound mind and sound body became essential in worship. The term *Muscular Christianity* was coined by the press to describe the work of authors Charles Kingsley and Thomas Hughes. Their "adventure novels replete with high principles and manly Christian heroes" sparked an interest in the social benefits of athleticism (Putney, 2001, pp. 12–13). Muscular Christianity's main focus was to address the concerns of boys directly, not abstractly, so that they could apply religion to their lives. The idea did not catch on quickly in the United States, but over time it has become one of the most notable tools employed in Evangelical Protestant outreach ministries.

The Young Men's Christian Association (YMCA) was started in England with strict religious ideals. Appalled at city lifestyles, George Williams created the YMCA as a place where men could fellowship together. The YMCA's initial activities in England were Bible studies, Christian readings, and prayer; all amusements were prohibited. The YMCA was designed to educate and promote Christian responsibilities in a world of temptation and self-indulgence (Baker, 2007, p. 47). Card playing, billiards, secular reading, and physical activity were forbidden by early English leaders, only to become a vital part of the American YMCAs.

In 1851, under the leadership of Captain Thomas Sullivan and the Reverend Lyman Beecher, Boston became the first city in the United States to open a YMCA. The Boston Association modeled its facilities after those of its English counterparts and emphasized the library and reading rooms where Bible classes could be held (Putney, 2001, p. 65). By 1856, there were over 50 YMCAs in the United States, from Georgia to California (Baker, 2007, p. 48). In time, American YMCAs began recruiting young men from all walks of life and employing a more secular approach than that of their English counterparts. Cards, secular novels, and athletic competition began to bring young men into the building, where leaders could preach the Word. Whereas English YMCAs acted as safe havens for Christian

young men, American YMCAs used popular activities to recruit and convert non-Christians. In 1860, the annual convention of the American YMCA decided that gymnasiums should be built at all YMCA locations, and by 1890 more than half of the 400 YMCAs in the United States had on-site gyms (Baker, 2007, p. 50), which were soon followed by bowling alleys, boxing rings, and swimming pools. Through use of these facilities, as well as camping trips and baseball leagues, the YMCA used sport and teamwork to expose young men to Muscular Christianity and "lead men to Christ."

It is difficult to underestimate the contribution to modern sport made by individuals associated with the YMCA. The term *bodybuilding* was first used in 1881 by Robert Roberts, a devout Baptist and gymnasium superintendent at the Boston YMCA, and William Morgan invented volleyball while serving as an instructor at the Holyoke, Massachusetts, YMCA in 1895 ("The YMCA in the United States," 2010). However, the YMCA's greatest contribution to sport came from James Naismith, a Presbyterian seminary graduate who was in residence at the YMCA Training School in Springfield, Massachusetts, when he developed the modern game of basketball in 1891 (Baker, 2007, p. 61). The sport's popularity grew exponentially over the years, and it has become the most popular organized YMCA sport. It has also become one of the leading evangelical tools for other Christian organizations such as Athletes in Action (AIA).

Sport remains integral to YMCA programming today. The "Y" sponsors leagues for baseball, soccer, tennis, football, basketball, volleyball, and gymnastics. The prevalence of these secular activities marks a dramatic change in the means employed by YMCAs today as compared with those at the time of their founding, but what of the ends? Although the mission of the YMCA remains "to put Christian principles into practice through programs that build healthy spirit, mind and body for all," there is little organized effort at the Y to proselytize today. According to William Baker (2007), "in 1888, most YMCA men agreed with Luther Gulick, who reminded

Although many players today do not know it, basketball was founded as a YMCA sport—part of an attempt to link Christian faith with sport through an idea of Muscular Christianity. Interestingly, this game is being played a few hundred yards from the first YMCA building ever built on a college campus in North America in 1883. The wood frame building is on the campus of Hanover College in Indiana, but it is now an Environmental Science Education Center—having become secularized like the YMCA itself. The sport spawned by the "Y" is alive and well on campus.

them that the gymnasium should always be a means to the end 'of leading men to Christ'" (p. 55). More than 100 years later, that end has long been lost at the YMCA as the organization has internally secularized.

Evangelical Protestant Sports Ministries

As the YMCA decreased its emphasis on sport as a method of increasing the religious sensibilities of young men, other organizations arose to fill the void. Professional athletes began openly sharing their testimonies, a practice that opened the door for many prominent religious leaders to use famous athletes in their efforts to attract young people to meetings. Billy Sunday, the famous evangelist, had begun his career in professional sports in the early 20th century. The first of several evangelical sports organizations, Sports Ambassadors (SA), was established in 1952. SA accelerated the use of athletes as spokespersons for Christ by organizing exhibition games in order to draw large crowds; during halftime, players would share their personal testimonies (Ladd & Mathisen, 1999, p. 129).

As the popularity of sports grew on college campuses, so did the opportunity for ministry. The Fellowship of Christian Athletes (FCA) was created in 1954 as a student–athlete Christian ministry and grew into an organization of summer camps and retreats designed to promote Christian ideals among high school and collegiate athletes (Ladd & Mathisen, 1999). Campus Crusade founder Bill Bright envisioned a more evangelical Christian ministry that he hoped would travel the world and preach the Gospel through sport. This approach became the focal point of its offshoot, Athletes in Action (AIA), founded in 1966 and intentionally positioned to the theological "right" of FCA. Today, AIA has a presence on nearly 100 U.S. college campuses and 35 professional sports teams. It fields summer teams in baseball, basketball, soccer, tennis, volleyball, wrestling, track and field, power lifting, and sports medicine to promote the Christian message and personal testimonies of Christian athletes.

As Mathisen (2006) observed, the founding of SA meant that "an entirely new genre of religious organizational forms was created, with sport occupying an essential presence" (p. 299). SA, FCA, and AIA are the "Big 3" Protestant sports ministries. Unlike in the formative years of the YMCA, which used religion to legitimize sport, the roles are now reversed: These organizations use sport to legitimize religion, which suggests the increasing social significance of sport and the relatively decreasing social significance of religion.

Religion and Spirituality at the Individual Level

In a secularized society, people can choose *whether* to be religious and, if so, *how* they are religious. They can choose to make religion part of their identity as an athlete or to make athletics part of their religious identity—or not. As the examples provided earlier suggest, many individuals do visibly choose to connect religion and sport. Still, for every Bob Cousy, who made the sign of the cross before shooting free throws, how many Catholic professional basketball players do not? Moreover, of all individuals who shoot free throws, how many make a religious sign prior to shooting? Unfortunately, efforts to measure this sport/faith relationship scientifically have lagged behind its media visibility.

Some smaller-scale studies suggest a link. For example, Storch and his colleagues (Storch, Bravata, and Storch, 2004) compared 57 intercollegiate athletes and 169 nonathlete undergraduates at the University of Florida and found that athletes had higher levels of conventional religious faith. There are good reasons to expect this finding. Sport shares many of the same values as certain religions, particularly evangelical Christianity (Mathisen, 2006). Overman (1997) highlighted the connection between sport and the Protestant ethic of success, self-discipline, and hard work. The Protestant ethic—with its emphasis on delayed gratification—is especially conducive to participation in organized sporting competition rather than in free and expressive play (Coakley, 2007, p. 538).

How generalizable are the findings of Storch and his colleagues? That is a question taken up by David Yamane and Teresa Blake using representative data on college students collected by UCLA's Higher Education Research Institute. Using multiple regression models that control for several other variables, Yamane and Blake (2008) found that college athletes are *less* religious than college students in general. It is possible that there is a value conflict between sport and religion (competitive aggression versus cooperation and compassion) that could lead to

lower levels of religious commitment, but the authors believe that sport is better understood as a *secular competitor* to religion. Religion and sport, in this view, are part of a zero-sum game (only one can win), since both require investments of time and energy. Indeed, the most visible manifestations of religion in sport—wearing religious symbols or making religious gestures—require no real religious commitment by the individuals in question (think Barry Bonds). Generally, when push comes to shove for college athletes, sport wins out over religion.

Of course, in the contemporary United States, it is important to distinguish between religion and spirituality. Whereas religion is sometimes associated with inherited tradition and dogmatic beliefs, spirituality is seen as a quality of an individual, particularly as relates to his or her personal experience. Spirituality is seen as a more primary and pure relationship to the divine than is "religion." This distinction raises the possibility that college athletes are perhaps spiritual but not religious. If so, this might be accounted for in two ways. First, one can more easily be spiritual on one's own terms, making it easier for athletes to accommodate spirituality in their busy lives. Second, for some, athletics itself can be a form of spiritual practice (Hoffman, 1992), and their peak or flow experiences during performance (Jackson & Csikszentmihalyi, 1999) may sensitize them to a spiritual dimension of life in general.

Yamane and Blake (2008) find no general support for this position. Their analyses of the UCLA data suggest that college athletes are actually *less* spiritual than students in general. Why? Perhaps, according to Mathisen (2006), because modern sport is the antithesis of play, "If play is free, sport is highly structured; if play is outside the ordinary, sport has become worklike; and, importantly, if play is intrinsic, sport is extrinsic" (p. 288). Therefore, as Mathisen (2006) noted, it may be precisely *outside* of formally organized sporting activity that we may see the true "signals of transcendence" (p. 287) that Peter Berger spoke of when examining *play*. Kliever (2001) also emphasized the notion that a modern ethic of work in sport trumps the ethic of play. Play, then, may be closer to spirituality than is sport.

Although college athletes appear generally to be less religious and spiritual than their fellow college students, one important exception emerges in Yamane and Blake's (2008) analyses. Evangelical Protestant athletes are *more* religious *and* spiritual than college students in general. This is not surprising given the history of "Muscular Christianity" in the United States, as well as the popularity of groups such as AIA and FCA on college campuses (Cherry, DeBerg, & Porterfield, 2001, p. 27). The theology of Muscular Christianity allows evangelical Protestants to sacralize sporting events and also provides a framework that allows athletes to negotiate conflicts between sport and their religious beliefs (Coakley, 2007, p. 556). This theology gets activated by AIA and FCA ministers and groups that provide strong systems of social support for belief and practice—"plausibility structures" (Berger, 1967), as we called them in Chapter 8—that are key to sustaining religiosity.

Of course, more systematic studies on more general populations are necessary in order to draw conclusions with certainty. For now, we would provisionally say that religion and sport in the lives of individuals are separate spheres of existence that connect at some times and for some people. However, as we would expect from the theory of secularization, there appears to be no inherent connection between them today.

Thus, to characterize the relationship between religion and sport in modern America as a "fusion" is an overstatement (Price, 2005). At certain times and places, religion and sport do have a close relationship. Still, this is not the same as saying that sport has a constitutionally religious dimension, as in Mayan ball games or sumo wrestling under Shinto. The idea that sporting activity is fundamentally religious, or that religious authorities can regulate sporting activity, is a thing of the past. The secularization of sport as a social institution is thus a reality. It is also a reality that sport instills values into young people that can have a pervasive and life-long impact—rather *like* religion.

INVISIBLE RELIGIONS

As we discussed in Chapter 5, several scholars have stressed the individualization of religion—the way in which each individual in modern society constructs his or her own meaning system by drawing from many religious and secular philosophies. Perhaps the most important work developing this thesis is that by Thomas Luckmann (1967). Luckmann used an extraordinarily broad definition of religion, referring to religion as the "symbolic universes of meaning" that infuse all of life with a sense of transcendent purpose. He emphasized worldview as an elementary and universal manifestation of religion (Luckmann, 1967). In this respect, Luckmann's definition of religion is similar to other functional definitions (Geertz, 1966; Yinger, 1970). However, rather than limiting religion to macrosystems of meaning—meaning systems that address death, suffering, and injustice—he sought to understand worldview at all levels of generality and specificity. He insisted that "no single interpretive scheme performs the religious function. It is rather the worldview as a whole, as a unitary matrix of meaning" that defines one's identity and serves as one's religious orientation (Luckmann, 1967, pp. 55–56). In essence, he pointed to personal identity as "a form of religiosity" (Luckmann, 1967, p. 70). A person's sense of identity—his or her values, attitudes, dispositions, and sense of self-worth—are part of his or her religiosity because all these are related to feelings about what makes life worth living. These are "invisible" forms of religion in that they do not have the social manifestations one normally associates with religion.

This is certainly a broad definition. Many social scientists have objected that it makes everything religious—or makes nothing at all specifically religious. They make an important point. The aspect of this definition that we find intriguing is that Luckmann has defined religiosity in a way very compatible with that of a number of 20th-century theologians. Richard Niebuhr, Paul Tillich, and a number of other modern theologians have strongly resisted the idea that one's faith or one's religiosity is expressed primarily through cognitive beliefs. Rather, they insisted that one's faith is most fully manifested in everyday assumptions, in actions, and even in personality structure. Hence, Niebuhr and Tillich sought to discover one's "real" center of worth by exploring the issue of what one ultimately trusts. Given the fact that some of the most widely acknowledged theologians (whose trade is meaning systems) have defined religiosity similarly to Luckmann, his formulation deserves our attention.

Luckmann believed that as society has become increasingly complex, and as institutions have specialized their sphere of influence, traditional religions influence a decreasing range of human behavior and thinking. Combined with this is the tendency of traditional religions to freeze their systems of belief so as to make them seem more eternal, absolute, and unchanging. At the same time, technological, political, and economic changes have continued to occur; indeed, in the modern world, change occurs at ever increasing rates. Luckmann maintained that this has caused traditional forms of religion to become irrelevant to the everyday experiences of the common person. He denied that this represents a decline of religiosity. The common person is as religious as ever, but the religiosity of the laity has taken on new forms. Luckmann insisted that claims of a decline in religiosity are due to the fact that sociologists have usually asked questions that measure only traditional religiosity (formal affiliation and attendance, belief in traditional doctrines, frequency of prayer, and so forth).

In the modern world, people derive their sense of meaning by drawing on a wide range of religious and secular philosophies. Each of these competes for the loyalties of the citizen, who is basically a consumer at the marketplace (an idea consistent with rational choice theory of religion). The product that each philosopher is selling is a worldview—with its own center of worth or system of values and its own definition of what makes life worth living. Popular religious

tracts, *Playboy* magazine, psychological theories expressed in best-selling books and magazines, and underlying themes and values in popular television programs can all affect a person's sense of the meaning of life and one's individual "philosophy of life."

Other organizations, social movements, or businesses also compete in the philosophy-of-life marketplace. Libertarianism is a political movement that exalts the rights of the individual to seek his or her own self-interests without interference. The prime formulator of libertarianism was the late Ayn Rand, whose newsletter was faithfully read by believers and whose public addresses packed houses with enthusiastic followers. Rand stressed individual initiative and the survival of the fittest and believed that altruism was the worst sort of vice. Selfishness, if one followed the logic of her argument, was the most exalted virtue and would ultimately lead to the best type of society. At the opposite end of the political spectrum, Marxism offers a coherent outlook on life and a constellation of values that promises to bring a better life in the future through collective action and collective consciousness. Each of these social movements offers a philosophy of life and a set of values that compete with traditional religions in defining the meaning and purpose of life.

Even business enterprises, like Amway Corporation, seek to motivate by stressing the primacy of financial independence, the ultimate value of free enterprise economics, and the rewards of close friendship with other distributors. In fact, the regular Amway weekend regional rallies can be analyzed as plausibility structures (see Chapter 8) that operate to reinforce the believability of the values and outlook presented by the corporation. Privatization of religion involves each individual developing a personalized meaning system or philosophy of life by drawing from many sources in modern life, including secular media, the traditional religions, and popular televangelism programs.

Talcott Parsons and Robert Bellah viewed the privatization process as a good and healthy sign, while Peter Berger pointed to the phenomenon as evidence of a decline in religion (discussed in Chapter 13). While Luckmann did not see the process as indicative of a decline in religion, neither did he view it as a particularly healthy trend. When individuals must construct their own meaning systems, those systems may seem less eternal and less compelling. The individual may therefore experience anomie or normlessness. Further, those who do construct a sustainable meaning system often develop one that is so privatized that it offers meaning only to the individual—ignoring the larger social structure. Because many privatized meaning systems in modern society exalt the autonomy of the individual (self-realization, individual social and geographic mobility, etc.), the locus of meaning is in the individual biography (Luckmann, 1967). With this locus of meaning, individuals are not likely to make sacrifices on behalf of the larger society. If this orientation continued indefinitely, the needs of the society itself would go unmet. For this reason, the privatization of religiosity could be unhealthy in the long run for the larger society.

Hence, Luckmann insisted that religiosity is not declining in the modern world; it is undergoing transformation. An alternative form of religiosity has been developing—a form that does not look like religion to many people because it lacks the institutional structures and the conventional dogmas characteristic of traditional religions.

Luckmann's thesis has drawn a great deal of attention. Several attempts have been made to measure the relative influence of traditional religious views and other "popular" meaning systems in personal philosophies of life. The results are mixed: Richard Machalek and Michael Martin (1976) found evidence to support Luckmann's thesis regarding invisible religions; Hart Nelsen and his colleagues (Nelsen, Everett, Mader, & Hamby, 1976) did not. William Bainbridge and Rodney Stark (1981) studied lay attitudes toward traditional religious doctrines and found that they may not be as impotent and irrelevant to the average citizen as Luckmann implied. On the other hand, Robert Wuthnow

(1973) found substantial variations in the personal theologies of seminary students—many departing significantly from traditional Christian theology. In any case, most social scientists would grant that the meaning systems of most Americans seem to be somewhat eclectic, with traditional religiosity, patriotism, and other value systems converging. At the present time, we do not know for sure whether this phenomenon is any more common in the modern world than it was in past eras.

Readers may find it interesting and worthwhile to reflect on their own sense of meaning and their own system of values. Do all your values evolve out of a traditional religion? Most of them? Some of them? What other sources have affected your outlook on life? What about the sense of meaning and the personal values of your friends and acquaintances? Does it make sense to you to refer to personalized systems of meaning as a form of religiosity? Why or why not? Is it essential for a meaning system to address the meaning of suffering and death in order for it to be called religion? These are important issues that have divided sociologists in their approaches to studying religion and in their generalizations about religious trends in this country.

Critical Thinking: Do you think religion is taking "invisible" forms? If it *is* happening, is this a destructive trend for religion and for society? Give your rationale for your answer.

QUASI-RELIGIOUS MOVEMENTS

Loosely integrated societies, in which the intensity of commitment to cultural tradition is low, are more likely than tightly integrated ones to generate cults and other nontraditional social movements (Stark & Roberts, 1982). So it is not surprising that in the United States—a pluralistic and rather loosely integrated society—there are many new religious and quasi-religious movements. Some of these movements hold rather esoteric beliefs; others are based on concepts from popular psychology literature and movements. Among the quasi-religious movements that we discuss in this section, none attracts a large following, and most do not attempt to articulate a comprehensive worldview that explains the meaning of death, suffering, and injustice. Nonetheless, they have collectively affected a significant segment of the North American population, especially in urban areas. Some of these kinds of movements elaborate their ideologies, develop theodicies, and evolve into full blown NRMs.

One such quasi-religious orientation is astrology, a set of beliefs about impersonal forces in the universe that profoundly influence human life on Earth. These forces can be "read" or predicted through an understanding of the stars. The zodiacal sign under which a person is born is thought to influence significantly (or even determine) one's personality structure and one's thinking processes. Astrology is not a new phenomenon, nor is it limited to any particular age group. It is not organized around a particular group of people (there is no "faith community"), there is no ordained clergy or other sanctioned leadership hierarchy, and there is no formal doctrine. Yet certain principles and beliefs, which are transmitted through books and word of mouth, are common to those who believe in astrology.

A surprisingly high number of Americans believe in astrology. According to a study by the National Science Foundation, 6% of the respondents reported that they read their horoscopes daily, and 7% of Americans reported that they believed astrology was very scientific (National Science Foundation, 2000). Many others follow the horoscopes printed in newspapers, know their zodiac sign and the characteristics of persons under that sign, and "half-believe" in the efficacy of astrology (i.e., they are not fully converts, but they remain open to astrology and believe that there is probably something valid about it). Roughly 44% of Americans check their horoscopes at least occasionally (National Science Foundation, 2000). In Canada, Bibby

found that 5% of the population firmly believe in astrology, and another 44% remain open to the possibility that it is true. He found that fully 77% of those polled read their horoscopes at least occasionally (Bibby, 1979).

Astrology is sometimes integrated into the worldviews of persons who are members of mainline religious groups. Their religiosity is a synthesis or blend of a traditional religion and astrology. For others who are not active in traditional religious groups, astrology may play a more significant role in their overall worldview. One study found that astrology seems to serve as an alternative to conventional religion for some people—especially marginal or subjugated members of society: the poorly educated, nonwhites, females, the unemployed, the overweight, the unmarried, the ill, and the lonely (Wuthnow, 1976a).

Another form of quasi-religious movement has focused more on the development of untapped human capabilities. Transcendental Meditation (TM), Silva Mind Control, and Scientology, are examples of this type of movement. TM is a meditation technique that bears some resemblance to yoga. Its advocates insist that it is not a new religion, and practitioners include both people who are active in traditional religious groups and those who are by traditional measures nonreligious. Some persons use this technique as a means of relaxation; others seek to tap the "cosmic consciousness" that is the ultimate source of energy in the universe.

TM was started in the late 1960s by a Hindu teacher, Maharishi Mahesh Yogi. It involves chanting a *mantra*—a word, phrase, or sound which is given to each recruit. A mantra is not to be shared with others, but chanting one and concentrating exclusively on it offers one a channel to inner bliss. TM masters maintain that such social problems as war, poverty, crime, and racism would disappear if everyone would engage in TM. They maintain that if everyone were in tune with the cosmic consciousness, most serious social problems would not exist. People would be more relaxed and more able to fulfill their cosmic purposes (Needleman, 1970). TM claims it is not a religion but only a discipline of meditation. Still, the organization's webpage describes them in terms that have quasi-religious overtones:

> The Transcendental Meditation technique is based on the ancient Vedic tradition of enlightenment in India . . . handed down by Vedic masters from generation to generation for thousands of years. About 50 years ago, Maharishi—the representative in our age of the Vedic tradition—introduced Transcendental Meditation to the world, restoring the knowledge and experience of higher states of consciousness at this critical time for humanity. When we teach the Transcendental Meditation technique today, we maintain the same procedures used by teachers thousands of years ago for maximum effectiveness. (TM, 2010)

While TM was founded by a Hindu leader, it is used by some Christians and by persons unaffiliated with any religious group. The organization claims 5 million practitioners worldwide (Bickerton, 2003).

Scientology is a highly organized and fully institutionalized movement related to popular psychology and parapsychology (belief in clairvoyance, telepathic communication, and psychic healing) and uses an educational model. It offers to help people become "clear" and to maximize their human potential. The more courses one takes, the higher one moves in the stratification system of the group. This, of course, enhances instrumental commitment, for one makes a financial investment (the courses are not cheap) and begins to rise in the system of respectability and esteem within the group (Bainbridge, 1978; Bainbridge & Stark, 1980; Wallis 1977). The Silva Method is a similar training-based meditation movement that offers to help people develop their psychic powers (extrasensory perception) and move people to a new stage of human evolution (Silva Method, 2010). Some observers view these movements as essentially business enterprises (the courses are substantial sources of income). However, Scientology, in particular, is very explicit about its claim to being a religious movement, stressing that its philosophy is a synthesis of western and eastern spirituality with modern science (Scientology, 2010).

Some writers have attempted to formulate an eclectic philosophy that purports to synthesize these diverse orientations into a unitary—but very diffuse—movement called the New Age movement. Proponents such as actress Shirley MacLaine and writer Marilyn Ferguson (1987) have claimed that a compelling new religion is emerging from these movements and from scientific sources. The most systematic statement for this perspective is Ferguson's book *The Aquarian Conspiracy*. New Age theology is very mystical (intuitive) in its source of knowledge, but it also blends popular psychology, "holistic medicine," process philosophy, and—from the natural and social sciences—general systems theory. Some advocates of New Age thought are theistic, but not all adherents believe in a deity. The New Age movement does seem to have a strong supernatural spirituality as a common core. Ferguson insists that this emerging form of spirituality has no organization but is an unplanned and uncoordinated groundswell. Because it has no organization and takes so many forms, the New Age movement seems to be largely in the eye of the beholder. Because its advocates are very process oriented (denying any absolute answers in life) and because they so fully embrace humanistic values, the New Age theology has become a favorite point of attack for conservative Christians.

Another form of quasi-religious movement is focused generally on science fiction and more specifically on the TV and film series *Star Trek*. *Star Trek* fan clubs have many characteristics that resemble religious bodies (Jindra, 1994). Michael Jindra (1994) argued that "without its institutional and confessional forms, we often fail to recognize religion in our own society" (p. 31), yet the *Star Trek* cult involves an origin myth, a set of beliefs, an organization to support and foster the beliefs, a network of fans who cultivate social bonds with one another, and rituals performed at collective gatherings. At one national *Star Trek* convention, a child was even playfully "baptized" into the pseudochurch.

Even the television program's writer and director, Nicholas Meyer, has said that *Star Trek* "has evolved into a sort of secular parallel to the Catholic Mass" (quoted in Jindra, 1994, p. 32). The program is built on a positive view of the future and is based on rational decision making, egalitarian values, a universalism of all living beings in the universe, and belief in the ultimate progress of humanity. In one sense, the movement deeply expresses both an American and a modernist worldview. Much like many religious groups, the most profoundly committed fans also cope with stigma for their beliefs. Yet many of these superfans suggest that Star Trek provides models and myths that influence them profoundly in everyday decisions and in making sense of the meaning of life.

The quasi-religious movements discussed here are only illustrative of a number of such movements that offer inner peace, ultimate fulfillment, spiritual expansion, or insights into the "truth" about human existence. Meher Baba, Spiritual Scientists, Association for Research and Enlightenment, Spiritual Frontiers Fellowship, and I Ching are only a few of the many other religious movements one might explore. Whether some of these are actually new religions may be debatable, but they do seem to represent a form of spirituality that might influence traditional religiosity or serve as an alternative mode of religion. For example, a person may use TM simply for stress reduction, to get into deeper touch with his or her inner self, or to tap a cosmic source of energy. Yet that person may not consider this activity "prayer" and may not respond positively to other traditional measures of religiosity (doctrinal orthodoxy, worship attendance, etc.). Likewise, a person who believes in astrology may not be religious in traditional ways, but he or she may use astrology as a worldview that offers to explain the meaning (or at least the cause) of events. Any empirical studies of religiosity that attempt to explain current trends and likely patterns for the future must take into account the possibility that religiosity is not declining but is changing in both form and substance.

The nature of religiosity in North America does seem to be in transition. Worldviews of both Canadians and Americans, including those who are active in traditional religious groups, appear to be somewhat more this-worldly (or secular, if you prefer that term). There also seems to be less willingness to assent to traditional doctrines. This may very well be a sign of an increase in the privatization of religiosity and an increase in syncretistic worldviews—patching together a meaning of life from diverse sources. At this point, we have little comparative data for firm generalizations. The meaning systems of people have, no doubt, always been characterized by a good deal of syncretism, but because of changes in access to the mass public due to television and other mass media, self-help groups, and others espousing their own philosophy of life probably have more influence on common citizens than in earlier eras. This may account for more individuality in meaning systems.

Whether this is a trend that will have unfortunate consequences for the society as a whole remains to be seen. The trend may have negative effects for the established religious communities: If fewer people feel committed to the theology that traditional religions espouse, it could involve a decrease in commitment to those organizations. On the other hand, privatization may bestow other offsetting benefits to the society—and perhaps even to religious organizations. Predictions at this point are highly speculative. The recent downturn in traditional forms of religiosity may be the beginning of a pattern, but at present the attendance levels at conventional faith communities are much higher than they were in 1776—when only 10% of the American population belonged to a church (Finke & Stark, 1992; Stark & Finke, 1988)—and are higher than other Global North countries in the modern world.

Our own interpretation is that religion is not in a declining phase—but readers should remember that we use a broad definition of religion. Systems of meaning that offer to explain the meaning of human events through a worldview, an ethos, and a system of symbols are not likely to disappear. However, certain traditional views and certain established religious institutions may well decline in their influence. Whether the reader views this change of religiosity as a decline in religion will be determined largely by his or her operational definition of religion.

> ***Critical Thinking:*** In our society there are a wide range of quasi-religious movements affecting Americans. Their core values often come from conflicting sources: traditional Judeo-Christian theology, or patriotism, or pop psychology, or playboy philosophy, or consumerism, or astrology. What are the consequences of this for religion? Why?

SUMMARY

In this chapter, we highlighted the benefits of "thinking outside the (God) box" for understanding religion in modern society. Looking only at religious organizations proper—churches, temples, sects, denominations, and congregations—we get a different view than if we look at the many ways that religion has changed its relationship with our social institutions and taken on a more fluid form in our culture.

Our first focus here was on religion and the media. These two institutions have long had a close relationship, with religious leaders and groups using the media to reach far beyond the walls of their congregations. The new media based on computer technology and the Internet at first blush would seem to pose revolutionary new possibilities for religion, but thus far, both on the production and consumption sides, action has been fairly traditional.

A second focus was on religion and sport. As the historical connection between religion and sport was broken in modern society, action moved to the organizational level, with parachurch groups like the YMCA taking center stage and negotiating the secular environment. At the individual level, there is no integral connection between religion and sport; as in other matters of faith in modern society, athletes can choose to integrate these two identities or not.

A third line of thinking emphasizes what Thomas Luckmann called "invisible" expressions of religiosity. Individuals may create their own systems of meaning, compiled from the philosophies expressed in popular religious tracts, best-selling pop psychology books, popular television programs, or independent political organizations of the far left or far right. Such invisible religion is often vulnerable because it lacks a plausibility structure, and it often is relevant only to one individual.

Our fourth and final consideration was quasi-religious movements. Astrology, TM, Silva Mind Control, Scientology, and the New Age movement are among these. They each attempt to explore supernatural forces and/or philosophies of life that can bring greater fulfillment.

Some scholars believe that these alternative forms of religion mean that religion is not so much declining in North America as changing in significant ways. New forms are emerging and old forms are taking on modified form. Religion in North America clearly does adjust and adapt to the larger society of which it is a part. As we will see in the next chapter, religion is also affecting and being affected by globalization.

15

RELIGION AND GLOBALIZATION

Here are some questions to ponder as you read this chapter:

- What is *globalization*, and why is it relevant to the study of religion?
- How might religion influence global events?
- How does the process of globalization influence religion?
- How might the existence of a religious transnational organization influence the power or legitimacy of a religious community within a nation?
- How do transnational migration patterns influence religion?
- How might a religious group contribute to conflict and warfare, even though the meaning system stresses a commitment to peace?

---

 consistent theme throughout this text has been that religion is a complex system that is involved in complex interrelationships with other structures and processes of the society. The meaning system, for example, is composed of an interplay of rituals, myths, and symbols, all of which reinforce a worldview and a sacred set of moods and motivations. That meaning system, in turn, is only one subsystem of the larger system we call religion. This larger system is made up of meaning, belonging, and structural (institutional) systems, all of which are interrelated in complex and sometimes contradictory ways. We found, for example, that the vested interests of officials in the institutional system may cause them to behave in ways that run counter to the meaning system—as occurs when denominations refuse to promote clergywomen to large and successful churches or when clergy are not promoted when they speak out against racism or other forms of inequity and bigotry.

Religion, in turn, is part of a larger system, the nation, and is in interaction with economic, health care, political, and educational institutions. Sometimes these interactions are mutually supportive and integrative, and sometimes they are conflictual.

The nation as a macro system, in turn, relates to yet a larger system: the world system of interdependent nations, involving transnational division of labor and mutual dependence for needed resources and products. (See Figure 3.5 in Chapter 3 on page 64) In this web of networks, subsystems can impact larger systems in important ways: For example, economic collapse in a country can affect not only that nation but the interaction of that nation with other nations. The processes in a more macro system can also shape processes in smaller systems: A multinational trade treaty or decisions made by the United Nations may profoundly affect the internal workings of a country's economy and therefore the economic circumstances of a family.

This chapter will briefly explore the role of religion in the largest of networks, as the entire world increasingly takes on the characteristics of a social system. We will find that even at this level, what religions *say* may differ in important ways from what they actually *do*. We will find that religion can affect global processes, but it is also often affected by them.

The scope of the impact of globalization on religion is sometimes indirect and subtle, but it is also sometimes extensive. For example, Wuthnow (1991) pointed out that Christianity in the United States has undergone an increasing split between those adherents who are quite well educated and those who have much less schooling. The increased educational level of many Americans—with large numbers of college educated citizens in the late 1950s and thereafter—was caused largely by government policies and huge outlays of funds to support higher education. Why did the government make these decisions in the focusing of resources? These actions were largely in response to the Cold War and space race with the Soviet Union and to the increasing role of the United States in high-tech international trade. Competition in the global markets required more people with university training. A side effect of that education, however, was a modification of the worldview and religious sensibilities of much of the citizenry. Wuthnow (1988) believed that a conflict—between those with high levels of education and those with less—has transformed religion in the latter half of the 20th century.

The more highly educated have become less particularistic, more open to other traditions and to the insights of science as they pertain to religion, and more willing to entertain alternative interpretations of scriptural passages based on biblical criticism. These modifications have led to conflicts between people who have attended universities and their neighbors on the same pew who have less formal education. So changes in the educational level of U.S. citizenry have influenced religion and religious communities. This brings us back to the open systems model and challenges us to recognize that some of the inputs that impact religion are from global forces—not just ones within national borders (see Figure 15.1). Further, as we shall see later in this chapter, religion may have *outputs* that impact international relations and global processes.

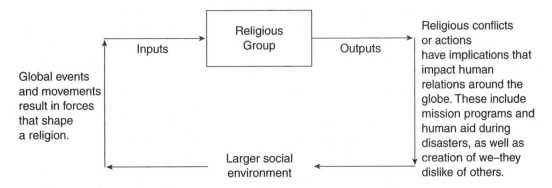

The idea that religious groups might impact international events suddenly became more real to many Americans on 9/11, when the World Trade Center was attacked in New York City. Religion had emboldened 19 men in an extremist Islamic movement to believe that their suicide would ensure a place in the afterlife. That religious outlook had also been used to create an intense sense of "us" versus "them." The result was catastrophic for the United States, for its citizens' sense of security and safety, and for international relations. It resulted in reallocation of financial priorities for a number of countries and a number of businesses (such as airline companies). Whole new security industries and technologies have been developed in response. The world is a different place, and religion played a role in the transformation.

Figure 15.1 Open Systems in a Global Social Environment

Critical Thinking: What are some other ways in which various nations, industries, and international relationships have been affected by the attacks on the World Trade Center and the Pentagon on 9/11? How was religion involved in *initiating* these changes? How have faith communities been *affected* by this event?

THE GLOBALIZATION PROCESS

During the past century and a half, and more especially during the past 50 years, societies around the world have been undergoing a radical transformation; each society and nation has become less an isolated and autonomous unit. The process, called globalization, has involved the entire world becoming "a single sociocultural place" (Robertson, 1989). This change involves several interdependent processes: (1) a structural interdependence of nation–states; (2) a synthesis and cross-fertilization of cultures as societies borrow ideas, technologies, artistic concepts, mass media procedures, and definitions of human rights from one another; (3) a change in socialization to a broader inclusiveness of others as being "like us" and to a sense of participation in a global culture; and (4) an increase in individualism, accompanied by a decrease in traditional mechanisms of control, such as strong clan and family ties (Robertson, 1989).

One scholar points out that the change of the millennium (January 1, 2000) was perhaps the first time that people around the globe all celebrated the same event. Because of mass media, the change to the third millennium on the Western calendar was acknowledged with celebrations in each of the 24 time zones, with those in other time zones watching. This, said Casanova (2001), is surely the first collective celebratory event of a virtual gathering community around the entire planet. This might be considered evidence of our becoming a single socioculture place.

There are several theories of global interdependence and transformation. One of the most widely discussed perspectives in the 1980s and 1990s has been "world systems theory," developed in large part by Immanuel Wallerstein (1974, 1979, 1984). This perspective, sometimes also referred to as "dependency theory," is based on a Marxian or conflict model. Affluent industrial societies based on a capitalistic economy are referred to as the *core* of the world economy. Interestingly, the most powerful countries in the world have formed a policy and consultation group—calling themselves the Group of 7 (or the G7—the United States, Japan, Germany, Canada, France, Great Britain, and Italy). Recently, Russia was added, and the group is now the G8. The G8 uses its collective power to regulate global economic policies to ensure stability (and thereby ensure that members' interests are secure). The G8 has the power to control world markets through the World Trade Organization, the World Bank, and the International Monetary Fund (Brecher, Costello, & Smith, 2006; Stein, 2006; Weidenbaum, 2006; Weller & Hersh, 2006).

Poor, nonindustrialized societies are viewed as part of the *periphery*. These peripheral countries typically have narrow, export-oriented economies, based largely on providing raw materials to the industrialized countries. They import manufactured products and in the process develop enormous debts to the richer nations. As a result, they become highly dependent on the developed countries. They remain poor not because of lack of resources or hard work but because they are exploited by the "haves" who control capital and technology (Shannon, 1989). These nations responded to the creation of the G7 with an organization of their own that they called the G77. They are attempting to create collective unity so they will have some power to determine their own destinies (Brecher et al., 2006; Hayden, 2006). The map below shows where the G8 and G77 nations are located. As you can see, the G8 nations are all located above the 15th degree of north latitude, so the current term used for the most powerful and less affluent countries are the Global North and the Global South.

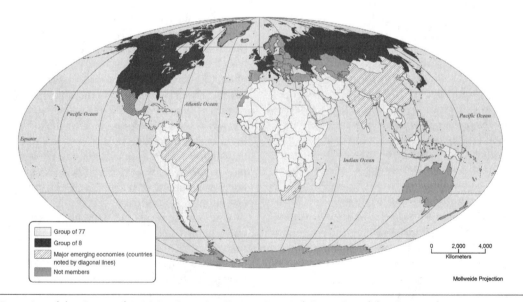

Countries of the Group of 8, Major Emerging Economies, and Countries of the Group of 77

Although Wallerstein's world systems theory says little directly about religion, it assumes that secularization is an unstoppable force and that religion is a relic destined for decline. Thus, religion does not receive much attention by these theorists (Robertson, 1992; Simpson, 1991). The prime movers of social behavior are thought to be economic. Like most neo-Marxian theories, when religion *is* discussed, it is often viewed simply as a mechanism by which the affluent preserve their power and keep the poor in their place. Ironically, as we shall see, some religious movements—especially within the Roman Catholic Church—use world systems theory to explain poverty in poor countries, despite the theory's indifference to religious variables.

An alternative view of globalization is set forth by Roland Robertson, whose perspective takes cultural values and norms more seriously as determinants of social processes. Robertson is convinced that religion is an important player in the global transformation process. Since he gave more consideration to the role of religion, we will explore his sociocultural theory in more depth in this chapter. However, those interested in global inequality issues should continue to be aware of the insights offered by Wallerstein relative to how power is used to serve the self-interests of the "haves" of the world.

According to Robertson, globalization has been influenced by religion and has impacted religion in important ways; it has been both a dependent and independent variable relative to religion. The topic of globalization is itself a complex issue, and as with other issues in this book, we can only illustrate some of the issues and demonstrate how sociological perspectives can be applied to the globalization process. Our discussion here will explore the emergence of global theologies, the impact of globalization on traditional religions, and the role of religions in international politics.

Critical Thinking: Provide examples—from economics, socialization, or other dimensions of life—of the impact of globalization on your own life.

THE EMERGENCE OF GLOBAL THEOLOGIES

Robert Wuthnow (2009) insisted that the Christian tradition has always been cross-national in character. Certainly Judaism and Islam have long histories of spanning cultures and regions before any "nation," in the modern sense, even existed. Religious groups are probably the first transcultural or transnational institutions and religious individuals among the first migrants (Levitt, 2003). Still, one interesting religious development during the past 120 years is the increase of global perspectives in the theologies and ethical systems of major world religions. Contemporary religious groups often stress the idea that the world is a single place and that all humans are children of the same God. Rather than stressing differences based on ethnicity or race, contemporary world religions often combat bigotry with affirmations of the fundamental similarity of people everywhere. For the thousand years prior to the 18th century, remarkably few efforts had been made to reach out to persons in other cultures (Hunter, 1983; Lee, 1992; Meyer, 1988; Vidler, 1961). Then in the 18th and 19th centuries, outreach programs were focused mostly on proselytizing and conversions. In the 20th century, the World Council of Churches, as well as many Christian denominations, sponsored a variety of programs aimed at health care, hunger relief, and amelioration of suffering caused by natural disasters. While these examples draw upon Christianity, global inclusiveness and cross-boundary compassion is a modern world trend in other world faiths as well.

Religions often construct images of the oneness of the world in their myths (Simpson 1991), the modern imagery of the "global village" being one example advanced by the World Council of Churches. The image itself creates a view of a closed, close-knit, interdependent, communal society (Turner, 1991a). Perhaps it is because of unifying images that religion often serves as a basis for unity and integration, even when language differences might separate. Roman Catholics speaking different languages and hailing from different parts of the globe may feel unified by their commonly held Catholic symbols. Members of the same religious tradition often feel a bond and are inclined to cooperate with others perceived as "like us" when the two might just as easily have adopted a posture of animosity. Religion serves as a master status in this situation rather than nationality being primary. George Thomas (2001) points out that theorists can no longer assume, as they did only a few decades ago, that the nation was the defining reference group for all people in virtually all circumstances. Globalization has changed that, and religion can now play a different role, at least in some cases.

Some social scientists believe this change in the direction of accommodation and tolerance of other religious traditions is itself a result of increased global interdependence (Lee, 1992; Meyer, 1988; Thomas, 1991). As we are forced to trade and interact with people in other parts of the world who control resources we need, it soon becomes clear that judgmental attitudes implying their inferiority—and an accompanying self-righteous posture regarding our own moral and spiritual values—are dysfunctional. Thus, since about 1875, many Christian theologians and ecclesiastical leaders have offered conciliatory attitudes toward other Christian churches and toward other religions. Lee (1992) wrote this:

> Rather than authority based solely in the Bible, there was now a new stress on the authority of personal religious experience. Instead of viewing the world as static, liberals accommodated new scientific views by attributing divine purposes to the newly evident evolutionary dynamism of nature and history. (p. 131)

Likewise, liberal theologies suggest that God may speak to people through other religious traditions, though Christianity may be depicted in these theologies as the fullest and most complete expression of God's Truth. Christ may not be the *only* path to Truth. The faith is depicted as "different only in degree from other religions" (Cauthen, 1962, p. 24).

More recent liberal formulations even hesitate to assert that Christianity is the fullest and most complete expression, for that implies a superiority/inferiority posture that continues to offend members of other traditions. Many theologians and church leaders now simply assert that Christ is the way they personally came to know God, either affirming or remaining mute on the matter of whether other faiths are equally valid paths to God. Indeed, Thomas (2001) pointed out that in a world cultural context, proselytizing (aggressively trying to convert others) "is viewed as a type of cultural supremacy and even as an act of aggression against an ethno-cultural group" (p. 527). Forcing one's own religion into another culture comes to be viewed as a form of cultural colonialism.

It is noteworthy that these very liberal or "tolerant" religions are usually strongest in precisely those areas characterized by cultural diversity and most involved in international transactions. Urban areas, coastal trade centers, and other "crossroads" of cultures and people are more likely to generate such outlooks than rural areas where the people rarely encounter much heterogeneity.

This shift in theology seems to be a result of globalization, but it has also resulted in counter-theologies and reactions that will be discussed later. Still, the trend toward greater tolerance on the part of the mainstream denominations was clear. The change in the position of the Roman Catholic Church during the 20th century was dramatic. In 1910, Pope Pius X required all Roman Catholic clergy and teachers to sign an "antimodernist oath." By the 1950s, Pope Pius XII acknowledged truth and goodness outside of Christianity, and in an encyclical, *Evangelii Praecones*, asserted that rather than attacking other religions as evil when evangelizing, the gospel should be assimilated into whatever is "good and honorable and beautiful" in existing religions.

The Second Vatican Council (1962–1965) opened the door even farther to viewing other religions as "colleagues with similar problems" rather than as enemies (Sheard, 1987). It is in this new context that the indigenous religion among Pueblo Indians in the Southwest of the United States—once brutally suppressed with acts of terrorism and torture and rituals strictly forbidden by church and state—is now becoming officially blended into

Roman Catholic Christianity. Developments in Zuni Pueblo, the largest of the 19 New Mexican Indian pueblos, are a case in point.

Catholicism first encountered the Zuni in 1540 in the form of Spanish explorer Francisco Vasquez de Coronado and his 2,000-person expedition in search of the Seven Cities of Gold (Flint, 2008). When Coronado, representing the Spanish crown and the Catholic Church which legitimated it, arrived at Zuni pueblo he read aloud and had translated a legal summons called the *Requerimiento*:

> We ask and require . . . that you acknowledge the Church as the ruler and superior of the whole world, and the high priest called Pope, and in his name the King and Queen [of Spain] . . . our lords, in his place, as superiors and lords and kinds of these islands and this mainland. . . . If, [however,] you do not do [what I ask] or you maliciously delay it, I assure you that, with the help of God, I will attack you mightily. I will make war [against] you everywhere and in every way I can. And I will subject you to the yoke and obedience of the Church and His Majesty. I will take your wives and children, and I will make them slaves . . . I will take your property. I will do all the harm and damage to you that I can. (quoted in Flint, 2008, pp. 70, 109)

Although the Zuni responded by firing arrows at Coronado's party, they were eventually subdued. Thus, near the center of the pueblo stands the old Catholic mission church, Our Lady of Guadalupe, which was built in 1629 using slave labor. Catholic missionaries fled after the pueblo revolts of 1680 but periodically returned, including in 1923 when Franciscan missionaries built a new church in Zuni pueblo. Over the years, the old mission church near the traditional dance plaza fell into disrepair.

After Vatican II, the Zuni tribe entered into an agreement with the Catholic Church (and the National Park Service) to excavate and restore Our Lady of Guadalupe. During the restoration, a Zuni Catholic artist, Alex Seowtewa, learned from his father and other pueblo elders that earlier in the

church's history it had kachina paintings on its walls. Kachinas are spirit-beings that are integral to the traditional religions of the pueblo Indians and they believe are critical to rainfall. In 1970, Seowtewa (and his son Ken) received permission from the Catholic parish priest to paint murals of kachina dancers on the upper walls of the church. The murals represent the seasonal cycles of Zuni ceremonial observances (Kennedy & Simplicio, 2009, p. 68). For his work, Seowtewa is sometimes called the Zuni Michelangelo, after the famed painter of the ceiling of the Sistine Chapel. As with Michelangelo, this work was done with the blessing and cooperation of the Catholic officials who are now much more open to the inculturation (adaptation to local religions and cultures) of the Catholic faith (Francis, 2000).[1] A few of the panels from the beautiful paintings on the walls of the Roman Catholic mission church can be seen at www.hanksville.org/voyage/misc/Zmission.html.

Critical Thinking: What might be some consequences of a synthesis of religions— as represented in the blending of Zuni kachinas alongside paintings of Jesus inside a Christian sanctuary? Is the broadening of religion to be more inclusive a step that will have mostly positive or negative results? Why?

Sometimes the impulse to modify the group's theology or system of ethics can come from internal dynamics within a religion. As the membership composition of a religious group changes, with a marked increase in membership in one region, there may be theological ramifications. The Roman Catholic Church, an international organization whose historic development and population base was European, is now undergoing transition from a Global North (industrialized, largely capitalistic, and powerful countries) to a Global South (poor, nonindustrialized nations)

[1]Significantly, it is mostly Zuni Kiva leaders and kachina dancers who feel that their religion is being compromised and that a sacrilege is being perpetrated by its official incorporation into the Roman Catholic Church (Haederle, 1991; Keith, 1999).

entity. Catholicism is experiencing growth in South America and Africa, causing the population makeup of the membership of this transnational church to shift (Budde, 1992; Della Cava, 2001).

As its membership majority has shifted from the Northern to the Southern Hemisphere, many Catholic theologians and priests have supported one or another version of liberation theology. Liberation theology tends to be highly critical of capitalistic economic systems and to view the world through the lens of conflict theory, championing the cause of those exploited by the capitalistic world system. The major cause of poverty in the Global South is not identified as slow adoption of industrialization but as neocolonial *exploitation* by the affluent nations of the world. Such direct attacks on capitalism as a system, attribution of sinister motives to the policy makers of G8 countries, and legitimation of movements antagonistic to the interests of affluent capitalists are bound to create tensions between the church and Western leaders.

This has created an interesting dilemma for the Roman Catholic Church; it is placed in a position of choosing between defending the self-interests of the poor—the new and very sizable constituency of the church—or sacralizing the interests of the affluent capitalistic countries that have been its historic constituency. Even within a single country, like Brazil, the church has found that liberation theology resulted in an expansion of Catholicism's constituency among the poor, but an alienation of middle class members who offer leadership skills and financial resources (Della Cava, 2001). Further, the "we" within the transnational Roman Catholic organization now includes people from a vast array of cultures, races, and regions of the globe. The church is itself increasingly multicultural. While Roman Catholicism is the most vivid example of these internal transformations, other Christian denominations are also experiencing substantial ethnic diversity (Della Cava, 2001).

The trend toward interreligious accommodation, the inclination to define other religious groups as fellow travelers, the desire for "dialogue" and discovery of common ground with other religions, and internal cultural diversity all compel established religious groups to accept pluralism as a positive value. Indeed, whether within the church, temple, or mosque or beyond its walls, embracing pluralism appears to be a critical first step toward accommodation of diverse peoples and alternative social systems. Those who believe that globalization is an unstoppable trend would interpret increased religious tolerance as the inevitable wave of the future.

On the other hand, a few evangelical Christian groups have tried to develop their own version of globalized theology, and these models do not emphasize pluralism and diversity. In a study of the conservative faith movement in Sweden and the United States, Coleman found that specialization of roles by various national groups was emphasized. Interdependence was recognized, but there was no particular expectation of equity, unification, and homogenization of peoples and cultures. The point that this study makes is that "globalization cannot be regarded as an internally homogeneous, inexorable phenomenon" (Coleman, 1993, p. 371).

Globalization will no doubt take many forms and result in a wide array of efforts to define its meaning and delineate the appropriate form of its evolution. Religious groups will likely make many attempts to define what global interdependence should look like. The one generalization that seems fairly certain is that religions can no longer ignore global interdependence and the fact that the world is to some extent becoming "a single sociocultural place." Theologies of the major world religions will likely be permanently affected by the expansion of vision to include the entire world.

Interestingly, Catalina Romero (2001) raised the opposite problem. Denominations typically establish policies in an effort to address specific problems. The problems are specific to a particular place and time, so the religious officials are necessarily thinking locally. However, what happens when the solution is broadcast globally? What are the consequences? The Roman Catholic Church has faced unexpected results when the Pope has said something in Rome or in Poland, but the circumstances were very different in South America or in Africa. Pope John Paul II's strong language espousing liberation and justice in Eastern Europe was interpreted in Peru as support for a movement calling for rich countries to forgive the debt of poor countries. Sometimes decisions and polices in one locale can have unanticipated interpretations and impacts elsewhere, especially if ecclesiastical leaders do not think globally before they speak (Romero, 2001).

IMPACT OF GLOBALIZATION ON TRADITIONAL RELIGIONS

The interdependence of the world has long been a factor in shaping religious movements. Robert Wuthnow (1980) argued that much religious change, including even the Protestant Reformation, has been caused by international processes. We do not have space for a long-term overview here, so we will limit the discussion to the modern world. Globalization in the past century has affected religion in a number of ways. In this section we will explore two processes that illustrate how religion is a *dependent* variable (a consequence): (1) modernization and secularization of social structures and cultures and (2) demographic migration patterns.[2]

Diffusion of Modernization and Religious Response

The ability of a society to function effectively in the interdependent world system seems to require "modernization." As societies modernize (which typically means becoming more Westernized) religious institutions are often relegated to a less encompassing role in social life. As we saw in Chapter 13, this

process of institutional differentiation is definitive of societal-level secularization. It also has organizational and individual-level correlates: Religious organizations begin to provide services that are perceived not as mandatory but "optional" (Turner, 1991b). Joseph Tamney (1992) described this as "selective religiosity." Official religious pronouncements compete with efforts by the state to define acceptable social behavior, and the individual is free to select which pronouncements to follow.

This monk, in traditional monastic dress, works on the most modern symbol of technology—a laptop computer. The reach of modernization—including into religious groups—has been extensive.

As governments become more concerned about "quality of life" issues and expand the scope of their concerns, they begin to encroach on the traditional realm of religion (Shupe & Hadden, 1989; Thomas, 1991). Religious groups traditionally had monopolistic control over education, stability of marriage ties, grief adjustment following death, and the provision of charities to help the destitute. However, in the modern world, those services are increasingly provided

---

[2]We treated a third process, the introduction of advanced communications and mass media technologies, in Chapter 14.

by "helping" or "service" professionals certified by governments (Turner, 1991b). Indeed, Wuthnow (1991) claimed that around the globe, as government spending to enhance the life-quality, health, and education of the citizenry increases, religious affiliation usually decreases. Moreover, governments increasingly become embroiled in disputes over issues of morality, justice, and meaning as they promulgate norms governing a very wide range of behaviors (Shupe & Hadden, 1989). National governments have displaced many of the social contributions of religious communities.

In addition, as the world globalizes, international nongovernmental organizations (INGOs) begin to expand to fulfill some of these same functions. The Economic and Social Council of the United Nations, in fact, has a branch that provides training, information, and advisory services for more than 3,200 agencies (www.un.org/esa/coordination/ngo/). Beginning with the first INGO—Anti-Slavery International, established in 1839—this sector of global civil society has grown dramatically (Keck & Sikkink, 1998). In 1854, there were only six INGOs, but by 1945 there were over a thousand. The Union of International Associations claims there were over 60,000 globally by 2007 (Davies, 2008). As John Boli and George Thomas (1999) argued, these INGOs play a key role in "constructing world culture." They addressed issues of poverty, education, peace, women's rights, reproductive health, and other areas of traditional interest to religious organizations. Not surprisingly, then, many INGOs are religiously-affiliated—over 3,000 of the more than 60,000 INGOs listed by the Union of International Organizations are connected to a faith tradition (Boli & Brewington, 2007). Still, that is only 5% of all INGOs. Moreover, the operating procedures that govern INGOs as a whole are secular, and there is also an emerging field of (secular) international law that defines boundaries of acceptable behavior in everything from commerce to human rights (Berman, 1991; Lechner, 1991a).

Moreover, the utilitarian and rational pursuit of self-interest in the political and economic realms has divorced them from religious norms. Turner (1991b) pointed out that distribution of resources is no longer tied directly to honor, loyalty, and obedience to traditional authority, as it once was. The result is that adultery, divorce, adolescent disobedience of parents, premarital sexual activity, and other traditionally forbidden religio-moral behaviors are not accompanied by economic sanctions. Religious norms lose their clout, and the realm of religious influence becomes more narrowly focused.

Another way to conceptualize this is to think of rational-utilitarian ties between people, ones stressing mutual self-interests and negotiation of resources in the interaction, as "contractual" relations. This is in contrast to "covenantal" relations: emotive-expressive ties based on bonding, shared values, and sense of commitment to specific other persons. The latter represent familial and religious relationships, while the former are more characteristic of bureaucratic, governmental, and economic affiliations. David Bromley and Bruce Busching (1988) and Bryan Turner (1991b) have developed arguments that the covenantal realm is shrinking in the face of expanding contractual ties. This is a global trend and it provides a profound threat to most traditional religions.

A number of scholars point out that the rise of religious fundamentalism and other legalistic, literalistic, and rigid revitalization movements in the 20th century is not a process unique to a single religious tradition or society. Therefore, it cannot be interpreted simply in light of local or national events, nor in terms of characteristics of a given religion. Its cause, they assert, is global (Antoun, 2008; Marty, 2001; Robertson, 1989; Thomas, 1991; Turner, 1991a; Wuthnow, 1980). Whether the revivalism or "fundamentalism" is HAMAS (Islamic Resistance Movement of Palestine) or the Wahhabi Islamic movement in Saudi Arabia, the Hasidic or *haredi* groups within Judaism, Soka Gakki in Japanese Buddhism, extremist fundamentalist groups within Christianity, or the Malaysian Dukway sect, the cause appears to be *reaction against* global modernization (Antoun, 2008; Chua,

2006; Davidman, 1990; Eitzen & Zinn, 2006; Marty & Appleby, 1992, 1994; Shupe & Hadden, 1989). About these revitalization movements around the world, religious historian Martin Marty (2001) wrote this:

Members of fundamentalist movements almost everywhere think they are old-time religion. But they're not. Fundamentalism comes after the Enlightenment in the West. It comes after the worldwide spread of technology and industry and along with the earliest stages of global economies. It is late-modern, and some people would say it is postmodern. Fundamentalism is a movement to fight back, and modernity is the enemy. (p. 44)

These movements, regardless of the religious tradition, have certain prevailing characteristics (Antoun, 2008; Heilman, 1992; Marty, 2001).

1. *Radicalism*—efforts to revitalize an idealized culture depicted in the past, a golden age or a golden place. That age is often frozen on paper in the form of scripture and usually depicts the ideas and social arrangements set forth by the founder of the group.

2. *Scripturalism*—absolutist interpretations of *selective* sections of scripture (many passages being ignored or *not* interpreted literally) and the belief that God speaks solely and decisively through the scriptures. The scriptures can be cited for "proof" in almost any kind of argument—moral, supernatural, literary, or scientific.

3. *Traditionism*—the need to demonstrate that there is nothing in the archaic and holy past that does not have direct application to the way we live today (Antoun, 2008).

4. *Oppositionism*—extreme hostility to secularization or modernization. This may well be the defining characteristic of these movements, but this opposition is also selective. Aspects of technology—such as computers, the internet, and cell phones—may be adopted and utilized to further the religious tradition.

5. *Totalism*—a posture that identifies religious issues as all-or-nothing conflicts where no compromise is tolerated. Often the most extreme

hatred is directed toward others of the same tradition who fail to define the key issues the same way or fail to recognize that a win–lose war is at stake.

6. *Puritanism*—the quest for purity in an impure world. While many other kinds of religious movements also have this passion for purity, the drive for internal purity and for militant opposition to the "unclean" is a feature of fundamentalist movements. Of course, various traditions have different ideas about the core elements of spiritual and moral purity.

Scholars believe that the formation of such movements in diverse traditions and places around the world is rooted in several features of the globalization process.

First, the pluralism and relativity of a diverse world is threatening to traditions that always protected the absoluteness of norms and values (Robertson, 1989; Thomas, 1991). Bryan Turner (1991b) pointed out that the whole idea that alternative lifestyles might be tolerable is offensive to those who are very certain that they alone know the Truth (p. 162).

Second, fear of economic dependence on other peoples—some of whom are very different and are viewed as undependable—stimulates a desire to reassert autonomy and, at least symbolically, to proclaim one's uniqueness (Robertson, 1989). Fundamentalism or rigid revivalism is in large part an attempt to reestablish independence from the world system.

Third, fundamentalism represents a reaction against the institutional differentiation that is a characteristic of modernization and secularization. Ironically, the blending of Islamic fundamentalism with governmental functions is much more a 20th-century phenomenon than a historic one (Turner, 1991a, 1991b). It represents a desire to merge institutions in reaction to the Western pattern of differentiating institutions. Shi'ite fundamentalists (Muslims) do not want religion removed from other spheres of life, and they react by forging an even closer marriage between religion and other societal functions. In essence, these are nativistic movements reacting against changes that they identify as part of neocolonialism. They are convinced—with good cause—that the acceptance of globalism involves *ethnocide*, the death of their traditional culture. Fundamentalism is an attempt to defend their culture.

Fourth, fundamentalisms are sometimes counteractions against religious reforms themselves. As Iran underwent modernization in the mid-20th century, official Islam in that country underwent reformulation, becoming liberated from folk versions and emphasizing a universalistic monotheism. It also embraced aspects of secularization, including changing roles for women and economic reform that eventually left Iran highly dependent on the United States and other Western nations. Yet modernization of the country also resulted in some economic dislocations and a degree of "cultural wobble." As conservatives rejected modernism in favor of traditionalism, they also rejected the new interpretation of Islam. Rather than returning to polytheistic folk versions of the faith, conservatives forged a literalistic and uncompromising interpretation of Islam (Turner, 1991a, 1991b), sometimes called Islamicism. Official Islam as it was taught by the elites was seen as having sold out to westernization; it was depicted as a traitor religion that abandoned traditional roots.

While the examples used here have been mostly of Islamic fundamentalism, the argument of many social scientists is that these movements are occurring across religious traditions, and that they represent reclaiming of authority over a sacred tradition. This religious tradition "is to be reinstated as an antidote for a society [astray] from its cultural moorings" (Shupe & Hadden, 1989, p. 111). The ultimate cause of the straying is the process of globalization.[3]

George Thomas (2001) pointed out that even the UN's Declaration of Human Rights has language that is offensive to some religious people. The UN document, drafted after World War II and the horrors of the Nazi holocaust against Jews, Gypsies, and homosexuals, recognizes the human right to choose one's religion. The very notion that one chooses one's religion is itself a form of modernization, for it stresses both individualism and freedom of choice (Thomas, 2001). Choice relativizes religion—one has a right to choose Religion A or Religion B. Indeed, relativization of cultural values is central to the globalization process and is an aspect of contemporary life that fundamentalists vehemently reject in the religious sphere (Campbell, 2005).

The polemical conflicts that have resulted from high religious boundaries can also create problems for religious researchers, especially if one comes from another religious tradition identified as "the enemy." The "Doing Research on Religion" feature discusses one example of a researcher caught in an awkward dilemma between his own heritage and the people and religious tradition that he found fascinating.

DOING RESEARCH ON RELIGION

David Buchman is an anthropologist interested in the Sufi movement within Islam. When he went to the Middle East—to Yemen—to do field research on this movement, he was faced with a dilemma. He was a Jew trying to establish rapport and do participant observation on Muslims who completely

---

[3]Note that, rather than viewing secularization as an unstoppable force, Robertson, Hadden, Shupe, and several other scholars cited in this chapter suggest that secularization often results in *renewed religious fervor* and the birth of *new types of religious movements*.

distrusted Israel in particular and Jews in general. Transnational research often causes dilemmas, but this situation was potentially dangerous, and it was resolved in a most interesting manner.

Is a Jew Among Arabs Like a Sufi Among Yemenis?

From May 1995 to August 1997 I conducted field research on the contemporary situation of Sufism in Yemen, concentrating on the beliefs and practices of a particular Sufi order located in the capital. Sufism is the mystical dimension of Islam, organized into orders or informal groups of disciples who follow the teachings of their spiritual master. Sufis say that their interpretation of Islam goes back to the founding of the religion itself, with the Prophet Muhammad being the first Sufi master. Less sympathetic interpretations posit Sufism as a harmless non-Muslim accretion at best and as a corruption of Islam at worst. Socially, throughout Islamic history, Sufis were treated in two opposed ways, which, depending on the politics of the time, tended to swing back and forth like a pendulum: sometimes they were persecuted as corruptors of Islam and sometimes highly promoted as the only holders of the inner dimension of the tradition. When persecuted, Sufis responded by hiding their teachings and practices from those outside the order to keep away the persecutors. When encouraged, the Sufis were more publicly open.

Because I am Jewish, before going to Yemen I needed to make a decision about how I would identify myself to people. Unfortunately, because of the Palestinian/Israeli conflict, many Arab Muslims do not distinguish between being Jewish and pro-Israeli, especially if the Jewish person is from the United States, a defender and financial backer of the Israeli state. I had two concerns. First, I was afraid that by stating I was from Jewish descent, I would be negatively stereotyped. I assumed, perhaps wrongly, that people would label me pro-Zionist, anti-Arab, or anti-Muslim, and so an initial antagonism and suspicion would be set-up between me and Yemenis before people knew my opinions. The second concern was ethical. I could say that I was of some other heritage, and thereby skirt at least a certain type of initial prejudice. It is difficult to justify such a stance morally and ethically. My research demands becoming close enough to people to talk about intimate religious experiences, and this requires utmost trust and sincerity between individuals. I could not be considered a trustworthy and sincere person if I denied my heritage, regardless of my current beliefs. Since I would be lying to them, why should I expect them to be honest with me? Moreover, the consequences of a discovered lie would make me even more of a suspicious character than not lying at all.

Various advisors gave me three answers: lie, tell the truth, or lie at first and then, maybe, tell the truth. Those who opted for the truth argued that ethics in the field is all important for an anthropologist. I should be honest about my own beliefs and background in order to expect people to be honest with me. More practically, I was told that if I became a successful anthropologist, I would have a life-long relationship with the people with whom I worked. If I was branded a liar, they might not want to continue the affiliation.

Those who recommended prevarication varied as to whether I should do so blatantly or subtly. Most justified the need for falsehood through recourse to personal safety. Being marked a Jew from the United States would make me a target of violence and/or a political pawn in a kidnapping scheme. Rather, I should conceal my background to some and—to trusted others—tell the truth, but

(Continued)

(Continued)

only if the issue arose. Another suggested that even in my lie I could be subtle, saying I was Protestant. In my mind I could justify such a statement by interpreting the word to mean one who protests the negative stereotypes of the questioner. So while the lie would have been outward in form, in its inward meaning it would remain true. I opted to follow the last piece of advice believing that I would be more candid about my past and current religious beliefs once I became close to certain Yemenis.

When I first started conducting research with the members of the Sufi order, I did not tell them I was of Jewish heritage. During my first year of research I attended their gatherings, accompanied them on pilgrimages to saints' tombs, and conducted informal and formal interviews. As time passed I found myself less and less observing and more and more participating in their group religious rituals. I was gaining their trust. After the initial questioning, they never asked me again; so the problem of my identity was solved at the beginning of fieldwork, or so I thought.

At the end of the first year, I felt extremely guilty about my lie, because I had become good friends with the head disciple. I decided finally to tell him about myself, preparing to receive an angry response. His reaction was unexpected. He smiled, saying, "Yes, I know. So what?" He and almost all the disciples suspected my Jewish background all along, but since I did not center my research on political discussions, they did not make an issue of it. Still, what he said next gave me insight not only into how well I unexpectedly handled things, but also into the nature of Sufism in Yemen itself. He insisted that I continue to conceal my heritage to Yemenis outside the order. All the disciples will keep the secret, he said.

This was shocking. Now that I revealed my identity why should I not tell it to all? As it turns out, what I was asked to do, was what the Sufis themselves have been doing for centuries: hiding those practices and beliefs from those in power who would use politics to persecute. My fieldwork revealed that due to opposition from an anti-Sufi Islamic political party in power, the order was going into hiding: the master of the Sufi order fled the country because of religious opposition to his teachings, their meetings were becoming more and more secretive, and their teachings were less and less explicitly "Sufi" because of hostile visitations from outsiders. Just as Sufism in Yemen was becoming more and more concealed, so also my Jewish identity should be kept hidden: it was not politically expedient to announce such things.

Another explanation as to why Sufis find it easy to keep such secrets is found in the Sufi interpretation of Islam itself. Sufis say that there are degrees and levels of interpretation within the religion. The outward meaning of Islam is for all, while the inward meaning, or 'realities,' (haqa'iq) should be kept hidden from those not worthy or strong enough to understand correctly. If certain realities or interpretations are spoken to those not intellectually prepared, such teachings will corrupt instead of bringing people closer to God. It is incumbent then that true realities of things be kept hidden from those without appropriate spiritual maturity. In this context, keeping secrets is not an unethical act, but a spiritual exercise.

My confession did not make the Sufis look upon me as a person of dubious character, as I expected. On the contrary, the revelation made me closer to them in that we all became members of a hidden society in which my religious heritage was the shared secret, a 'reality' to be kept hidden from those who would corrupt. No one spoke about it to outsiders. Just as I became a member of their Sufi group, learning their secrets and practicing their rituals, so, after I confessed, they become members of a newly formed social group based upon the concealing of my secret of being Jewish.

In sum, an important self-identity issue for me during my fieldwork was whether or not I should be honest in informing people of my Jewish heritage. I decided to conceal my background not knowing that my personal situation paralleled that of the Sufis. The Sufis and I became mirror images reflecting each other, and we saw ourselves in the cultural situation of the other. A Jew among Arabs is like a Sufi among Yemenis; both must conceal their true realities until the pendulum swings in the other direction—at which time revelation of truths will make all closer to God.

David Buchman

Hanover College

We must add that worldwide revivalistic movements are not the only type of religious reaction attributed to global modernism. An alternative response to institutional differentiation and the narrowing scope of religious influence, according to Robertson (1989) and Turner (1991a, 1991b), is privatization of religion. Since religions with a universal message are not linked closely to a specific ethnic group or geographic region, and since religious institutions have been moved to the periphery of social life, individuals are freer to formulate their own theology, their own meaning system. Thus, global trends toward diversity and institutional autonomy may be linked to the rise in individualization of religion, as well as to the formation of fundamentalist movements. In the face of global interdependence, liberals may tend toward privatization of religion, while conservatives tend to institute movements proclaiming absolute and infallible answers to questions of meaning and morality.

> **Critical Thinking:** Why is pluralism so threatening to so many religions around the world? Why is pluralism central to globalization? Is acceptance of pluralism important to world peace? Why or why not?

World Population Patterns and Religious Consequences

Another way in which global trends may influence religion is through demographic changes, especially the massive human geographic mobility that has taken place since the 19th century. Of course, migration and its relationship to religion is nothing new. As Peggy Levitt (2003) put it, "You know, Abraham was really the first immigrant" (p. 847). Protestant and Catholic Christians fled England to found colonies in North America in which they could freely practice their faith; French Huguenots migrated to Switzerland and Holland for more tolerance. More recently, changes in U.S. immigration laws in 1965 allowed for increased immigration into North America. Much of this immigration was from non-European regions of the world, especially Asia. With these new immigrants came many new religious traditions. Gordon Melton, a specialist on new religious movements (NRMs), insists that part of the increase in the number of NRMs in North America is simply a function of high global immigration rates (Melton, 1993).

Improved communication technologies have brought new awareness of other parts of the world and advanced transportation technologies have allowed more people to move to new locations than at any time in human history. Robert Wuthnow and Stephen Offutt (2008) reviewed some of the global data on migration: 7.5% of the British population and 11% of the French population are immigrants (p. 215). Sixty-six percent of those living in Macao, 58% in Kuwait, and 40% in Bahrain and Hong Kong are immigrants, while 54% of Mozambiquans and 41% of Botswanans said their parents work in South Africa. The population of El Salvador is about

7 million, barely more than twice the number of Salvadorans living abroad (including 2.5 million living in the United States). We analyze the religious consequences of this migration by looking at two broad categories of phenomena: (1) glocalization and (2) transnational religious connections.

Glocalization

The term *glocalization* was brought into the sociology of religion by globalization theorist Roland Robertson (1992, 1995). It derives from a Japanese concept, *dochakuka*, which roughly translates as "global localization," and that was used to characterize the marketing process of tailoring global products to local tastes (Robertson, 1992, pp. 173–174). Applied to the sociology of religion, glocalization focuses our attention on the reality that the standardizing processes of globalization do not simply wipe out particular local cultures, nor do global migrants simply abandon their cultures of origin. Globalization and glocalization go hand in hand producing both homogenization (McDonald's in India!) and hetereogeneity (they serve a Maharaja Mac, a Big Mac with lamb or chicken, and the McAloo Tikki, a vegetarian burger, because most Indians do not eat beef). Thus, "globalization is always also *glocalization*, the global expressed in the local and the local as the particularization of the global" (Beyer, 2007, p. 98).

Glocalization occurs because when religious individuals migrate, they bring their religions with them into a new sociocultural setting. Giulianotti and Robertson (2007) identified four different glocalization strategies that are developed by these individuals relative to those new sociocultural environments:

- *Relativization:* The social actors seek to preserve their prior cultural institutions, practices, and meanings within the new environment—maintaining differentiation from the host culture.
- *Accommodation:* The social actors absorb pragmatically the practices, institutions, and meanings associated with other societies, in order to maintain key elements of the local culture.

- *Hybridization:* The social actors synthesize local and other cultural phenomena to produce distinctive, hybrid cultural practices, institutions, and meanings.
- *Transformation:* The social actors come to favor the practices, institutions, or meanings associated with other cultures. Transformation may produce fresh cultural forms or, more extremely, the abandonment of the local culture in favor of alternative and/or hegemonic cultural forms. (p. 135)

Relativization involves migrants preserving their local cultural practices and maintaining some separation from the host culture. In her work on religious transnationalism, Peggy Levitt (2004) described the re-creation of the Devotional Associates of Yogeshwar or the Swadhyaya movement, in Lowell, Massachusetts, by immigrants from Gujarati, India. She explained that Swadhyaya members have maintained the founder's teachings, albeit in a new sociocultural setting.

> For example, all followers are supposed to "devote time to God" by participating in Yogeshwar Krishi. In India, these are cooperative farms or fishing enterprises, created by the movement. Members donate their labor each month and then distribute their earnings to the poor. To do Yogeshwar Krishi in Chicago, Swadhayees formed a small company that made ink refills for pens where they work in addition to their full-time jobs. In Massachusetts, groups of families get together to assemble circuit boards on contract from computer companies.
>
> The Gujarati migrants maintained their local religious practices as a way of insulating themselves from "what they perceived as inferior Western values," and in general "they wanted to remain socially apart." (Levitt, 2004, p. 13)

Accommodation is similar to relativization, in terms of the maintenance of prior local culture, but differs in the level of engagement with the host culture. Accommodating glocalization involves some absorption of the host culture. In a study of Tamil Hindus from Sri Lanka living in Germany, Martin Baumann and Kurt Salentin (2006) found that the immigrant experience "favors as well as demands the adoption of organizational patterns found in the host society" (p. 309). Thus, although Hindu religiousness in South Asia is largely noncongregational, in Germany it is much more centered on the congregation as a communal place of worship. The same congregational pattern has been found among immigrant religions in Houston, Texas, and, as was discussed at length in Chapter 8, elsewhere in the United States, reflecting the de facto congregationalism of American religion (Yang & Ebaugh, 2001). An opposite process of accommodation takes place among Muslims in France, according to Caitlin Killian (2007). Based on interviews with immigrant African Muslim women in France, Killian argued that they have accommodated to the secularized culture of France by transforming Islam from a publicly celebrated community of believers to a privately practiced religion of the heart.

Hybridization as a glocalizing strategy involves a synthesis of the local and host cultures into a (relatively) new set of cultural forms. Various religious syncretisms fall under this umbrella, including ones like voodoo and Santeria in the Caribbean that result from forced migration (the export of African religions via slavery) or the distinctively African forms of Christianity that result from missionary activity on that continent (importation of Christianity into Africa). Anthropologist Karen Richman (2005) has written about the fascinating dynamics that take place within Haitian Catholicism: "a complex, syncretic blend of European and African-Creole religious ideologies and practices, centered around the material reality of spiritual affliction, sorcery, and magic" (p. 165). It came about due to the combination of African religions carried by slaves and the Catholicism of the French colonialists. Historically the Haitians who have immigrated to the United States have continued to practice their creolized Catholicism (McAlister, 1998), but Richman (2005) found that many more recent immigrants to Palm Beach County in South Florida have been converting to evangelical Protestantism, creating yet another new mix: a "modern, ascetic

cloak worn by the new converts" underneath which "spiritual healing, sorcery, and magic remain at the heart of their syncretic practices" (p. 165).

Transformational glocalization occurs when migrants favor the religious practices of the receiving culture over their local culture. Fenggang Yang's study of Chinese converts to Christianity in the United States provides a good example of this modality. Yang (1998) found that the environment of far-reaching change in China and the subsequent "chain migration"—running from one strange place to another—fundamentally altered the religious needs and outlooks of the population. These changes, he found, meant that the message of evangelical Christianity is more compatible with their needs and perspectives than had been the case with previous cohorts of Chinese people. As a consequence, they have been more willing to abandon their traditional religions and embrace Protestant Christianity.

Just as globalization is a multifaceted phenomenon, so too is glocalization. In fact, it is probably more accurate to refer to glocalizations

This is a Christian festival in Peru, but mixed in with this are celebrations of various other heroes and figures from local native religion. The synthesis is part of the glocalization that occurs as a world religion blends with local customs and beliefs.

(plural), in light of this discussion of four different approaches to glocalization by which the global becomes local. Our discussion so far has focused on the various glocalization projects that are undertaken by migrants to negotiate the relationship between their local cultures of origin and their new host cultures. However, migrants do not simply move and leave behind the world from which they came. The growing field of transnationalism highlights the ongoing connections people have with their communities of origin and how those origins continue to influence religious development in the new locale.

Transnational Religious Connections

A second consequence of increasing migration for religion is what Wuthnow and Offutt (2008) called "transnational religious connections." The idea recognizes that people who migrate often continue to have ties to the home community, communicating with, remitting money, visiting, and even retiring to the home country. Some have called them "transmigrants" rather than "immigrants" for this reason (Basch, Schiller, & Blanc, 1994). Moreover, some flows of people across national borders are intentionally short term, such as missionaries, short-term volunteers, and even tourists. As these people flow back and forth in their transnational religious connections, several changes may take place.

Since family members may no longer be united by common nationality or territorial bond, religion may become a more important element of belonging and connectedness. Heidi Larson (2000) found this among Hindu, Muslim, and Sikh immigrants living in England.

Further, the religious practices, rituals, and beliefs of a home country may be influenced as people back home hear of the experiences and religious celebrations of their geographically mobile kin. Peggy Levitt (1998) found that with the Dominican Republic located in geographical proximity to the United States, there is a fairly high level of migration and visitation back and forth. Dominican family members now living in Boston not only return occasionally for visits, but "adopt-a-parish" ties have encouraged communication and visits by priests and others. The clergy in the Dominican Republic have begun to adopt ideas of ministry (a more "professional" relationship to the congregation) and more rationalized (bureaucratic) procedures of running the church as an organization. Rituals have sometimes been affected, too, as immigrants have communicated ideas and practices back to the home Dominican village.

The result has been local-level globalization—expanded perspectives and practices even in an isolated island village. At the same time, family members now in Boston maintain connections to the home community, and this prevents complete assimilation and cultural absorption of the immigrants into the mainstream of the larger American society.[4] Both churches are influenced by the transnational migration (Levitt, 1998).

Not all transnational movements of people are intended to be permanent or semipermanent. Some are short term by design, and yet they can have important religious implications. Barak Kalir (2009), for example, studied religious conversion among Chinese temporary workers in Israel brought in mostly from the villages of Fujian to do construction work. The visas for these workers in Israel are limited to 5 years. Despite the fact that they knew they would return to their homeland in due time, many still converted to evangelical Protestantism during their stays. Actually, as Kalir explained, it is not

---

[4]Transnationalism within religious organizations limits pluralism; for example, Spanish speaking churches in places like the United States are often pan-ethnic, including people from several nationalities. Thus, the culture of any one country is diluted (Levitt, 1998). However, it also preserves pluralism with a constant infusion of ideas and immigrants from other places. Because of the importance of cultural preservation to people and because of the dynamic character of culture, we might conclude with John Simpson (1996) that the world will never be *entirely* a *single* sociocultural place.

despite the fact that they would be returning to China that these migrant workers converted, it was *because* they would be returning. "Finding Jesus in the Holy Land" allows temporary migrant workers to be involved in a religious tradition that they associate with modernity and progress (and hence high status) and a religious organization that affords them opportunities to cultivate practical skills and network connection.

Thus, conversion to Christianity for these migrant workers increases their cultural capital (knowledge, skills, education), symbolic capital (honor, prestige, recognition), and social capital (connections to social networks), which can be deployed to their advantage when they return to their rural, uneducated, lower-class origins. Kalir's (2009) insight into this transnational religious connection is only possible because he shifted his "analytical gaze from the host country [to] consider more pertinently the effects that a conversion to Christianity carries for Chinese migrants when they return back home" (p. 143).

Professional religious workers—aka missionaries—continue to be a significant transnational connection in the religious field. According to Wuthnow and Offutt (2008), in 2001 there were 42,787 American missionaries working full-time in other countries, an increase of 16% over the previous decade (p. 216). The International Mission Board of the Southern Baptist Convention, for example, had a budget of nearly $300 million in 2005 and a staff of 500 full-time employees (Wuthnow & Offut, 2008, p. 217). Missionary activities do not exist only at the denominational level; they are connected to local congregations as well. Seventy-four percent of church members in the United States report that their congregation supported a foreign missionary in the previous year, 40% said their congregation has a committee with an

international focus, and 20% have a full-time staff member in their congregation responsible for global ministries and missions (Wuthnow & Offutt, 2008, p. 217). Thus, the average member of a congregation in the United States has considerable exposure to transnational connections through their local congregations.

Last, even people who go abroad for short periods of time can find themselves affected religiously. In the U.S. context, these are known as "short-term mission trips." Wuthnow and Offutt (2008) estimated that 1.6 American church members participate in such trips annually, with a median length of 8 days (p. 218). Although this does not necessarily yield permanent transnational connections, the individuals who take them can have transformative religious experiences. Using longitudinal survey day from the National Survey of Youth and Religion, scholars found that adolescents who had reported not participating in a "mission trip or religious service project" in 2002 but that had participated by 2005 (17% of the adolescents surveyed) had also increased religious involvement and solidified religious beliefs over that time span (Trinitapoli & Vaisey, 2009).[5] The mission trip, it seems, had a significant effect on the adolescents' religiosity, at least in the short term.

Related to short-term missions are pilgrimages. A pilgrimage can be defined as "a journey undertaken for religious purposes that culminates in a visit to a place considered to be the site or manifestation of the supernatural" (Tomasi, 2002, p. 3). As with mission trips, these can be domestic or transnational, but the best known pilgrimage sites entail transnational travel for most pilgrims. Millions of people every year engage in this transnational religious practice, including some 2.5 million Muslims making the annual *hajj* to Mecca, 1.8 million individuals (Christians and Jews) visiting the Holy Land of

---

[5]This operationalization of mission trips is broader than what would be preferable for the current discussion of transnationalism as it includes both international and domestic travel; however, the effects reported we would suspect would be *greater* for those whose mission trips were international than for those who engaged in domestic service projects.

Israel, and well over 100,000 pilgrims traveling annually to Kashmir to see the Amarnath cave shrine and worship Lord Shiva (Wuthnow & Offutt, 2008, p. 220). As with mission trips, we can ask what the religious effect is of these pilgrimages. Sociologist Matthew Loveland (2008) has investigated this for American Catholics. Owing to its sacramental imagination, Catholicism has a host of pilgrimage sites, among the best known being the Vatican (Italy) and the various shrines to Marian apparitions in Fatima, Portugal (4 million visitors annually), in Lourdes, France (5 million), and in Mexico City (10 million). Using data from a national survey of U.S. Catholics, Loveland found that those Catholics who had undertaken a pilgrimage (15%) were transformed by the experience, resulting in more orthodox religious beliefs. They were significantly more likely than the Catholics who had not been on a religious pilgrimage to support the church hierarchy's teachings against the use of contraception and for a celibate priesthood.

Religion, then, can be influenced in important ways by global trends. The emergence of global theologies and the global diffusion of modernization, the concurrent secularization of social structures and cultures, and demographic migration patterns have all impinged on religion. All three subsystems of religion can be affected: (1) the institutional structure and its tie to other institutions in the society, (2) the sense of religious belonging as a source of personal identity, and (3) the meaning system, which tries to make sense of people's experiences and establish norms of behavior. Before we conclude this look at religion in a global context, however, we need to consider ways that religious groups and movements may impinge on international relations.

Critical Thinking: Based on your reading of this section, what are the most important ways that you think religion has been *affected by* globalization? *Why* are these the primary influences on religion?

THE ROLE OF RELIGION IN INTERNATIONAL POLITICS

Religion is not always just a *dependent* variable in globalization issues. Sometimes religion is the *cause* of global processes. Religious groups themselves are often multinational conglomerates of sorts. Some religious organizations have international constituencies. Others may have memberships concentrated in one nation but sponsor benevolent and evangelical programs that span the globe. By reaching into other parts of the world, religious organizations develop vested interests, sympathies, and responsibilities to adherents that may bring it into conflict with one or more governments.

Diffusion of Western Religion and Consequences in International Politics

Religious mission programs not only affect the members of the faith community but they also have an impact on the destination site. In some cases, mission programs have contributed to economic vitality in an area. Sometimes denominations—through the Heifer Project and other programs—have provided well-drilling equipment and pumps for irrigation, animals, and other agricultural assistance that help people feed their families. In 1992, the United Methodist Church sponsored the building and staffing of Africa University in Zimbabwe—an effort to provide education and training for development in countries across the continent. Many other denominations could be cited for similar projects and programs to improve people's standards of living. The dilemma, of course, is that these very programs assume and enhance westernization of the cultures; indigenous ways of life are often undermined. Missionaries respond that ethnocide—the destruction of the cultures of another people, especially indigenous or minority people—has already occurred, and the only question is whether organizations in the more technologically developed countries are going to help poor countries to cope with the new global environment. Moreover, there is evidence that at least some missionaries have helped to revitalize and to preserve aspects of indigenous cultures. The role of missionaries has been complex (Sanneh, 1991).

Missionaries themselves have been predominantly from the United States, especially since early in the 20th century. In 1911, 11% of all Protestant missionaries were American. By 1925, half were from the United States, and by 1980, the figure had increased to 80% (Walls, 1991). Of course, Roman Catholics and other Christian groups were also sending missionaries, but Protestant Christians from the United States have been especially active in this enterprise. Interestingly, African and Latin American churches are now starting to send missionaries to Europe and the United States, since religious revivals have been so vibrant in the Southern Hemisphere (Jenkins, 2002; Wuthnow 1991).

Still, there is a concern that goes beyond that of Christian mission programs undermining traditional culture and beliefs. Howes (1991) documented the fact that the mission programs to Latin America sponsored by U.S. churches have often been aligned with U.S. international interests and may have done more for the United States than for the target country. Clearly the U.S. government has intentionally tried to use (even manipulate) religiously sponsored mission programs to achieve objectives that primarily served capitalism, business interests, and U.S. national security. Whether the programs helped the people in the target countries was of secondary interest (Howes, 1991). In any case, concern about the ethics of changing other cultures and of being intrusive with North American culture has

caused mainline denominations to cut back on evangelistic orientations to missions.[6]

Religious movements and institutions can have an impact on international processes in other ways, too. American televangelists have sometimes expanded their "markets" by broadcasting programs abroad. In doing so they diffuse a very Americanized version of Christianity. When this message is not warmly received by indigenous governments and is prohibited, these evangelists sometimes jam the message into the country with a high-kilowatt station located in a nearby country. As Jeffrey Hadden (1990, 1991) pointed out, these "pirate broadcasters" assume that God's law and purposes (which they are sure they alone know) stand above the laws of mortals. They will not be dissuaded by local officials or national leaders if they feel they are called to evangelize a country.

Often these televangelists are Americans— and rather ethnocentric and culturally insensitive Americans at that. According to Hadden, the bad feelings generated by this aggressive promulgation of a religious tradition can create diplomatic tensions and conflicts. Other highly sensitive negotiations between the target country and the United States may be jeopardized (Hadden, 1991). Thus, aggressive religious entrepreneurs from the United States may interfere with sensitive negotiations or with the national interests of their own home countries.

Of course, religious groups from Global North countries are also frequently involved in benevolence programs directed toward impoverished areas of the world. These programs have the potential to enhance goodwill between countries, but occasionally the manner in which assistance is provided can be demeaning to the recipient

---

[6]By contrast, Pentecostal or charismatic forms of both Protestantism and Catholicism have spread rapidly. At the turn of the 20th century, the number of Pentecostals and Charismatic Catholics in Latin America were negligible. By 1970s, they were still less than 5% of the population combined. By 2005, however, they had exploded to 28% of the population, including 13% Pentecostals and 14% charismatics (mostly Catholics). Seventy-three percent of all Protestants in Latin America are Pentecostal (Pew Research Center Forum on Religion & Public Life, 2006). Della Cava (2001) reported that the charismatic movement has also made major inroads in Latin American Roman Catholicism, in many places replacing the quasi-Marxian liberation theology movements. Thus, mission programs, at the very least, have changed the type of religiosity in some places.

nation or people. If the latter occurs, the net result could be hostile feelings rather than ones of mutual respect and appreciation.

Whether the concern is international bigotry, prospects for economic cooperation and trade, a struggle for human rights and justice, or a desire for world peace, religious groups are likely to be important players in the outcome. Religion affects international relations and is affected by global trends. As in all other social processes, religion is both a dependent and independent variable.

Transnational Versus Autocephalous Religious Organizations

We have already discussed the fact that the Roman Catholic Church is an international body that must be sympathetic to the circumstances of people in many nations. It cannot align itself with the economic and political interests of one nation or set of nations without risking the alienation of tens of millions of members in another part of the world. There are other ways in which multinational religious organizations can be involved in social and economic policies of a nation, as well.

In Poland, during the 1970s and 1980s, the citizenry often felt colonized by a foreign power—the Soviet Union. The Communist government in Poland was extremely unpopular and the people had no real way to make the government more responsive to their needs. In a survey conducted in 1983, only 3% of the respondents felt that the government and the Communist Party represented the interests of the Polish people. By contrast, 60% identified the Catholic Church and the Pope as protecting Polish interests (Tamney, 1992, p. 33).

Joseph Tamney (1992) reported that the Catholic Church became a real power broker in negotiations between unions and the government. Indeed, bishops in Poland sometimes took on a role akin to a union leader and in so doing became major players in the political future of the country. Since those bishops were ultimately responsible to the Pope and had a source of support external to the nation, they experienced a good deal of independence. Clergy who were supported by a state church, one financed with tax dollars, would not likely have the same freedom to challenge the government. The global nature of the Catholic Church influenced the role bishops could play in the Polish situation.

The locus of control of a religious hierarchy is an important issue. On the one hand, a religious structure may involve an organization that is external to any one nation and includes representation from many countries. Alternatively, a religious body may be *autocephalous*—that is, they are not *organizationally* linked to congregations in other countries; the religious hierarchy and membership is contained within a single nation. Transnationalism can influence the power of religious officials in our complex global environment. On the other hand, in an autocephalous religious community, the vitality of the community may be connected more closely to its functions in the mobilization of political power of a particular state or for affirming ethnic identity (Turner, 1991b). Poland is an especially intriguing case in that a single religion tended to be identified with national pride and culture, but that religion was global in its power base. That combination provided an exceptionally strong base for challenging governmental policies (Tamney, 1992).

Religion, Local, or Ethnic Identity and World Peace

This role of religion in solidifying ethnic identity may be even more important for religions in pluralistic societies, where supernatural sanction of one's own culture helps to fend off anomie. Nonetheless, the conclusion that a confluence of religion and ethnic group interests is important to religious vigor raises serious questions about the capacity of religion to unite people across these boundaries. Can a religion that advocates tolerance and diversity and downplays we–they polemics have vitality? In a local community seeking divine legitimation of its own values, religious ambivalence and nonspecificity is not appreciated.

Religious groups often are faced with a dilemma (Robertson, 1989). Church officials may choose to embrace the emerging global culture, acknowledge and celebrate global economic interdependence, and foster greater tolerance and open-mindedness toward those who are "different." The alternative course would be to intensify the solidarity between the religion and the national or ethnic group and to stress the differences between "us" and "them." The latter course is likely to enhance the numbers of adherents. Although virtually all religions give lip service to a desire for world peace, and although world peace requires greater tolerance and understanding between people, an institution is actually rewarded with increased numbers and additional resources if it takes the path of identifying with ethnic pride. The institutional reward system may actually work *against* openness to diversity. Despite meaning system proclamations in support of peace and tolerance of others, the drive for institutional survival may entice religious leaders to follow a path leading to popularity and membership growth. Those religions that do not choose this path may dwindle in numbers and risk their continued viability.

There is some evidence that religion does enhance patriotism and national fervor (e.g., Eisinga, Felling, & Peters, 1991), often in the process creating we/they distinctions against outsiders. Political scientist Ted Jelen pointed out that there is great danger to a religious system if it sacralizes the nation and its political structures, for in theological terms, this can easily become idolatry. The nation or the land or the political entity becomes sacred in itself and begins to be the primary sense of worth for the citizenry (Jelen, 1995). See the "Illustrating Sociological Concepts" feature for a discussion of this issue in Wales.

Critical Thinking: Religions—more than any other institutions—claim to support peace, yet your authors suggest that institutional rewards come to those religions that ignore in practice what they preach regarding peace. Does this argument make sense? Why or why not? What evidence supports your view?

Churches often grow in membership—they are rewarded institutionally—if they celebrate patriotism and sacralize national pride and loyalty. This enhanced we–they thinking does not lend itself well to peace advocacy and may run counter to claims of worshipping the "Prince of Peace." This huge flag hung in the Crystal Cathedral in California during a Sunday worship service.

(Text continued on p. 401)

ILLUSTRATING SOCIOLOGICAL CONCEPTS

Religion and Nationalism in Wales:

The Struggle for Local Identity and Global Perspective

Keith A. Roberts

The connection between religion and nationalism in Wales is both intense and ambiguous. The primary religious affiliation in Wales has long been Nonconformity, which includes all forms of Protestantism that are *not* Church of England. (Nonconformist congregations are called chapels or "Capels"; the word "churches" in Wales refers to Church of England congregations.) As I interviewed Welsh Nonconformists—clergy and laypeople, I found that a close link between religion and nationalism was justified on several grounds: the historical tie that has existed for centuries, preservation of language, celebration of peoplehood, and the demands of social justice. My respondents felt the chapels should support Welsh nationalism, though many were very clear that they supported a **cultural nationalism** (preservation of the language, the cultural uniqueness, the passion for music and poetry, and the sense of peoplehood in Wales). They were opposed to congregations or denominations being involved in any kind of **political nationalism** (struggles over power or distribution of resources, efforts to gain a measure of political independence from England, or support for the state and the actions of the state). The role of the chapels in preserving the language—and therefore the literature, history, and culture of the Welsh—was seen as a justice issue, for England has long tried to obliterate Welsh language and culture. Nonconformist worship services have been conducted in Welsh for centuries and helped the language to survive. Even the translation of the Bible into Welsh in 1567 may well have played a role in saving the language (Davies, 1993). Without Nonconformity, the long legacy of Welsh culture would have been abolished. Indeed, part of the attraction of Nonconformity was probably the fact that it represented something *other than* the Church of England. Nonconformist chapels were part of a larger protest against the large, powerful, and often oppressive neighbor to the east—England.

Yet this was an uneasy and sometimes uncomfortable alliance. There is a very clear sense in Wales that a confounding of the Christian community and the nation was a very slippery slope. The Welsh clergy in my interviews identified several dangers.

Worship of the Nation. A number of the Nonconformists I interviewed characterized the primary danger of a close religion/national identity link as that of idolatry: worshiping the nation or making the health and security of the nation one's "ultimate concern." A laywoman in Llanberis put the issue forcefully: "Patriotism is the antithesis of religion, is it not? There are dangers of patriotism becoming an idolatry–it becomes a false god we worship rather than the true God of all humanity." A retired congregational pastor on the coast made a similar comment: "There is a deep awareness that God and God alone is to be worshipped in the Welsh chapels. No other loyalty can enter the sanctuary. None whatsoever. We would never *dream,* for example, of having a flag in church—not even to wrap a casket during a funeral. Any symbol of the nation inside a church would be considered blasphemy."

Xenophobia. As one nonconformist minister in south Wales put it, "There is always the danger of the blend of nationalism and religion turning into xenophobia [fear and hatred of the outsider]. It is an extremely powerful combination, and potentially extremely dangerous." Noel Davies,

General Secretary of Churches Together in Wales, told me, "There is a fine line between the church helping the people understand national responsibility and the development of xenophobia. The former can so easily slide into the latter . . . We see this all around the world."

Welsh Language as a Possible Source of Division Within Chapels. A retired congregationalist minister told me: "In my last chapel I was stationed in south Wales. I did Welsh-only services because there were already English-only services being conducted in the town at other chapels. In so doing, monoglots [single language people] in my chapel felt left out because I was speaking only in Welsh during worship. So the language becomes a source of tension and conflict when members who are Anglophone monoglots [English speakers only] feel excluded. On the other hand, affirmation of the language is an essential part of Welshness and fostering it is an important role of the church . . . It was a real dilemma for me."

Linking religion and nationalism was viewed as extremely precarious, bordering on blasphemy. Yet it persisted. The question was, how did Welsh Nonconformists make sense of the dilemma and justify continuing links, for they clearly thought the links between religion and patriotism in Ireland, South Africa, the United States, and elsewhere were deeply offensive. Why was it any less so in Wales?

I found five arguments or constructions that justified this close tie in the minds of my informants. The contention of my informants was that there were special characteristics of Welsh culture or Welsh nationalism that made for a compatible fit between *Cymru cenedlaetholdeb* (Welsh nationalism) and Christian theology.

1. Universalism of the Church and the Internationalism of Plaid Cymru. The theological inclusiveness of Nonconformity fits very well, I was told, with the profound internationalism of the Welsh nationalist movement. The founders of the modern Cymru (Welsh) movement, such as Saunders Lewis, had a strong international orientation. As a Roman Catholic, Lewis was deeply influenced by the global perspective of the church. Denzel Morgan at the University of Wales, Aberystwyth, told me "The Welsh national party was founded by people whose experiences were formed by leaving Wales. Saunders Lewis [founder of Plaid Cymru, the Welsh nationalist political party] spent a number of years in France, for example. So Wales was viewed as being within a brotherhood of nations."

   Historian Neil Evans added, "The Brits often accused the Welsh of being parochial, but by knowing more about other countries than the Brits did, Welsh nationalists were able to outflank them and combat this charge of being parochial. Plaid Cymru looked to other small nations for support and even for models of nationalism and liberation. So they were very much in touch with international trends in Ireland, Scotland, Denmark, and other small countries."

2. The Justice Issue and Defensive Nationalism. Welsh nationalism is a defensive nationalism—very mild in nature but geared to preserve an oppressed and endangered culture. The church should stand by the oppressed, I was repeatedly told, and should preserve that which is of value in any culture. While Christianity should never allow itself to be identified with an aggressive nationalism or a group that benefits from great power and privilege, the responsibility of the church is different when a people and a culture are threatened.

(Continued)

(Continued)

One religious leader said "England is—and I use this in a soft respect—but England is the oppressor. The English are a threat to our language because they come in insisting that we ought to speak their language and that they own the territory. We are expected to adjust to the British newcomer. There is not such an expectation by immigrants from other parts of the world because they expect to adapt to our life and to learn our language and our customs. The English offend the Welsh because of their arrogance."

The fact is that Wales has no standing army and no international power. Therefore the Welsh Nationalism movement seems innocuous to many leaders; the usual dangers of mixing religion and national power are less relevant.

3. Nonconformist and Plaid Cymru Pacifism. One congregational pastor in North Wales expressed a theme I heard often: "There is a very strong sense among Nonconformists that if a Christian community is not pacifist and the nationalism is not committed to pacifism, then clearly there cannot be compatibility between nationalism and Christianity." Tudor Jones, the evangelical Welsh scholar and passionate nationalist, expressed the same sentiment: "Nationalism means so many different things. It has manifested itself in frightening ways . . . But Welsh nationalism is moderate nationalism—not embedded in a notion of war." If he were living in any country with a tradition of dominating any other country or with a tradition of using war as a means of problem-solving, he insisted that he would reject any notion that the church should be aligned with nationalism.

Others, like Noel Davies, the head of the Welsh ecumenical movement, said something else I heard on many occasions: "Welsh nationalism is *not* patriotism. Patriotism has a British connotation. Patriotism has a militaristic tone. Most of us would never use the word patriotism for Welshness. Patriotism is making the nation into a God. On the other hand, nationalism is building a community with a distinctive sense of peoplehood. Nationalism, as we understand it, is pacifist. Patriotism is always militaristic in its final outcomes." A Congregationalist summed up a consistent message in Wales, "Nationalism that is going to have any connection to Christianity must be pacifist and it must be internationalist. If it does not have that, it cannot be embraced by the church."

4. Cross-Cutting Loyalties and Interests. Cross-cutting group membership can reduce we/they thinking or the resultant xenophobia. Because nonconformist chapels are autocephalous churches with religious and national identities being parallel and coterminous, there is potential for the creation of xenophobia. Yet Wales is in a real cross-roads location. It is only a few miles from other countries, and travel to other places and friendships across national lines are easy to establish and are common. It is much less possible in Wales to live physically isolated from people who are from different parts of the world than is true in many parts of the U.S. and Canada. Moreover, the Welsh people often feel deep solidarity with other small nations who feel crushed by the large nations. So although the churches in Wales are autocephalous, the Welsh sense of Christianity is global. The Welsh do seem to have a strong sense of multiple group membership, and those multiple groups are cross-national in character.

Many Welsh clergy, laypeople, and scholars view any connection between religion and nationalism with some caution, feeling that religion can only embrace a very moderate nationalism. That nationalism must (1) focus on preservation of a culture rather than on loyalty to a state, (2) be internationalist in outlook, and (3) be pacifist. If any of these factors is missing, religious embracement of patriotism or nationalism is seen as dangerous.

Sociologically, the interesting point is that a religious tradition needs to affirm the peoplehood of an ethnic group or a nation, and religious groups are particularly likely to affirm this if that group is oppressed or powerless. The faith must relate to the basic identity of the people and their culture. On the other hand, close association of a religion and a nation can create bigotry and prejudice against those who are different and may justify massive killing through warfare against "those people" who are obviously not favored by God. If it does the latter, it may undermine some of the basic principles of the faith. The dilemma is very real.

Critical Thinking: Does the rationale developed by Welsh nonconformists for how religion might embrace nationalism—without creating jingoistic nationalism and bigotry—make sense? Is religion and patriotism a mix that can easily lead to we–they thinking and even warfare? Why do you think as you do?

Toward a Global Civil Religion?

In Chapters 3 and 13, we discussed Durkheim's idea, elaborated in recent years by Robert Bellah, that societies—especially pluralistic ones—need a common core of sacred symbols, values, and beliefs to enhance social integration and to provide an overarching system of meaning. In his essay on civil religion, Bellah (1970b) expressed his expectation that a global civil religion would eventually emerge, alluding to the possibility of "American civil religion becoming simply one part of a new civil religion of the world" (p. 186). The world has no such universal system, no myths, rituals, and symbols that powerfully unite the people of the globe, though the spread of human rights and international law gives some prospect that this could be a possible future (Casanova, 2001; Cristi & Dawson, 2007, p. 285), even in the face of bewildering religious diversity across the globe. There are roughly 2,200 million Christians, 1,388 million Muslims, 875.7 million Hindus, 385.6 million Buddhists,

652 million people who belong to ethnic folk religions, and this is just a few of the groups. The religious diversity and composition of the world are set forth in Figure 15.2.

Not only does the globe lack a comprehensive and unifying meaning system but the planet is actually populated by religions that have historically each claimed exclusive access to divine truth. Bryan Turner (1991a) wrote, "It is difficult to imagine how one can have several universalistic, global, evangelistic religions within the same world political space. How can one have mutually exclusive households within the world cultural system?" (p. 179). In fact, concern about the uncivil nature of some forms of religion dates back at least to Jean Jacques Rousseau, who stated in *The Social Contract* that some religions are contrary to order and peace within a nation: "It is impossible to live at peace with those we regard as damned. . . . the man who makes bold to say 'outside the church is no salvation' should be driven from the state" (Rousseau, 1954, pp. 221, 223).

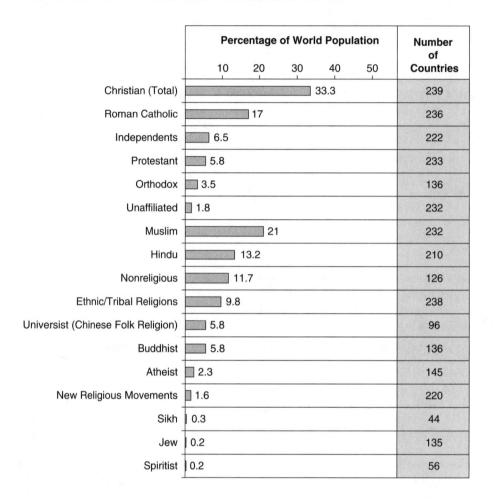

Figure 15.2 Religious Membership Around the Globe

SOURCE: Reprinted with permission from *Encyclopedia Britannica Almanac* 2008. Copyright © 2008 by Encyclopedia Britannica, Inc.

NOTE: Bahai, Confusianism, Jainism, and Taoism each comprise about .2% of the world's population. Shintoism and Zoroastrianism each makes up just .1%.

It is not just social coherence that is at stake. In the midst of a pluralistic world, with extraordinarily powerful self-interests inclining nations and societies to conflict, world organizations such as the World Court and the United Nations lack the legitimated authority that civil religion can provide (Berman, 1991; Meyer, 1980; Robertson, 1989, 1991; Simpson, 1991). Without some sacralization of their authority, such bodies find it very difficult to make their policies binding.

Of course, many other sociologists insist that the primary "glue" that unites most modern societies is not common beliefs and values anyway. Economic interdependence, technological linkages between social units, and bureaucratic processes provide the primary coherence, they argue (Turner 1991b; Wallerstein 1974, 1984). For these scholars, a global civil religion is unnecessary.

> *Critical Thinking:* Is a core set of common values and beliefs important to an integrated and well-functioning society? Does *global* harmony and genuine cooperation require some global religion? If so, how will this affect existing world religions? If not, what does the lack of importance of global religion say about the role and the vitality of religion in the future?

SUMMARY

Religion is not only itself a complex social system, with internal processes that are sometimes integrative and sometimes contradictory and conflictive, but it is one part of the larger national and global macro systems. Religion can both affect and be affected by globalization processes.

Globalization involves several interdependent processes: structural interdependence of nation–states, synthesis and cross-fertilization of cultures as a modern secularized global culture emerges, a change in socialization to a broader inclusiveness of others as being "like us," and an increase in individualism and reduction of traditional mechanisms of social control.

Theologies of world religions have been influenced by this globalization trend. As people have traded and become increasingly interdependent, liberal theologies have emerged that view other religions as colleagues with similar problems and allow the possibility that other traditions may also offer insights into divine truth. Further, shifts in the membership composition of religious organizations with global constituencies can cause a shift in theological views and ethics, as occurs with the anticapitalist liberation

theology of Roman Catholics in Global South countries. The Catholic Church finds itself in a dilemma as to whether it should sacralize the social system of its traditional constituency, the affluent Western world, or support its rapidly growing population in the Southern Hemisphere. In any case, global perspectives seem to be emerging in the theologies of many groups, as diverse groups try to make sense of this new and compelling social reality.

Globalization impacts religion in a number of ways. The emerging global culture is a highly secularized one stressing a rational-utilitarian outlook on the world and calling for institutional differentiation of religion from other spheres. This means that "covenantal relations," characterized by love, commitment, and sacrifice on behalf of others, are waning as the self-interested utilitarian "contractual relations" of government and economics expand. In reaction to this and to the compromises to modernism made by liberal theologies, a pattern of rigid and uncompromising revivalistic movements has developed in religions around the world. These revivalistic or "fundamentalist" religions attempt to reclaim sacred authority over a society and culture, combatting ethnocide of their culture by the press of global homogeneity. This means that fundamentalist movements are a result of worldwide trends, not just local factors.

The movement of people around the globe has increased. It is both affected by and affects religion. Sociologists studying transnational migration have used the term *glocalization* to describe the need for migrant communities to negotiate the relationship between global and the local and have described various strategies of glocalization employed by migrants (relativization, accommodation, hybridization, and transformation). Transnational religious connections are formed by "transmigrants"—immigrants, temporary migrants, missionaries, and pilgrims—and their effect on religious communities and individuals.

Many scholars believe that social systems need either a unifying religious system or a "civil religion." The entire world has no single set of sacred symbols, beliefs, and values that unify all people and that legitimate the growing number of global institutions designed to reduce conflicts and meet human needs. These scholars feel that the lack of any legitimating and solidifying sacred system is a problem in the prospect for world peace.

No other institutions are so explicit about their commitment to peace as religious ones. Christians, for example, claim to follow the Prince of Peace, and most other world religions give voice to the need for peace. Yet the often exclusive claim to truth and the incredibly powerful we–they sentiments they generate often belie their presumed commitment. If it is peace they want, religious groups often only want it on *their* terms. Once again in this text we learn that if we want to understand religion in society we must look beyond the official creeds. Just as religion may verbally oppose racism, yet support it in other ways, so also religion may undermine the conditions that would lead to peace even as it proclaims commitment to it.

Religion is, indeed, a complex phenomenon. It is composed of many parts and is itself one component of yet larger systems. It may contribute to integration or to schism, it may give voice to one set of values yet undermine those values in other ways, it may elicit the most noble and caring qualities of persons, or it may foster the most hateful and narrow-minded forms of bigotry. Whatever else one may conclude about religion, it is a multifaceted, dynamic, and always interesting part of our social system and of our most intimate personal lives. Its role in the emerging world system is no less intriguing and important.

EPILOGUE

The Sociological Perspective on Religion: A Concluding Comment

Sociology seeks to understand the social processes of religion. Because the discipline is limited to empirical investigation, it does not address the truth or falsity of a religious system. The researcher adopts a posture of at least temporary agnosticism. Nonetheless, explanations of cause and effect are offered with only empirically identifiable causes being noted. Hence, the effect is apparently to explain away any supernatural causes. Sociology offers only one lens or one vantage point for understanding religious processes, and it tends to operate within the confines of that vantage point. Other disciplines and other vantage points can offer other insights.

Furthermore, the interpretations of causality that are presented in this book represent the current understandings of sociologists, and the social sciences remain ever open to new data and new perspectives. Social scientists must always be willing to accept a new interpretation if it seems more plausible and if the data support it. The discipline will continue to develop, and new data may support or disprove theories about religious behavior that

now seem plausible. Methods of data collection and data interpretation are constantly being challenged and reassessed. The social scientist must always remain open to new data and fresh insights. That is the nature of science as a mode of inquiry.

Regardless, the meaning of a transcendent dimension of life for individuals can never be fully grasped through objective, scientific study. Margaret Poloma (1982b) insisted that scientific empiricism is itself a worldview and that the determinism of the social sciences can dull one's sensitivity to the mystical, intuitive, symbolic–imagistic side of life. Milton Yinger (1970) stated the case eloquently:

> No one would claim that the analysis of paint, painter, and patron exhausts the meaning of art; we are becoming cautious about making equivalent claims for the analysis of religion. The scientist must realize that propositions derived from objective study do not exhaust the meaning of things. (p. 2)

Social psychologist James Vander Zanden (1987) pointed out that our five senses are limited to certain sensations. We know that there are wavelengths of light that we cannot see and frequencies of sound we cannot hear. Yet we do not deny that they exist. Scientists have now demonstrated that some creatures have sensory capabilities not even available to humans. Some birds, for example, are sensitive to changes in barometric pressures and to the earth's magnetic fields; in fact, some arctic species have magnetite in their heads, which some scientists think is how these birds know the location of the earth's poles. Most birds can see ultraviolet rays and some sharks find their prey through sensitivity to magnetic fields given off by fish buried in the sand. Many snakes can *see* infrared rays that we only experience as heat. Vander Zanden concluded that we must never forget the limitations of our own empirical senses: Scientists must approach all forms of human knowledge with a healthy dose of humility.

Indeed, Canadian sociologist of religion Reginald Bibby (2002) pointed to the myopic view that a strictly empiricist assumption can entail:

> The old analogy still applies: limiting religion's potential fullness to science's eyes is like searching for a lost key only under a light because the lighting is better there, or like searching for handwritten material on a computer because it's easier than going through the endless number of file cabinet drawers. (p. 228)

With Milton Yinger, Thomas O'Dea, Reginald Bibby and others, we remain convinced that neither theism nor atheism are inherently unsophisticated. Both are assumptions about the nature of the world that are of interest to sociologists.

The purpose of this text has been neither to destroy the faith of believers nor to make believers out of skeptics. The purpose has been to help readers gain insight into the complexity of religion and the relationship between religion and society. In the process, the views of readers (including their worldviews) may have been changed or modified. Sometimes the insights of social scientists are unsettling, for they challenge our assumptions about reality. That is the nature of the search for truth; the seeker must be willing to follow the data wherever they lead. Certainly sociology does not offer the *whole* truth about religion or about any aspect of human life, but sociological investigation can *contribute* to a *holistic* understanding of human experience—including religious experience. As we suggested in Chapter 2, sociological study of religion cannot be proven to be directly beneficial to religious faith, but surely ignorance is more harmful in the long run than is disquieting knowledge. Likewise for nonbelievers, ignorance of this phenomenon—which is so important to many people—leaves a gap that prevents a holistic understanding of the human experience.

Future contributions to the understanding of religion will likely require two characteristics: (1) uncompromising academic rigor and (2) an honest recognition of the limitations of our current knowledge. Certainly, there is much that we do not know about religion and about religious behavior. We only hope that our own fascination with the sociology of religion has been contagious to the readers and that this brief introduction will serve only as the beginning of a continuing inquiry.

BIBLIOGRAPHY

Allport, Gordon W. 1966. The religious context of prejudice. *Journal for the Scientific Study of Religion* (fall), 447–457.

Altman, Dennis. 1982. *The homosexualization of America.* New York: St. Martin's Press.

Ammerman, Nancy. 1987. *Bible believers: Fundamentalists in the modern world.* New Brunswick, NJ: Rutgers University Press.

Ammerman, Nancy. 1990. *Baptist battles: Social change and religious conflict in the Southern Baptist Convention.* New Brunswick, NJ: Rutgers University Press.

Ammerman, Nancy. 1997. *Congregation and community.* New Brunswick, NJ: Rutgers University Press.

Ammerman, Nancy. 2003. Religious identities and religious institutions. In M. Dillon (Ed.), *Handbook of the sociology of religion* (pp. 207–224). Cambridge: Cambridge University Press.

Ammerman, Nancy. 2005. *Pillars of faith: American congregations and their partners.* Berkeley: University of California Press.

Ammerman, Nancy. 2006. Denominationalism/ congregationalism. In H. R. Ebaugh (Ed.), *Handbook of religion and social institutions* (pp. 353–372). New York: Springer.

Ammerman, Nancy. 2009. Congregations: Local, social, and religious. In P. Clarke (Ed.), *Oxford handbook of the sociology of religion* (pp. 562–580). New York: Oxford University Press.

Ammerman, Nancy; Carroll, Jackson; Dudley, Carl; & McKinney, William. 1998. *Studying congregations: A handbook.* Nashville, TN: Abingdon Press.

Anderson, Charles. 1970. *White protestant Americans.* Englewood Cliffs, NJ: Prentice Hall.

Anthony, Francis-Vincent; Hermans, Chris A. M.; & Sterkens, Carl. 2010. A comparative study of mystical experience among Christian, Muslim, and Hindu students in Tamil Nadu, India. *Journal for the Scientific Study of Religion, 49,* 264–277.

Anti-Defamation League. 2009. *Audit of anti-Semitic incidents, 2008.* Retrieved from www.adl.org/ main_Anti_Semitism_Domestic/2008_Audit.htm

Antoun, Richard T. 2008. Understanding fundamentalism: Christian, Islamic, and Jewish movements (2nd ed.). Lanham, MD: Rowman & Littlefield.

Arnett, Jeffrey J. 2004. *Emerging adulthood: The Winding road from late teens through the twenties.* New York: Oxford University Press.

Assimeng, Max. 1987. *Saints and social structures.* Jema, Ghana: Ghana Publishing Co.

Associated Press. 2009. *Family spat divides television empire* (January 31).

Association of American Publishers. 2010. *Industry Statistics 2009* (April 7). Retrieved from publishers. org/main/IndustryStats/indStats_02.htm

Austin, Allen D. (Ed.). 1997. *African Muslims in antebellum America* (Rev. ed.). New York: Routledge.

Baer, Hans A. 1984. *The black spiritual movement: A religious response to racism.* Knoxville: University of Tennessee Press.

Bagby, Ihsan. 2003. Imams and mosque organization in the United States: A study of mosque leadership and organizational structure in American mosques. In P. Strum & D. Tarantolo (Eds.),

Muslims in the United States, (pp. 113–134). Washington, DC: Woodrow Wilson International Center for Scholars.

Bainbridge, William Sims. 1978. *Satan's power: Ethnography of a deviant psychotherapy cult.* Berkeley: University of California Press.

Bainbridge, William Sims, & Stark, Rodney. 1979. Cult formation: Three compatible models. *Sociological Analysis* (winter), 283–295.

Bainbridge, William Sims, & Stark, Rodney. 1980. Scientology: To be perfectly clear. *Sociological Analysis* (summer), 128–316.

Bainbridge, William Sims, & Stark, Rodney. 1981. The consciousness reformation reconsidered. *Journal for the Scientific Study of Religion* (March), 1–15.

Bainton, Roland. 1960. *Christian attitudes toward war and peace: A historical survey and critical re-evaluation.* New York: Abingdon Press.

Baker, Joseph. 2008. An investigation of the sociological patterns of prayer frequency and content. *Sociology of Religion, 69,* 169–185.

Baker, Joseph, & Smith, Buster. 2009. The nones: Social characteristics of the religiously unaffiliated. *Social Forces, 87,* 1251–1263.

Baker, William. 2007. *Playing with God: Religion and modern sport.* Cambridge, MA: Harvard University Press.

Balch, Robert W. 1980. Looking behind the scenes in a religious cult: Implications for the study of conversion. *Sociological Analysis* (summer), 137–143.

Balch, Robert W., & Taylor, David. 1976. Salvation in a UFO. *Psychology Today* (October), 58–66, 106.

Balch, Robert W., & Taylor, David. 1977. Seekers and saucers. *American Behavioral Scientist* (July/August), 839–860.

Balch, Robert W., & Taylor, David. 2002. Making sense of the Heaven's Gate suicides. In D. G. Bromley & J. G. Melton (Eds.), *Cults, religion, and violence* (pp. 209–228). New York: Cambridge University Press.

Ballantine, Jeanne, & Hammack, Floyd. 2008. *The sociology of education: A systematic analysis* (6th ed.). Upper Saddle River, NJ: Prentice Hall.

Ballantine, Jeanne, & Roberts, Keith A. 2011. *Our social world.* Thousand Oaks, CA: Sage/Pine Forge.

Baltzell, E. Digby. 1964. *The Protestant establishment.* New York: Random House.

Bandura, Albert. 1977. *Social learning theory.* New York: General Learning Press.

Bankston, Carl, & Zhou, Min. 2000. De facto congregationalism and social mobility in Laotian and Vietnamese immigrant communities: A study of religious institutions and economic change. *Review of Religious Research, 41,* 453–470.

Bao, Wan-Ning; Whitbeck, Les; Hoyt, Danny; & Conger, Rand. 1999. Perceived parental acceptance as a moderator of religious transmission among adolescent boys and girls. *Journal of Marriage and Family, 61,* 362–374.

Barker, Eileen. 1984. *The making of a Moonie: Brainwashing or choice?* New York: Basil Blackwell.

Barlow, Philip L. 1991. *Mormons and the Bible: The place of the Latter-Day Saints in American religion.* New York: Oxford University Press.

Barnouw, Victor. 1982. *An introduction to anthropology: Ethnology* (4th ed.). Homewood, IL: Dorsey Press.

Baroja, Julio Caro. 1964. *The world of the witches.* Chicago: University of Chicago Press.

Barrett, David; Kurian, George; & Johnson, Todd (Eds.). 2001. *World Christian encyclopedia* (2 vols.). New York: Oxford University Press.

Barro, Robert; Hwang, Jason; & McCleary, Rachel. 2010. Religious conversion in 40 countries. *Journal for the Scientific Study of Religion, 49,* 15–36.

Bartkowski, John P. 2001. Breaking walls, raising fences: Masculinity, intimacy, and accountability among the Promise Keepers. In R. H. Williams (Ed.), *Promise Keepers and the new masculinity: Private lives and public morality* (pp. 33–53). Lanham, MD: Lexington Books.

Bartkowski, John P., & Read, Jen'nan Ghazal. 2003. Veiled submission: Gender, power, and identity among Evangelical and Muslim women in the United States. *Qualitative Sociology, 26,* 71–92.

Basch, Linda; Schiller, Nina Glick; & Blanc, Cristina Szanton. 1994. *Nations unbound: Transnational projects, postcolonial predicaments, and deterritorialized nation-states.* London: Gordon and Breach.

Batson, C. Daniel. 1977. Experimentation in psychology of religion: An impossible dream. *Journal for the Scientific Study of Religion* (December), 413–418.

Batson, C. Daniel; Schoenrade, Patricia; & Ventis, W. Larry. 1993. *Religion and the individual: A social-psychological perspective.* New York: Oxford University Press.

Baumann, Martin, & Salentin, Kurt. 2006. Migrant religiousness and social incorporation: Tamil Hindus from Sri Lanka in Germany. *Journal of Contemporary Religion, 21,* 297–323.

Bayer, Alan E. 1975. Sexist students in American colleges: A descriptive note. *Journal of Marriage and the Family* (May), 391–396.

Beck, Marc. 1978. Pluralist theory and church policy positions on racial and sexual equality. *Sociological Analysis* (winter), 338–350.

Becker, Howard. 1932. *Systematic sociology.* New York: Wiley.

Beckford, James A. 1985. *Cult controversies: The societal response to the New Religious Movements.* London: Tavistock.

Bell, Daniel. 1976. *The cultural contradictions of capitalism.* New York: Basic Books.

Bellah, Robert N. 1970a. Christianity and symbolic realism. *Journal for the Scientific Study of Religion* (summer), 89–96.

Bellah, Robert N. 1970b. Civil religion in America. *Beyond belief: Essays on religion in a post industrial world* (pp. 168–215). New York: Harper & Row.

Bellah, R. N. 1970c. Religious evolution. *Beyond belief: Essays on religion in a post industrial world* (pp. 20–50). New York: Harper & Row.

Bellah, Robert N. 1975. *The broken covenant: American civil religion in time of trial.* New York: Seabury Press.

Bellah, Robert N.; Madsen, Richard; Sullivan, William M.; Swidler, Ann; & Tipton, Steven M. 1985. *Habits of the heart: Individualism and commitment in American life.* Berkeley: University of California Press.

Bengtson, Vern; Copen, Casey; Putney, Norella; & Silverstein, Merril. 2009. A longitudinal study of the intergenerational transmission of religion. *International Sociology, 24,* 325–345.

Benokraitis, Nijole V. 2010. *Marriages and families: Changes, choices, and constraints* (12th ed.). Englewood Cliffs, NJ: Prentice Hall.

Benz, Ernest. 1964. On understanding non-Christian religions. In L. Schneider (Ed.), *Religion, culture, and society* (pp. 3–9). New York: Wiley.

Berger, Helen A. 1999. *A community of witches: Contemporary neo-paganism and witchcraft in the United States.* Columbia, SC: University of South Carolina Press.

Berger, Peter L. 1967. *The sacred canopy.* Garden City, NY: Doubleday.

Berger, Peter L. 1974. Some second thoughts on substantive versus functional definitions of religion. *Journal for the Scientific Study of Religion* (June), 125–133.

Berger, Peter L. 1979. *The heretical imperative.* Garden City, NY: Anchor Press.

Berger, Peter L. 1981. The class struggle in American religion. *The Christian Century* (February), 194–200.

Berger, Peter L., & Luckmann, Thomas. 1966. *The social construction of reality.* Garden City, NY: Doubleday.

Berman, Harold J. 1991. Law and religion in the development of a world order. *Sociological Analysis* (spring), 27–36.

Berry, Brewton, & Tischler, Henry L. 1978. *Race and ethnic relations* (4th ed.). Boston: Houghton Mifflin.

Beyer, Peter. 2000. Secularization from the perspective of globalization. In W. H. Swatos Jr. & D. V. A. Olson (Eds.), *The secularization debate* (pp. 81–93). Lanham, MD: Rowman & Littlefield.

Beyer, Peter. 2007. Globalization and glocalization. In J. Beckford & N. J. Demerath III (Eds.), *Sage handbook of the sociology of religion* (pp. 88–117). Los Angeles: Sage.

Beyerlein, Kraig. 2004. Specifying the impact of conservative Protestantism on educational attainment. *Journal for the Scientific Study of Religion, 42,* 505–518.

Beyerlein, Kraig, & Chaves, Mark. 2003. The political activities of religious congregations in the United States. *Journal for the Scientific Study of Religion, 42,* 229–246.

Bibby, Reginald W. 1978. Why conservative churches really are growing: Kelley revisited. *Journal for the Scientific Study of Religion* (June), 129–138.

Bibby, Reginald W. 1979. Religion and modernity: The Canadian case. *Journal for the Scientific Study of Religion* (March), 1–17.

Bibby, Reginald W. 1987a. *Fragmented gods: The poverty and potential of religion in Canada.* Toronto: Irwin.

Bibby, Reginald W. 2002. *Restless gods: The renaissance of religion in Canada.* Toronto: Stoddard Publishing Co.

Bibby, Reginald W., & Brinkerhoff, Merlin B. 1973. The circulation of the saints: A study of people who join conservative churches. *Journal for the Scientific Study of Religion* (September), 273–283.

Bibby, Reginald W., & Brinkerhoff, Merlin B. 1983. The circulation of the saints revisited. *Journal for the Scientific Study of Religion* (September), 253–262.

Bibby, Reginald W., & Brinkerhoff, Merlin B. 1992. On the circulatory problems of saints: A response to Perrin and Mauss. *Review of Religious Research* (December), 170–175.

Bibby, Reginald W., & Brinkerhoff, Merlin B. 1994. Circulation of the saints, 1966–1990: New data, new reflections. *Journal for the Scientific Study of Religion* (September), 273–280.

Bickerton, Ian. 2003. Bank makes an issue of mystic's mint. *Financial Times* (February 8), p. 9.

Billig, Michael. 1995. *Banal nationalism.* Thousand Oaks, CA: Sage.

Bird, Phyllis. 1974. Images of women in the Old Testament. In R. R. Ruether (Ed.), *Religion and sexism* (pp. 41–88). New York: Simon & Schuster.

Blanchard, Dallas A., & Prewitt, Terry J. 1993. *Religious violence and abortion: The Gideon project.* Gainesville: University Press of Florida.

Bloch, Jon P. 2001. The new and improved Clint Eastwood: Change and persistence in Promise Keepers self-help literature. In R. H. Williams (Ed.), *Promise Keepers and the new masculinity: Private lives and public morality* (pp. 11–31). Lanham, MD: Lexington Books.

Bogan, Jesse. 2009. *America's biggest megachurches* (June 26). Retrieved from www.forbes.com/2009/06/26/americas-biggest-megachurches-business-megachurches.html

Boli, John, & Brewington, David. 2007. Religious organizations. In P. Beyer & L. Beaman (Eds.), *Religion, globalization, and culture* (pp. 203–232). Leiden: Brill.

Boli, John, & Thomas, George. 1999. *Constructing world culture: International nongovernmental organizations since 1875.* Stanford, CA: Stanford University Press.

Bonilla-Silva, Eduardo. 2003. *Racism without racists: Color-blind racism and the persistence of racial inequality in the United States.* New York: Rowman & Littlefield.

Borg, Marcus. 2001. *Reading the Bible again for the first time.* San Francisco: HarperCollins.

Bowles, Samuel, & Gintis, Herbert. 1976. *Schooling in capitalist America.* New York: Basic Books.

Brasher, Brenda. 1998. *Godly women: Fundamentalism and female power.* New Brunswick, NJ: Rutgers University Press.

Brasher, Brenda. 2004. *Gimme that online religion.* New Brunswick, NJ: Rutgers University Press.

Breault, Kevin. 1989a. New evidence on religious pluralism, urbanism, and religious participation. *American Sociological Review, 54,* 1048–1053.

Breault, Kevin. 1989b. A reexamination of the relationship between religious diversity and religious adherents. *American Sociological Review, 54,* 1056–1059.

Brecher, Jeremy; Costello, Tim; & Smith, Brendan. 2006. Globalization and social movements. In D. S. Eitzen & M. B. Zinn (Eds.), *Globalization: The transformation R-5 references of social worlds* (pp. 330–347). Belmont, CA: Wadsworth.

Brightman, Edgar Sheffield. 1940. *A philosophy of religion.* New York: Greenwood Press.

Bromley, David G. 1991. Satanism: The new cult scare. In J. T. Richardson, J. Best, & D. G. Bromley (Eds.), *The Satanism scare* (pp. 49–72). New York: Aldine de Gruyter.

Bromley, David G. 1997. Remembering the future: A sociological narrative of crisis episodes, collective action, culture workers, and countermovements. *Sociology of Religion* (summer), 105–140.

Bromley, David G. 2009. New religions as a specialist field of study. In P. Clarke (Ed.), *The Oxford handbook of the sociology of religion* (pp. 723–741). New York: Oxford University Press.

Bromley, D. G., & Busching, Bruce C. 1988. Understanding the structure of contractual and covenantal social relations: Implications for the sociology of religion. *Sociological Analysis* (December), 15S–32S.

Bromley, David G., & Melton, J. Gordon. (Eds.). 2002. *Cults, religion, and violence.* New York: Cambridge University Press.

Bromley, David G., & Shupe, Anson D., Jr. 1979. *Moonies in America: Cult, church, and crusade.* Beverly Hills, CA: Sage.

Bromley, David G., & Shupe, Anson D., Jr. 1981. *Strange gods: The great American cult scare.* Boston: Beacon Press.

Bruce, Steve. 1995. The truth about religion in Britain. *Journal for the Scientific Study of Religion* (December), 417–430.

Bruce, Steve. 1999. *Choice and religion: A critique of rational choice theory.* Oxford: Oxford University Press.

Bruce, Steve. 2002. *God is dead: Secularization in the West.* Oxford: Blackwell.

Budde, Michael L. 1992. *The two churches: Catholicism and capitalism in the world system.* Durham, NC: Duke University Press.

Bullough, Vern L. 1973. *The subordinate sex: A history of attitudes toward women.* Urbana: University of Illinois Press.

Bunt, Gary. 2009. Religion and the Internet. In P. Clarke (Ed.), *Oxford handbook of the sociology of religion* (pp. 705–720). Oxford: Oxford University Press.

Burawoy, Michael. 1987. The limits of Wright's analytical Marxism and an alternative. *Berkeley Journal of Sociology, 32,* 51–72.

Burgess, Ernest W. 1916. *The function of socialization in social evolution.* Chicago: University of Chicago Press.

Burke, Kenneth. 1935. *Permanence and change.* New York: New Republic.

Burrow, Rufus, Jr. 1999. *Personalism: A critical introduction.* St. Louis, MO: Chalice Press.

Burrow, Rufus, Jr. 2006. *God and human dignity: The personalism, theology, and ethics of Martin Luther King, Jr.* Notre Dame, IN: Notre Dame Univ. Press.

Burstein, Paul. 2007. Jewish educational and economic success in the United States: A search for explanations. *Sociological Perspectives, 50,* 209–228.

Burtchaell, James. 1998. *The dying of the light: The disengagement of colleges and universities from their Christian churches.* Grand Rapids, MI: Eerdmans.

Byrne, Donn, & McGraw, Carl. 1964. Interpersonal attraction toward negroes. *Human Relations* (August), 201–213.

Byrne, Donn, & Wong, Terry J. 1962. Racial prejudice, interpersonal attraction, and assumed dissimilarity of attitudes. *Journal of Abnormal and Social Psychology* (October), 246–253.

Campbell, Colin. 1983. Romanticism and the consumer ethic: Intimations of a Weber-style thesis. *Sociological Analysis, 44,* 279–296.

Campbell, Ernest Q., & Pettigrew, Thomas F. 1959. *Christians in racial crisis.* Washington, DC: Public Affairs Press.

Campbell, George V. P. 2005. *Everything you know seems wrong: Globalization and the relativization of tradition.* Lanham, MD: University Press of America.

Cantrell, Randolph; Krile, James; & Donohue, George. 1983. Parish autonomy: Measuring denominational differences. *Journal for the Scientific Study of Religion, 22,* 276–287.

Caplow, Theodore; Bahr, Howard; & Chadwick, Bruce. 1983. *All faithful people: Change and continuity in Middletown's religion.* Minneapolis: University of Minnesota Press.

Carden, Maren Lockwood. 1969. *Oneida.* New York: Harper & Row.

Carus, Carl G. 1925. In T. Lessing (Ed.), *Symbolik der menschlichen gestalt.* Celle: N. Kampmann. (Original work published 1853)

Casanova, Jose. 2001. Religion, the new millenium, and globalization. *Sociology of Religion* (winter), 415–441.

Cauthen, Kenneth. 1962. *The impact of American religious liberalism.* New York: Harper & Row.

Chang, Patricia M. Y. 1997. Female clergy in the contemporary Protestant Church: A current assessment. *Journal for the Scientific Study of Religion* (December), 565–573.

Chaves, Mark. 1993. Denominations as dual structures: An organizational analysis. *Sociology of Religion* (summer), 147–169.

Chaves, Mark. 1994. Secularization as declining religious authority. *Social Forces* (March), 749–774.

Chaves, Mark. 1997. *Ordaining women: Culture and conflict in religious organizations.* Cambridge, MA: Harvard University Press.

Chaves, Mark. 2004. *Congregations in America.* Cambridge, MA: Harvard University Press.

Chaves, Mark, & Anderson, Shawna. 2008. Continuity and change in American congregations: Introducing the second wave of the National Congregations study. *Sociology of Religion* (winter), *69,* 415–440.

Chaves, Mark, & Cann, David. 1992. Regulation, pluralism, and religious market structure: Explaining religion's vitality. *Rationality and Society, 4,* 272–290.

Chaves, Mark, & Cavendish, James. 1997. Recent changes in ordination conflicts: The effect of a social movement on interorganizational controversy. *Journal for the Scientific Study of Religion* (December), 574–584.

Chaves, Mark, & Gorski, Phillip S. 2001. Religious pluralism and religious participation. *Annual Review of Sociology, 27,* 261–281.

Chaves, Mark; Konieczny, Mary Ellen; Beyerlein, Kraig; & Barman, Emily. 1999. The national congregations study: Background, methods, and selected results. *Journal for the Scientific Study of Religion, 38*(4), 458–476.

Chaves, Mark, & Sutton, John. 2004. Organizational consolidation in American Protestant denominations, 1890–1990. *Journal for the Scientific Study of Religion, 43,* 51–66.

Chaves, Mark, & Tsitsos, William. 2001. Congregations and social services: What they do, how they do it, and with whom. *Nonprofit and Voluntary Sector Quarterly, 30,* 660–683.

Cherlin, Andrew. 1999. *Public and private families.* Boston: McGraw-Hill.

Cherry, Conrad; DeBerg, Betty; & Porterfield, Amanda. 2001. *Religion on campus.* Chapel Hill: University of North Carolina Press.

Chesnut, R. Andrew. 2010. Conservative Christian competitors: Pentecostals and charismatic Catholics in Latin America's new religious economy. *SAIS Review, 30,* 91–103.

Christerson, Brad; Edwards, Korie L.; & Emerson, Michael O. 2005. *Against all odds: The struggle for racial integration in religious organizations.* New York: NYU Press.

Chua, Amy. 2006. Globalizing hate. In S. D. Eitzen & M. B. Zinn (Eds.), *Globalization: The transformation of social worlds* (pp. 234–238). Belmont, CA: Wadsworth.

Cimino, Richard, & Lattin, Don. 2002. *Shopping for faith: American religion in the new millennium.* San Francisco: Jossey-Bass.

Clinebell, Howard J. 1965. *Mental health through Christian community.* Nashville, TN: Abingdon Press.

Coakley, Jay. 2007. *Sports in society: Issues and controversies.* Boston: McGraw-Hill.

Cobb, Jennifer. 1998. *Cybergrace: The search for God in the digital world.* New York: Random House.

Cohen, Jere. 2002. *Protestantism and capitalism: The mechanisms of influence.* New York: Aldine de Gruyter.

Cohen, Steven M. 1983. *American modernity and Jewish identity.* New York: Tavistock.

Cohn, Norman. 1964. Medieval millenarism: Its bearing on the comparative study of millenarian movements. In L. Schneider (Ed.), *Religion, culture, and society* (pp. 168–181). New York: Wiley.

Coleman, Simon. 1993. Conservative Protestantism and the world order: The faith movement in the United States and Sweden. *Sociology of Religion* (winter), 353–373.

Collett, Jessica, & Lizardo, Omar. 2009. A power-control theory of gender and religiosity. *Journal for the Scientific Study of Religion, 48,* 213–231.

Condran, John G., & Tamney, Joseph B. 1985. Religious "nones": 1957 to 1982. *Sociological Analysis* (winter), 415–424.

Cone, James H. 1969. *Black theology and black power.* New York: Seabury Press.

Cone, James H. 1970. *Liberation.* Philadelphia: Lippincott.

Cone, James H. 1972. *The spirituals and the blues.* Westport, CT: Greenwood Press.

Cornwall, Marie. 1987. The social bases of religion: A study of factors influencing religious belief and commitment. *Review of Religious Research* (September), 44–56.

Coser, Lewis A. 1954. *The functions of social conflict.* New York: Free Press.

Coser, Lewis A. 1967. *Continuities in the study of social conflict.* New York: Free Press.

Cowan, Douglas. 2007. Religion on the Internet. In J. Beckford & N. J. Demerath III (Eds.), *Sage handbook of the sociology of religion.* Thousand Oaks, CA: Sage.

Cox, Edwin. 1967. *Sixth form religion.* London: SCM Press.

Cox, Harvey. 1964. *On not leaving it to the snake.* New York: Macmillan.

Cox, Harvey. 2009. *The future of faith.* New York: HarperCollins.

Craig, Steve. 2004. How America adopted radio: Demographic differences in set ownership reported in the 1930–1950 U.S. Censuses.

Journal of Broadcasting & Electronic Media, 48, 179–195.

Cristi, Marcela, & Dawson, Lorne L. 2007. Civil religion in America and in global context. In J. Beckford & N. J. Demerath III (Eds.), *Handbook of sociology of religion* (pp. 251–276). London: Sage.

Daly, Mary. 1970. Women and the Catholic Church. In R. Morgan (Ed.), *Sisterhood is powerful* (pp. 124–138). New York: Vintage Books.

Darley, J. M., & Batson, C. Daniel. 1973. From Jerusalem to Jericho: A study of situational and dispositional variables in helping behavior. *Journal of Personality and Social Psychology* (July), 100–108.

Darnell, Alfred, & Sherkat, Darren. 1997. The impact of Protestant fundamentalism on educational attainment. *American Sociological Review, 62,* 306–316.

Davidman, Lynn. 1990. Accommodation and resistance to modernity: A comparison of two contemporary Orthodox Jewish groups. *Sociological Analysis* (Spring), 35–51.

Davidman, Lynn. 1991. *Tradition in a rootless world: Women turn to Orthodox Judaism.* Berkeley: University of California Press.

Davidson, James D. 1977. Socio-economic status and ten dimensions of religious commitment. *Sociology and Social Research* (July), 462–485.

Davidson, James, & Pyle, Ralph. 2006. Social class. In H. R. Ebaugh (Ed.), *Handbook of religion and social institutions* (pp. 185–206). New York: Springer.

Davie, Grace. 2000. *Religion in modern Europe: A memory mutates.* Oxford: Oxford University Press.

Davies, John. 1993. *A history of Wales.* London: Penguin Books.

Davies, Thomas Richard. 2008. *The rise and fall of transnational civil society: The evolution of international non-governmental organizations since 1839.* (Working Paper CUTP/003) (April). City University London, Centre for International Politics.

Davis, James A.; Smith, Tom W.; & Marsden, Peter V. 2003. General Social Surveys, 1972–2002: [cumulativefile]. 2nd ICPSR version. Chicago, IL: National Opinion Research Center; Storrs, CT: Roper Center for Public Opinion Research, University of Connecticut; Ann Arbor, MI: Inter-university Consortium for Political and Social Research.

Davis, Kingsley. 1949. *The human society.* New York: Macmillan.

Dawson, Lorne. 1990. Self-affirmation, freedom, and rationality: Theoretically elaborating "active" conversions. *Journal for the Scientific Study of Religion* (June), 141–163.

Dawson, Lorne. 2009. Church-sect-cult: Constructing typologies of religious groups. In P. Clarke (Ed.), *The Oxford handbook of the sociology of religion* (pp. 525–544). New York: Oxford University Press.

Dawson, Lorne, & Cowan, Douglas. 2004. *Religion online: Finding faith on the Internet.* New York: Routledge.

deGobineau, C. A. 1855/1970. Essay on the inequality of the human races. In M. Biddiss (Ed.), *Gobineau: Selected political writings.* London: Jonathon Cape.

Della Cava, Ralph. 2001. Transnational religions: The Roman Catholic Church in Brazil and the Orthodox Church in Russia. *Sociology of Religion* (winter), 535–550.

Demerath, N. J., III. 1965. *Social class in American Protestantism.* Chicago: Rand McNally.

Demerath, N. J., III. 2000. The varieties of the sacred experience: Finding the sacred in a secular grove. *Journal for the Scientific Study of Religion* (March), 2–11.

Dempewolff, Judith Ann. 1974. Some correlates of feminism. *Psychological Reports* (April), 671–676.

Denton, Melinda; Pearce, Lisa; & Smith, Christian. 2008. *Religion and spirituality on the path through adolescence, Research Report Number 8.* National Study of Youth and Religion, University of North Carolina at Chapel Hill.

Deutscher, Irwin. 1966. Words and deeds: Social science and social policy. *Social Problems* (winter), 235–254.

Deutscher, Irwin. 1973. *What we say, what we do.* Glenview, IL: Scott, Foresman.

DeYoung, Curtiss; Emerson, Michael O.; Yancey, George; & Kim, Karen Chai. 2003. *United by faith: The multiracial congregation as an answer to the problem of race.* New York: Oxford University Press.

Dillon, Michele. 2007. Age, generation, and cohort in American religion and spirituality. In J. Beckford

& N. J. Demerath III (Eds.), *The Sage handbook of the sociology of religion* (pp. 526–548). Thousand Oaks, CA: Sage.

Dillon, Michele, & Wink, Paul. 2007. *In the course of a lifetime: Tracing religious belief, practice, and change.* Berkeley: University of California Press.

DiMaggio, Paul, & Powell, Walter. 1983. The iron cage revisited: Institutional isomorphism and collective rationality in organizational fields. *American Sociological Review, 48,* 147–160.

Dobbelaere, Karel. 1981. *Secularization: A multidimensional concept.* Beverly Hills, CA: Sage.

Dobbelaere, Karel. 2000. Toward an integrated perspective of the processes related to the descriptive concept of secularization. In W. H. Swatos Jr. & D. V. A. Olson (Eds.), *The secularization debate* (pp. 21–39). Lanham, MD: Rowman & Littlefield.

Douglas, Mary. 1966. *Purity and danger.* London: Routledge and Kegan Paul.

Douglas, Mary. 1968. Pollution. In D. Sills (Ed.), *International encyclopedia of the social sciences* (Vol. XII, pp. 336–341). New York: Macmillan and Free Press.

Douglass, Jane Dempsey. 1974. Women and the continental reformation. In R. R. Ruether (Ed.), *Religion and sexism* (pp. 292–318). New York: Simon & Schuster.

Downton, James V., Jr. 1979. *Sacred journeys: The conversion of young Americans to Divine Light Mission.* New York: Columbia University Press.

Downton, James V., Jr. 1980. An evolutionary theory of spiritual conversion and commitment: The case of Divine Light Mission. *Journal for the Scientific Study of Religion* (December), 381–396.

DuBois, W. E. B. 1903. *The souls of black folks.* Chicago: A. C. McClurg.

Durkheim, Emile. 1995. *The elementary forms of the religious life* (K. E. Fields, Trans.). New York: Free Press. (Original work published 1912)

Dynes, Russell R. 1955. Church-sect typology and socioeconomic status. *American Sociological Review* (October), 555–560.

Edwards, Jonathan. 1966. Sinners in the hands of an angry God. In O. E. Winslow (Ed.), *Jonathan Edwards: Basic writings* (pp. 156–160). New York: New American Library.

Eichler, Margrit. 1972. *Charismatic and ideological leadership in secular and religious millenarian movements: A sociological study.* PhD dissertation, Duke University. Ann Arbor, MI.

Eisinga, Rob; Felling, Albert; & Peters, Jan. 1991. Christian beliefs and ethnocentrism in Dutch society: A test of three models. *Review of Religious Research* (June), 305–319.

Eitzen, D. Stanley, & Zinn, Masine Baca. 2006. *Globalization: The transformation of social worlds.* Belmont, CA: Wadsworth.

Eliade, Mircea. 1959. *The sacred and the profane: The nature of religion* (W. R. Trask, Trans.). New York: Harcourt, Brace World.

Elinsky, Rachel. 2005. Religious publishing for the red state consumers and beyond. *Publishing Research Quarterly, 21,* 11–29.

Emerson, Michael O. (with Woo, Rodney). 2006. *People of the dream: Multiracial congregations in the United States.* Princeton, NJ: Princeton University Press.

Emerson, Michael O., & Kim, Karen Chai. 2003. Multiracial congregations: A typology and analysis of their development. *Journal for the Scientific Study of Religion, 42,* 217–227.

Emerson, Michael O.; Mirola, William; & Monahan, Susanne. 2011. *Religion matters: What sociology teaches us about religion in our world.* Boston: Allyn & Bacon.

Emerson, Michael O., & Smith, Christian. 2000. *Divided by faith: Evangelical religion and the problem of race in America.* New York: Oxford University Press.

Erickson, Joseph A. 1992. Adolescent religious development and commitment: A structural equation model of the role of family, peer group, and educational influences. *Journal for the Scientific Study of Religion, 31,* 131–152.

Erikson, Kai. 1966. *Wayward Puritans.* New York: Wiley.

Estus, Charles, & Overington, Michael A. 1970. The meaning and end of religiosity. *American Journal of Sociology* (March), 760–778.

Fabian, Stephen Michael. 1992. *Space-time of the Bororo of Brazil.* Gainesville: University Press of Florida.

Faith and Leadership. 2010. *Clergy women make connections.* Duke Divinity School, Durham, NC. Retrieved from www.faithandleadership

.com/features/articles/clergy-women-make-connections?page=full&print=true

Fallding, Harold. 1980. An overview of mainline Protestantism in Canada and the United States of America. In C. H. Jacquet, Jr. (Ed.), *Yearbook of American and Canadian churches, 1980.* (pp. 249–257). Nashville, TN: Abingdon Press.

Fanfani, Amintore. 1936. *Catholicism, Protestantism, and capitalism.* New York: Sheed and Ward.

Farber, I. E.; Harlow, Harry F.; & West, Louis Jolyon. 1951. Brainwashing, conditioning, and D. D. D. (debility, dependency, and dread). *Sociometry* (December), 271–283.

Farley, John E. 2010. *Majority-minority relations* (6th ed.). Upper Saddle River, NJ: Prentice Hall.

Fauset, Arthur H. 1944. *Black gods of the metropolis.* Philadelphia: University of Pennsylvania Press.

Ferguson, Marilyn. 1987. *The Aquarian conspiracy.* Los Angeles: J. P. Tarcher.

Ferraro, Kenneth, & Kelley-Moore, Jessica. 2002. Religious consolation among men and women: Do health problems spur seeking? *Journal for the Scientific Study of Religion, 39,* 220–244.

Festinger, Leon; Riecken, Henry W.; & Schachter, Stanley. 1956. *When prophecy fails.* New York: Harper & Row.

Fialka, John. 2003. *Sisters: Catholic nuns and the making of America.* New York: St. Martin's Press.

Finke, Roger. 1990. Religious deregulation: Origins and consequences. *Journal of Church and State, 32,* 609–626.

Finke, Roger. 1997. The consequences of religious competition: Supply-side explanations for religious change. In L. A. Young (Ed.), *Rational choice theory and religion: Summary and assessment* (pp. 45–64). New York: Routledge.

Finke, Roger, & Stark, Rodney. 1988. Religious economies and sacred canopies: Religious mobilization in American cities, 1906. *American Sociological Review* (February), 41–49.

Finke, Roger, & Stark, Rodney. 1989. Evaluating the evidence: Religious economies and sacred canopies. *American Sociological Review, 54,* 1054–1056.

Finke, Roger, & Stark, Rodney. 1992. *The churching of America, 1776–1990: Winners and losers in our religious economy.* New Brunswick, NJ: Rutgers University Press.

Finke, Roger, & Stark, Rodney. 1998. Religious choice and competition. *American Sociological Review, 63,* 761–766.

Finke, Roger, & Stark, Rodney. 2001. The new holiness clubs: Testing church-to-sect propositions. *Sociology of Religion* (summer), 175–189.

Finke, Roger, & Stark, Rodney. 2005. *The churching of America, 1776–2005: Winners and losers in our religious economy* (2nd ed.). New Brunswick, NJ: Rutgers University Press.

Flint, Richard. 2008. *No settlement, no conquest: A history of the Coronado Entrada.* Albuquerque: University of New Mexico Press.

Flory, Richard, & Miller, Donald. 2008. *Finding faith: The spiritual quest of the post-boomer generation.* New Brunswick, NJ: Rutgers University Press.

Form, William. 2000. Italian Protestants: Religion, ethnicity, and assimilation. *Journal for the Scientific Study of Religion* (September), 307–320.

Francis, Mark. 2000. *Shape a circle ever wider: Liturgical inculturation in the United States.* Chicago: Liturgy Training Publications.

Frank, Mark G., & Gilovich, Thomas. 1988. The dark side of self- and social perception: Black uniforms and aggression in professional sports. *Journal of Personality and Social Psychology, 54*(1), 74–85.

Frankl, Razelle. 1987. *Televangelism: The marketing of popular religion.* Carbondale: Southern Illinois University Press.

Frazier, E. Franklin. 1957. *Negroes in the United States* (Rev. ed.). New York: Macmillan.

Frazier, E. Franklin. 1963. *The Negro church in America.* New York: Schocken Books.

Fredman, Ruth Gruber. 1981. *The Passover seder.* Philadelphia: University of Pennsylvania Press.

Freese, Jeremy. 2004. Risk preferences and gender differences in religiousness: Evidence from the World Values Survey. *Review of Religious Research, 46,* 88–91.

Freston, Paul. 2007. Evangelicalism and fundamentalism: The politics of global popular protestantism. In J. Beckford & N. J. Demerath III (Eds.), *Sage handbook of the sociology of religion* (pp. 205–226). Thousand Oaks, CA: Sage.

Fukuyama, Yoshio. 1961. The major dimensions of church membership. *Review of Religious Research* (spring), 154–161.

Fulton, Aubyn S.; Gorsuch, Richard L.; & Manard, Elizabeth A. 1999. Religious orientation, antihomosexual sentiment, and fundamentalism among Christians. *Journal for the Scientific Study of Religion* (March), 14–22.

Gaddy, Gary. 1984. The power of the religious media: Religious broadcast use and the role of religious organizations in public affairs. *Review of Religious Research* (June), 289–302.

Gaertner, Samuel L., & Dovidio, John F. 1986. The aversive form of racism. In J. F. Dovidio & S. L. Gaertner (Eds.), *Prejudice, discrimination, and racism: Theory and research* (pp. 61–89). San Diego, CA: Academic Press.

Gallup, George, Jr., & Lindsay, D. Michael. 1999. *Surveying the religious landscape: Trends in U.S. beliefs.* Harrisburg, PA: Morehouse Publishing.

The Gallup Organization. 1987. *The Gallup report: Religion in America.* Princeton, NJ: Author.

The Gallup Organization. 2000. *Gallup poll topics: A-Z.* Retrieved from http//www.gallup.com/poll/indicators/indreligion.asp

Geertz, Clifford. 1957. Ritual and social change: A Javanese example. *American Anthropologist* (February), 32–54.

Geertz, Clifford. 1958. Ethos, world view and the analysis of sacred symbols. *The Antioch Review* (winter), 421–437.

Geertz, Clifford. 1966. Religion as a cultural system. In M. Banton (Ed.), *Anthropological approaches to the study of religion* (pp. 1–46). London: Tavistock.

Geertz, Clifford. 1968. *Islam observed: Religious development in Morocco and Indonesia.* Chicago: University of Chicago Press.

Geertz, Clifford. 1973. Thick description: Toward an interpretive theory of culture. In *The interpretation of cultures: Selected essays* (pp. 3–30). New York: Basic Books.

Gerbner, George; Gross, Larry; Hoover, Stewart; Morgan, Michael; Signorilli, Nancy; Cotugno, Harry, et al. 1984. *Religion and television.* Philadelphia: University of Pennsylvania and The Gallup Organization.

Gill, Sam D. 1982. *Native American religions.* Belmont, CA: Wadsworth.

Gill, Sam D. 2004. *Native American religions* (2nd ed.) Belmont, CA: Wadsworth.

Gillis, Chester. 1999. *Roman Catholicism in America.* New York: Columbia University Press.

Giulianotti, Richard, & Robertson, Roland. 2007. Forms of glocalization. *Sociology, 41,* 133–152.

Glock, Charles Y., & Stark, Rodney. 1966. *Christian beliefs and anti-Semitism.* New York: Harper & Row.

Gmelch, George J. 1971. Baseball magic. *Transaction* (June), pp. 39–41, 54.

Goode, Erich. 1967. Some critical observations on the church-sect typology. *Journal for the Scientific Study of Religion* (April), 69–77.

Goodstein, Laurie. 2010. Lutherans offer warm welcome to gay pastors. *New York Times* (July 25). Retrieved from www.nytimes.com/2010/07/26/us/26lutheran.html

Goodstein, Laurie, & Sussman, Dalia. 2010. Catholics criticize Pope on abuse scandal, but see some hope. *New York Times* (May 4). Retrieved from www.nytimes.com/2010/05/05/us/05poll.html

Gorski, Phillip S. 2000. Historicizing the secularization debate: Church, state, and society in late medieval and early modern Europe, CA 1300 to 1700. *Social Forces* (February), 138–167.

Gorski, Phillip S. 2003. *The disciplinary revolution: Calvinism and the rise of the state in early modern Europe.* Chicago: University of Chicago Press.

Gorski, Phillip. 2010. *A neo-Weberian theory of American civil religion* (January 8). Retrieved from http://blogs.ssrc.org/tif/2010/01/08/a-neo-weberian-theory-of-american-civil-religion/

Gorsuch, Richard L., & Aleshire, Daniel. 1974. Christian faith and ethnic prejudice: A review and interpretation of research. *Journal for the Scientific Study of Religion* (September), 281–307.

Graham, Billy. 1983. The future of T.V. evangelism. *TV Guide* (March), 4–11.

Greeley, Andrew M. 1972. *The denominational society.* Glenview, IL: Scott, Foresman.

Greeley, Andrew M. 1974. *Ethnicity in the United States.* New York: Wiley.

Greeley, Andrew M. 1995. *Religion as poetry.* New Brunswick, NJ: Transaction Publishers.

Greeley, Andrew M. 2000. *The Catholic imagination.* Berkeley: University of California Press.

Green, Robert W. (Ed.). 1959. *Protestantism and capitalism: The Weber thesis and its critics.* Boston: Heath.

Greil, Arthur L., & Rudy, David R. 1984. What have we learned from process models of conversion?

An examination of ten case studies. *Sociological Focus* (October), 305–323.

Gremillion, Joseph, & Castelli, Jim. 1987. *The emerging parish: The Notre Dame study of Catholic life since Vatican II.* San Francisco: Harper & Row.

Griffin, Glenn A.; Gorsuch, Richard L.; & Davis, Andrea Lee. 1987. A cross-cultural investigation of religious orientation, social norms, and prejudice. *Journal for the Scientific Study of Religion* (September), 358–365.

Griffith, R. Marie. 2000. *God's daughters: Evangelical women and the power of submission.* Berkeley: University of California Press.

Groothius, Douglas. 1997. *The soul in cyber-space.* Grand Rapids, MI: Baker Books.

Gross, Thomas K., & English, Carey W. 1998. Bible thumping hits TV. *U.S. News and World Report* (January 26), 47.

Guest, Matthew. 2009. The reproduction and transmission of religion. In P. Clarke (Ed.), *The Oxford handbook of the sociology of religion* (pp. 652–670). New York: Oxford University Press.

Gunnoe, M. J., & Moore, K. A. 2002. Predictors of religiosity among youth aged 17–22: A longitudinal study of the national survey of children. *Journal for the Scientific Study of Religion, 17,* 359–379.

Gusfield, J. R. 1963. *Symbolic crusade: Status politics and the American temperance movement.* Chicago: University of Chicago Press.

Guttmann, Allen. 1978. *From ritual to record: The nature of modern sports.* New York: Columbia University Press.

Hadaway, Christopher Kirk; Marler, Penny L.; & Chaves, Mark. 1993. What the polls don't show: A closer look at U.S. church attendance. *American Sociological Review* (December), 741–752.

Hadden, Jeffrey K. 1969. *The gathering storm in the churches.* Garden City, NY: Doubleday.

Hadden, Jeffrey K. 1987. Religious broadcasting and the new Christian right. *Journal for the Scientific Study of Religion* (March), 1–24.

Hadden, Jeffrey K. 1990. The globalization of American televangelism. *International Journal of Frontier Missions* (January), 1–10.

Hadden, Jeffrey K. 1991. The globalization of American televangelism. In R. Robertson & W. R. Garrett (Eds.), *Religion and global order* (pp. 221–244). New York: Paragon.

Hadden, Jeffrey K., & Shupe, Anson D., Jr. 1988. *Televangelism: Power and politics on God's frontier.* New York: Holt.

Haederle, Michael. 1991. Indian artist blends Catholic and Zuni traditions in murals. Los Angeles Times (December 25).

Halaby, Charles. 2004. Panel models in sociological research: Theory into practice. *Annual Review of Sociology, 30,* 507–544.

Hall, John R. 2002. Mass suicide and the Branch Davidians. In D. G. Bromley & J. G. Melton (Eds.), *Cults, religion, and violence* (pp. 149–169). New York: Cambridge University Press.

Hall, Mitchell. 1990. *Because of their faith: CALCAV and religious opposition to the Vietnam War.* New York: Columbia University Press.

Hamberg, Eva. 2009. Unchurched spirituality. In P. Clarke (Ed.), *Oxford handbook of the sociology of religion* (pp. 742–757). Oxford: Oxford University Press.

Hamilton, Malcolm. 2009. Rational choice theory: A critique. In P. Clark (Ed.), *The Oxford handbook of the sociology of religion* (pp. 116–133). New York: Oxford University Press.

Hammond, Phillip. 1992. *Religion and personal autonomy: The third diestablishment in America.* Columbia: University of South Carolina Press.

Hangen, Tona. 2002. *Redeeming the dial: Radio, religion, and popular culture in America.* Chapel Hill: University of North Carolina Press.

Hannan, Michael, & Tuma, Nancy. 1979. Methods of temporal analysis. *Annual Review of Sociology,* 5, 303–328.

Hargrove, Barbara. 1979. *The sociology of religion.* Arlington Heights, IL: AHM Publishing.

Harrison, Rebecca. 2007. African Anglicans to snub pro-gay rights U.S. bishop. *Reuters.* Retrieved from www.reuters.com/article/idUSL0885356520070111

Hartley, Eugene L. 1946. *Problems in prejudice.* New York: King's Crown Press.

Hawley, John Stratton. 1986. Images of gender in the poetry of Krishna. In C. W. Bynum, S. Harrell, & P. Richman (Eds.), *Gender and religion* (pp. 231–256). Boston: Beacon Press.

Hayden, Tom. 2006. Seeking a new capitalism in Chiapas. In D. S. Eitzen & M. B. Zinn (Eds.), *Globalization: The transformation of social worlds* (pp. 328–354). Belmont, CA: Wadsworth.

Hayes, Bernadette C. 1995. The impact of religious identification on political attitudes. *Sociology of Religion* (summer), 177–194.

Heilman, Samuel. 1992. *Defenders of the faith: Inside ultra-Orthodox Jewry.* New York: Schocken Books.

Helland, Christopher. 2000. Online-religion/religion-online and virtual communitas. In J. K. Hadden & D. E. Cowan (Eds.), *Religion on the Internet: Research prospects and promises* (pp. 205–223). London: JAIPress/Elsevier Science.

Helland, Christopher. 2007. *Turning cyberspace into sacred space: Examining the religious revolution occurring on the World Wide Web.* Retrieved from video.google.com/videoplay?docid= -6186132236261787419&q=type%3Agoogle+en gEDU

Henley, Nancy M., & Pincus, Fred. 1978. Interrelationship of sexist, racist and homosexual attitudes. *Psychological Reports* (February), 83–90.

Herberg, Will. 1955. *Protestant-Catholic-Jew.* Garden City, NY: Doubleday.

Herbrechtsmeier, William. 1993. Buddhism and the definition of religion: One more time. *Journal for the Scientific Study of Religion* (March), 1–18.

Herskovits, Melville. 1958. *The myth of the Negro past.* Boston: Beacon Press.

Hertzke, Allen D. 1988. *Representing God in Washington: The role of religious lobbies in the American polity.* Knoxville: University of Tennessee Press.

Hesselbart, Susan. 1976. A comparison of attitudes toward women and attitudes toward blacks in a Southern city. *Sociological Symposium* (fall), 45–68.

Hesser, Gary, & Weigert, Andrew J. 1980. Comparative dimensions of liturgy: A conceptual framework and feasibility application. *Sociological Analysis* (fall), 215–229.

Hoffman, Shirl. 1992. Evangelicalism and the revitalization of religious ritual in sport. In S. J. Hoffman (Ed.), *Sport and religion* (pp. 111–125). Champaign, IL: Human Kinetics.

Hoge, Dean R., & Carroll, Jackson W. 1978. Determinants of commitment and participation in suburban Protestant churches. *Journal for the Scientific Study of Religion* (June), 107–128.

Hoge, Dean R., & Roozen, David A. 1979. Understanding church growth and decline, 1950–1978. New York: Pilgrim Press.

Homans, George C. 1941. Anxiety and ritual: The theories of Malinowski and Radcliffe-Brown. *American Anthropologist* (April), 164–172.

Hood, Ralph W., Jr.; Morris, R. J.; & Watson, P. J. 1989. The differential report of prayer experience and religious orientation. *Review of Religious Research, 31,* 39–45.

Hoover, Stewart. 2009. Religion and the media. In P. Clarke (Ed.), *Oxford handbook of the sociology of religion* (pp. 688–704). Oxford: Oxford University Press.

Hoover, Stewart; Clark, L. S.; & Rainie, L. 2004. *Faith online.* Washington, DC: Pew Internet and American Life Project.

Hostetler, John A. 1993. *Amish society* (4th ed.). Baltimore: Johns Hopkins University Press.

Hout, Michael, & Fischer, Claude. 2002. Why more Americans have no religious preference: Politics and generations. *American Sociological Review, 67,* 165–190.

Hout, Michael, & Fischer, Claude. 2009. *The politics of religious identity in the United States, 1974–2008* (August 9). Paper presented at the annual meeting of the American Sociological Association, San Francisco, CA.

Hout, Michael; Greeley, Andrew; & Wilde, Melissa. 2001. The demographic imperative in religious change in the United States. *American Journal of Sociology, 107,* 468–500.

Howe, Neil, & Strauss, William. 1993. *13th gen: Abort, retry, ignore, fail?* New York: Vintage Books.

Howe, Neil, & Strauss, William. 2000. *Millennials rising: The next great generation.* New York: Vintage Books.

Howes, Graham. 1991. "God Damn Yanquis"— American hegemony and contemporary Latin American Christianity. In W. C. Roof (Ed.), *World order and religion* (pp. 83–97). Albany: State University of New York Press.

Hudson, Winthrop S. 1949. Puritanism and the spirit of capitalism. *Church History* (March), 3–17.

Human Rights Campaign. 2003. *Answers to questions about marriage equality.* Washington, DC: Human Rights Campaign, Family Net Project.

Hummer, Robert; Rogers, Richard; Nam, Charles; & Ellison, Christopher. 1999. Religious involvement and U.S. adult mortality. *Demography, 36,* 273–285.

Hunt, Larry L., & Hunt, Janet G. 1977. Black religion as both opiate and inspiration of civil rights militance: Putting Marx's data to the test. *Social Forces* (September), 1–14.

Hunt, Stephen. 2007. Religion as a factor in life and death through the life-course. In J. Beckford & N. J. Demerath III (Eds.), *The Sage handbook of the sociology of religion* (pp. 608–629). Thousand Oaks, CA: Sage.

Hunter, James Davidson. 1983. *American evangelicalism.* New Brunswick, NJ: Rutgers University Press.

Iannaccone, Laurence R. 1982. Let the women be silent. *Sunstone* (May/June), 38–45.

Iannaccone, Laurence R. 1990. Religious practice: A human capital approach. *Journal for the Scientific Study of Religion, 29*(3), 297–314.

Iannaccone, Laurence R. 1992. Religious markets and the economics of religion. *Social Compass, 39,* 123–131.

Iannaccone, Laurence R. 1994. Why strict churches are strong. *American Journal of Sociology* (March), 1180–1211.

Iannaccone, Laurence R. 1995. Voodoo economics? Reviewing the rational choice approach to religion. *Journal for the Scientific Study of Religion* (March), 76–88.

Illinois diocese votes to split from church. 2008. *New York Times* (July 26). Retrieved from www.nytimes.com/2008/11/09/us/09quincy.html?ref=v_gene_robinson

Jackson, Lynne M., & Hunsberger, Bruce. 1999. An intergroup perspective on religion and prejudice. *Journal for the Scientific Study of Religion* (December), 509–523.

Jackson, Susan, & Csikszentmihalyi, Mihaly. 1999. *Flow in sports: The keys to optimal experiences and performances.* Champaign, IL: Human Kinetics.

Jacobs, Janet. 1987. Deconversion from religious movements. *Journal for the Scientific Study of Religion* (September), 294–308.

James, Jonathan. 2010. *McDonaldization, masala McGospel and om economics: Televangelism in contemporary India.* New Delhi: Sage Publications India.

James, William. 1958. *Varieties of religious experience.* New York: New American Library. (Original work published 1902).

Jansen, B. J., Tapia, A., & Spink, A. 2010. Searching for salvation: An analysis of US religious searching on the World Wide Web. *Religion, 40,* 39–52.

Jelen, Ted G. 1995. Religion and the American political culture: Alternative models of citizenship and discipleship. *Sociology of Religion* (fall), 271–284.

Jenkins, J. Craig. 1977. Radical transformation of organizational goals. *Administrative Science Quarterly* (December), 568–586.

Jenkins, Philip. 2000. *Mystics and messiahs: Cults and new religions in American history.* New York: Oxford University Press.

Jenkins, Philip. 2002. The next Christendom: The coming of global Christianity. New York: Oxford University Press.

Jindra, Michael. 1994. Star Trek fandom as a religious phenomenon. *Sociology of Religion* (Spring), 27–51.

Johnson, Benton. 1963. On church and sect. *American Sociological Review* (August), 539–549.

Johnson, Martin, & Mullins, Phil. 1992. Mormonism: Catholic, Protestant, different? *Review of Religious Research* (September), 51–62.

Johnson, Paul E. 1959. *Psychology of religion* (Rev. ed.). Nashville, TN: Abingdon Press.

Johnstone, Ronald L. 2001. *Religion in society: A sociology of religion* (6th ed.). Englewood Cliffs, NJ: Prentice Hall.

Johnstone, R. 2007. *Religion in society: A sociology of religion.* Upper Saddle River, NJ: Pearson Prentice Hall.

Jordan, Winthrop D. 1968. *White over black.* Baltimore: Penguin.

Judah, J. Stillson. 1974. *Hare Krishna and the counterculture.* New York: Wiley.

Kahoe, Richard D. 1974. The psychology and theology of sexism. *Journal of Psychology and Theology* (fall), 284–290.

Kalir, Barak. 2009. Finding Jesus in the Holy Land and taking him to China: Chinese temporary migrant workers in Israel converting to evangelical Christianity. *Sociology of Religion, 70,* 130–156.

Kalton, Graham. 1983. *Introduction to survey sampling.* Beverly Hills, CA: Sage.

Kanter, Rosabeth Moss. 1972. *Commitment and community.* Cambridge, MA: Harvard University Press.

Karabel, Jerome. 2005. *The chosen: The hidden history of admission and exclusion at Harvard, Yale, and Princeton.* New York: Houghton Mifflin.

Kay, William K. 2009. Pentecostalism and religious broadcasting. *Journal of Beliefs and Values, 30.3,* 245–254.

Keck, Margaret, & Sikkink, Kathryn. 1998. *Activists beyond borders: Advocacy networks in international politics.* Ithaca, NY: Cornell University Press.

Keith, Katherine Drouin. 1999. Zuni murals connect two cultures. *National Catholic Reporter* (March 26).

Kelley, Dean M. 1972. *Why conservative churches are growing.* New York: Harper & Row.

Kelley, Dean M. 1978. Comment: Why conservative churches are still growing. *Journal for the Scientific Study of Religion* (June), 165–172.

Kelsey, George D. 1965. *Racism and the Christian understanding of man.* New York: Scribner.

Kennedy, Tom, & Simplicio, Dan. 2009. First contact at Hawikku (Zuni): The day the world stopped. In M. Weigle (Ed.), *Telling New Mexico: A history* (pp. 61–72). Santa Fe: Museum of New Mexico Press.

Kephart, William M., & Zellner, William W. 1998. *Extraordinary groups: An examination of unconventional life-styles* (6th ed.). New York: St. Martin's Press.

Kilbourne, Brook, & Richardson, James T. 1989. Paradigm conflict, types of conversion, and conversion theories. *Sociological Analysis* (spring), 1–21.

Killian, Caitlin. 2007. From a community of believers to an Islam of the heart: "Conspicuous" symbols, Muslim practices, and the privatization of religion in France. *Sociology of Religion, 68,* 305–320.

Kinder, Donald R., & Sears, D. O. 1981. Symbolic racism versus racial threats to the good life. *Journal of Personality and Social Psychology* (40), 414–431.

King, Martin Luther, Jr. 2010. *Stride toward freedom: The Montgomery story.* Boston: Beacon Press.

King, Pamela E., Furrow, James L., & Roth, Natalie. 2002. The influence of families and peers on adolescent religiousness. *The Journal for Psychology and Christianity, 2,* 109–120.

Kirkpatrick, L. A. 1993. Fundamentalism, Christian Orthodoxy, and intrinsic religious orientation as predictors of discriminatory attitudes. *Journal for the Scientific Study of Religion* (March), 256–268.

Kliever, Lonnie. 2001. God and games in modern culture. In J. L. Price (Ed.), *From season to season: Sports as American religion* (pp. 39–48). Macon, GA: Mercer University Press.

Kluckhohn, Clyde. 1972. Myths and rituals: A general theory. In W. A. Lessa & E. Z. Vogt (Eds.), *Reader in comparative religion: An anthropological approach* (3rd ed., pp. 93–105). New York: Harper & Row.

Knickmeyer, Ellen. 2008. Turkey's Gul signs head scarf measure." *The Washington Post* (February 23). Retrieved from www.washingtonpost.com/wp-dyn/content/article/2008/02/22/AR2008022202988.html

Knox, Robert. 1850. *Races of men.* Philadelphia: Lea and Blanchard.

Knudsen, Dean D.; Earle, John R., & Schriver, Donald W., Jr. 1978. The conception of sectarian religion: An effort at clarification. *Review of Religious Research* (fall), 44–60.

Koch, Bradley. 2009. *The Prosperity Gospel and economic prosperity: Race, class, giving, and voting.* Unpublished doctoral dissertation, Department of Sociology, Indiana University, Bloomington, IN.

Koch, Jerome R., & Curry, Evans W. 2000. Social context and the Presbyterian gay/lesbian debate: Testing open systems theory. *Review of Religious Research* (December), 206–214.

Kohn, Melvin L. 1989. *Class and conformity: A study in values* [Midway Reprint Edition]. Chicago: University of Chicago Press.

Kosmin, Barry A., & Keysar, Ariela. 2006. *Religion in a free market: Religious and non-religious Americans.* Ithaca, NY: Paramount Market Publishing.

Kosmin, Barry A., & Keysar, Ariela. 2009. *Summary report. American Religious Identification Survey (2008)* (March). Hartford, CT: Trinity College. Retrieved from www.americanreligionsurvey-aris.org/

Kosmin, Barry A., & Lachman, Seymour P. 1993. *One nation under God: Religion in contemporary American society.* New York: Harmony.

Kovel, Joel. 1970. *White racism: A psychohistory.* New York: Pantheon Books.

Kox, William; Meeus, Wim; & Harm't, Hart. 1991. Religious conversion of adolescents: Testing the Lofland and Stark model of religious conversion. *Sociological Analysis* (fall), 227–240.

Krause, Neal. 2001. Social support. In R. H. Binstock & L. K. George (Eds.), *Handbook of aging and the social sciences* (pp. 272–294). New York: Academic Press.

Krause, Neal. 2006. Aging. In H. R. Ebaugh (Ed.), *Handbook of religion and social institutions* (pp. 139–160). New York: Springer.

Kraybill, Donald B. 2001. *The riddle of Amish culture* (Rev. ed.). Baltimore: John Hopkins University Press.

Kurien, Prema. 1998. Becoming American by becoming Hindu: Indian Americans take their place at the multicultural table. In R. S. Warner & J. Wittner (Eds.), *Gatherings in diaspora: Religious communities and the new immigration* (pp. 37–70). Philadelphia: Temple University Press.

LaBarre, Weston. 1972. *The Ghost Dance.* New York: Dell.

Ladd, Thomas, & Mathisen, James. 1999. *Muscular Christianity: Evangelical Protestants and the development of American sport.* Grand Rapids, MI: Baker Books.

Lambert, Yves. 2000. Religion in modernity as a new axial age: Secularization or new religious forms? In W. H. Swatos Jr. & D. V. A. Olson (Eds.), *The secularization debate* (pp. 95–125). Lanham, MD: Rowman & Littlefield.

Larsen, Elena. 2000. *Wired churches, wired temples: Taking congregations and missions into cyberspace.* Washington, DC: Pew Internet & American Life Project.

Larson, Heidi. 2000. "We don't celebrate Christmas, we just give gifts": Adaptions to migration and social change among Hindu, Muslim, and Sikh children in England. In S. K. Houseknecht & J. G. Pankhurst (Eds.), *Family, religion, and social change in diverse societies* (pp. 283–302). New York: Oxford University Press.

Lawson, Ronald. 1995. Broadening the boundaries of church-sect theory. *Journal for the Scientific Study of Religion* (December), 652–672.

Lawson, Ronald. 1996. Onward Christian soldiers: Seventh-Day Adventists and the issue of military service. *Review of Religious Research* (March), 193–218.

Lawson, Ronald. 1998. Sect-state relations: Accounting for the differing trajectories of Seventh-day Adventists and witnesses. *Sociology of Religion* (winter), 351–377.

Lawton, Leora, & Bures, Regina. 2001. Parental divorce and the "switching" of religious identity. *Journal for the Scientific Study of Religion, 40,* 99–111.

Laythe, Brian; Finkel, Deborah; & Kirkpatrick, Lee A. 2001. Predicting prejudice from religious fundamentalism and right wing authoritarianism. *Journal for the Scientific Study of Religion, 40* (March), 1–10.

Lazerwitz, Bernard; Winter, J. Alan; Dashefsky, Arnold; & Tabory, Ephraim. 1998. *Jewish choices: American Jewish denominationalism.* Albany, NY: SUNY Press.

Leach, Edmund R. 1972. Ritualization in man in relation to conceptual and social development. In W. A. Lessa & E. Z. Vogt (Eds.), *Reader in comparative religion* (3rd ed., pp. 333–337). New York: Harper & Row.

Lechner, Frank L. 1991a. Religion, law, and global order. In R. Robertson & W. R. Garrett (Eds.), *Religion and global order* (pp. 263–280). New York: Paragon.

Lechner, Frank L. 1991b. The case against secularization: A rebuttal. *Social Forces, 69,* 1103–1119.

Lechner, Frank L. 2003. Defining religion: A pluralistic approach for the global age. In A. Greil & D. Bromley (Eds.), *Defining religion: Investigating the boundaries between the sacred and the secular* (pp. 67–83). Oxford: Elsevier.

Lechner, Frank L. 2007. Rational choice and religious economies. In J. Beckford & N. J. Demerath III (Eds.), *The Sage handbook of the sociology of religion* (pp. 81–97). Thousand Oaks, CA: Sage.

Lee, Richard Wayne. 1992. Christianity and the other religions: Interreligious relations in a shrinking world. *Sociological Analysis* (summer), 125–139.

Lehman, David. 2010. Rational choice and the sociology of religion. In B. Turner (Ed.), *The New Blackwell companion to the sociology of religion* (pp. 181–200). Malden, MA: Blackwell.

Lehman, Edward C., Jr. 1980. Patterns of lay resistance to women in ministry. *Sociological Analysis* (winter), 317–338.

Lehman, Edward C., Jr. 1981. Organizational resistance to women in ministry. *Sociological Analysis* (summer), 101–118.

Lehman, Edward C., Jr. 1985. *Women clergy: Breaking through gender barriers.* New Brunswick, NJ: Transaction Books.

Lehman, Edward C., Jr. 1987a. Sexism, organizational maintenance, and localism: A research note. *Sociological Analysis* (fall), 274–282.

Lehman, Edward C., Jr. 1987b. Research on lay member's attitudes toward women in ministry: An assessment. *Review of Religious Research* (June), 319–329.

Lehne, Gregory K. 1995. Homophobia among men: Supporting and defining the male role. In M. Kimmel & M. Messner (Eds.), *Men's lives* (3rd ed., pp. 325–336). Boston: Allyn & Bacon.

Lenski, Gerhard. 1963. *The religious factor* (Rev. ed.). Garden City, NY: Doubleday.

Leslie, Michael. 2003. *International televangelism/ American ideology: The case of the 700 Club* (May 2–4). Paper presented at the International Conference on Television in Transition, MIT, Boston.

Levine, Saul V. 1984. *Radical departures: Desperate detours to growing up.* San Diego: Harcourt Brace Jovanovich.

Levitt, Mairi. 1995. Sexual identity and religious socialization. *British Journal of Sociology, 46,* 529–536.

Levitt, Peggy. 1998. Local level global religion: The case of the U.S.-Dominican migration. *Journal for the Scientific Study of Religion* (March), 74–89.

Levitt, Peggy. 2003. You know, Abraham was really the first immigrant: Religion and transnational migration. *International Migration Review, 37,* 847–873.

Levitt, Peggy. 2004. Redefining the boundaries of belonging: The institutional character of transnational religious life. *Sociology of Religion, 65,* 1–18.

Lewis, Amanda. 2003. *Race in the schoolyard.* New Brunswick, NJ: Rutgers University Press.

Lieberson, Stanley. 2000. *A matter of taste: How names, fashions, and culture change.* New Haven, CT: Yale University Press.

Lightfoot, Neil R. 2003. *How we got the Bible.* Grand Rapids, MI: Baker Books.

Lincoln, C. Eric. 1968. *Is anybody listening to black America?* New York: Seabury Press.

Lincoln, C. Eric. 1994. *The black Muslims in America* (3rd ed.). Grand Rapids, MI: William B. Eerdmans.

Lincoln, C. Eric., & Mamiya, Lawrence H. 1990. *The black church in the African American experience.* Durham, NC: Duke University Press.

Lindner, Eileen W. (Ed.). 2000. *Yearbook of American and Canadian churches.* Nashville, TN: Abingdon Press.

Lindner, Eileen W. (Ed.). 2010. *Yearbook of American and Canadian churches.* Nashville, TN: Abingdon Press.

Lipman-Blumen, Jean. 1972. How ideology shapes women's lives. *Scientific American* (January), 33–42.

Lipset, Seymour. 1960. *Political man.* Garden City, NY: Doubleday.

Lockhart, William H. 2001. "We are all one life," but not of one gender ideology: Unity, ambiguity, and the Promise Keepers. In R. H. Williams (Ed.), *Promise Keepers and the new masculinity: Private lives and public morality* (pp. 73–92). Lanham, MD: Lexington Books.

Lofland, John. 1977. *Doomsday cult* [Enlarged ed.]. New York: Irvington.

Lofland, John, & Skonovd, Norman. 1981. Conversion motifs. *Journal for the Scientific Study of Religion* (December), 373–385.

Lofland, John, & Stark, Rodney. 1965. Becoming a world saver: A theory of conversion to a deviant perspective. *American Sociological Review, 30*(6), 863–874.

Loftus, Jeni. 2001. America's liberalization in attitudes toward homosexuality, 1973–1998. *American Sociological Review, 66,* 762–782.

Los Angeles diocese elects second gay Episcopal bishop, highlighting an Anglican split. 2009. *New York Times* (December 6). Retrieved from www .nytimes.com/2009/12/07/us/07episcopal.html?_r=1&ref=v_gene_robinson

Loveland, Matthew. 2003. Religious switching: Preference development, maintenance, and change. *Journal for the Scientific Study of Religion, 42,* 147–157.

Loveland, Matthew. 2008. Pilgrimage, religious institutions, and the construction of orthodoxy. *Sociology of Religion, 69,* 317–334.

Luckmann, Thomas. 1967. *The invisible religion*. New York: Macmillan.

Luker, Kristin. 2008. *Salsa dancing into the social sciences: Research in an age of info-glut*. Cambridge, MA: Harvard University Press.

Machalek, Richard, & Martin, Michael. 1976. Invisible religions: Some preliminary evidence. *Journal for the Scientific Study of Religion* (December), 311–322.

MacWilliams, Mark. 2002. Virtual pilgrimages on the Internet. *Religion, 32*, 315–335.

Maguire, Brendan, & Weatherby, Georgie Ann. 1998. The secularization of religion and television commercials. *Sociology of Religion, 59*, 171–178.

Mahler, Jonathan. 2005. The soul of the new exurb. *New York Times Magazine* (March 27). Retrieved from www.nytimes.com/2005/03/27/magazine/327MEGACHURCH.html

Malinowski, Bronislaw. 1931. Culture. In E. R. A. Seligman & A. Johnson, *Encyclopaedia of the Social Sciences* (Vol. IV, pp. 621–645). New York: Macmillan.

Malinowski, Bronislaw. 1944. *A scientific theory of culture and other essays*. Chapel Hill: The University of North Carolina Press.

Malinowski, Bronislaw. 1948. *Magic, science, and religion and other essays*. New York: Free Press.

Man, John. 2002. *Gutenberg: How one man remade the world with words*. New York: John Wiley.

Mannheim, Karl. 1927. The problem of generations. *Essays in the sociology of knowledge* (pp. 98–110). London: RKP.

Manning, Christel. 1999. *God gave us the right: Conservative Catholic, Evangelical Protestant, and Orthodox Jewish women grapple with feminism*. New Brunswick, NJ: Rutgers University Press.

Marett, Robert R. 1914. *The threshold of religion*. London: Methuen & Co.

Marler, Penny Long, & Hadaway, C. Kirk. 2002. "Being religious" or "being spiritual" in America: A zero-sum proposition? *Journal for the Scientific Study of Religion, 41*(2), 289–300.

Marsden, George. 1991. *Understanding fundamentalism and evangelicalism*. Grand Rapids, MI: Eerdmans.

Marsden, George. 1994. *The soul of the American university: From Protestant establishment to established nonbelief*. New York: Oxford University Press.

Marti, Gerardo. 2009. Affinity, identity, and transcendence: The experience of religious racial integration in diverse congregations. *Journal for the Scientific Study of Religion, 48*, 54–69.

Martin, David A. 1969. *The religious and the secular: Studies in secularization*. London: Routledge & Kegan Paul.

Martin, David A. 1978. *A general theory of secularization*. Oxford, England: Basil Blackwell.

Martin, David A. 1993. *Tongues of fire: The explosion of Protestantism in Latin America*. New York: Blackwell.

Martin, Patricia Yancey; Osmond, Marie Withers; Hesselbart, Susan; & Wood, Meredith. 1980. The significance of gender as a social and demographic correlate of sex role attitudes. *Sociological Focus* (October), 338–396.

Martin, Todd F., White, James M., & Perlman, Daniel. 2003. Religious socialization: A test of the channeling hypothesis of parental influence on adolescent faith maturity. *Journal of Adolescent Research, 18*, 169–187.

Martin, William. 1972. *Christians in conflict*. Chicago: Center for the Scientific Study of Religion, University of Chicago.

Martins, Stuart, & Magida, Arthur. 2003. *How to be a perfect stranger: The essential religious etiquette handbook*. Woodstock, VT: Skylight Paths Publishing.

Marty, Martin. 1958. *The new shape of American religion*. New York: Free Press.

Marty, Martin E. 1972. Ethnicity: The skeleton of religion in America. *Church History* (March), 5–21.

Marty, Martin E. 2001. The logic of fundamentalism. *Boston College Magazine* (fall), 44–46.

Marty, Martin, & Appleby, R. Scott. (Eds.). 1992. *The glory and the power: The fundamentalist challenge to the modern age*. Boston: Beacon Press.

Marty, Martin, & Appleby, R. Scott. (Eds.). 1994. *Accounting for fundamentalism: The dynamic character of movements*. Chicago: University of Chicago Press.

Marx, Gary. 1967. *Protest and prejudice*. New York: Harper & Row.

Marx, Karl. 1977. *Contribution to the critique of Hegel's "Philosophy of Right."* (J. O'Malley, Ed.). Cambridge: Cambridge University Press. (Original work published 1844)

Marx, Karl. 2008. *The 18th Brumaire of Louis Bonaparte.* LaVergne, TN: Wildside Press. (Original work published 1852)

Maslow, Abraham. 1964. *Religions, values, and peak experiences.* Columbus: Ohio State University Press.

Mason, Karen, & Bumpass, Larry L. 1975. U.S. women's sex role ideology, 1970. *American Journal of Sociology* (March), 1212–1219.

Mathisen, James A. 1987. Thomas O'Dea's dilemmas of institutionalization: A case study and reevaluation after twenty-five years. *Sociological Analysis* (winter), 302–318.

Mathisen, James A. 2006. Sport. In H. R. Ebaugh (Ed.), *Handbook of religion and social institutions* (pp. 285–303). New York: Springer.

Mauss, Armand L., & Barlow, Philip L. 1991. Church, sect, and scripture: The Protestant Bible and Mormon Sectarian retrenchment. *Sociological Analysis* (winter), 397–414.

McAlister, Elizabeth. 1998. The Madonna of 115th Street revisited: Vodou and Haitian Catholicism in the age of transnationalism. In R. S. Warner & J. Wittner (Eds.), *Gatherings in diaspora: Religious communities and the new immigration* (pp. 123–160). Philadelphia: Temple University Press.

McCallion, Michael; Maines, David; & Wolfel, Susan. 1996. Policy as practice: First holy communion as a contested situation. *Journal of Contemporary Ethnography, 25,* 300–326.

McCloud, Sean. 2007. Liminal subjectivities and religious change: Circumscribing Giddens for the study of contemporary American religion. *Journal of Contemporary Religion* (October), *22,* 295–303, 309.

McCutcheon, Alan L. 1988. Denominations and religious intermarriage: Trends among white Americans in the twentieth century. *Review of Religious Research* (March), 213–227.

McGaw, Douglas B. 1979. Commitment and religious community: A comparison of a charismatic and a mainline congregation. *Journal for the Scientific Study of Religion* (June), 146–163.

McGuire, Meredith B. 2002. *Religion: The social context* (5th ed.). Belmont, CA: Wadsworth.

McLaughlin, Elenor Commo. 1974. Equality of souls, inequality of sexes: Women in medieval theology. In R. R. Ruether (Ed.), *Religion and Sexism* (pp. 213–266). New York: Simon & Schuster.

McLeod, Jay. 1995. *Ain't no makin it* (2nd ed.). Boulder, CO: Westview.

McMurry, Mary. 1978. Religion and women's sex-role traditionalism. *Sociological Focus* (April), 81–95.

McPherson, Miller; Smith-Lovin, Lynn; & Cook, James. 2001. Birds of a feather: Homophily in social networks. *Annual Review of Sociology, 27,* 415–444.

McRoberts, Omar M. 2003. *Streets of glory: Church and community in a black urban neighborhood.* Chicago: University of Chicago Press.

Mead, Frank; Hill, Samuel; & Atwood, Craig. (Eds.). 2005. *Handbook of denominations in the United States* (12th ed.). Nashville, TN: Abingdon Press.

Meier, Harold C. 1972. Mother centeredness and college youths' attitudes towards social equality for women: Some empirical findings. *Journal of Marriage and the Family* (February), 115–121.

Melton, J. Gordon. 1993. Another look at new religions. *Annals of the American Academy of Political and Social Science* (May), 97–112.

Menschung, Gustav. 1964. The masses, folk belief, and universal religion. In L. Schneider (Ed.), *Religion, culture, and society* (pp. 269–272). New York: Wiley.

Merton, Robert. 1968. *Social theory and social structure.* New York: Free Press.

Metropolitan Community Church. 2010. *History of MCC.* Retrieved from http://ufmcc.com/overview/history-of-mcc/

Meyer, John W. 1980. The world polity and the authority of the national-state. In A. Bergson (Ed.), *Studies of the modern world-system* (pp. 109–137). New York: Academic Press.

Meyer, John W. 1988. *The evolution of a world religious system: Some research design ideas.* Paper presented at the annual meeting of the Association for the Sociology of Religion, Atlanta, GA.

Miller, Alan, & Hoffman, John. 1995. Risk and religion: An explanation of gender differences in religiosity. *Journal for the Scientific Study of Religion, 34,* 63–75.

Miller, Alan, & Stark, Rodney. 2002. Gender and religiousness: Can socialization explanations be saved? *American Journal of Sociology, 107,* 1399–1423.

Miller, Donald E. 1997. *Reinventing American Protestantism.* Berkeley: University of California Press.

Mitchell, Barbara. 2006. *The boomerang age: Transitions to adulthood in families.* New Brunswick, NJ: Transaction Publishers.

Mixon, Stephanie Litizzette; Lyon, Larry; & Beatty, Michael. 2004. Secularization and national universities: The effect of religious identity on academic reputation. *Journal of Higher Education, 75,* 400–419.

Moberg, David. 1980. Prison camp of the mind. In G. Gaviglio & D. E. Raye (Eds.), *Society as it is* (pp. 318–330). New York: Macmillan.

Moberg, David. 1984. *The church as a social institution* (2nd ed.). Grand Rapids, MI: Baker Book House.

Morin, Stephen F., & Garfinkle, Ellen M. 1978. Male homophobia. *Journal of Social Issues, 34*(1), 29–47.

Morris, Aldon. 1986. *The origins of the civil rights movement: Black communities organizing for change.* New York: Free Press.

Mosse, George L. 1985. *Toward the final solution: A history of European racism.* Madison: University of Wisconsin Press.

Mullen, Mike. 1994. Religious polities as institutions. *Social Forces, 73,* 709–728.

Murvar, Vatro. 1975. Toward a sociological theory of religious movements. *Journal for the Scientific Study of Religion* (September), 229–256.

Musick, March, & Wilson, John. 1995. Religious switching for marriage reasons. *Sociology of Religion* (fall), 257–270.

Muslims aim to end televangelism. 1998. *Christianity Today* (March 2), *42*(3), 78.

Myers, Scott M. 1996. An interactive model of religiosity inheritance: The importance of family context. *American Sociological Review, 61,* 858–866.

Myrdal, Gunar. 1944. *An American dilemma.* New York: Harper & Row.

National Conference of State Legislatures. 2011. Common-law marriage. Retrieved from www.ncsl.org/default.aspx?tabid=4265

National Science Foundation. 2000. Frequency of reading astrology reports, by selected characteristics, 1985–1999. Washington, DC: Congressional Information Service.

Needleman, Jacob. 1970. *The new religions.* Garden City, NY: Doubleday.

Neihardt, John G. 1961. *Black Elk speaks.* Lincoln: University of Nebraska Press.

Neitz, Mary Jo. 1990. Steps toward a sociology of religious experience: The theories of Mihaly Csikszentmihalyi and Alfred Schutz. *Sociological Analysis* (spring), 15–33.

Neitz, Mary Jo. 2000. Queering the Dragonfest: Changing sexualities in a post-patriarchal religion. *Sociology of Religion* (winter), 369–391.

Nelsen, Hart M.; Everett, Robert F.; Mader, Paul Douglas; & Hamby, Warren C. 1976. A test of Yinger's measure of nondoctrinal religion: Implications for invisible religion as a belief system. *Journal for the Scientific Study of Religion* (September), 263–268.

Nelsen, Hart M.; Madron, Thomas W.; & Yokley, Raytha L. 1975. Black religion's Promethean motif: Orthodoxy and militancy. *American Journal of Sociology* (July), 139–146.

Nelsen, Hart M., & Nelsen, Anne Kuesener. 1975. *Black church in the Sixties.* Lexington: University Press of Kentucky.

Nelsen, Hart M., & Snizek, William E. 1976. Musical pews: Rural and urban models of occupational and religious mobility. *Sociology and Social Research* (April), 279–289.

Nelsen, Hart M.; Yokley, Raytha; & Nelsen, Anne. (Eds.). 1971. *The black church in America.* Lexington: University Press of Kentucky.

Nelson, Geoffrey. 1968. The concept of cult. *Sociological Review* (November), 351–363.

Nelson, Mary. 1975. Why witches were women. In Jo Freeman (Eds.), *Women: A feminist perspective* (pp. 335–350). Palo Alto, CA: Mayfield.

Nelson, Timothy. 2005. *Every time I feel the spirit: Religious experience and ritual in an African American church.* New York: New York University Press.

Nelson, Timothy J. 2009. At ease with our own kind: Worship practices and class segregation in American religion. In W. Mirola & S. McCloud, *Religion and class in America: Culture, history, and politics* (pp. 45–68). Leiden, Netherlands: Brill.

Nesbitt, Paula D. 1997. Clergy feminization: Controlled labor or transformative change?" *Journal for the Scientific Study of Religion* (December), 585–598.

New media. 2011. *Wikipedia.* Retrieved from http://en.wikipedia.org/wiki/New_media

Newcomb, Theodore, & Svehla, George. 1937. Intra-family relationships in attitude. *Sociometry, 1,* 180–205.

Newport, Frank. 1979. The religious switcher in the United States. *American Sociological Review* (August), 528–552.

Niebuhr, H. Richard. 1957. *The social sources of denominationalism.* New York: Meridian Books. (Original work published 1929)

Niebuhr, H. Richard. 1960. Faith in God and in gods. In *Radical monotheism and Western culture* (pp. 114–126). New York: Harper & Row.

Noel, Donald L. 1968. A theory of the origins of ethnic stratification. *Social Problems* (fall), 157–172.

Nottingham, Elizabeth K. 1971. *Religion: A sociological view.* New York: Random House.

O'Dea, Thomas F. 1957. *The Mormons.* Chicago: University of Chicago Press.

O'Dea, Thomas F. 1961. Five dilemmas in the institutionalization of religion. *Journal for the Scientific Study of Religion* (October), 30–39.

O'Dea, Thomas F. 1966. *The sociology of religion.* Englewood Cliffs, NJ: Prentice Hall.

O'Dea, Thomas F. 1968. Sects and cults. *International encyclopedia of the social sciences* (Vol. 14, pp. 130–136). New York: Macmillan.

Olsen, Marvin. 1978. *The process of social organization: Power in social systems* (2nd ed.). New York: Holt, Rinehart & Winston.

Olson, Daniel V. A. 1998. Religious pluralism in contemporary U.S. counties. *American Sociological Review, 63,* 759–761.

Olson, Laura; Cadge, Wendy; & Harrison, James. 2006. Religion and public opinion about same-sex marriage. *Social Science Quarterly, 87,* 340–360.

Orru, Marco, & Wang, Amy. 1992. Durkheim, religion, and Buddhism. *Journal for the Scientific Study of Religion* (March), 47–61.

Ostling, Richard. 1987. Raising eyebrows and the dead. *Time* (July 13).

Oswald, Ramona Faith. 2000. A member of the wedding? Heterosexism and family ritual. *Journal of Social and Personal Relationships* (June), 349–368.

Oswald, Ramona Faith. 2001. Religion, family, and ritual: The production of gay, lesbian, and transgendered outsiders-within. *Review of Religious Research* (December), 39–50.

Otto, Rudolf. 1923. *The idea of the holy.* London: Oxford University Press.

Overman, S. 1997. *The influence of the Protestant ethic on sports and recreation.* Aldershot, UK: Averbury.

Ozorak, Elizabeth W. 1989. Social and cognitive influences on the development of religious beliefs and commitment in adolescence. *Journal for the Scientific Study of Religion, 28,* 448–463.

Pagels, Elaine. 1979. *The Gnostic Gospels.* New York: Vintage Books.

Palfrey, John, & Gasser, Urs. 2008. *Born digital: Understanding the first generation of digital natives.* New York: Basic Books.

Parker, Everett; Barry, David; & Smythe, Dallas. 1955. *The television-radio audience and religion.* New York: Harper.

Parker, Robert Allerton. 1935. *A yankee saint: John Humphrey Noyes and the Oneida community.* New York: Putnam.

Parsons, Talcott. 1964. Christianity in modern industrial society. In E. Tiryakian (Ed.), *Sociological theory, values, and sociocultural change* (pp. 233–270). Glencoe, IL: Free Press.

Parvey, Constance F. 1974. The theology and leadership of women in the New Testament. In R. R. Ruether (Ed.), *Religion and sexism* (pp. 117–149). New York: Simon & Schuster.

Peck, Janice. 1993. *The Gods of televangelism: The crisis of meaning and the appeal of religious television.* Crosskill, NJ: Hampton.

Peek, Lori. 2005. Becoming Muslim: The development of a religious identity. *Sociology of Religion, 66,* 215–242.

Pelikan, Jaroslav. 2003. *Credo: Historical and theological guide to creeds and confessions of faith in the Christian tradition.* New Haven, CT: Yale University Press.

Perkins, H. Wesley. 1983. Organized religion as opiate or prophetic stimulant: A study of American and English assessments of social justice in two urban settings. *Review of Religious Research* (March), 206–224.

Perkins, H. Wesley. 1985. A research note on religiosity as opiate or prophetic stimulant among students in England and the United States. *Review of Religious Research* (March), 269–280.

Perrin, Robin D. 1989. American Religion in the post-Aquarian age: Values and demographic factors in church growth and decline. *Journal for the Scientific Study of Religion* (March), 75–89.

Perrin, Robin D.; Kennedy, Paul; & Miller, Donald E. 1997. Examining the sources of conservative church growth. *Journal for the Scientific Study of Religion* (March), 71–80.

Perrin, Robin D., & Mauss, Armand L. 1991. Saints and seekers: Sources of recruitment to the Vineyard Christian Fellowship. *Review of Religious Research* (December), 97–111.

Perry, Troy D. 2010. *The Lord is my shepherd, and he knows I'm gay (excerpts)*. Retrieved from ufmcc.com/overview/history-of-mcc/

Petersen, Larry R., & Donnenwerth, Gregory V. 1998. Religion and declining support for religious beliefs about gender roles and homosexual rights. *Sociology of Religion* (winter), 353–371.

Petterson, Thorlief. 1986. The American uses and gratifications of T.V. worship services. *Journal for the Scientific Study of Religion* (December), 391–409.

Petts, Richard, & Knoester, Chris. 2007. Parents' religious heterogamy and children's well being. *Journal for the Scientific Study of Religion, 48,* 373–389.

Pew Research Center for the People & the Press. 2009. *Majority continues to support civil unions* . Retrieved from http://people-press.org/reports/

Pew Research Center Forum on Religion & Public Life. 2005. *Myths of the modern megachurch* (May 23). Retrieved from http://pewforum.org/Christian/Evangelical-Protestant-Churches/Myths-of-the-Modern-Megachurch.aspx

Pew Research Center Forum on Religion & Public Life. 2006. *Overview: Pentecostalism in Latin America.* Retrieved from http://pewforum.org/Christian/Evangelical-Protestant-Churches/Overview-Pentecostalism-in-Latin-America.aspx

Pew Research Center Forum on Religion & Public Life. 2007a. *Changing faiths: Latinos and the transformation of American religion* (April 25). Retrieved from www.pewforum.org/Changing-Faiths-Latinos-and-the-Transformation-of-American-Religion.aspx

Pew Research Center Forum on Religion & Public Life. 2007b. *Muslim Americans: Middle class and mostly mainstream* (May 22). Retrieved from www.pewresearch.org/pubs/483/muslim-americans

Pew Research Center Forum on Religion & Public Life. 2008a. *A portrait of American Catholics on the eve of Pope Benedict's visit to the U.S* (March 27). Retrieved from www.pewforum.org

Pew Research Center Forum on Religion & Public Life. 2008b. *U.S. religious landscape survey.* Retrieved from http://religions.pewforum.org

Pew Research Center Forum on Religion & Public Life. 2009. *A religious portrait of African Americans* (January 30). Retrieved from http://pewforumorg/A-Religious-Portrait-of-African-Americans.aspx

Pharr, Suzanne. 1997. *Homophobia: A weapon of sexism.* Berkeley: Chardon Press.

Poloma, Margaret M. 1982a. *The charismatic movement: Is there a new Pentecost?* Boston: Twayne.

Poloma, Margaret M. 1989. *The assemblies of God at the crossroads: Charisma and institutional dilemmas.* Knoxville: University of Tennessee Press.

Poloma, Margaret M. 1982b. "Toward a Christian sociological perspective: religious values, theory and methodology." *Sociological analysis* (summer): 95–108.

Poloma, Margaret M. 2005. Charisma and structure in the Assemblies of God: Revisiting O'Dea's five dilemmas. In D. A. Roozen & J. R. Nieman (Eds.), *Church, identity, and change: Theology and denominational structures in unsettled times* (pp. 45–96). Grand Rapids, MI: Eerdmans.

Pope, Liston. 1942. *Millhands and preachers.* New Haven, CT: Yale University Press.

Pope, Liston. 1957. *Kingdom beyond caste.* New York: Friendship.

Popenoe, David. 1993. American family decline, 1960–1990: A review and appraisal. *Journal of Marriage and the Family, 55,* 527–555.

Postmes, T., & Branscombe, N. R. (Eds.) 2010. *Rediscovering social identity.* London: Psychology Press.

Pratt, James Bissett. 1964. Objective and subjective worship. In L. Schneider (Ed.), *Religion, culture, and society* (pp. 143–156). New York: Wiley.

Price, Jammie, & Dalecki, Michael G. 1998. The social basis of homophobia: An empirical illustration. *Sociological Spectrum* (April–June), 143–159.

Price, Joseph L. 2005. An American apotheosis: Sports as popular religion. In B. D. Forbes & J. H. Mahan (Eds.), *Religion and popular culture in America* (Rev. ed., pp. 195–212). Berkeley: University of California Press.

Priest, Robert J., & Nieves, Alvaro L. (Eds.). 2007. *This side of heaven: Race, ethnicity, and Christian faith.* New York: Oxford University Press.

Purvis, Sally. 1995. *The stained-glass ceiling: Churches and their women pastors.* Louisville, KY: Westminster/John Knox.

Putney, Clifford. 2001. *Muscular Christianity: Manhood and sports in Protestant America 1880–1920.* Cambridge, MA: Harvard University Press.

Pyle, Ralph. 2006. Trends in religious stratification: Have religious group socioeconomic distinctions declined in recent decades? *Sociology of Religion, 67,* 61–79.

Quinley, Harold E. 1974. *The prophetic clergy: Social activism among Protestant ministers.* New York: Wiley.

Radcliffe-Brown, A. R. 1939. *Taboo.* Cambridge, MA: Harvard University Press.

Radde-Antweiler, Kerstin. 2008. "Virtual Religion": An approach to a religious and ritual topography of second life. *Online—Heidelberg Journal of Religions on the Internet, 3.1,* 174–211.

Ramo, Joshua Cooper. 1996. Finding God on the web. *Time* (December 16). Retrieved from www.time.com/time/magazine/article/0,9171,985700,00.html

Read, Jen'nan Ghazal, & Bartkowski, John. 2000. To veil or not to veil? A case study of identity negotiation among Muslim women in Austin, Texas. *Gender and Society, 14,* 395–417.

Redekop, Calvin. 1974. A new look at sect development. *Journal for the Scientific Study of Religion* (September), 345–352.

Regnerus, Mark; Smith, Christian; & Smith, Brad. 2004. Social context in the development of adolescent religiosity. *Applied Developmental Science, 8,* 27–38.

Regnerus, Mark, & Uecker, Jeremy. 2006. Finding faith, losing faith: The prevalence and context of religious transformations during adolescence. *Review of Religious Research, 47,* 217–237.

Regnerus, Mark, & Uecker, Jeremy. 2007. Religious influences on sensitive self-reported behaviors: The product of social desirability, deceit, or embarrassment? *Sociology of Religion, 68,* 145–163.

Religion: Air worship. 1931. *Time* (February 9). Retrieved from www.time.com/time/magazine/article/0,9171,741032–1,00.html

Religion: Radio religion. 1946. *Time* (January 21). Retrieved from www.time.com/time/magazine/article/0,9171,934406,00.html

Richardson, James T. 1985. Paradigm conflict in conversion research. *Journal for the Scientific Study of Religion* (June), 163–719.

Richardson, James T.; Best, Joel; & Bromley, David G. (Eds.). 1991. *The Satanism scare.* New York: Aldine de Gruyter.

Richman, Karen. 2005. The Protestant ethic and the dis-spirit of vodou. In K. Leonard, A. Stepick, M. Vasquez, & J. Holdaway (Eds.), *Immigrant faiths: Transforming religious life in America* (pp. 165–187). Lanham, MD: AltaMira Press.

Robbins, Thomas, & Anthony, Dick. 1978. New religions, families, and brainwashing. *Society* (May/June), 77–83.

Robbins, Thomas, & Lucas, Phillip Charles. 2007. From "cults" to new religious movements: Coherence, definition, and conceptual framing in the study of new religious movements. In J. Beckford & N. J. Demerath III (Eds.), *The Sage handbook of the sociology of religion* (pp. 227–247). Thousand Oaks, CA: Sage.

Roberts, Keith A. 1992. Ritual and the transmission of a cultural tradition: An ethnographic perspective. In J. Carroll & W. C. Roof (Eds.), *Beyond establishment: Protestant identity in a post Protestant age* (pp. 74–98). Louisville, KY: Westminster/John Knox.

Robertson, H. M. 1959. *Aspects of the rise of economic individualism: A criticism of Max Weber and his school.* New York: Kelley & Millman. (Original work published 1933)

Robertson, Roland. 1989. Globalization, politics, and religion. In J. A. Beckford & T. Luckmann (Eds.), *The changing face of religion* (pp. 1–9). London: Sage.

Robertson, Roland. 1991. Globalization, modernization, and postmodernization: The ambiguous position of religion. In R. Robertson & W. R. Garrett (Eds.), *Religion and global order* (pp. 281–291). New York: Paragon.

Robertson, Roland. 1992. *Globalization: Social theory and global culture.* London: Sage.

Robertson, Roland. 1995. Glocalization: Time-space and homogeneity-heterogeneity. In M. Featherstone, S. Lash, & R. Robertson (Eds.), *Global modernities* (pp. 25–44). London: Sage.

Rodriguez, Eric M., & Ouellette, Suzanne C. 2000. Gay and lesbian Christians: Homosexual and religious identity integration in the members and participants of a gay-positive church. *Journal for the Scientific Study of Religion, 39*(3), 333–347.

Rokeach, Milton. (Ed.). 1968. The nature of attitudes. *Beliefs, attitudes, and values.* San Francisco: Jossey-Bass.

Rokeach, Milton; Smith, Patricia W.; & Evans, Richard I. 1960. Two kinds of prejudice or one?

In M. Rokeach (Ed.), *The open and closed mind* (pp. 132–168). New York: Basic Books.

Romero, Catalina. 2001. Globalization, civil society and religion from a Latin American perspective. *Sociology of Religion* (winter), 475–490.

Roof, Wade Clark. 1978. *Commitment and community.* New York: Elsevier.

Roof, Wade Clark. 1993. *A generation of seekers: The spiritual journeys of the baby boom generation.* San Franscisco: Harper.

Roof, Wade Clark. 1999. *Spiritual marketplace: Baby boomers and the remaking of American religion.* Princeton, NJ: Princeton University Press.

Roof, Wade Clark. 2009. Generations and religion. In P. Clarke (Ed.), *Oxford handbook of the sociology of religion* (pp. 616–634). Oxford: Oxford University Press.

Roof, Wade Clark, & McKinney, William. 1987. *American mainline religion: Its changing shape and future.* New Brunswick, NJ: Rutgers University Press.

Rosenfeld, Michael J. 2008. Racial, educational, and religious endogamy in the United States: A comparative historical perspective. *Social Forces* (September), 87, 1–32.

Rosewicz, Barbara. 1985. At Jerusalem church, people often ignore tenth commandment. *Wall Street Journal* (April 5), pp. 1, 5.

Rothenberg, Paula S. 2008. *White privilege* (3rd ed.). New York: Worth.

Rousseau, Jean-Jacques. 1954. *The social contract* (W. Kendall, Trans.). Chicago: Henry Regnery Company. (Original work published 1762)

Ruether, Rosemary Radford. 1974c. Misogynism and virginal feminism in the fathers of the church. In R. R. Ruether (Ed.), *Religion and sexism* (pp. 150–183). New York: Simon & Schuster.

Ruether, Rosemary Radford. 1974. The persecution of witches. *Christianity and Crises* (December 23), 291–295.

Ruether, Rosemary Radford. 1975. *New woman, new earth.* New York: Seabury Press.

Ryder, Norman B. 1965. The cohort as a concept in the study of social change. *American Sociological Review, 30,* 843–861.

Samuelsson, Kurt. 1961. *Religion and economic action* (E. G. French, Trans.). New York: Basic Books. (Original work published 1957)

Sanneh, Lamin. 1991. The yogi and the commissar: Christian missions and the new world order in Africa. In W. C. Roof (Ed.), *World order and religion* (pp. 173–192). Albany: State University of New York Press.

Sargeant, Kimon Howland. 2000. *Seeker churches: Promoting traditional religion in a nontraditional way.* New Brunswick, NJ: Rutgers University Press.

Scheepers, Peer; Gijsberts, Merove; & Hello, Evelyn. 2002. Religion and prejudice against ethnic minorities in Europe: Cross-national tests on a controversial relationship. *Review of Religious Research* (March), 242–265.

Schneider, Louis. 1970. *Sociological approach to religion.* New York: Wiley.

Schoenherr, Richard. 2002. In D. Yamane (Ed.), *Goodbye father: The celibate male priesthood and the future of the Catholic Church.* New York: Oxford University Press.

Schwadel, Philip. 2008. Poor teenagers' religion. *Sociology of Religion, 69*(2), 125–149.

Schwadel, Philip. 2010. Period and cohort effects on religious nonaffiliation and religious disaffiliation: A research note. *Journal for the Scientific Study of Religion, 49,* 311–319.

Scientology. 2010. *Scientology is a religion.* Retrieved from www.scientology.org

Sewell, William, & Hauser, Robert. 1975. *Education, occupation, and earnings. Achievement in the early career.* New York: Academic Press.

Shannon, Thomas Richard. 1989. *An introduction to the world-system perspective.* Boulder, CO: Westview Press.

Shannon, William. 1963. *The American irish.* New York: Macmillan.

Sharot, Stephen. 1991. Judaism and the secularization debate. *Sociological Analysis* (fall), 255–275.

Sheard, Robert B. 1987. *Interreligious dialogue in the Catholic Church since Vatican II.* Lewiston, NY: Mellon.

Sherkat, Darren E. 1997. Embedding religious choices: Integrating preferences and social choices into rational choice theories of religious behavior. In L. A. Young (Ed.), *Rational choice theory and religion: Summary and assessment* (pp. 65–85). New York: Routledge.

Sherkat, Darren E. 2003. Religious socialization: Sources of influence and influences of agency. In M. Dillon (Ed.), *Handbook of the sociology of religion* (pp. 151–178). New York: Cambridge University Press.

Sherkat, Darren E. 2010. Religion and verbal ability. *Social Science Research, 39,* 2–13.

Sherkat, Darren E., & Ellison, Christopher G. 2001. Recent developments and current controversies in the sociology of religion. *Annual Review of Sociology,* (27), 363–386.

Sherkat, Darren E., & Wilson, John. 1995. Preferences, constraints, and choices in religious markets: An examination of religious switching and apostasy. *Social Forces* (March), 993–1026.

Shields, David L. 1986. *Growing beyond prejudices: Overcoming hierarchical dualism.* Mystic, CT: Twenty-Third Publishing.

Shipps, Jan. 1985. *Mormonism: The story of a new religious tradition.* Urbana: University of Illinois Press.

Shupe, Anson D., Jr. 1995. *In the name of all that's holy: A theory of clergy malfeasance.* New York: Praeger.

Shupe, Anson D., Jr. 1998. *Wolves within the fold: Religious leadership and abuses of power.* New Brunswick, NJ: Rutgers University Press.

Shupe, Anson D., Jr., & Bromley, David G. 1978. Witches, Moonies, and evil. *Society* (May/June), 75–76.

Shupe, Anson, D., Jr., & Hadden, Jeffrey K. 1989. Is there such a thing as global fundamentalism? In J. K. Hadden & A. Shupe (Eds.), *Secularization and fundamentalism reconsidered* (pp. 109–122). New York: Paragon.

Shupe, Anson D., Jr.; Stacey, William A.; & Darnell, Susan. 2000. *Bad pastors: Clergy misconduct in modern America.* New York: New York University Press.

Siegel, Karolynn, & Shrimshaw, Eric W. 2002. The perceived benefits of religious and spiritual coping among older adults living with HIV/AIDS. *Journal for the Scientific Study of Religion, 41,* 87–101.

Sikkink, David. 1999. The social sources of alienation from public schools. *Social Forces 78,* 51–86.

Sikkink, David, & Hill, Jonathan. 2006. Education. In H. R. Ebaugh (Ed.), *Handbook of religion and social institutions* (pp. 41–66). New York: Springer.

Silva Method. 2010. The Silva Method. Retrieved from www.silvamethod.com

Simon, Stephanie. 2007. It's Easter; shall we gather at the desktops? *Los Angeles Times* (April 8).

Retrieved from articles.latimes.com/2007/apr/08/nation/na-virtual8

Simpson, John H. 1991. Globalization and religion: Themes and perspectives. In R. Robertson & W. R. Garrett (Eds.), *Religion and global order* (pp. 1–18). New York: Paragon.

Simpson, John H. 1996. "The Great reversal": Selves, communities, and the global system. *Sociology of Religion* (summer), 115–125.

Skeggs, Beverly. 2004. *Class, self, culture.* London: Routledge.

Sklare, Marshall, & Greenblum, Joseph. 1979. *Jewish identity on the suburban frontier* (2nd ed.). Chicago: University of Chicago Press.

Smelser, Neil J. 1962. *Theory of collective behavior.* New York: Free Press.

Smith, Carole R.; Williams, Lev; & Willis, Richard H. 1967. Race, sex, and belief as determinants of friendship acceptance. *Journal of Personality and Social Psychology* (February), 127–137.

Smith, Christian. 1996a. *Disruptive religion: The force of faith in social movement activism.* New Brunswick, NJ: Rutgers University Press.

Smith, Christian. 1996b. *Resisting Reagan: The U.S. Central America peace movement.* Chicago: University of Chicago Press.

Smith, Christian. 1998. *American evangelicalism: Embattled and thriving.* Chicago: University of Chicago Press.

Smith, Christian. 2003. *Moral, believing animals: Human personhood and culture.* New York: Oxford University Press.

Smith, Christian (with Denton, Melinda Lundquist). 2005. *Soul searching: The religious and spiritual lives of American teenagers.* New York: Oxford University Press.

Smith, Christian (with Snell, P.). 2009. *Souls in transition: The religious and spiritual lives of emerging adults.* New York: Oxford University Press.

Smith, Christian, & Faris, Robert. 2005. Socioeconomic inequality in the American religious system: An update and assessment. *Journal for the Scientific Study of Religion, 44,* 95–104.

Smith, Tom W. 1992. Are conservative churches growing? *Review of Religious Research* (June), 305–329.

Snow, David A., & Phillips, Cynthia L. 1980. The Lofland-Stark conversion model: A critical reassessment. *Social Problems* (April), 430–447.

Sommerville, C. John. 2002. Stark's age of faith argument and the secularization of things: A commentary. *Review of Religious Research* (fall), 361–372.

Southwold, Martin. 1982. True Buddhism and village Buddhism in Sri Lanka. In J. Davis (Ed.), *Religious organization and religious experience* (pp. 137–152). London: Academic Press.

Spickard, James V. 1998. Rethinking religious social action: What is "rational" about rational choice theory? *Sociology of Religion* (summer), 99–115.

Spinoza, Baruch. 2009. *A theologico-political treatise.* Gloucestershire, UK: Dodo Press.

Spiro, Melford. 1966. Religion: Problems of definition and explanation. In M. Banton (Ed.), *Anthropological approaches to the study of religion* (pp. 85–126). London: Tavistock.

Spiro, Melford. 1978. *Burmese supernaturalism.* Philadelphia: Institute for the Study of Human Issues.

Spitz, Bob. 1992. Mrs. Gere goes to Hollywood. *Redbook, 179*(5), 110–116.

Staples, Clifford L., & Mauss, Armand L. 1987. Conversion or commitment: A reassessment of the Snow and Machalek approach to the study of conversion. *Journal for the Scientific Study of Religion* (June), 133–147.

Starhawk. 1989. *Spiral dance: A rebirth of the ancient religion of the Great Goddess.* San Francisco: HarperSanFrancisco.

Stark, Rodney. 1972. The economics of piety: Religious commitment and social class. In G. W. Thielbar & S. D. Feldman (Eds.), *Issues in social inequality* (pp. 483–503). Boston: Little, Brown.

Stark, Rodney. 1984. The rise of a New World faith. *Review of Religious Research* (September), 18–27.

Stark, Rodney. 1986. The class basis of early Christianity: Inferences from a sociological model. *Sociological Analysis* (fall), 216–225.

Stark, Rodney. 1992. Do Catholic societies really exist? *Rationality and Society,* 4, 261–271.

Stark, Rodney. 1995. Reconstructing the rise of Christianity: The role of women. *Sociology of Religion* (fall), 229–244.

Stark, Rodney. 2000a. Religious effects: In praise of "idealistic humbug." *Review of Religious Research* (March), 289–310.

Stark, Rodney. 2000b. Secularization, R.I.P. In W. H. Swatos, Jr., & D. V. A. Olson, *The secularization debate* (pp. 41–66). Lanham, MD: Rowman & Littlefield.

Stark, Rodney, & Bainbridge, William Sims. 1979. Of churches, sects, and cults: Preliminary concepts for a theory of religious movements. *Journal for the Scientific Study of Religion* (June), 119–121.

Stark, Rodney, & Bainbridge, William Sims. 1985. *The future of religion: Secularization, renewal, and cult formation.* Berkeley: University of California Press.

Stark, Rodney, & Bainbridge, William Sims. 1996. *A theory of religion.* New Brunswick, NJ: Rutgers University Press.

Stark, Rodney, & Finke, Roger. 1988. American religion in 1776: A statistical portrait. *Sociological Analysis* (spring), 39–51.

Stark, Rodney, & Finke, Roger. 2000. *Acts of faith: Explaining the human side of religion.* Berkeley: University of California Press.

Stark, Rodney, Finke, Roger, & Iannaccone, Laurence R. 1995. Pluralism and piety: England and Wales, 1951. *Journal for the Scientific Study of Religion* (December), 431–444.

Stark, Rodney, & Glock, Charles Y. 1968. *American piety: The nature of religious commitment.* Berkeley: University of California Press.

Stark, Rodney, & Glock, Charles Y. 1969. Prejudice and the churches. In C. Y. Glock & E. Siegelman, *Prejudice U.S.A.* (pp. 70–95). New York: Praeger.

Stark, Rodney, & Iannaccone, Laurence R. 1994. A supply–side reinterpretation of the "secularization" of Europe. *Journal for the Scientific Study of Religion* (September), 230–252.

Stark, Rodney, & McCann, James. 1993. Market forces and Catholic commitment: Exploring the new paradigm. *Journal for the Scientific Study of Religion, 32,* 111–124.

Stark, Rodney, & Roberts, Lynne. 1982. The arithmetic of social movements: Theoretical implications. *Sociological Analysis* (spring), 53–67.

Stark, Werner. 1967. *Sectarian religion. Vol. 2, The sociology of religion: A study of Christendom.* New York: Fordham University Press.

Stein, Nicholas. 2006. No way out. In D. S. Eitzen & M. B. Zinn (Eds.), *Globalization: The transformation of social worlds* (pp. 293–299). Belmont, CA: Wadsworth.

Steinberg, Stephen. 1965. Reform Judaism: The origin and evolution of a "church movement." *Journal for the Scientific Study of Religion* (October), 117–129.

Stevens, Mitchell. 2003. *Kingdom of children: Culture and controversy in the homeschooling movement.* Princeton, NJ: Princeton University Press.

Stoll, David. 1990. *Is Latin America turning Protestant? The politics of evangelical growth.* Berkeley: University of California Press.

Stolzenberg, Ross; Blair-Loy, Mary; & Waite, Linda. 1995. Age and family life cycle effects on church membership. *American Sociological Review, 60,* 84–103.

Storch, E., Roberti, J., Bravata, E., & Storch, J. 2004. Strength of religious faith: A comparison of intercollegiate athletes and non-athletes. *Pastoral Psychology, 52,* 485–492.

Straus, Roger. 1976. Changing oneself: Seekers and the creative transformation of life experience. In J. Lofland (Ed.), *Doing social life* (pp. 252–273). New York: Wiley Interscience.

Straus, Roger. 1979. Religious conversion as a personal and collective accomplishment. *Sociological Analysis* (summer), 158–165.

Suchman, Mark C. 1992. Analyzing the determinants of everyday conversion. *Sociological Analysis,* S15–33.

Sullins, Paul. 2000. The stained glass ceiling: Career attainment for women clergy. *Sociology of Religion* (fall), 243–266.

Swatos, William H., Jr. 1998. Denomination/denominationalism. In W. H. Swatos (Ed.), *Encyclopedia of religion and society* (pp. 134–136). Walnut Creek, CA: AltaMira.

Takayama, K. Peter. 1975. Formal polity and change of structure: Denominational assemblies. *Sociological Analysis, 36,* 17–28.

Takayama, K. Peter, & Cannon, Lynn Weber. 1979. Formal polity and power distribution in American Protestant denominations. *Sociological Quarterly, 20,* 321–332.

Talmon, Yonina. 1965. The pursuit of the millennium: The relation between religion and social change. In W. A. Lessa & E. Z. Vogt (Eds.), *Reader in comparative religion* (2nd ed., pp. 522–537). New York: Harper & Row.

Tamadonfar, Mehren. 2001. Islam, law, and political control in contemporary Iran. *Journal for the Scientific Study of Religion* (June), 205–219.

Tamney, Joseph B. 1992. *The resilience of Christianity in the modern world.* Albany: State University of New York Press.

Tamney, Joseph B., & Johnson, Steven D. 1985. Consequential religiosity in modern society. *Review of Religious Research* (June), 360–378.

Tamney, Joseph B., & Johnson, Steven D. 1998. The popularity of strict churches. *Review of Religious Research* (March), 209–223.

Tawney, R. H. 1954. *Religion and capitalism.* New York: New American Library. (Original work published 1924)

Tedlin, Kent. 1978. Religious preference and pro/anti activism on the Equal Rights Amendment issue. *Pacific Sociological Review* (January), 55–66.

Thomas, Charles B., Jr. 1985. Clergy in racial controversy: A replication of the Campbell and Pettigrew study. *Review of Religious Research* (June), 379–390.

Thomas, George M. 1991. A world polity interpretation of U. S. religious trends since World War II. In W. C. Roof (Ed.), *World order and religion* (pp. 217–247). Albany: State University of New York Press.

Thomas, George M. 2001. Religions in global civil society. *Sociology of Religion* (winter), 515–533.

Thompson, Edward H., Jr. 1991. Beneath the status characteristic: Gender variations in religiousness. *Journal for the Scientific Study of Religion* (December), 381–394.

Thornton, Arland, & Freedman, Deborah. 1979. Changes in the sex role attitudes of women, 1962–1977: Evidence from a panel study. *American Sociological Review* (October), 831–842.

Thumma, Scott. 2001. *Megachurches Today 2000.* Hartford, CT: Hartford Institute for Religion Research. Retrieved from hirr.hartsem.edu/megachurch/faith_megachurches_FACT summary .html

Thumma, Scott, & Bird, Warren. 2008. Changes in American megachurches: Tracing eight years of growth and innovation in the nation's largest-attendance congregations (September). Retrieved from hirr.hartsem.edu/megachurch/megachurches_research.html

Thumma, Scott; Travis, Dave; & Bird, Warren. 2005. *Megachurches today 2005: Summary of research findings.* Retrieved from hirr.hartsem.edu/megachurch/megachurches_research.html

Tillich, Paul. 1957. *Dynamics of faith.* New York: Harper & Row.

Titiev, Mischa. 1972. A fresh approach to the problem of magic and religion. In W. A. Lessa & E. Z. Vogt (Eds.), *Reader in comparative religion: An anthropological approach* (3rd ed., pp. 430–433). New York: Harper & Row.

Tomasi, Luigi. 2002. Homo Viator: From pilgrimage to religious tourism via the journey. In W. Swatos & L. Tomasi (Eds.), *From medieval pilgrimage to religious tourism* (pp. 1–24). Westport, CT: Praeger.

Toolin, Cynthia. 1983. American civil religion from 1789 to 1981: A content analysis of presidential inaugural addresses. *Review of Religious Research* (September), 39–48.

Transcendental Meditation. 2010. The TM technique. Retrieved from www.tm.org

Trevor-Roper, Hugh. 1967. Witches and witchcraft. *Encounter* (June), 13–34.

Trible, Phyllis. 1979. Eve and Adam: Genesis 2–3 reread. In C. P. Christ & J. Plaskow, *Womanspirit rising* (pp. 74–83). New York: Harper & Row.

Trinitapoli, Jenny. 2005. Congregation-based services for elders: An examination of patterns and correlates. *Research on Aging, 27,* 241–264.

Trinitapoli, Jenny, & Vaisey, Stephen. 2009. The transformative role of religious experience: The case of short-term missions. *Social Forces, 88,* 121–146.

Tschannen, Olivier. 1991. The secularization paradigm: A systematization. *Journal for the Scientific Study of Religion* (December), 395–415.

Turner, Bryan S. 1991a. Politics and culture in Islamic Fundamentalism. In R. Robertson & W. R. Garrett (Eds.), *Religion and global order* (pp. 161–181). New York: Paragon.

Turner, Bryan S. 1991b. *Religion and social theory.* London: Sage.

Turner, Jonathan H., & Maryanski, Alexandra. 1979. *Functionalism.* Menlo Park, CA: Benjamin/Cummings.

Tylor, Edward B. 1958. *Primitive culture,* Vol. II. New York: Harper & Row. (Original work published 1873).

U.S. Bureau of Labor Statistics. 2006. *Women in the professions.* Retrieved from www.bls.gov/opub/ted/2006/dec/wk1/art02.htm

U.S. Bureau of Labor Statistics. 2009. *Labor force statistics from the Current Population Survey: Women in the labor force.* Retrieved from www.bls.gov/cps/wlf-intro-2009.htm

U.S. Census Bureau. 2010. *Mean earnings by highest degree earned.* Retrieved from www.census.gov/compendia/statab/cats/education/educational_attainment.html

U.S. Flag Code. 2007. U.S. Flag Code (Federal Flag Code Amendment Act of 2007). Retrieved from www.suvcw.org/flag.htm

Van Ausdale, Debra, & Feagin, Joe R. 1996. Using racial and ethnic concepts: The critical case of very young children. *American Sociological Review* (October), 779–793.

van Biema, David, & Chu, Jeff. 2006. Does God want you to be rich? *Time* (September 18), pp. 48–56.

Van der Leeuw, Gerardus. 1963. *Religion in essence and manifestation* (Vol. I.). New York: Harper Torchbooks.

Vander Zanden, James W. 1983. *American minority relations* (4th ed.). New York: Knopf.

Vander Zanden, James W. 1987. *Social psychology* (4th ed.). New York: Random House.

Vidler, Alec R. 1961. *The church in an age of revolution, 1989 to the present day.* London: Hodder & Stoughton.

Vogt, Evon Z. 1952. Water witching: An interpretation of a ritual pattern in a rural American community. *Scientific Monthly* (September), pp. 175–186.

Vrame, Anton. 2008. Four types of "orthopraxy" among Orthodox Christians in America." In A. Papanikolaou & E. H. Prodromou (Eds.), *Thinking through faith: New perspectives from Orthodox Christian scholars* (pp. 279–308). Crestwood, NY: St. Vladimir's Seminary Press.

Wallace, Anthony F. C. 1972. Revitalization movements. In W. A. Lessa & E. Z. Vogt (Eds.), *Reader in comparative religion: An anthropological approach* (3rd ed., pp. 503–512). New York: Harper & Row.

Wallace, Ruth A. 1992. *They call her pastor: A new role for Catholic women.* Albany: State University of New York Press.

Wallerstein, Immanuel. 1974. *The modern world system.* New York: Academic Press.

Wallerstein, Immanuel. 1979. *The capitalist world economy.* New York: Cambridge University Press.

Wallerstein, Immanuel. 1984. *The politics of the world economy.* New York: Cambridge University Press.

Wallis, Roy. 1977. *The road to total freedom: A sociological analysis of Scientology.* New York: Columbia University Press.

Wallis, Roy, & Bruce, Steve. 1991. Secularization: Trends, data, and theory. *Research in the Social Scientific Study of Religion, 3,* 1–31.

Walls, A. F. 1991. World Christianity, the missionary movement, and the ugly American. In W. C. Roof (Ed.), *World order and religion* (pp. 147–172). Albany: State University of New York Press.

Walton, Jonathan L. 2009. *Watch this! The ethics and aesthetics of black televangelism.* New York: New York University Press.

Ward, Martha C., & Edelstein, Monica. 2006. *A world full of women* (4th ed.). Boston: Pearson Education.

Warner, R. Stephen. 1993. Work in progress toward a new paradigm for the sociological study of religion in the United States. *American Journal of Sociology* (March), 1044–1093.

Warner, R. Stephen. 2005. *A church of our own: Disestablishment and diversity in American religion.* New Brunswick, NJ: Rutgers University Press.

Warner, W. Lloyd. 1953. *American life: Dream and reality.* Chicago: University of Chicago Press.

Washington, Joseph R., Jr. 1964. *Black religion.* Boston: Beacon Press.

Washington, Joseph R., Jr. 1972. *Black sects and cults.* Garden City, NY: Doubleday.

Watt, William Montgomery. 1996. *A short history of Islam.* Oxford: Oneworld Publications.

Waugh, Earle. 1994. Reducing the distance: A Muslim congregation in the Canadian north. In J. Wind & J. Lewis (Eds.), *American congregations: Portraits of twelve religious communities* (pp. 572–611). Chicago: University of Chicago Press.

Weaver, Horace R. 1975. *Getting straight about the Bible.* Nashville, TN: Abingdon Press.

Weber, Max. 1946. *From Max Weber: Essays in sociology.* H. H. Gerth & C. W. Mills (Eds. & Trans.). New York: Oxford University Press.

Weber, Max. 1947. *The theory of social and economic organization.* A. M. Henderson & T. Parsons (Eds. & Trans.). New York: Oxford University Press.

Weber, Max. 1949. *The methodology of the social sciences.* New York: Free Press.

Weber, Max. 1951. *The religion of China* (H. H. Gerth, Trans.). New York: Free Press. (Original work published 1920–1921)

Weber, Max. 1952. *Ancient Judaism.* H. H. Gerth & D. Martindale (Eds. & Trans.). New York: Free Press. (Original work published 1920–1921)

Weber, Max. 1958a. *The Protestant ethic and the spirit of capitalism* (T. Parsons, Trans.). New York: Scribner. (Original work published 1904–1905)

Weber, Max. 1958b. *The Religion of India.* H. H. Gerth & D. Martindale (Eds. & Trans.). New York: Free Press. (Original work published 1920–1921)

Weber, Max. 1963. *The sociology of religion* (E. Fischoff, Trans.). Boston: Beacon Press. (Original work published 1922)

Weidenbaum, Murray. 2006. Globalization: Wonderland or waste land? In D. S. Eitzen & M. B. Zinn (Eds.), *Globalization: The transformation of social worlds* (pp. 53–60). Belmont, CA: Wadsworth.

Weinberg, George. 1973. *Society and the healthy homosexual.* Garden City, NY: Doubleday.

Welch, Susan. 1975. Support among women for the issues of the women's movement. *Sociological Quarterly* (spring), 216–227.

Weller, Christian E., & Hersh, Adam. 2006. Free markets and poverty. In D. S. Eitzen & M. B. Zinn (Eds.), *Globalization: The transformation of social worlds* (pp. 69–73). Belmont, CA: Wadsworth.

Welter, Barbara. 1976. The feminization of American religion, 1800–1860. *Dimity convictions.* Athens: Ohio University Press.

Whitehead, Andrew. 2010. Sacred rites and civil rights: Religion's effect on attitudes toward same-sex unions and the perceived cause of homosexuality. *Social Science Quarterly, 91,* 63–79.

Wilcox, W. Bradford. 2006. Family. In H. R. Ebaugh (Ed.), *Handbook of religion and social institutions* (pp. 97–120). New York: Springer.

Williams, Gwyneth I. 2001. Masculinity in context: An epilogue. In R. H. Williams (Ed.), *Promise Keepers and the new masculinity: Private lives and public morality* (pp. 105–114). Lanham, MD: Lexington Books.

Williams, Michael A. 1986. Uses of gender imagery in ancient Gnostic texts. In C. W. Bynum,

S. Harrell, & P. Richman (Eds.), *Gender and religion* (pp. 196–227). Boston: Beacon Press.

Williams, Peter W. 1980. *Popular religion in America: Symbolic change and the modernization process in historical perspective.* Englewood Cliffs, NJ: Prentice Hall.

Williams, Rhys, & Vashi, Gira. 2007. *Hijab* and American Muslim women: Creating the space for autonomous selves. *Sociology of Religion, 68*(3), 269–287.

Wilmore, Gayraud S. 1972. *Black religion and black radicalism.* Garden City, NY: Doubleday.

Wilson, Bryan R. 1959. An analysis of sect development. *American Sociological Review* (February), 3–15.

Wilson, Bryan R. 1970. *Religious sects.* New York: McGraw-Hill.

Wilson, Bryan R. 1976. *Contemporary transformations of religion.* Oxford: Clarendon Press.

Wilson, Bryan R. 1982. Religion in sociological perspective. New York: Oxford University Press.

Wilson, Gerald L.; Keyton, Joann; Johnson, David; Geiger, Cheryl; & Clark, Johanna C. 1993. Church growth through member identification and commitment: A congregational case study. *Review of Religious Research* (March), 259–272.

Wimberley, Ronald C.; Hood, Thomas C.; Lipsey, C. M.; Clelland, Donald; & Hay, Marguerite. 1975. Conversion in a Billy Graham crusade: Spontaneous event or ritual action? *The Sociological Quarterly* (spring), 162–170.

Winter, J. Alan. 1977. *Continuities in the sociology of religion.* New York: Harper & Row.

Witten, Marsha. 1995. *All is forgiven: The secular message in American Protestantism.* Princeton, NJ: Princeton University Press.

Wolfe, Alan. 2003. *The transformation of American religion: How we actually live our faith.* New York: Free Press.

Wood, James R. 1970. Authority and controversial policy: The churches and civil rights. *American Sociological Review* (December), 1057–1069.

Wood, James R. 1981. *Leadership in voluntary organizations: The controversy over social action in protestant churches.* New Brunswick, NJ: Rutgers University Press.

Wood, James R., & Bloch, Jon P. 1995. The role of church assemblies in building a civil society: The case of the United Methodist General Conference's

debate on homosexuality. *Sociology of Religion* (summer), 121–136.

Wright, Derek, & Cox, Edwin. 1967. A study of the relationship between moral judgment and religious belief in a sample of English adolescents. *Journal of Social Psychology* (June), 135–144.

Wrong, Dennis H. 1961. The oversocialized conception of man in modern sociology. *American Sociological Review, 26*(2), 183–193.

Wuthnow, Robert. 1973. New forms of religion in the seminary. In C. Y. Glock (Ed.), *Religion in sociological perspective* (pp. 187–203). Belmont, CA: Wadsworth.

Wuthnow, Robert. 1976a. Astrology and marginality. *Journal for the Scientific Study of Religion* (June), 157–168.

Wuthnow, Robert. 1976b. *The consciousness reformation.* Berkeley: University of California Press.

Wuthnow, Robert. 1980. World order and religious movements. In A. Bergson (Ed.), *Studies of the modern world-system* (pp. 57–75). New York: Academic Press.

Wuthnow, Robert. 1981. Two traditions of religious studies. *Journal for the Scientific Study of Religion* (March), 16–32.

Wuthnow, Robert. 1987. The social significance of religious television. *Review of Religious Research* (December), 125–134.

Wuthnow, Robert. 1988. *The restructuring of American religion: Society and faith since World War II.* Princeton, NJ: Princeton University Press.

Wuthnow, Robert. 1991. International realities: Bringing the global picture into focus. In W. C. Roof (Ed.), *World order and religion* (pp. 19–37). Albany, NY: State University of New York Press.

Wuthnow, Robert. 1994a. *God and Mammon in America.* New York: The Free Press.

Wuthnow, Robert. (Ed.). 1994b. *I come away stronger: How small groups are shaping American religion.* Grand Rapids, MI: Eerdmans.

Wuthnow, Robert. 1996. *Poor Richard's principle.* Princeton, NJ: Princeton University Press.

Wuthnow, Robert. 1998. *After heaven: Spirituality in America since the 1950s.* Berkeley: University of California Press.

Wuthnow, Robert. 1999. *Growing up religious: Christians and Jews and their journeys of faith.* Boston: Beacon Press.

Wuthnow, Robert. 2009. *Boundless faith: The global outreach of American churches.* Berkeley: University of California Press.

Wuthnow, Robert, & Evans, John. (Eds.). 2002. *The quiet hand of God: Faith-based activism and the public role of mainline Protestantism.* Berkeley: University of California Press.

Wuthnow, Robert, & Offutt, Stephen. 2008. Transnational religious connections. *Sociology of Religion, 69,* 209–232.

Yamane, David. 1997. Secularization on trial: In defense of a neosecularization paradigm. *Journal for the Scientific Study of Religion* (March), 109–122.

Yamane, David. 1998. Spirituality. In W. H. Swatos (Ed.), *Encylopedia of religion and society.* Walnut Creek, CA: AltaMira.

Yamane, David, & Blake, Teresa. 2008. *Sport and sacred umbrellas on campus: The religiosity and spirituality of college athletes.* Unpublished manuscript, Department of Sociology, Wake Forest University, Winston-Salem, NC.

Yamane, David, & MacMillen, Sarah. 2006. *Real stories of Christian initiation: Lessons for and from the RCIA.* Collegeville, MN: Liturgical Press.

Yamane, David; Mellies, Charles; & Blake, Teresa. 2010. Playing for whom? Sport, religion, and the double movement of secularization in America. In E. Smith (Ed.), *Sociology of sport and social theory* (pp. 81–94). Champaign, IL: Human Kinetics.

Yamane, David, & Polzer, Megan. 1994. Ways of seeing ecstasy in modern society: Experiential-expressive and cultural-linguistic views. *Sociology of Religion, 55,* 1–25.

Yancey, George. 1999. An examination of the effects of residential and church integration on racial attitudes of whites. *Sociological Perspectives* (42), 279–304.

Yang, Fenggang. 1998. Chinese conversion to evangelical Christianity: The importance of social and cultural contexts. *Sociology of Religion* (fall), 273–257.

Yang, Fenggang. 1999. *Chinese Christians in America: Conversion, assimilation, and adhesive identities.* College Station, PA: PSU Press.

Yang, Fenggang, & Ebaugh, Helen Rose. 2001. Transformations in new immigrant religions and their global implications. *American Sociological Review, 66,* 269–288.

Yinger, J. Milton. 1961. Comment. *Journal for the Scientific Study of Religion* (October), 40–44.

Yinger, J. Milton. 1969. A structural examination of religion. *Journal for the Scientific Study of Religion* (spring), 88–100.

Yinger, J. Milton. 1970. *The scientific study of religion.* New York: Macmillan.

Yinger, J. Milton. 1977. A comparative study of the substructure of religion. *Journal for the Scientific Study of Religion* (March), 67–86.

The YMCA in the United States. 2010. *YMCA of the USA.* Retrieved from http://www.ymca.net/history/

Young, Lawrence. 1997. *Rational choice theory and religion: Summary and assessment.* New York: Routledge.

Yu, Eui-Young. 1988. The growth of Korean Buddhism in the United States, with special reference to Southern California. *The Pacific World: Journal of the Institute of Buddhist Studies* (n.s.), *4,* 82–93.

Zablocki, Benjamin. 1971. *The joyful community: An account of the Bruderhof.* Baltimore: Penguin Books.

Zaechner, R. C. (Ed.). 1967. *The concise encyclopedia of living faith.* Boston: Beacon Press.

Zhai, J. E.; Ellison, Christopher G.; Glenn, N.; & Marquardt, E. 2007. Parental divorce and religious involvement among young adults. *Sociology of Religion, 68*(2), 125–144.

Zuckerman, Phil. 2003. *Invitation to the sociology of religion.* New York: Routledge.

CREDITS

Chapter 1

Photo 1.1, page 5. Photo by Elise Roberts

Photo 1.2, page 10. © Istockphoto.com/James Pauls

Chapter 2

Photo 2.1, page 29. Photo by Keith Roberts

Chapter 3

Photo 3.1, page 48. Photo by Elise Roberts

Photo 3.2, page 55. © Used by permission of the United States Holocaust Memorial Museum

Chapter 4

Photo 4.1, page 68. Photo by Keith Roberts

Photo 4.2, page 77. © Antoine Gyori/Sygma/Corbis

Photos 4.3, page 83. Photo by Keith Roberts

Photos 4.4, page 83. Photo by Keith Roberts

Chapter 5

Photo 5.1, page 95. © Istockphoto.com/Jacob Hellbach

Photo 5.2, page 104. © Brand X Pictures/Brand X Pictures/Thinkstock

Photo 5.3, page 108. © istockphoto.com/Christopher Futcher

Chapter 6

Photo 6.1, page 132. Photo by Keith Roberts

Photo 6.2, page 141. Photo by Keith Roberts

Photo 6.3, page 145. NA

Chapter 7

Photo 7.1, page 154. © Brigham Young Family Association

Photo 7.2, page 160. Photo by Keith Roberts

Photo 7.3, page 167. Photos by Keith Roberts

Chapter 8

Photo 8.1, page 191. Photo by Keith Roberts

Photo 8.2, page 192. © Istockphoto.com

Photo 8.3, page 203. © Corbis

Photo 8.4, page 206. Photo by Keith Roberts

Chapter 9

Photo 9.1, page 225. © Brooks Kraft/Sygma/Corbis

Photo 9.2, page 227. © Kevin Lamarque/Reuters/Corbis

Photo 9.3, page 232. © Toby Melville/Reuters/Corbis

Chapter 10

Photo 10.1, page 244. © William Campbell/Sygma/Corbis

Photo 10.2, page 248. Photo courtesy of St. Joseph's Indian School, Chamberlain, SD

Photo 10.3, page 255. Photo by Keith Roberts

Chapter 11

Photo 11.1, page 273. Photo by Keith Roberts

Photo 11.2a, page 288. © Brian Snyder/Reuters/Corbis

Photo 11.2b, page 288. © Dan Habib/Concord Monitor/Corbis

Chapter 12

Photo 12.1, page 301. Photo by Keith Roberts

Photo 12.2, page 319. Bob Adelman/Magnum Photos

Chapter 13

Photo 13.1, page 334. Photo by Keith Roberts

Photo 13.2, page 341. Photo by Keith Roberts

Photo 13.3, page 344. Photo by Keith Roberts

Chapter 14

Photo 14.1, page 353. © Jack Hollingsworth/Photodisc/Thinkstock

Photo 14.2, page 358. N/A

Photo 14.3, page 366. © Courtesy of Hanover College

Chapter 15

Photo 15.1, page 383. © istockphoto.com/Richard Stamper

Photo 15.2, page 391. Photo by Virginia Caudill

Photo 15.3, page 397. © Vince Streano/Corbis

Name Index

SUBJECT INDEX

A

Activism, 293, 303–305, 309–311, 319–322
Affiliation, 113–114
   class, 228–233
   denominational membership, 135–137
   educational level, 228–229, 231, 237
   gender, 262–263
   global, 402
   race, 237–240
   See also church growth and decline,
      denominations, retention of members,
      switching
African Americans
   affiliation, 238–240
   minister's leadership role, 318–319
   religion in community, 312–313
   spirituals, 316–317
   See also African Methodist Episcopal church,
      Baptist churches—historically black,
      liberation theology—African American,
      National of Islam, slave religion
African Methodist Episcopal (AME) church,
   188, 202–203, 312
Age effects, 100–101
Agnostics, 145–146, 262–263
American Religious Identification Survey (ARIS),
   113–114, 239
American Society of Muslims, 188
Amish, 56–57, 109, 225
   Old Order, 172, 225, 271
Anabaptists, 136, 207, 231
Anomie (normlessness), 50, 86, 126,
   327, 370, 396
Anti-Semitism, 53, 251

Anxiety, 14, 45–46, 76, 224, 327
   primary and secondary, 46
Apocalypticism, *see* eschatological worldviews
Art, 248, 271
Asceticism, 107, 166, 173–174
   See also Protestant Ethic
Asians, 238–240, 258
   See also immigration
Assemblies of God, 114, 141–142, 162–164, 206,
   229, 300
Astrology, 120, 176, 371, 373–374
Atheists, 114, 135, 145–146, 228, 238–240, 253,
   262–263
Athletes in Action, 365–366, 368
Attendance (religious services), 28, 97–99, 205–206,
   233, 304–305, 334–335, 373
Attitudes, social and political, 301–302
Authority, 56–58, 115, 119, 139, 161, 326
   charismatic, 152–154, 337
   See also denominations—polity structure,
      secularization
Autocephalous organizations, 396, 400

B

Baby boomers, 16, 115, 116, 208–210, 212
   See also generations
Baptist churches, 187–189, 196–197, 205–206
   American, 142, 229, 274–275, 300
   Historically black, 229, 238, 240, 312–313
   See also Southern Baptist Convention
Baptists, 114, 136
Bigotry. *See* prejudice
Brainwashing, 123–125
Bruderhof, 5, 109, 155, 172

443

SAGE Research Methods Online
The essential tool for researchers

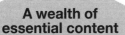